Differential Equations:

A Visual Introduction for Beginners

Ninth printing

By Dan Umbarger
www.mathlogarithms.com

Illustrations by Al Diaz and Mark Lewkowicz

Dedication

To my wife, Bramleta, who gave me three wonderful children: Jimmy, Terri, and Keelan.

In Appreciation

It is difficult to have all the skill sets necessary to write a book such as this. I have skills in lead-ins, progressions and sequencing, presentation, motivation, reinforcing, and organization, but I needed to reach out to others to supplement my knowledge base. Llewellyn Smith of Arlington (wyzant.com) was my primary math tutor. Brandlyn Lee, a graduate student in physics at UTD and Jason Burnside of wyzant.com helped me with much of the physics. Kirk Singleton of Colorado Springs (guru.com) was my "MATLAB go to guy" for most of my wondrous MATLAB slope fields and particular solutions. Al Diaz (guru.com) and Mark Lewkowicz of creationnetwork.com did all my wonderful illustrations. A former student, William Jameson, Harvard 2016, Dr. Alan Graves, Steve Vaughn of wyzant.com, Chris Reid of varsitytutors.com, Jeffrey Pair of Jeffrey's Tutoring, and especially Ms. Erin Brownlee of North Dakota State University all helped me to proofread my manuscript. Matthew Stein of ETS Tutoring and a graduate student in physics at SMU assisted me with a proof. Brian Clausing of Ohio helped with proofreading and directed me with a link to the wonderful position paper by Dr. Gian-Carlo Rota (deceased) of M.I.T. that guided me in my choice of topics. Dr. Larry Green of Lake Tahoe JC helped me to understand the rough correspondence between types of differential-equation applications and their solution types. Dr. Davis Cope, an old college chum now at North Dakota State University, helped correct a serious misunderstanding about "exact" differential equations. Finally, to my relief and joy, the editor of my two previous books, LaTeX wizard John Morris (editide.us), again agreed to help me edit and lay out the book.

All errors are, of course, my own.

ISBN: 978-0-9833973-6-6 LCCN: 2015911559

8 9 10 11 12 13 14 15 16 17 18 19 20

Also by the author: *Explaining Logarithms* and *Explaining Bayes Theorem*, both at www.mathlogarithms.com, and *Twenty Key Ideas in Beginning Calculus* at www.calculusbook.net.

Contents

Foreword

I still remember him, my university differential-equations teacher, Dr. D. It was back in the mid-1970s. He was really smart. He had earned his PhD by the time he was 25 or so. Although he never said so, it was widely whispered that he was awarded a Phi Beta Kappa Key during his study. He was really smart! He was also unfailingly happy, congenial, professional and friendly. He posted office hours and honored them. When you walked into his office he seemed genuinely happy to see you.

It was a small university with small math classes, so student questions were always allowed and encouraged. I clearly recall one day a student asking Dr. D. how he went from one line of his work to the next. Dr. D.'s response was that there was only one step involved in his work. We all ganged up on him cajoling, begging and whining so he looked again at what he had done. He studied his work for a full 10 seconds before saying, with effort, "Well I suppose you could have —." After that another thought came to him, and another. By the time he was through, much to our collective glee and satisfaction, he had replaced the step that confused us with four or five steps. That experience was with me every day as I taught mathematics to high school students over a 30-yr+ career. Dr. D. was like the neighborhood kid who, when playing "hide and go seek," would count to 100 by 5s. He was so smart that he was totally oblivious to the fact that he had collapsed five steps of a process to one step. He honestly believed (until badgered by his students) that what he had done involved only one step.

What is my point? Clearly, I am not saying Dr. D. was dumb. Au contraire! The point here is quite the opposite. But many math students learning mathematics need it to be presented sequentially, in small increments, with lots of explanation, visuals and repetition.

In preparation for my work, I obtained several books and several video series. One of them started out with the slide in the text box at right being read to the viewer. This material was minutes 1-5 of that 25-hour video series!!!

> How to Solve Linear (Differential) Equations
> $y'(x) + P(x) \times y(x) = Q(x)$
>
> 1. Calculate the integrating factor,
>
> $$I(x) := e^{\int P(x)\,dx},$$
>
>   and multiply that by both sides.
> 2. This makes the left-hand side into
>
> $$e^{\int P(x)\,dx}y' + P(x)e^{\int P(x)\,dx}y = Iy' + I'y = (Iy)'.$$
> 3. We can then integrate both sides.

Well excuse me, Dr. Smart Man, but what are differential equations? What are they used for? You seem to be starting from the general and working to the specific. The benefits of teaching with that approach are well known. If I understand what you are saying I will be able to apply your instruction to hundreds, countless situations. However, the drawbacks are … how shall I say it politely … well … hmmm … *What the hell are you talking about*? Who do you think you are talking to? One of your colleagues who already knows what you are saying? What you are doing is not unlike a farmer throwing around seeds without first plowing or fertilizing or watering. Many of us learn by moving from specific to the general. By working repeatedly with all the pieces slowly, over time, people learn how they synthesize and interact. Only then are they ready for generalization. Just because a technique is efficient and "the best way to do something" doesn't mean it is the way you should teach. There is the matter of "readiness." The teacher and the student have shared responsibilities here.

You state that you are going to solve a "linear equation." I know how the term "linear equation" is used in algebra. I don't think your slide title is referring to those kind of "linear equations" but you never address that. You simply state that the linear equation you are working with is $y'(x) + P(x) \times y(x) = Q(x)$. That doesn't look like the linear equations I know about, $y = mx + b$ or $ax + by = c$. What are those $P(x)$ and $Q(x)$ you are talking about? I know about $f(x)$ and $g(x)$, but what's with the capital letters, P and Q? Is that capitalization significant? It's hard to be interested in what you are talking about. Also, I know what an integral is, but what is an "integrating factor"? Oh! There it is on pg. 724 of my 1973 calculus book in a chapter titled "Differential Equations." We never got that far. Are you assuming that people should already know about "integrating factors" before studying differential equations? That seems circular. I feel really dumb. I feel defeated. Where did that $I(x) = e^{\int P(x)\,dx}$ come from? Is it like Athena jumping out of the head of Zeus? Is this like a religion class where we should accept teaching on faith? Is the example you start with the easiest one you could think of? I have heard something about *separable* differential equations and that they are easier to work with. Wouldn't it be better to begin working with those separable kinds of differential equations ... just to get started and build up the student's readiness, vocabulary, and confidence? **I don't see any sort of *visuals*.** All you are doing is talk, talk, talk ... slide after slide. Have you ever heard that "A picture is worth 1000 words"? You are solving a lot of problems, one after the other but *Why?* How do those problems connect to the world? Are differential equations like non-Euclidean geometry back in the 1800s or like topology ... just mental exercises with no application? I have heard that differential equations are used in economics, business, engineering, chemistry, physics, etc. It would be interesting and motivating to me if I could make some sort of connection to the real world.

Once, when I was teaching high school math, the department head, Tom Hall, asked me to sit in on an interview for a teacher applying to teach at our high school. The applicant boasted that when he was in college he was considered a "calculus machine." We did not hire him. We did not want a "calculus machine" teaching our students. We wanted someone who could present, sequence, motivate, and explain. Today, even more than ever, with MATLAB, Maple, and Mathematica, calculators with integral and derivative buttons, and both computer and calculator capability to graph slope fields, it is clear that understanding and comprehension is more important than cranking out answers.

If you are a "calculus machine" who feels that working differential equations is all about solving equations, then this book is not for you. It will bore you. It will possibly offend you. I can state with great authority that there are hundreds of YouTube videos and dozens and dozens of books on and about differential equations. You should turn your attention to that material.

Some 25 years ago, I read an interesting book titled *They're Not Dumb, They're Different* by Sheila Tobias (Research Corporation, 1990). It was about a very tiny study of bright university postgraduates who, as freshmen had either avoided science classes or who had transferred out of them. As postgraduates, they were monetarily bribed to go back and take an introductory science class in either physics or chemistry (their choice). They were asked to keep notes on their reactions to the curriculum as they went through their class. Most of that book is not relevant to this book, but following are some quotes from or about the study participants that I believe are helpful in explaining this book's philosophy.

Eric: The greatest stumbling block to understanding was the lack of identifiable goals and the absence of linkage between concepts (pg. 29).

Jack: If he (the teacher) could tell us what's coming next, why we moved from projectile to circular motion ... I would find it easier to concentrate ... I always wanted to know how to connect the small parts of a large subject. In humanities classes, I searched for themes in novels, connections in history, and organizing principals in poetry (pg. 34) ... I never really knew where we were heading ... He (the teacher) knows the whole picture but we don't (pg. 38).

Michelle: My curiosity simply did not extend to the quantitative solution ... I was more interested in the "why" (pg. 40).

Tom: More attention was given to Avogadro's number than to Avogadro's insight ... how and why he came up with that number (pg. 43).

Stephanie: In the humanities and social sciences we are taught to ask "why" questions. In chemistry I felt we were only being taught to ask "how" (pg. 58).

I totally relate to all the statements and feelings above. I believe that the best teacher is one who builds slowly upon previous knowledge, uses anticipatory sets, plans schedules of reinforcement, connects to real-world applications, paces carefully, makes copious use of visuals whenever possible, evolves and organizes ideas, spirals concepts, and repeats, repeats, and repeats whenever possible. See my previous books at www.mathlogarithms.com and www.calculusbook.net. If you are a visual learner who would like to be gently and kindly exposed to the topic of differential equations, welcome, you have come to the right place. I have 30+ years' experience in teaching grades 5–12 mathematics and computer science. I cannot teach you the "real thing" but I can help you get ready for the "real thing" in a humane, and hopefully interesting, format.

Lastly, during the course of researching and writing this book, I came across a fascinating nine-page article by Gian-Carlo Rota (www.ega-math.narod.ru/Tasks/GCRota.htm). With his nearly 40 years' experience teaching differential equations at MIT using materials he had written, I felt that Dr. Rota had considerable authority and insight into curricular issues in that discipline.

Dr. Rota began by sharing the fact that he had read a reprint review copy of Cauchy's introductory course in differential equations. From that point forward he shared with the reader a great many insights and opinions I will "cherry pick" and present as follows.

1. "It was a surprise to discover how little the content of the course had changed since Cauchy."
2. "As I read Cauchy's textbook, I realized how much the material we now teach is obsolete."
3. "The order of presentation of the outworn topics has not been altered."
4. The current curriculum of any text "will list a number of disconnected tricks ... that are passed off as useful."
5. "Lecturers in the course [are] unaware of any applications of differential equations beyond those given in elementary texts."
6. The curriculum is unchanged because "vested interests dominate every nook and cranny of our society ... A revamped curriculum of elementary differential equations ... would require [teachers] to have to learn the subject anew ... [and] the fatuous, expensive, multi-colored textbooks that are now cornering the market would be forced out of print."
7. **"Teaching a subject of which no honest examples can be given is ... demoralizing."** Author's emphasis.
8. According to Dr. Rota it is still important for students to learn about: separation of variables, changes of variables, linear differential equations, the universal occurrence of the exponential function, stability, the relationship between trajectories and integrals of systems, phase plane analysis, Laplace transforms, and especially solving linear systems with constant coefficients.
9. Dr. Rota concludes with the following: A differential equations teacher should "teach concepts, not tricks ... What matters is [the student] getting a feeling for the importance of the subject and their coming out of the course with the conviction of the inevitability of differential equations, and with enhanced faith in the power of mathematics."

Having died in the late 1990s, Dr. Rota did not have access to the current versions of mathematical software: Mathematica, Maple, MatLab. One can only surmise that such software would only reinforce Dr. Rota in the opinions he espoused in this paper.

Chapter 0
Equations and Differentials, Review of Differential Calculus

It is traditional to start an introductory treatment of differential equations with a definition: "A differential equation is . . . ." If you would like to start there, please turn to Chapter 1: Getting Started.

Equations

An equation is a balancing of an equal amount of information or quantity on each side of a balancing point, although the information or quantities are generally presented in different forms.

Even though Bobby weighs three times as much as Mary, her force on the left side of the balance point has the same moment as Bobby's force on the right side of the balance point because her distance from the balance point is three times Bobby's distance from the balance point.

Algebraic Equations in One Variable

For most people, algebraic equations most commonly come to mind when thinking of equations. The goal in solving an algebraic equation is to find a value (or set of values; think polynomial equation) that, when substituted into the original equation, will result in equal quantities on both sides. Remember, the number of solutions ($f(x) = 0$) is equal to the degree of the polynomial.

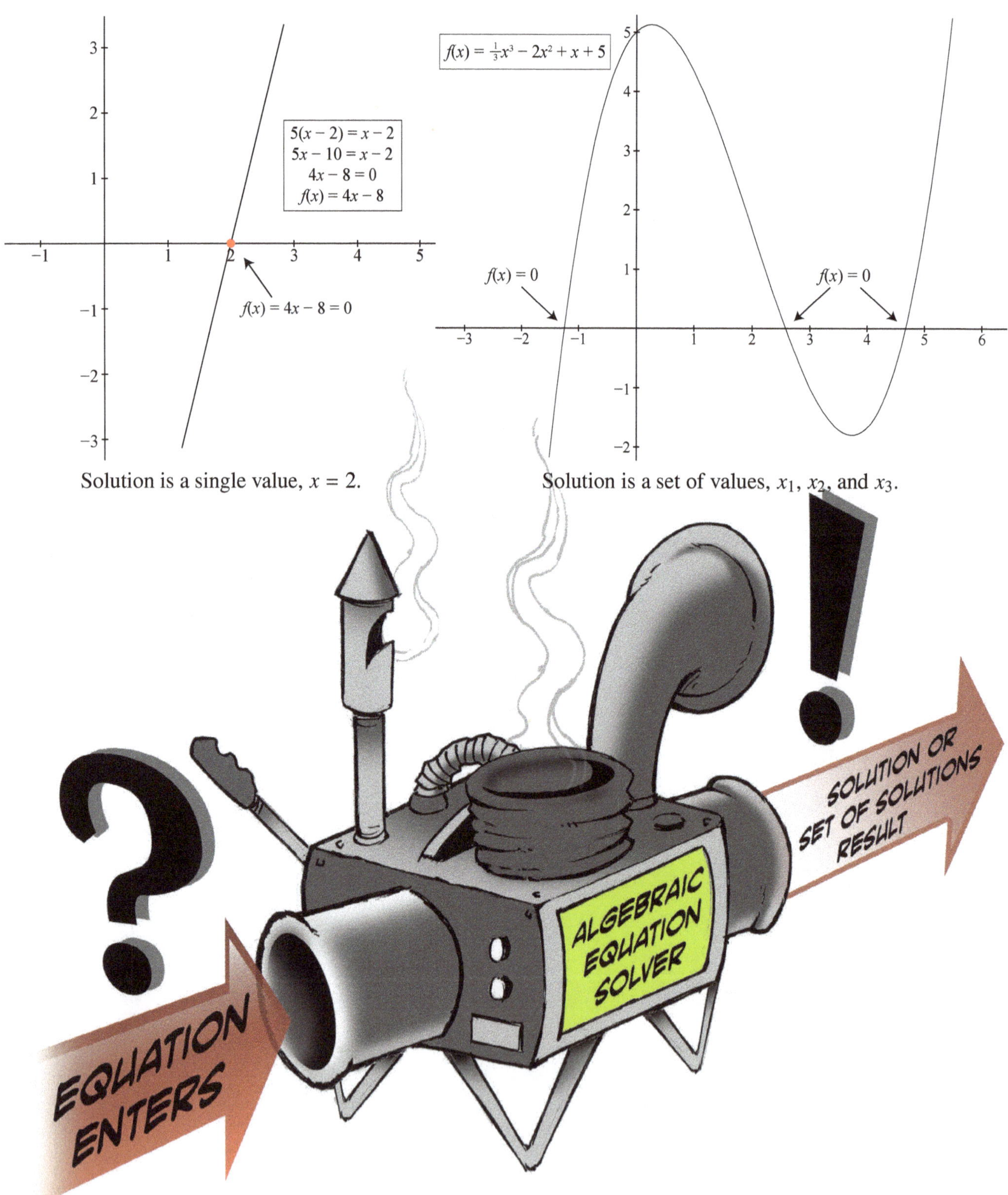

Solution is a single value, $x = 2$.

Solution is a set of values, x_1, x_2, and x_3.

Chemical Equations

Chemistry teaches the "Law of Conservation of Mass": In a chemical reaction, the number of atoms in the "reactants" of the chemical reaction must be equal to the number of atoms in the "product(s)" of a chemical reaction.

Propane	+	Oxygen	=	Carbon Dioxide	+	Water	
C_3H_8	+	O_2	$\longrightarrow$	CO_2	+	H_2O	unbalanced
C_3H_8	+	O_2	$\longrightarrow$	$3CO_2$	+	H_2O	carbon balanced
C_3H_8	+	O_2	$\longrightarrow$	$3CO_2$	+	$4H_2O$	carbon and hydrogen balanced
C_3H_8	+	$5O_2$	$\longrightarrow$	$3CO_2$	+	$4H_2O$	all balanced

The goal of balancing a chemical equation is to end up with the same number of atoms of all reactant elements as there are atoms in the product element(s). The fact that this process involves an equation is somewhat obscured because there was no "equal" sign involved. That would be confusing because using an equal sign here would give the appearance of violating the multiplicative property of equality.

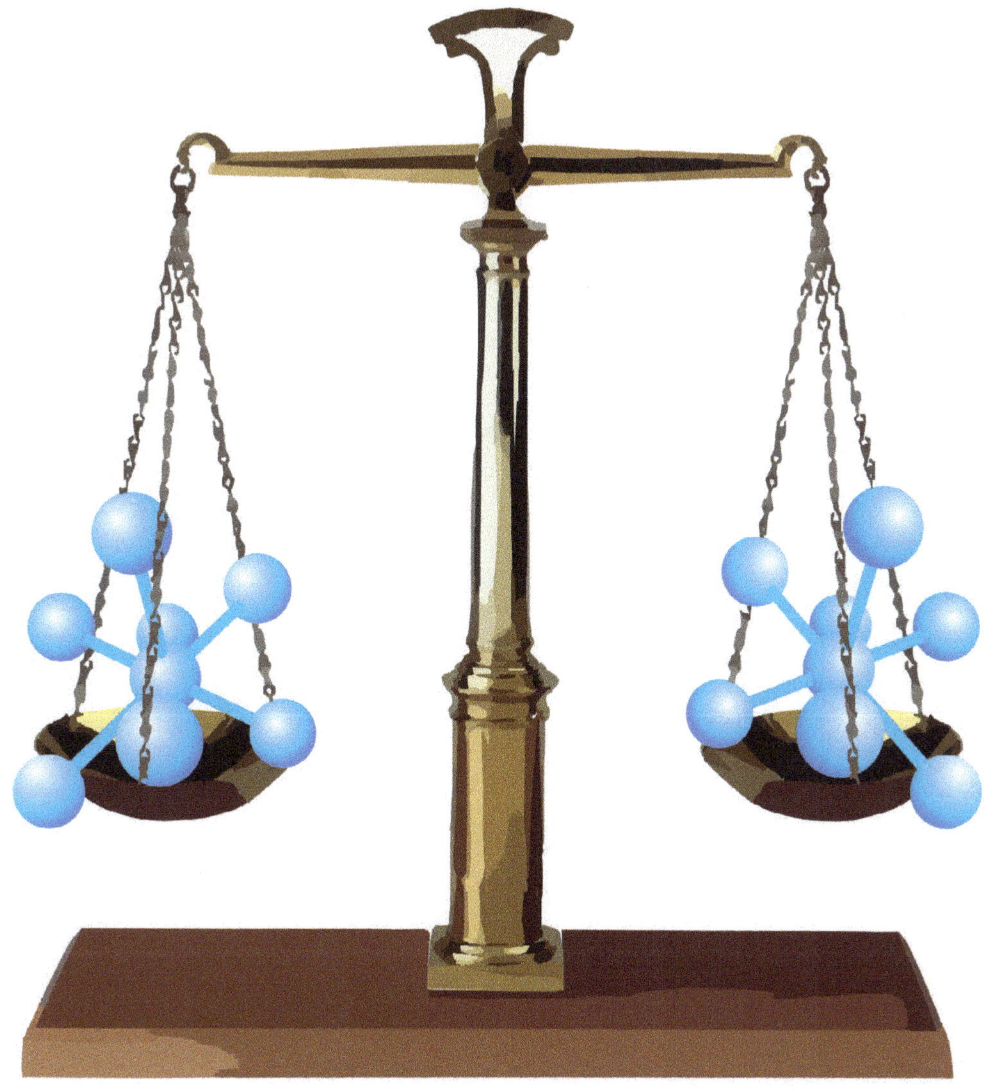

Derivatives

A derivative can be thought of as: 1) the slope of a tangent to a curve at a given point or 2) the instantaneous speed of a moving object at a given point in time. Formally, all derivatives are obtained by way of a definition which finesses the problem of indeterminate division by zero. However, in practice, theorems derived from that definition give us a derivative value. The derivative function machine will find the derivative of the specified function, $f(x)$.

Differential Equations

Definition 1: A "differential equation" is an equation involving an unknown function and one or more of its derivatives. Three different notations for the same equation are shown below. The symbols used in a differential equation class are confusing because of the history of the concepts. Calculus was developed independently by two different people (Sir Isaac Newton and Gottfried Leibniz) who each used different symbols. Today, we have names for the different forms: derivative form, differential form, and differential operator (modern) form, respectively.

$$y'' + y = x + 3 \qquad \frac{d^2y}{dx^2} + y = x + 3 \qquad D^2y + y = x + 3$$

The goal in solving an algebraic equation is to find a scalar (quantity) or set of scalars (think polynomial equation) that, when substituted into the equation, will balance the two sides of the equation. The goal in solving a differential equation is to find a function that, when substituted into the original differential equation, will balance the two sides of that equation.

Since we learned in beginning calculus class that $\frac{d}{dx}f(x) = f'(x)$ and $\int f'(x)\,dx = f(x)$, we should anticipate that a "differential equation solver" will somehow involve integration. Example: Solve the differential equation $\frac{dy}{dx} = 2x$. This differential equation is asking, "What function $f(x)$ has the derivative $2x$?" That is, if $f'(x) = 2x$, what is $f(x)$? Hopefully by the time you are finished with this book you will understand the following steps:

$$\frac{dy}{dx} = 2x$$
$$dy = 2x\,dx$$
$$\int dy = \int 2x\,dx$$
$$\int y^0\,dy = 2\int x^1\,dx$$
$$\frac{y^{0+1}}{0+1} = 2\frac{x^{1+1}}{1+1}$$
$$y = 2\frac{x^2}{2}$$
$$y = x^2$$
$$f(x) = x^2$$

ck: $y = x^2$
$$\frac{d}{dx}y = \frac{d}{dx}x^2$$
$$\frac{dy}{dx} = 2x^{2-1}$$
$$\frac{dy}{dx} = 2x \quad \text{ck}$$

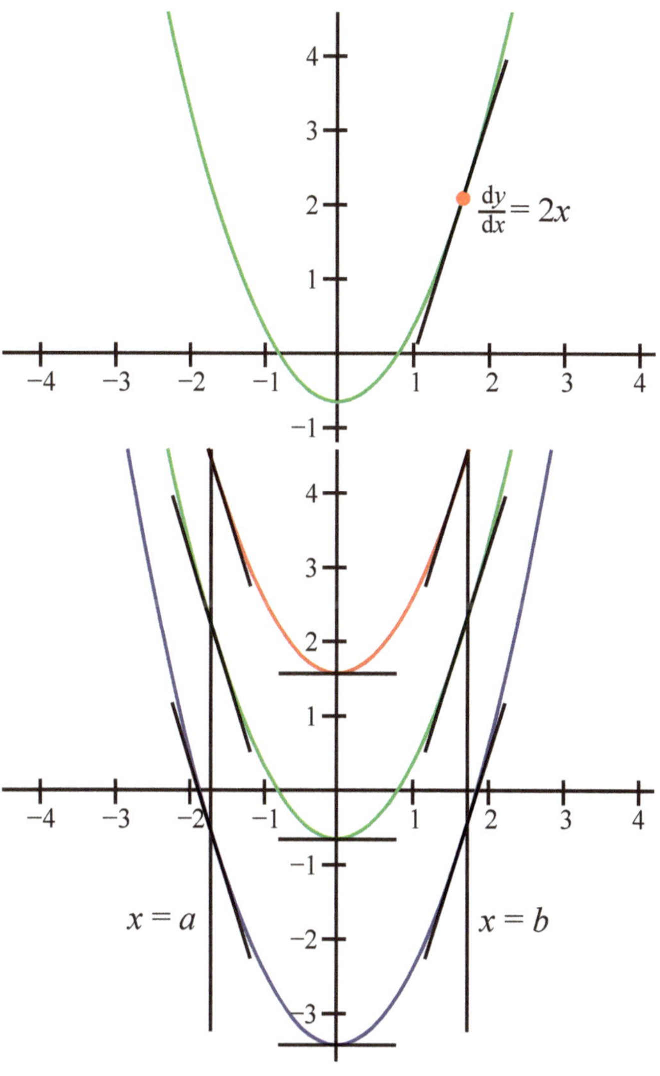

However, since a vertical shift does not change the slope of a curve at a given point x, more than one curve would have a slope of $2x$ at any given point x. We could indicate this by stating, "For $\frac{dy}{dx} = 2x$, the family of function solutions would be $x^2 + c$, where c represents any constant." The vertical lines here are called isoclines. An isocline is a line or curve that intersects a family of functions so that the slopes of those functions at the points of intersection are the same.

Hence the goal of solving a differential equation is to solve for a function ***or set of functions*** that, when substituted into the original equation, will cause both sides to be balanced or equal.

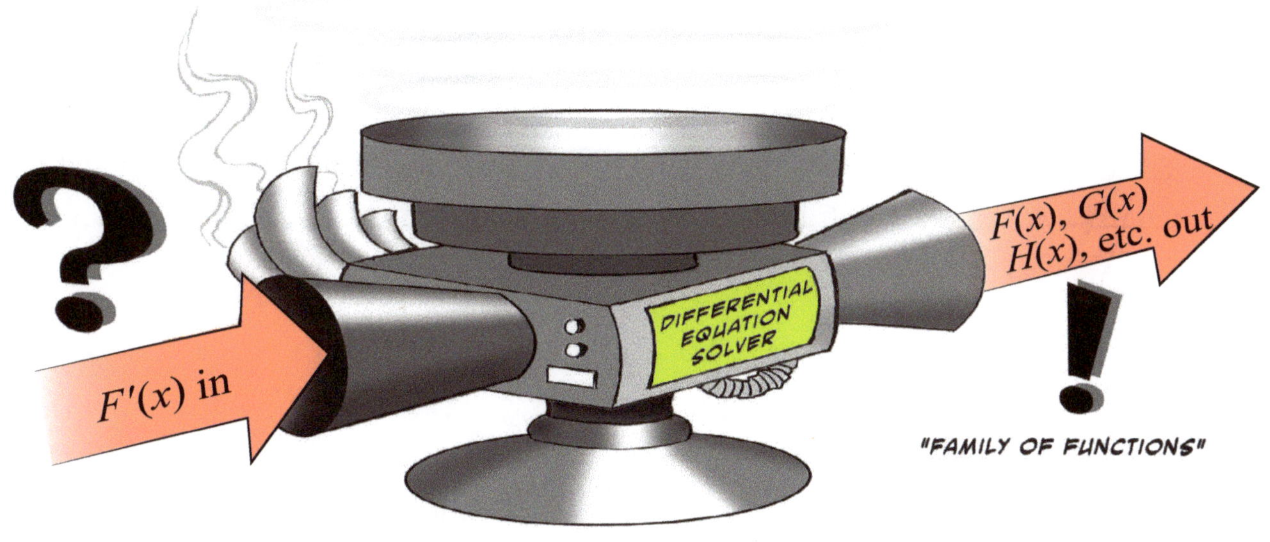

In Algebra, we study problems involving slope: Imagine a line passes through points (3, 5) and (−2, −1). What is the slope of that line? What is its equation? What is $f(6)$ of the unknown function passing through the two points?

1. Calculate m. $m = \frac{y_2 - y_1}{x_2 - x_1} = \frac{-1-5}{-2-3} = \frac{-6}{-5} = \frac{6}{5}$
2. Calculate b.

$$y = mx + b$$
$$5 = \frac{6}{5} \times 3 + b$$
$$\frac{25}{5} = \frac{18}{5} + b$$
$$b = \frac{7}{5}$$

3. Using values for m and b, we can determine the equation of the line passing through the two given points: $y = f(x) = \frac{6}{5}x + \frac{7}{5}$.
4. The y value of the point lying on this line whose x value is 6 would be

$$y = f(6) = \frac{6}{5} \times 6 + \frac{7}{5} = \frac{36}{5} + \frac{7}{5} = \frac{43}{5}.$$

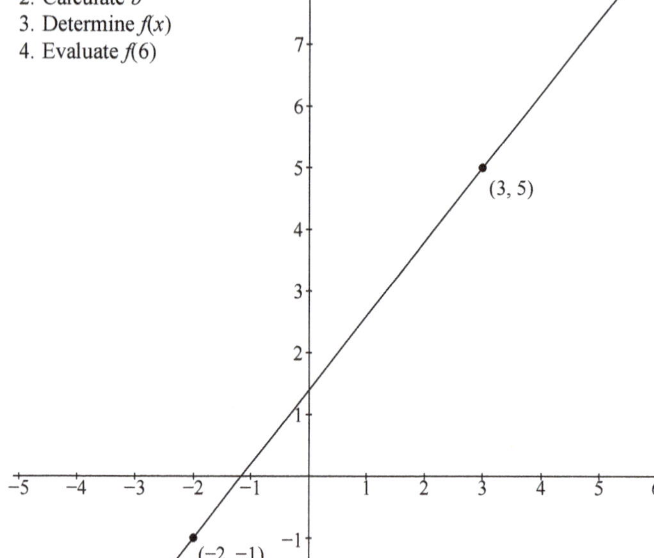

When scientists or engineers "solve" differential equations, they usually have information that allows the identification of which equation—among the infinite equations solving the differential equation—would be applicable for the specific situation. For example, for the equation $\frac{dy}{dx} = 2x$, the solution set is $f(x) = x^2 + c$ implying infinite solutions. Which (if any) of those infinite solutions will pass through the point (2, 1)? Substituting (2, 1) into the equation $y = x^2 + c$ we get $1 = 2^2 + c$, $1 = 4 + c$, therefore $c = -3$. The function $f(x) = x^2 - 3$ is the specific function whose derivative, $\frac{dy}{dx} = 2x$, is what we need. What is the y value of the unknown function when $x = 3$? Compute $y = 3^2 - 3 = 6$. The process of thinking of this curvilinear differential equation problem is very much like the linear algebraic one above except the slope is already given: $\frac{dy}{dx} = 2x$!!

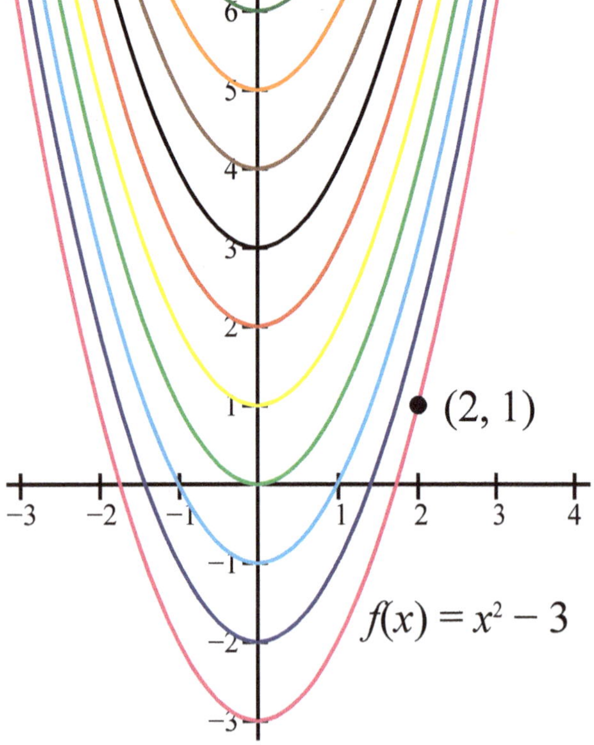

Family of Solutions for $\frac{dy}{dx} = 2x$

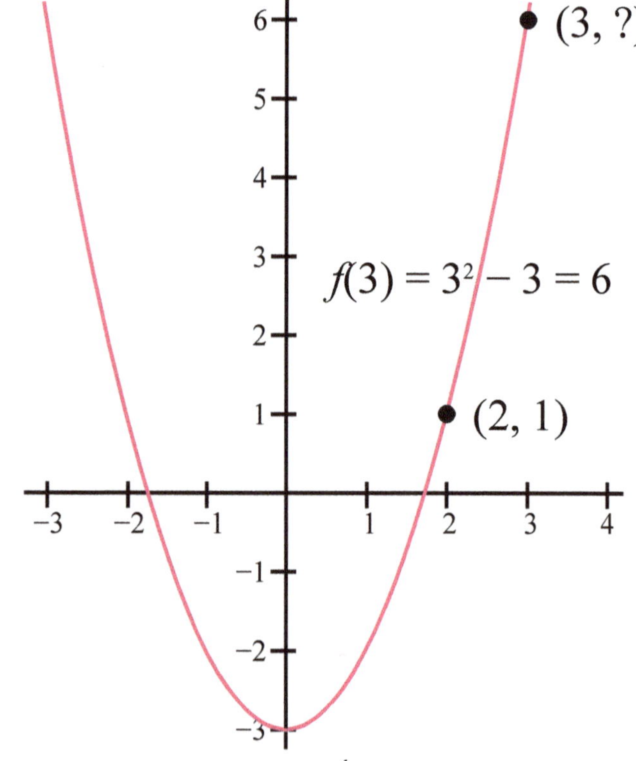

Specific Solution for $\frac{dy}{dx} = 2x$ and $f(2) = 1$

Differentials

"A differential is a small increment. The differential, dx, of the independent variable x, is any arbitrary increment in the value of x, and the corresponding differential dy is defined by the relation $dy = f'(x)\,dx$, where $f'(x) = \lim_{\Delta x \to 0} \frac{\Delta y}{\Delta x}$."* Some sources take liberties with the definition and define a "differential" as an infinitesimally small distance. The ratio $\frac{dy}{dx}$ is Leibniz's abbreviated form of $f'(x) = \lim_{\Delta x \to 0} \frac{\Delta y}{\Delta x}$.

Now, if you understood that, you should jump straight to Chapter 1. If you did not understand that definition and wish to do so, continue to work your way through this chapter. In my book *Twenty Key Ideas in Beginning Calculus,*† I spend three entire chapters (38 pages) talking about a concept called a derivative, why the concept of a derivative is useful, and why a derivative cannot be obtained using traditional algebraic techniques. In Chapter 4 (18 pages), I show how Newton and Leibniz used a definition to resolve the issue of how to calculate a derivative without fear of the indeterminate division of zero. Those 56 pages of material are compressed here into just a few pages. Please see my earlier book for more explanation and clarity.

Differential Calculus (a quick review)

A major goal in any beginning (differential) calculus class is the study of how to obtain the slope of a tangent to a curve at a given point. That skill, by itself, is not really very important, but it turns out to be vital to many, many very important real-world problems. By finding the slope of a tangent to a curve at a given point, one can indirectly obtain (infer) information about other important problems.

Given the graph $y = x^2$, how could you find the slope of the tangent at the point $(8, 64)$? The familiar slope formula, $m = \frac{y_2 - y_1}{x_2 - x_1}$, can only be applied when two distinct points are available. A classic technique in mathematics is to (temporarily) leave the problem one wishes to solve and to work on a more solvable problem that has similarities to the "unsolvable" problem. In the graphs below, we do not know how to find the slope of the tangent at the point $(8, 64)$, but we can find the slope of the secant line from $(0, 0)$ to $(8, 64)$. We improve our approximation of the tangent-line slope by moving the extra point from $(0, 0)$ toward the desired point of tangency $(8, 64)$, finding the slope of the secant line from (x, y) to $(8, 64)$ as $x \to 8$. (Integer x values result in easy arithmetic.) For x_1 values of 0, 4, 6, and 7, slope values 8, 12, 14, and 15 result.

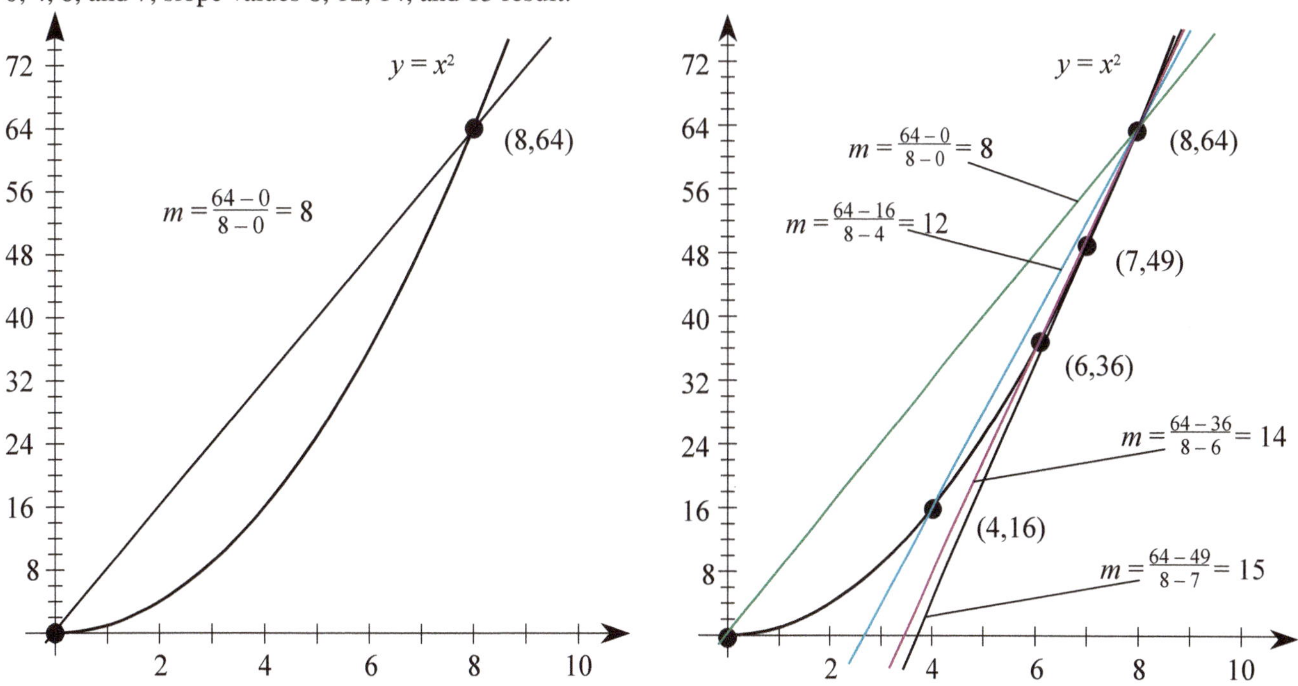

Dictionary of Mathematics, C. C. T. Baker, Hart Publishing, 1966.

†*Twenty Key Ideas in Beginning Calculus,* Dan Umbarger, www.mathlogarithms.com, 2011.

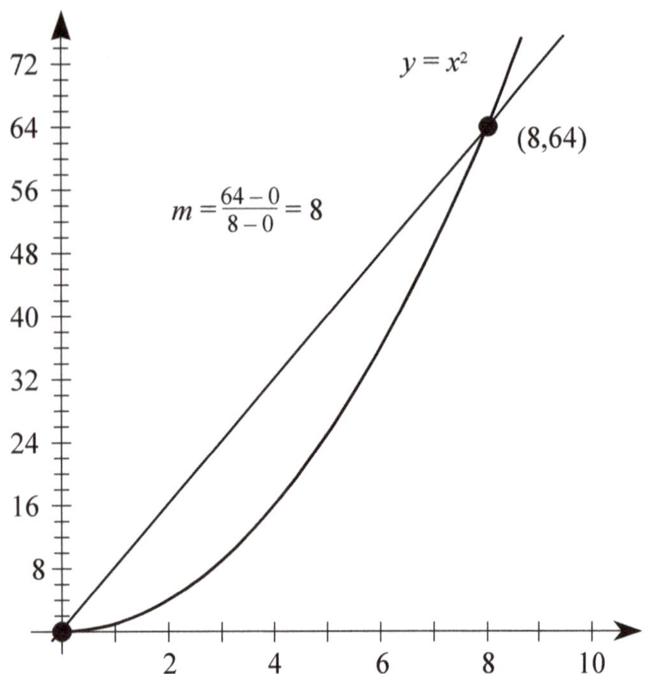

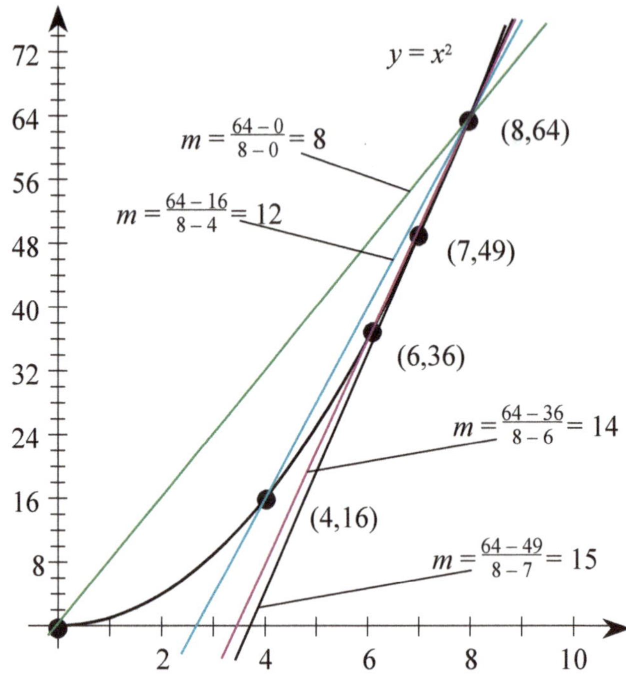

This same information can be shown in table form.

x (domain)	y (range)	$m = \frac{y_2-y_1}{x_2-x_1}$
0	0	$m = \frac{64-0}{8-0} = 8 \downarrow$
4	16	$m = \frac{64-16}{8-4} = 12 \downarrow$
6	36	$m = \frac{64-36}{8-6} = 14 \downarrow$
7	49	$m = \frac{64-49}{8-7} = 15 \downarrow$

Continuing this pattern with $x \to 8$, we get successively closer and closer to the desired slope at point $(8, 64)$.

x (domain)	y (range)	$m = \frac{y_2-y_1}{x_2-x_1}$
7.9	62.41	$m = \frac{64-62.41}{8-7.9} = 15.9 \downarrow$
7.99	63.8401	$m = \frac{64-63.8401}{8-7.99} = 15.99 \downarrow$
7.999	63.984001	$m = \frac{64-63.984001}{8-7.999} = 15.999 \downarrow$

Those diminishing horizontal and vertical distances are known as differentials. The slope of the tangent line to the curve $y = x^2$ appears to be 16. Before calculus, we could not say for sure, but it strongly looked as if the slope would not be greater than 16.

Using the formula $m = \frac{y_2-y_1}{x_2-x_1}$ to calculate the slope of a tangent poses two problems for the noncalculus student: 1) There is only one point available; so $x_2 = x_1$, resulting in an undefined division by zero. 2) Furthermore, since $y_2 = y_1$, the result in using the traditional slope value is $\frac{0}{0}$. Curiously and counterintuitively, this $\frac{0}{0}$ situation can be addressed. Stay tuned!

CONVERGING FROM LEFT

15.9 15.99 15.999 16

LimitMan

Newton and Leibniz both realized (independently) that seeking the slope at a single point in space is futile because an infinity of lines can be drawn through a point (implying infinite slopes; Fig. below left). However, through a given point on a continuous curve only one line could be drawn tangent to the curve (Fig. below right). There is something about the latter case that imparts predictability to the process of approximating the slope at that point even though it involves division by zero. The value $\frac{a \text{ where } a \neq 0}{b \text{ where } b \to 0}$ is undefined when b reaches 0. However, the value $\frac{a \text{ where } a \to 0}{b \text{ where } b \to 0}$ is indeterminate but not undefined. This distinction is very important!

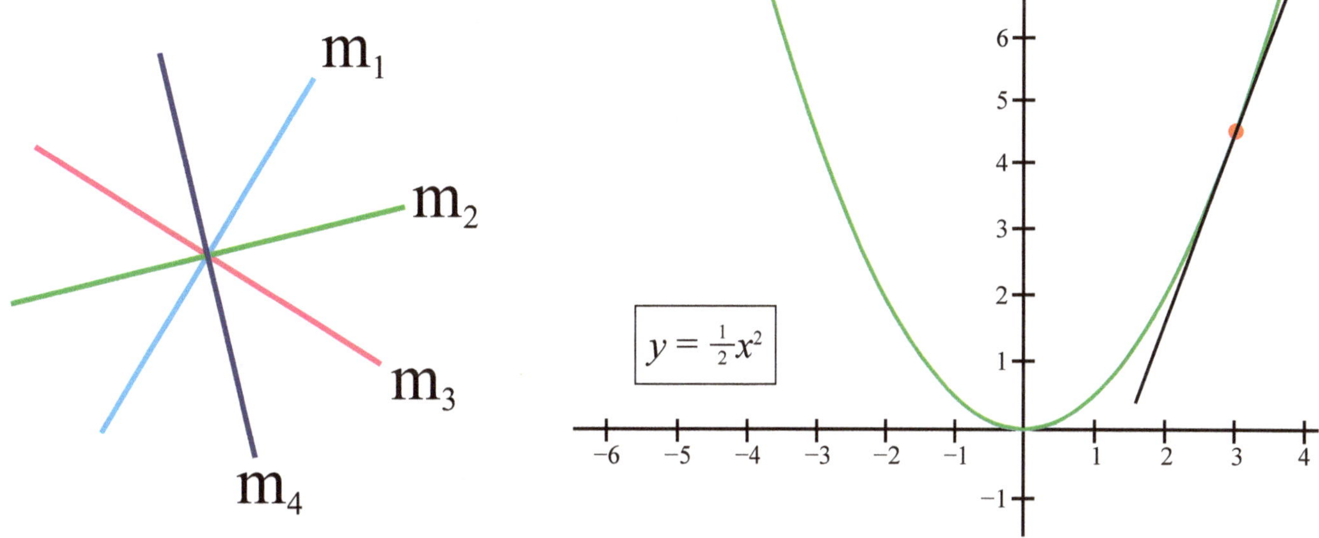

The solution to the indeterminate division-by-zero conundrum was clever and, in retrospect, easy. The community of mathematicians led by Newton and Leibniz made it go away with a definition that was consistent with (did not contradict) established mathematics. The definition is a bit cryptic to the uninitiated. Some images are helpful in understanding that definition.

In the graph below left, one sees a sequence of six points, p_1, p_2, p_3, p_4, p_5, and p_6 approaching point $(8, 64)$ from the left. The respective secant slopes from p_1 to $(8, 64)$, p_2 to $(8, 64)$, p_3 to $(8, 64)$, p_4 to $(8, 64)$, p_5 to $(8, 64)$, and p_6 to $(8, 64)$ are calculated using the two-point slope formula, $m = \frac{y_2 - y_1}{x_2 - x_1}$. As $p_n \to (8, 64)$, the slope of the secant line between points p_n and $(8, 64)$ approaches the slope of the tangent line at point $(8, 64)$. This is shown more abstractly in the graph below right where it is implied that there is an infinity of points approaching point $(8, 64)$.

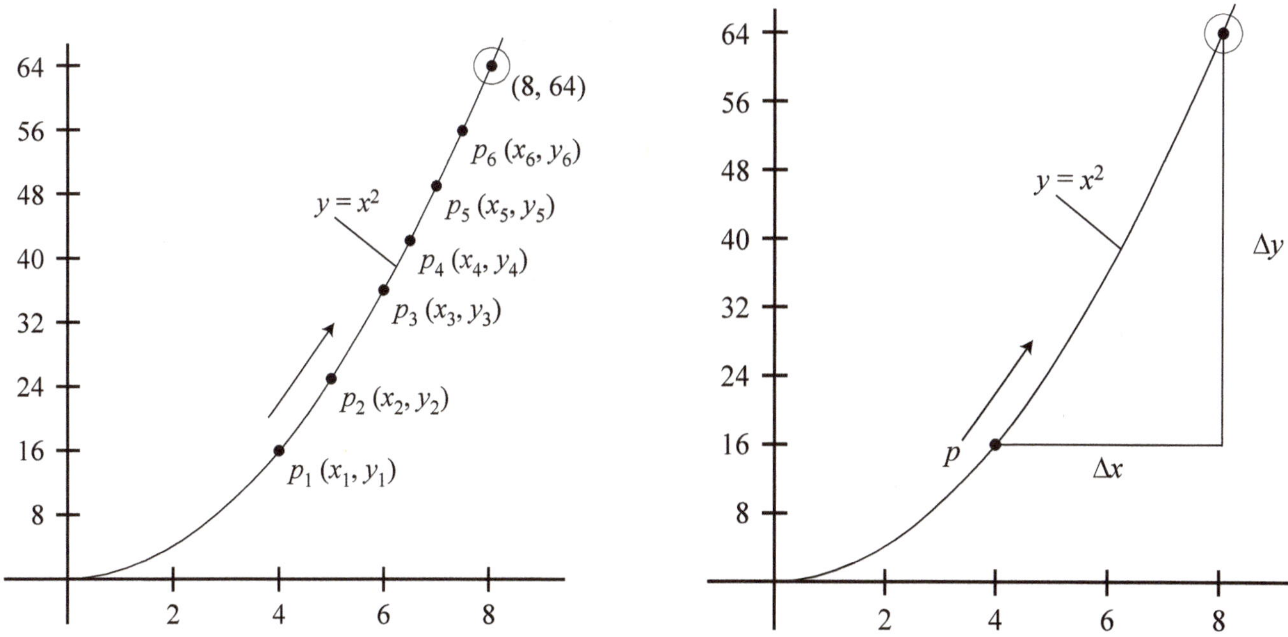

The "definition" Newton and Leibniz used to resolve the problem of needing two points to calculate a slope and only having one involved the limit operator:

$$f'(x) = \lim_{p \to x} \frac{f(x) - f(p)}{x - p}, \text{ in derivative form and}$$

$$\frac{\mathrm{d}y}{\mathrm{d}x} = \lim_{p \to x} \frac{y(x) - y(p)}{x - p}, \text{ in differential form.}$$

What this definition means is that as points p_n successively approach point x, the slope of the secant lines approaches the slope of the tangent line at x. The definitions shown above can be combined with algebra to obtain useful theorems.

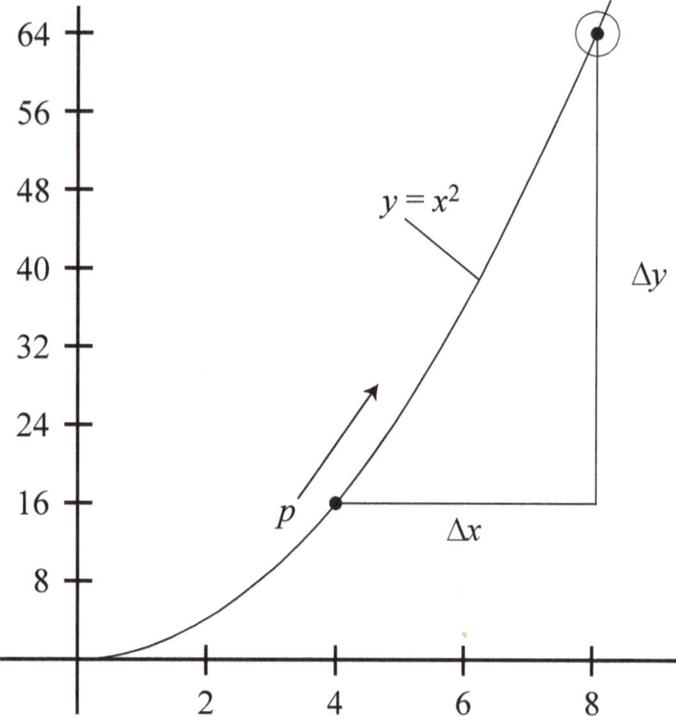

The use of a definition to resolve issues in math is not a new concept. $0! = 1$ and $b^0 = 1$ (for $b \neq 0$) are two such definitions you have used. Also, how were you taught to solve for y in the equation $x = b^y$? Were you taught to use the same process you would use to solve $3x + 7y = 4$ for y? No. You were taught to solve the equation $x = b^y$ for y by applying what is essentially a definition: $x = b^y$ is equivalent to $y = \log_b x$. Subject to consistency with all other mathematics, a mathematician can make up any definition that can help solve a problem. For example, we often define Δy as $y_2 - y_1$.

If $f(x) = x^n$, then $f'(x) = nx^{n-1}$ (derivative form) and if $y = x^n$, then $\frac{\mathrm{dy}}{\mathrm{dx}} = \mathrm{n}x^{n-1}$ (differential form).

Proof. The definition says that $f'(x) = \lim_{p\to x} \frac{f(x)-f(p)}{x-p}$. Hence for $f(x) = x^n$,

$$f'(x) = \lim_{p\to x} \frac{x^n - p^n}{x - p}$$

From algebra, we know that $x^2 - p^2 = (x-p)(x+p)$ $x^3 - p^3 = (x-p)(x^2 + xp + p^2)$ $x^4 - p^4 = (x-p)(x^3 + x^2p + xp^2 + p^3)$ Or in general $x^n - p^n = (x-p)(x^{n-1} + x^{n-2}p + x^{n-3}p^2 + \cdots + xp^{n-2} + p^{n-1})$

(Continuing the proof) $$= \lim_{p\to x} \frac{\cancel{(x-p)}(x^{n-1} + x^{n-2}p + x^{n-3}p^2 + \cdots + p^{n-1})}{\cancel{(x-p)}}, \quad x \neq p$$

No more denominator, no more division by zero.

$$\begin{aligned}
&= \lim_{p\to x}(x^{n-1} + x^{n-2}p + x^{n-3}p^2 + \cdots + p^{n-1}) \\
&= x^{n-1} + x^{n-2}x + x^{n-3}x^2 + \cdots + x^{n-1} \\
&= \underbrace{x^{n-1} + x^{n-1} + x^{n-1} + \cdots + x^{n-1}}_{n \text{ terms}} \\
&= nx^{n-1}
\end{aligned}$$

□

So if $f(x) = x^2$, then $f'(x) = 2 \times x^{2-1} = 2x$ using the derivative form. Alternatively, if $y = x^2$, then $\frac{\mathrm{dy}}{\mathrm{dx}} = 2 \times x^{2-1} = 2x$ using the differential form. The slope of the tangent to the curve $y = x^2$ is $2x$ ($f'(x) = \frac{\mathrm{dy}}{\mathrm{dx}} = 2x$).

$\frac{\mathrm{d}}{\mathrm{d}x} f(x) = \lim_{p\to x} \frac{f(x) - f(p)}{x - p}$ (definition) and, for $y = x^n$, $\frac{\mathrm{dy}}{\mathrm{dx}} = nx^{n-1}$ (theorem derived from the definition).

Before Newton and Leibniz, a branch of mathematics known as Euclidean geometry was the consistent synergism of undefined terms (points, lines, space, etc.), common notions (postulates), and theorems. After Newton and Leibniz, calculus became the consistent synergism of basic algebraic properties (commutative, associative, distributive, etc.), definitions (definition of derivative, definition of area under a curve), and easy-to-apply theorems (derivative of x^n; derivative of a polynomial sum, etc.) that derive from the properties and definitions. As with geometry, all the properties, definitions, and theorems in calculus are consistent (do not contradict each other).

When calculating ***slopes of secants***, we used the formula:

$$m = \frac{y_2 - y_1}{x_2 - x_1} = \frac{\Delta y \text{ where } \Delta y = y_2 - y_1}{\Delta x \text{ where } \Delta x = x_2 - x_1 \text{ but } x_2 \neq x_1}.$$

Calculating ***slopes of tangents*** we used the formula $m = \frac{\mathrm{dy} \text{ where } \mathrm{dy} = \Delta y \to 0}{\mathrm{dx} \text{ where } \mathrm{dx} = \Delta x \to 0}$ which culminates in the indeterminate form $\frac{0}{0}$.

The definition $f'(x) = \lim_{p\to x} \frac{f(x)-f(p)}{x-p}$ resolves the problem of the indeterminate value $\frac{0}{0}$: $m = \frac{\mathrm{dy} \text{ where } \mathrm{dy} = \Delta y \to 0}{\mathrm{dx} \text{ where } \mathrm{dx} = \Delta x \to 0}$ can be written as $m = \lim_{\mathrm{dx}\to 0} \frac{\mathrm{dy}}{\mathrm{dx}}$. (No more $\frac{0}{0}$!) As $\mathrm{dx} \to 0$ so, too, will $\mathrm{dy} \to 0$, but only rarely at the same rate.

Comparing the two slope formulas side by side, we get $m_{\text{secant}} = \frac{\Delta y \text{ where } \Delta y = y_2 - y_1}{\Delta x \text{ where } \Delta x = x_2 - x_1 \text{ and } x_2 \neq x_1}$ and $m_{\text{tangent}} = \lim_{\mathrm{dx}\to 0} \frac{\mathrm{dy}}{\mathrm{dx}}$.

The symbols $\mathrm{d}y$ and $\mathrm{d}x$ are examples of "differentials." In beginning calculus, the $\frac{\mathrm{d}y}{\mathrm{d}x}$ is always shown as a ratio. In differential equations, the numerator and denominator are frequently separated:

$$f'(x) = \frac{\mathrm{d}y}{\mathrm{d}x} \text{ is equivalent to } f'(x)\,\mathrm{d}x = \mathrm{d}y \text{ or } \mathrm{d}y = f'(x)\,\mathrm{d}x.$$

Compare the following definition of the differential originally shown on page 7.

> A differential is a small increment. The differential, $\mathrm{d}x$, of the independent variable x, is any arbitrary increment in the value of x, and the corresponding differential $\mathrm{d}y$ is defined by the relation $\mathrm{d}y = f'(x)\,\mathrm{d}x$, where[a]
>
> $$f'(x) = \lim_{\Delta x \to 0} \frac{\Delta y}{\Delta x}.$$
>
> [a] *Dictionary of Mathematics*, C. C. T. Baker, Hart Publishing, 1966.

with the slope of a tangent line shown above: $m_{\text{tangent}} = \lim_{\mathrm{d}x \to 0} \frac{\mathrm{d}y}{\mathrm{d}x}$. Hopefully the text since page 7 has helped you to better understand that definition.

One of the reasons that math is so difficult is that cryptic math symbols are often used to communicate very complex ideas. Unlike human languages, the symbols are always very abstract and concise. Here is a classic example from math education back in the precalculator 1960s. "The product of two numbers, $a \times b$, can be found by taking the antilogarithm of the sum of the logs of a and b." This was taught as: $a \times b = 10^{\log a + \log b}$. You see that the logarithm rule of multiplying $a \times b$ completely strips out the words in favor of symbols. All the necessary information is still there, but it is communicated tersely in symbols. ***Words and symbols are processed differently and by different parts of the brain. The brain interprets and processes symbols and pictures much more quickly than words. The old saying is, "A picture is worth 1,000 words." That is also true for symbols.***

The term *esoteric* means, "designed for and understood by the specially initiated alone; not communicated, or not intelligible to the general population." Technical material is inherently esoteric. Understanding mathematics requires knowledge of specialized vocabulary and specialized symbols. Specialized vocabulary and symbols common to groups of people allow those groups to communicate huge amounts of information quickly and efficiently. If you understand the two definitions that were previously presented in Chapter 0 and how they are interrelated,

1. "A differential is a small increment. The differential, $\mathrm{d}x$, of the independent variable x, is any arbitrary increment in the value of x, and the corresponding differential $\mathrm{d}y$ is defined by the relation $\mathrm{d}y = f'(x)\,\mathrm{d}x$, where $f'(x) = \lim_{\Delta x \to 0} \frac{\Delta y}{\Delta x}$, and

2. $f'(x) = \lim_{p \to x} \frac{f(x) - f(p)}{x - p}$, in the definition of the derivative from page 10," then Chapter 0 becomes pretty well superfluous because most of the information on all of those pages is communicated in those two definitions. However this is a "chicken and egg situation": most people cannot understand these esoteric symbols and definitions without extensive explanation. You have to be a member of the "brotherhood" of mathematicians to understand them.

Senior citizens who have told the same stories and jokes so many times that, for brevity and efficiency, they have given them all numbers.

Some people just can't tell a joke.

All of this long discussion is pretty well necessary to understand that there are subtle differences between the notations in the following two text boxes. In the text box below left, you separate the numerator from the denominator of the fraction $\frac{2}{3}$ by multiplying both sides by a scalar

In the text box below right, you separate the numerator and denominator of the derivative $\frac{\mathrm{d}y}{\mathrm{d}x}$ by multiplying both sides by a differential, $\mathrm{d}x$. ($\mathrm{d}x$ is not a scalar.)

$$\frac{2}{3}x = 10$$
$$3 \times \frac{2}{3}x = 10 \times 3$$
$$2x = 30$$

$$\frac{\mathrm{d}y}{\mathrm{d}x} = 10$$
$$\mathrm{d}x \times \frac{\mathrm{d}y}{\mathrm{d}x} = 10 \times \mathrm{d}x$$
$$\mathrm{d}y = 10\,\mathrm{d}x$$

Since, by definition, $\frac{\mathrm{d}}{\mathrm{d}x}f(x) = \lim_{p\to x}\frac{f(x)-f(p)}{x-p}$, all that limit stuff is implied whenever you see $\frac{\mathrm{d}y}{\mathrm{d}x}$. It is also implied when the numerator differential and the denominator are separated. So while $\frac{\mathrm{d}y}{\mathrm{d}x} = 10$ says that the limit of the original ratio of differentials as $p \to x$ would be 10, the equivalent form $\mathrm{d}y = 10 \times \mathrm{d}x$ means something like "a tiny, forever decreasing, y distance will be limited by a value which is 10 times a tiny, forever decreasing, x distance." ***For beginning students, the definition of $\frac{\mathrm{d}y}{\mathrm{d}x}$ can be thought of as the ratio of two differentials.*** Keep in mind that $\lim_{x\to p}$ is explicitly stated in the definition of $\frac{\mathrm{d}y}{\mathrm{d}x}$, but it is implied when the differentials are separated. If you are taking applied differential equations in a science or engineering class it would not really hurt you to think of the manipulations in the right text box as working with scalars. A mathematics professor might be upset with some of the above discussion as it ignores some nuances that mean a lot to them, but those nuances really are not important to the rest of us in applied settings.

Definition 2: A differential equation is an equation with an unknown function and one or more of its derivatives. Its derivatives can be processed using one or more integration techniques resulting in the unknown function. Finally, here are some examples of differential equations in three different forms.

Newton (prime) notation	Leibniz (differential) notation	Operator (modern) notation
$y' + y = x + 3$	$\frac{\mathrm{d}y}{\mathrm{d}x} + y = x + 3$	$\mathrm{D}y + y = x + 3$
$y'' + y = x + 3$	$\frac{\mathrm{d}^2y}{\mathrm{d}x^2} + y = x + 3$	$\mathrm{D}^2y + y = x + 3$

The purpose of a differential-equation class or book is to teach you how to solve for the function *y*—that is, *y*(*x*)—in equations such as those shown above. Put another way, you will be solving for a function, not a (scalar) variable as you did in algebra.

Chapter 1
Getting Started

A differential equation is an equation in one or more variables involving one or more of its own derivatives. A major goal of taking a class in differential equations is to solve for $f(x)$ if given $f'(x)$, $f''(x)$, etc. That is, somehow work backward and find the function whose derivative is given. For example, given $f'(x) = 2x$, what is $f(x)$?

Once we obtain the mystery $f(x)$, we can evaluate it for any desired x. What would be the function whose derivative, $\frac{dy}{dx}$, is $2x$? We know from differential calculus that the derivative of $y = x^2$ would be $2x$. Hence, if $\frac{dy}{dx} = 2x$ then $y = x^2$. (This can also be stated as: if $f'(x) = 2x$, then $f(x) = x^2$.)

But wait! A vertical shift of a function does not impact the slope of tangent lines! In the figure at right, the tangent slopes to the curves $y = x^2 - 1$ (green), $y = x^2 + 2$ (red), and $y = x^2 - 3$ (blue) are all equal when $x = a$, when $x = 0$, and when $x = b$. There is not a single functional solution (one unique function whose derivative is $2x$), but a set or family of them. We would indicate this by stating that for $\frac{dy}{dx} = 2x$ the set or family of function solution*s* is $x^2 + c$, where c represents any constant. The vertical lines here, $x = a$, $x = 0$, and $x = b$ are called isoclines. An isocline is notable because the slope of each function at each point on the isocline is the same. Hence, the goal of solving a differential equation is to solve for a function ***or family of functions*** which, when substituted into the original equation, will balance the two sides or make them equal.

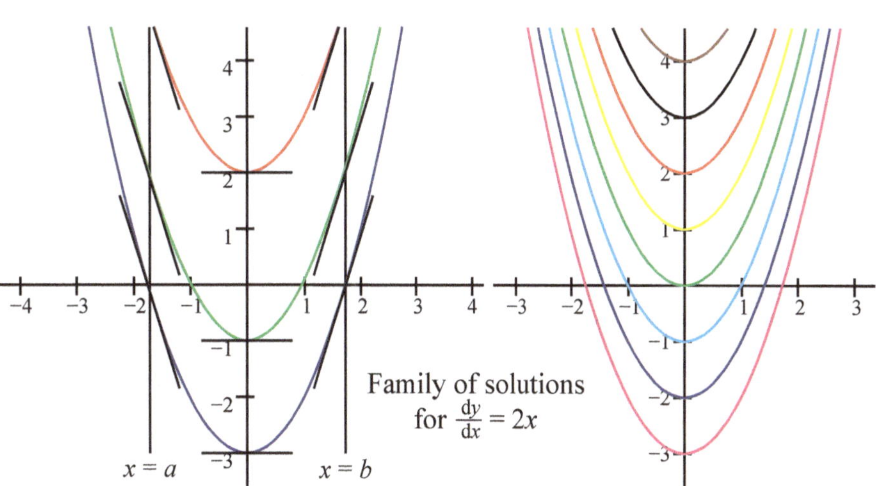

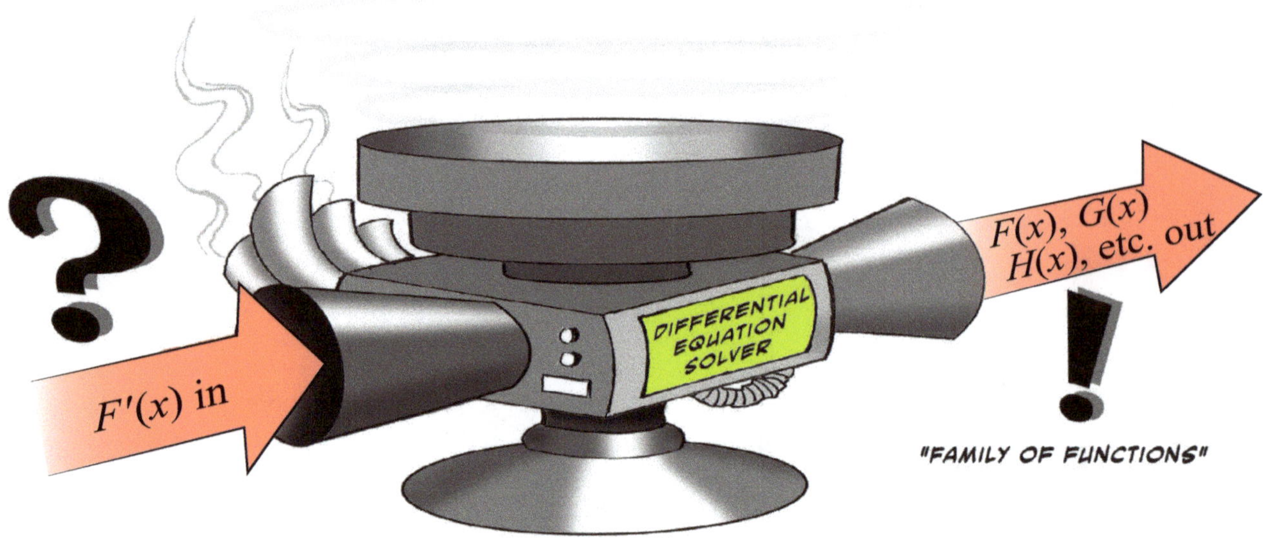

From that family of function solutions we often, using information given us, will identify the particular one that is appropriate to our situation and use it to evaluate for specific values of x. That is called solving an initial value problem (IVP). In differential calculus, we studied how to obtain a derivative function from a given function. In differential equations, we study how to obtain a function from a given differential.

In a famous French play *Le Bourgeois Gentilhomme* by Molière, a comical but buffoonish character, Monsieur Jourdain, is amazed to learn that he had been speaking prose all his life and didn't even know it.

"Par ma foi! Il y a plus de quarante ans que je dis de la prose sans que j'en susse rien, et je vous suis le plus obligé du monde de m'avoir appris cela."

As Monsieur Jourdain discovers, it is quite possible to speak prose without knowing that you are doing it. However, it is very, very difficult to learn abstract skills and abstractions without, ***at some conscious or unconscious level***, building upon earlier more primitive ones. Differential-equations teachers will not tell you the information that follows because they are so smart and they have internalized it so deeply that they assume that you have too. To be fair to them, the university curriculum requires that they move at a very fast pace, chop-chop. Nevertheless, it is a truism that mathematicians are not always "math educators." Be grateful if you have a teacher who can move comfortably between the two worlds. Thank him or her.

It turns out that you have been practicing many of the necessary skills to work with and solve differential equation problems since algebra and you did not even know it. Making the connections shown in the following page or so of text will level the steep learning curve leading to differential equations and perhaps give you the confidence to say, "Wow! I have been doing much of this work since algebra and I did not even know it!"

There are many, many coplanar lines that can be drawn with the same linear slope. Lines that have the same slope are said to be parallel. They do not intersect.

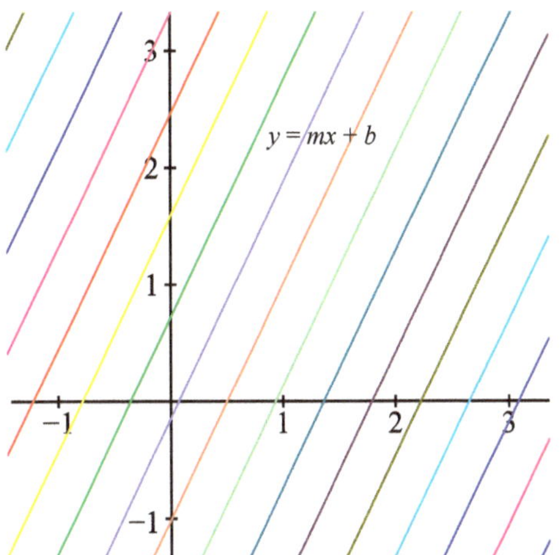

Euclid's famous fifth postulate suggests that only one of a set of parallel lines can pass through a given point.

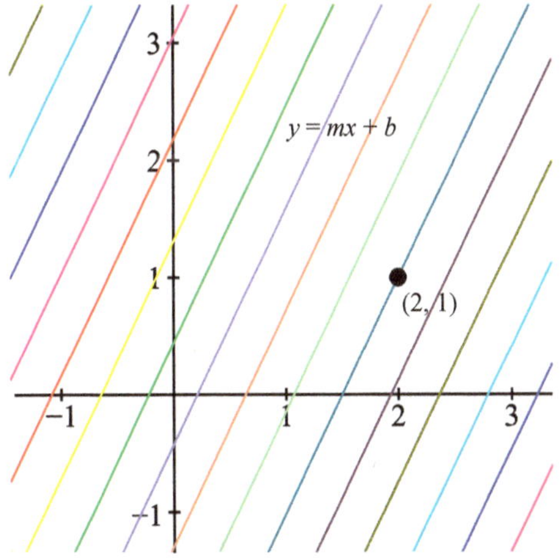

We use the above assumption to identify one of a family of parallel lines. Find the equation of a line, $y = mx + b$, that has a slope of 2 and passes through the point (2, 1).

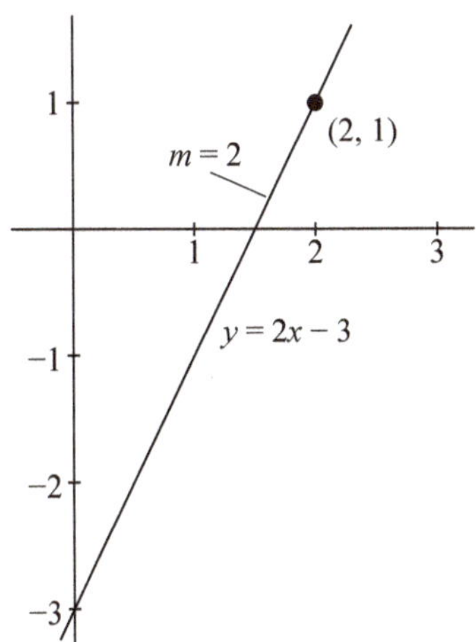

$$y = mx + b \rightarrow 1 = 2(2) + b \rightarrow b = -3$$

The line passing through (2, 1) with a slope of 2 is $f(x) = 2x - 3$.

There are many, many quadratic equations that can be drawn each with the same function slope. For example, $\frac{dy}{dx} = 2x$. Coplanar curves with the same slope can be said to be "parallel." They do not intersect.

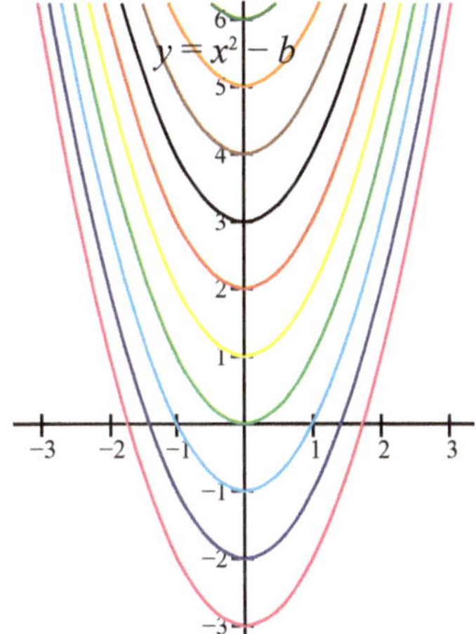

The "Theorem of Uniqueness" says that only one of a family of "parallel" curves can pass through a given point.

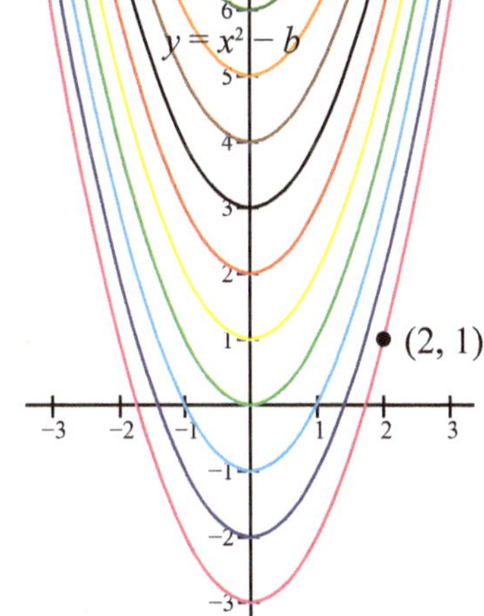

We use the above theorem to identify one of a family of "parallel" parabolas. Find the equation of the parabola, $y = x^2 + b$, that passes through the point (2, 1). Here the slope of $y = x^2 + b$ is $2x$ because $\frac{d}{dx}(x^2 + b) = 2x$.

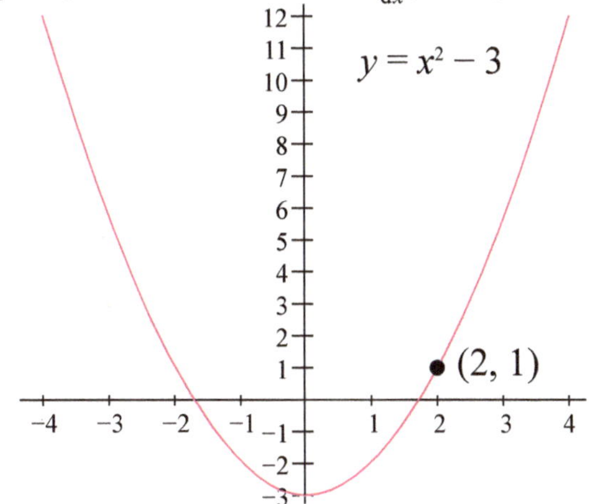

$$y = x^2 + b \rightarrow 1 = 2^2 + b \rightarrow b = -3$$

The parabola passing through (2, 1) with a slope of 2 is $f(x) = x^2 - 3$.

It is possible to find the equation of the line with given slope and passing through a given point. You could then find the value of y for any given x on that line: $y = f(x)$.

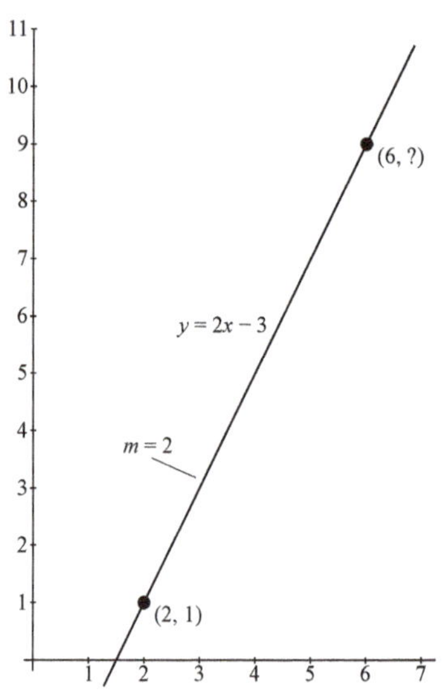

$y = 2x - 3 \rightarrow y = 2(6) - 3 \rightarrow y = 12 - 3 \rightarrow y = 9$

The line with slope 2 passing through (2, 1) also passes through (6, 9). Only one straight line does this.

It may be possible to find a specific function from a family of solutions to $\frac{dy}{dx}$ that passes through a specific point. ***This is called solving an initial value problem (IVP)***. You could then find the y for any given x on that function, $y = x^2 - 3$.

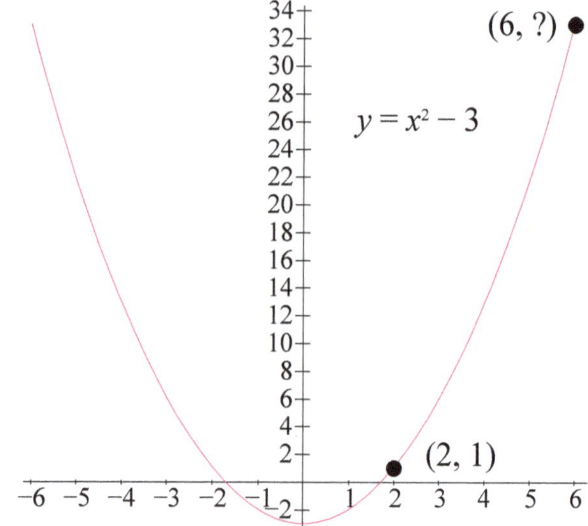

$y = x^2 - 3 \rightarrow y = 6^2 - 3 \rightarrow y = 36 - 3 \rightarrow y = 33$

The parabola with slope $2x$ passing through (2, 1) also passes through (6, 33). Only one parabola does this. The complicated existence and uniqueness theorem addresses this in a real differential-equations class.

Math is easier if you can make connections to previous topics! Differential equations can be seen as a curvilinear version of work done previously with linear functions.

Or, as Monsieur Jourdain would have said, “Par ma foi! Il y a plus de quarante ans que je dis de la prose sans que j’en susse rien, et je vous suis le plus obligé du monde de m’avoir appris cela.”

In Chapter 1, the differential equation we have been using was $\frac{dy}{dx} = 2x$ (or $f'(x) = 2x$). For that simple equation it was intuitive from beginning calculus that $f(x) = x^2$. Any differential equation of any consequence will not be solved by inspection. It is good to understand that working with differential equations is kind of a reverse process of differential calculus. Since the differential equation ($f'(x)$ or $\frac{dy}{dx} = 2x$) was obtained by differentiating $f(x)$, then perhaps you can anticipate that $f(x)$ will be obtained by integrating $f'(x)$.

For $f(x) = x^2 - 3$ the derivative $f'(x) = 2x$.

Therefore it follows that:

$\frac{dy}{dx} = 2x$	given differential equation
$dy = 2x\,dx$	multiply by dx
$\int dy = \int 2x\,dx$	integrate both sides
$\int y^0\,dy = 2\int x^1\,dx$	integration continued
$\frac{y^{0+1}}{0+1} = 2\frac{x^{1+1}}{1+1}$	integration continued
$y = 2\frac{x^2}{2}$	simplify
$y = x^2$	simplify again
$y = x^2 + c$	There are many functions with $\frac{dy}{dx} = 2x$.

If it is given that the function $y = x^2 + c$ passes through the point (2, 1), then

$1 = 2^2 + c$

$c = -3$

$y = x^2 + c$

$y = x^2 - 3$ equation of the parabola of form $y = x^2 + b$ and passing through (2, 1)

check: $\frac{d}{dx}(x^2 - 3) = 2x!$

When $x = 6$, $y = x^2 - 3 = 33$.

Of all the functions whose derivative is $\frac{dy}{dx} = 2x$, only one passes through the point (2, 1). That function also passes through the point (6, 33).

We are used to working with functions and function notation from both algebra and calculus. We have seen both symbols $f'(x)$ and $\frac{dy}{dx}$. However, up to now, that notation has mostly been used to evaluate a scalar or to indicate a function. For example, for $f(x) = x^2$, we have $f'(x) = 2x$ or $f'(5) = 10$. In differential equations, it will often be helpful to think of $f'(x)$, $\frac{dy}{dx}$, as an infinite set of tiny tangent segments, so tiny that each line segment is the length of a point. (Points don't have length ... use your imagination ... think of a computer screen pixel.) Commercial computer software is available to create *slope fields*. Because the brain has a tendency to "fill in" gaps, you can, with imagination, "see" a finite representation of that desired set of functions that results from solving the "differential function." The following is a progression of slope fields with $\Delta x = 1$, 0.6, 0.3, and 0.1.

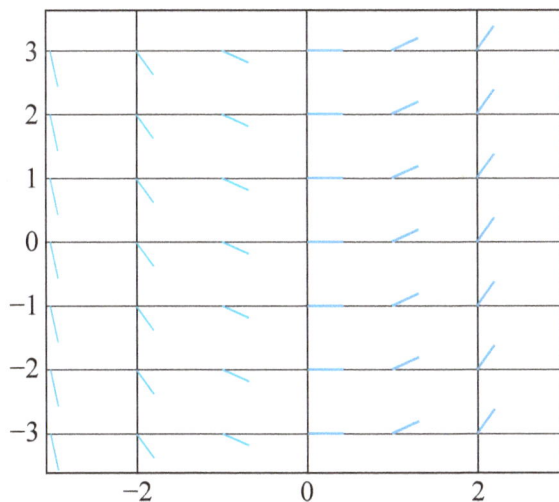

$\frac{dy}{dx} = 2x$, $\Delta x = 1$

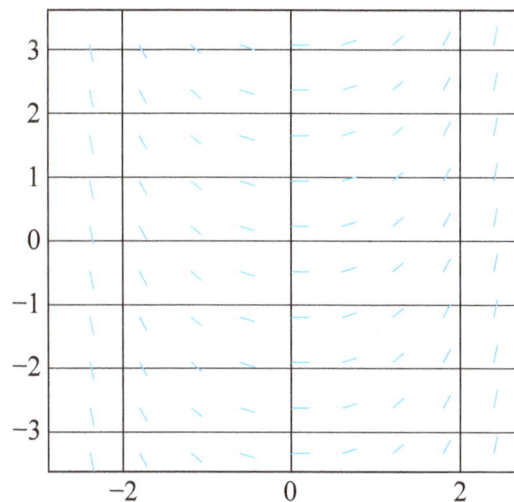

$\frac{dy}{dx} = 2x$, $\Delta x = 0.6$

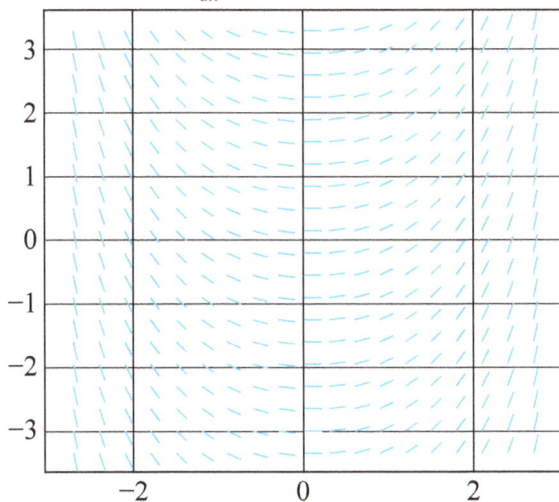

$\frac{dy}{dx} = 2x$, $\Delta x = 0.3$

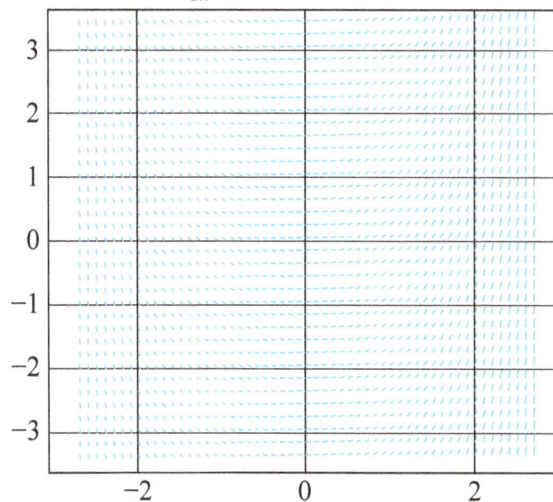

$\frac{dy}{dx} = 2x$, $\Delta x = 0.1$

Putting all these ideas together into one marvelous MATLAB screen we see the slope field family prescribed by the differential equation $\frac{dy}{dx} = f'(x) = 2x$. We know that there is a specific function, $f(x)$, somewhere in the slope field family of functions whose derivative is $2x$, that passes through the point (2, 1). Solving the differential equation $f'(x)$, we get $f(x) = x^2 + c$ where c indicates an integration constant, a vertical shift of $f(x) = x^2$. Using the information that the specific function we want passes through the point (2, 1), we solved for c by substituting the (x, y) values (2, 1) into $y = x^2 + c$ to find $c = -3$. So $y = x^2 + c$, for IVP (2,1), becomes $y = x^2 - 3$. Then we could find, for $x = 6$, $y = 6^2 - 3 = 33$!!

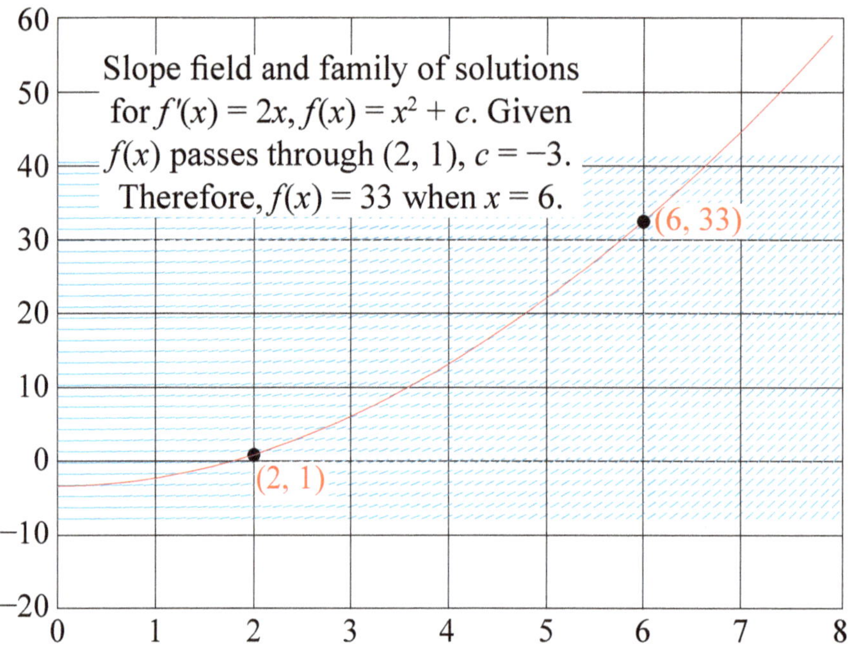

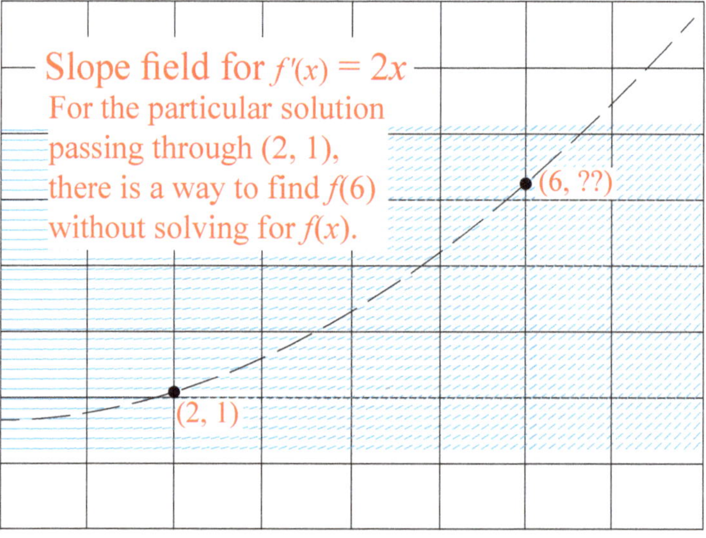

Slope field for $f'(x) = 2x$
For the particular solution
passing through (2, 1),
there is a way to find $f(6)$
without solving for $f(x)$.
(6, ??)
(2, 1)

Do you remember in calculus how you learned to successively approximate the slope of a tangent line by calculating the slope of approaching secant lines? (See Chapter 0.) Well a smart Swiss mathematician, Leonhard Euler, 1707–1783, figured out a way to successively approximate the solution to a differential equation for a given value of x. The genius of Euler—who is credited with this iterative method of solving for $f(x)$ if given $\frac{dy}{dx}$, that is $f'(x)$—was that 1) he could mentally visualize the entire slope field for $\frac{dy}{dx}$, that is every $f(x)$ whose derivative is $f'(x)$, and 2) he realized that he could use the known fact that (x, y) was on the unknown $f(x)$ to successively approach $f(z)$ for any z using algebra and trigonometry and the known $f'(x)$—i.e., $\frac{dy}{dx}$. Today, we have graphing calculators and computer software packages that will help us see what Euler could visualize in the 1700s. As the slope indicators become shorter and the tangent indicators become more numerous, each specific (particular) slope field approaches a particular solution, $f(x)$. Using the information we do have, $\frac{dy}{dx} = f'(x) = 2x$ and the fact that $f(x)$ passes through the point (2, 1), we can get an initial estimation of $f(6)$ without knowing the function $f(x)$.

In the figure above, we have a point (2, 1) and we know the slope of the unknown function at $x = 2$ is 4. That is, $f'(2) = 2 \times 2 = 4$. We wish to find the unknown point (6, ?) on the unknown function $f(x)$. With the two given x values (2 & 6), we can determine $\Delta x = 6 - 2 = 4$. If we could determine Δy, we could calculate $y + \Delta y$ and obtain the new y value. How can we determine the Δy value? We know from studying trigonometry that $\tan\theta = \frac{opp=y}{adj=x}$ hence $opp = (\tan\theta) \times adj$. We know from studying algebra that $m = \frac{\text{rise}=y}{\text{run}=x}$. So, $\tan\theta = m$.

1. The slope $= m = \frac{y}{x} = \frac{opp}{adj}$.
2. $opp = m \times adj$

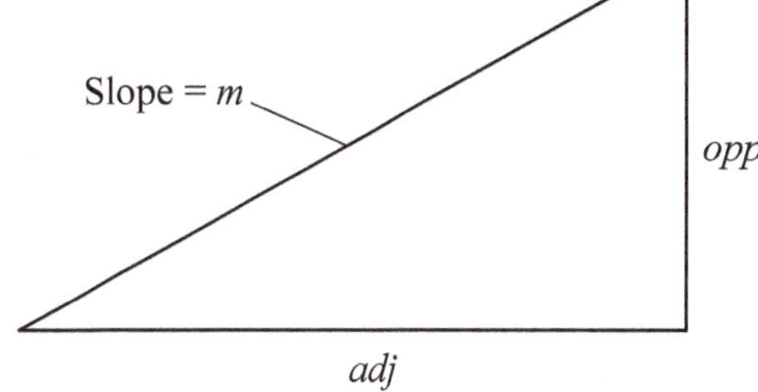

Substituting into $opp = m \times adj$

$$\Delta y = m \times \Delta x$$
$$\Delta y = 2x \times 4 \ldots\ m \text{ at point } (2, 1) \text{ is } 2x$$
$$\Delta y = 4 \times 4 = 16 \ldots\ m \text{ at point } (2, 1) = 4$$

New y = old $y + \Delta y$. $y = 1 + 16 = 17$. We estimate that $f(6)$ for the unknown function, $f(x)$, might be 17. Basically we have "extended" out (extrapolated) the tangent line until it reached an x value of 6.

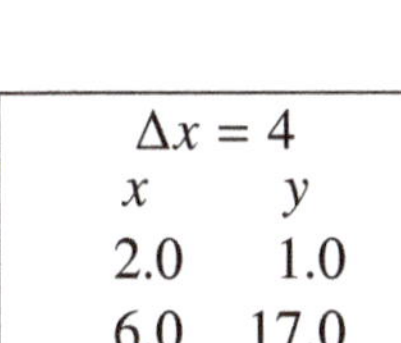

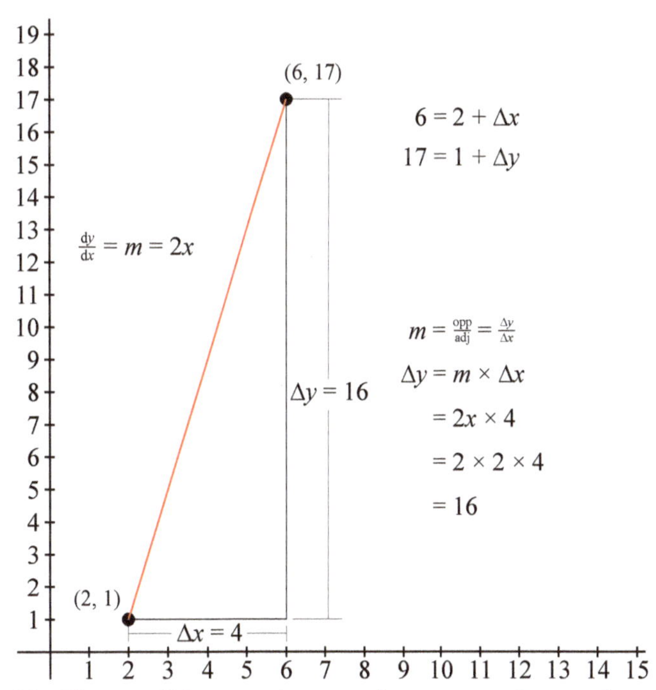

Recall from differential calculus (reviewed in Chapter 0) how we improved our secant slope estimate of a tangent slope in the example by decreasing Δx:

x (domain)	y (range)	$m = \frac{y_2-y_1}{x_2-x_1}$	
0	0	$m = \frac{64-0}{8-0} = 8$	↓
4	16	$m = \frac{64-16}{8-4} = 12$	↓
6	36	$m = \frac{64-36}{8-6} = 14$	↓
7	49	$m = \frac{64-49}{8-7} = 15$	↓
7.9	62.41	$m = \frac{64-62.41}{8-7.9} = 15.9$	↓
7.99	63.8401	$m = \frac{64-63.8401}{8-7.99} = 15.99$	↓
7.999	63.984001	$m = \frac{64-63.984001}{8-7.999} = 15.999$	↓
8	64	$m = \frac{64-64}{8-8} = \frac{0}{0} = 16???$ Indeterminate division by zero	

That same "successive approximation" technique can be used in differential equations to improve our estimation of $f(6)$. That is, given $f'(x) = \frac{dy}{dx} = 2x$ and the fact that the function we seek passes through point (2, 1), we can improve our original estimation of the unknown $f(x)$ at 6—$f(6) \sim 17$—by reducing the Δx we used to project out from (2, 1).

For the example above, change from $\Delta x = 4$ to $\Delta x = 1$.

$$\Delta x = 1, \quad \Delta y = 2x_n \times \Delta x \quad \left(\text{because } \frac{dy}{dx} = 2x, \text{ so } dy = 2x \times dx\right)$$

N	$x_n = x_{n-1} + \Delta x$	$\Delta y = 2 \times x_{n-1} \times \Delta x$	$y_n = y_{n-1} + \Delta y$	(x_n, y_n)
1	$x_1 = 2$ (given)	Not applicable	$y_1 = 1$ (given)	(2, 1)
2	$x_2 = 2 + 1 = 3$	$2 \times 2 \times 1 = 4$	$y_2 = 1 + 4 = 5$	(3, 5)
3	$x_3 = 3 + 1 = 4$	$2 \times 3 \times 1 = 6$	$y_3 = 5 + 6 = 11$	(4, 11)
4	$x_4 = 4 + 1 = 5$	$2 \times 4 \times 1 = 8$	$y_4 = 11 + 8 = 19$	(5, 19)
5	$x_5 = 5 + 1 = 6$	$2 \times 5 \times 1 = 10$	$y_5 = 19 + 10 = 29$	(6, 29)

By decreasing Δx, we get values closer and closer to the actual $f(6)$. The text box at right and the five text boxes below were generated by the computer program shown on the following page.

$\Delta x = 1$

x	y
2.0	1.0
3.0	5.0
4.0	11.0
5.0	19.0
6.0	29.0

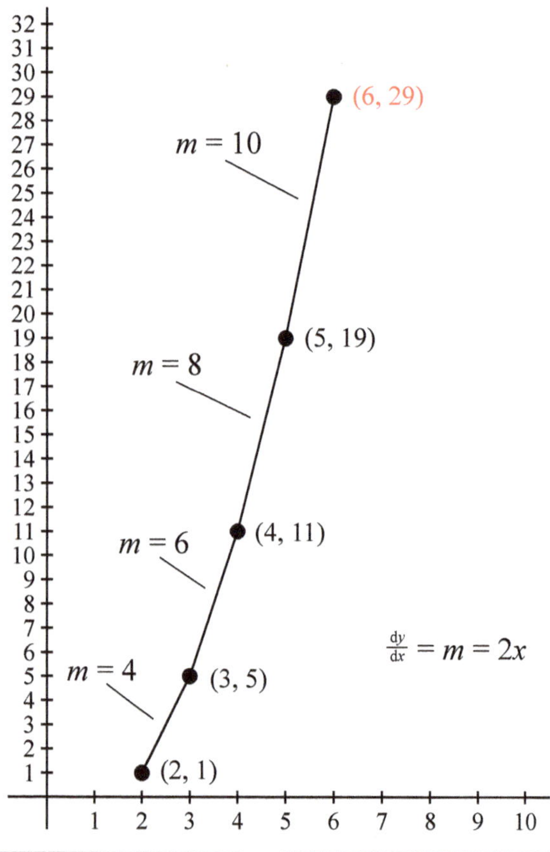

$\Delta x = 4$

x	y
2.0	1.0
6.0	17.0

$\Delta x = 1$

x	y
2.0	1.0
3.0	5.0
4.0	11.0
5.0	19.0
6.0	29.0

$\Delta x = 0.1$

x	y
2.0	1.00
2.1	1.40
2.2	1.82
⋮	⋮
5.7	29.12
5.8	30.26
5.9	31.42
6.0	32.66

$\Delta x = 0.01$

x	y
2.00	1.00
2.01	1.04
2.02	1.08
2.03	1.12
⋮	⋮
5.97	32.60
5.98	32.72
5.99	32.84
6.00	32.96

$\Delta x = 0.001$

x	y
2.000	1.000
2.001	1.004
2.002	1.008
2.003	1.012
2.004	1.016
⋮	⋮
5.997	32.960
5.998	32.972
5.999	32.984
6.000	32.996

Compare (6, 32.996) for $\Delta x = 0.001$, with the exact solution (6, 33) we got several pages back. Fortunately, these answers are close. Otherwise, there would be egg on the author's face!

```
public class EulersMethodSolvingAnODE
{
  public static void main(String args[])
  {
    double x = 2;
    double y = 1;
    double deltaX = 1; // Δx is run for 4, 1, 0.1, 0.01, and 0.001
    System.out.println("x y"); // column headers
    while (x <= 6)
    {
      System.out.println(x + ", " + y);
      // tan theta = opp/adj, so opp = tan theta * adj. Ergo y_n = y_n - 1 + dy/dx * Δ x
      y = y + (2 * x) * deltaX; // new y = old y + length side opposite theta
      x = x + deltaX; // new x = old x + deltaX
    } // end while
  } // end main
} // end JavaTemplate
```

There are software packages available that allow you to experiment with the ideas taught in a differential equations class. A jar file can be purchased from Cengage Publishers. One of the options allows the user to experiment with Euler's method for different differential equations.

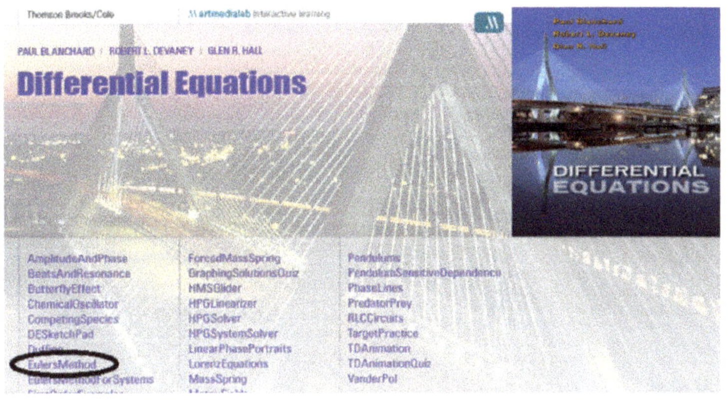

Below, passing through (0.24, 1.2), you see Euler's method for $\Delta x = 1, 0.5, 0.25, 0.125$, and the Runge–Kutta 4 algorithm applied to the differential equation $\frac{dy}{dt} = y^2 - 4t$.

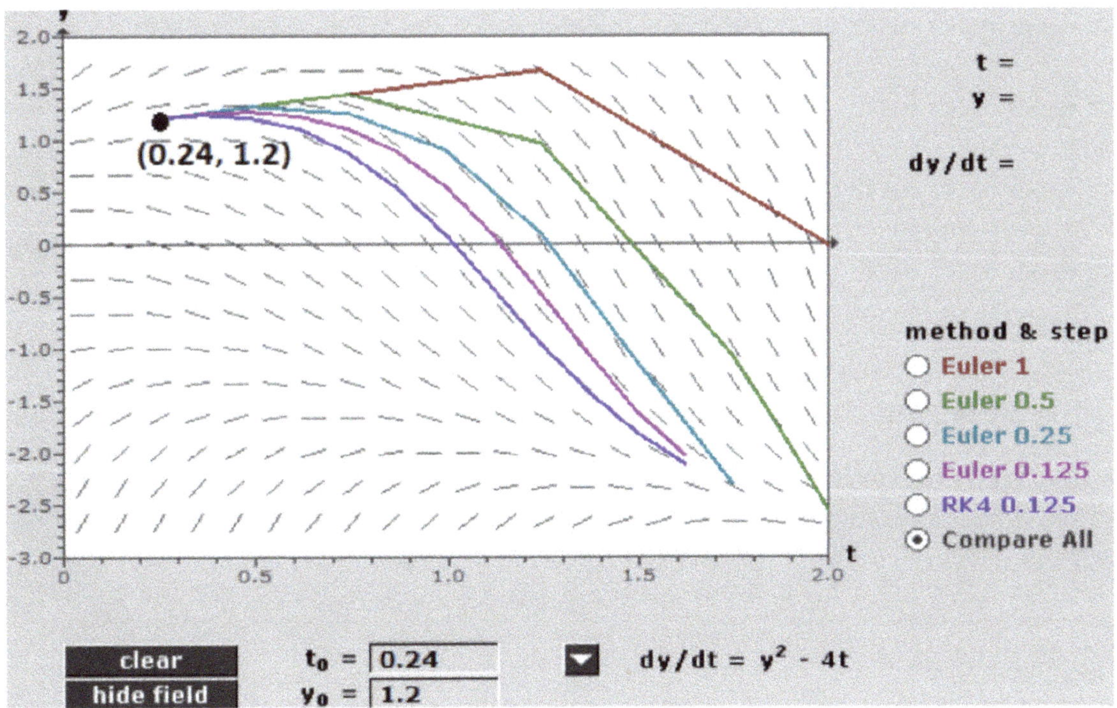

For people with a programming background it is not too much of a stretch to understand where the slope fields and the graphs of a particular function come from. Code based on Euler's method (or perhaps another iterative algorithm called the Runge–Kutta method) is used on the given differential equation, $\frac{dy}{dx}$ or $f'(x)$, to generate a set

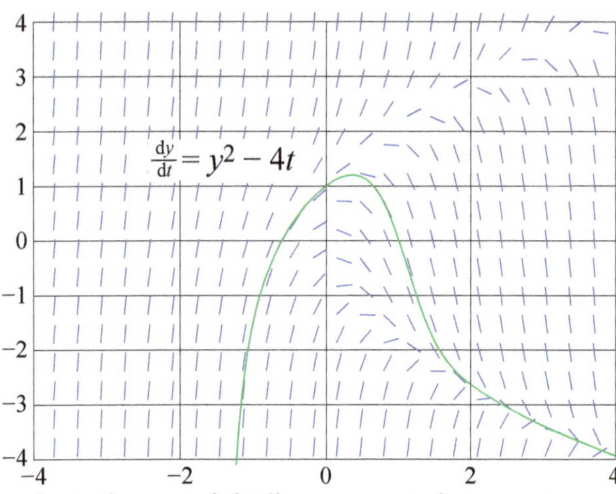

of ordered pairs of points which, as closely as desired, approximate points lying on the unknown function, $f(x)$. Then, using the slope information immediately available from the differential equation ($\frac{dy}{dx} = m$, right?) together with any desired delta value and simple trig function knowledge such as shown above in Euler's method, one could write code to identify end points of a tangent segment to the unknown $f(x)$. The code to draw straight-line segments between two points is well known and precoded in many languages (e.g., `g.drawLine` in Java). Voilà! There you have a tangent segment drawn to a point that is on, or very, very close to being on the unknown function curve. If you bring your delta close to zero, then the tangent-line slope segment approaches length "zero" which in "computerese" means one pixel. Then your graph would be a *very close visualization of the unknown solution function itself.* That function (shown above) can be thought of as a particular example of the family of solutions differentiated from the others by the fact that each of its tangent-slope segments are of length "one pixel." Look at the figure and reread this paragraph if it was not clear at first.

$\Delta x = 0.001$	
x	y
−1.000	−1.476
−0.999	−1.468
−0.998	−1.461
−0.997	−1.455
⋮	⋮
3.995	−3.937
3.996	−3.938
3.997	−3.939
3.998	−3.940
3.999	−3.941
4.000	−3.942

Chapter 1 Review

Chapter 1 starts out defining a differential equation as an equation in one or more variables involving one or more of its own derivatives. A major goal of taking a class in differential equations is to solve for $f(x)$ if given $f'(x)$, $f''(x)$, etc. That is, somehow work backward and find the function whose derivative is given. That paradigm is just the opposite of differential calculus when you are given a function and asked to find its derivative. **Because, in solving a differential equation, you are solving for a function, sometimes the $f(x)$ is abbreviated as f.**

Side-by-side examples of coplanar parallel linear lines and coplanar "parallel" parabolas were given. It was shown that, if given a point on a straight line as well as its slope, you could find the y value of a different point on that line if given its x value. Similarly, it is the case for the parabolas. The concepts *slope field* and *particular solution* were introduced as well as Euler's iterative technique for approximating an unknown point on an unknown function whose derivative and initial value are known.

Online Application

Visit demonstrations.wolfram.com/NumericalMethodsForDifferentialEquations for a great little application that will demonstrate some of the concepts from this chapter. "Numerical Methods for Differential Equations" from the Wolfram Demonstrations Project. Contributed by Edda Eich-Soellner.

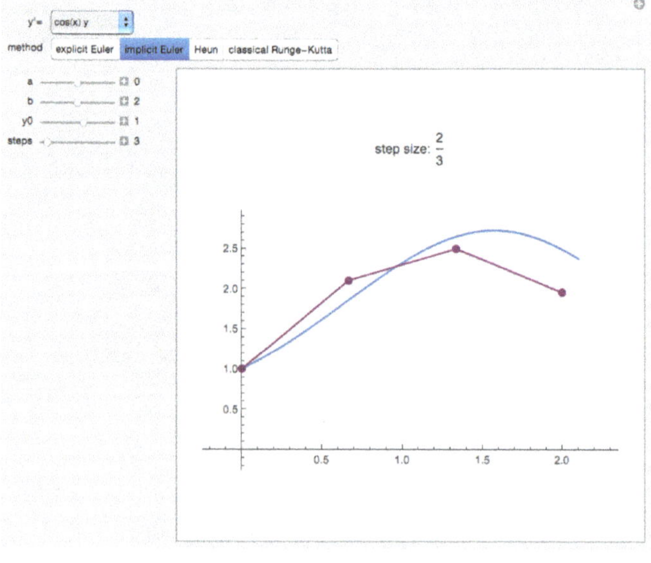

Chapter 2

Using Discrete Math to Solve Growth and Decay Problems

Author's note: Many of the ideas from Chapter 1 are reviewed here in a new context along with the four concepts listed below. If review and reinforcement bother you, perhaps you should jump ahead or perhaps these materials are not suited for you and you should switch to a presentation more appropriate to your level.

A large part of the beauty of mathematics is the synergism of ideas. That synergism, however, is part of what makes the discipline enigmatic to many. The category of math problems known as "differential growth and decay problems" is covered in Chapters 2 and 3. They are the synthesis of many ideas, but prominently these four: direct proportion (also known as direct variation, $y = kx$), geometric sequences ($ar^0, ar^1, ar^2, ar^3, \ldots, ar^{n-1}$), discrete versus continuous variables, and integration, $\lim_{n\to\infty} \sum_{i=1}^{n} f(x_i)\Delta x$.

Direct Variation, $y = kx$

The terms direct variation and directly proportional refer to a functional relationship between domain (x) and range (y) values controlled by a single constant (k). For example, $y = kx$. Here, as x gets larger, so does the y value. However, the rate of increase of the y values depends upon k (the constant of proportionality). As k gets larger, y gets larger faster, and as k gets smaller, the y value grows more slowly. This is shown in the three tables and graph at right.

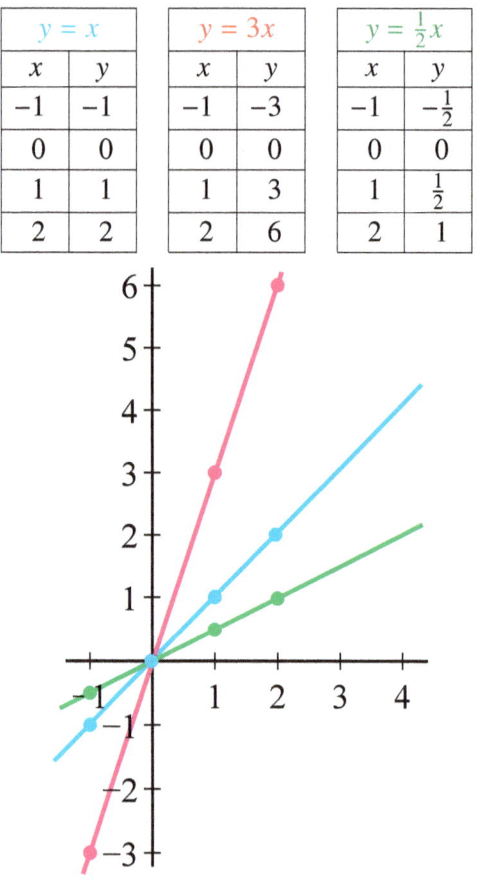

$y = x$	
x	y
−1	−1
0	0
1	1
2	2

$y = 3x$	
x	y
−1	−3
0	0
1	3
2	6

$y = \frac{1}{2}x$	
x	y
−1	$-\frac{1}{2}$
0	0
1	$\frac{1}{2}$
2	1

In Algebra 1, we often see problems such as the following. Those who work more hours get paid more. Suppose a person worked four hours and received \$85. How much would they get paid for working 11 hours? There are different ways to approach such a problem, but the following way is best for introducing the ideas in this chapter and the next. Let p equal the amount of pay and h equal the number of hours worked. Then the amount of pay could be indicated by the formula $p = r \times h$ where r is the hourly rate of pay. The r in the formula can also be thought of as a "growth factor" or a "constant of proportionality."

Step #1
$p = rh$
$85 = r \times 4$
$r = \$21.25 / \text{hr}$

Step #2
$p = rh$
$p = \$21.25 \times 11$
$p = \$233.75$

This sort of problem is called a *direct-variation* problem. That is, the pay varies directly with the number of hours worked. The pay is also dependent upon the person's hourly rate. In the problem condition here, you can use the fact (initial value problem, IVP) that the person works four hours for \$85 to determine the hourly rate. Then, you can use that information to establish a formula ($p = 21.25 \times h$) allowing you to calculate the person's pay for any desired number of hours. In beginning algebra, the work you do will usually result in a linear equation. In calculus the work you do will invariably be curvilinear. For many people, it is helpful to treat calculus curvilinear direct-variation problems as an advanced version of what you have already done in earlier math: algebraic linear direct variation. That way, what you are learning is not really a new concept, but an advanced version of a previous topic. This allows you to concentrate on the details and not be confused about the concept.

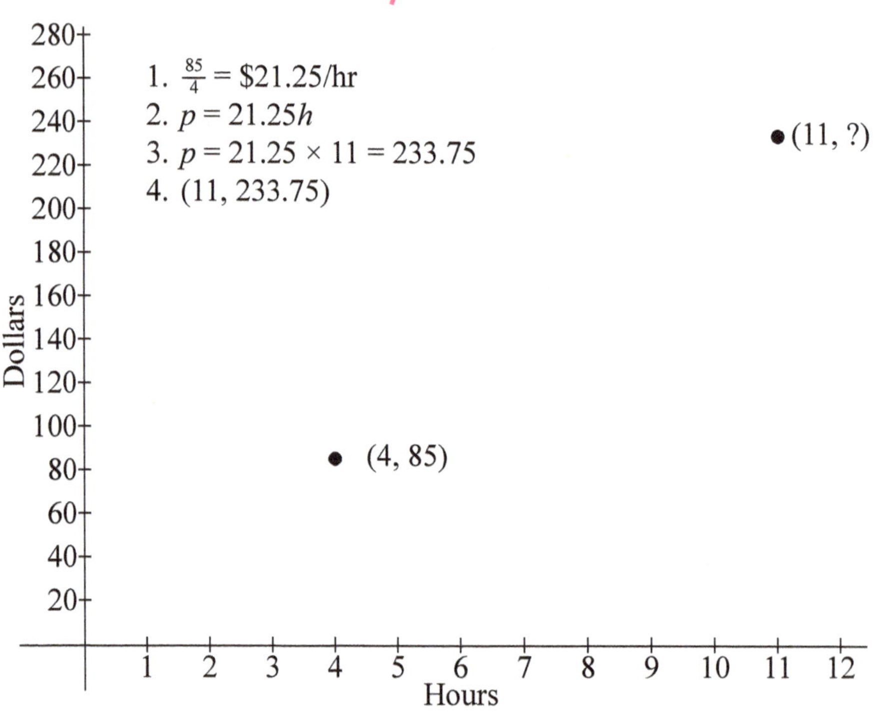

Biologists make different assumptions when they make predictions about wildlife populations. For example, one such assumption might be that the ***change of the population over time*** is directly proportional to the current population, $\frac{dp}{dt} = kp$. As time increases the population will increase at a rate determined by k which might represent the interaction of birth and death rates. The value of k will help determine the appearance of the family of solution functions. Following are slope fields for $\frac{dp}{dt}$ for $k = 2$ and $k = \frac{1}{3}$. (In both cases, if the point (t, p) were given, the indicated slope could be used to find $p(t)$ allowing you to find the unknown function's value for any t value).

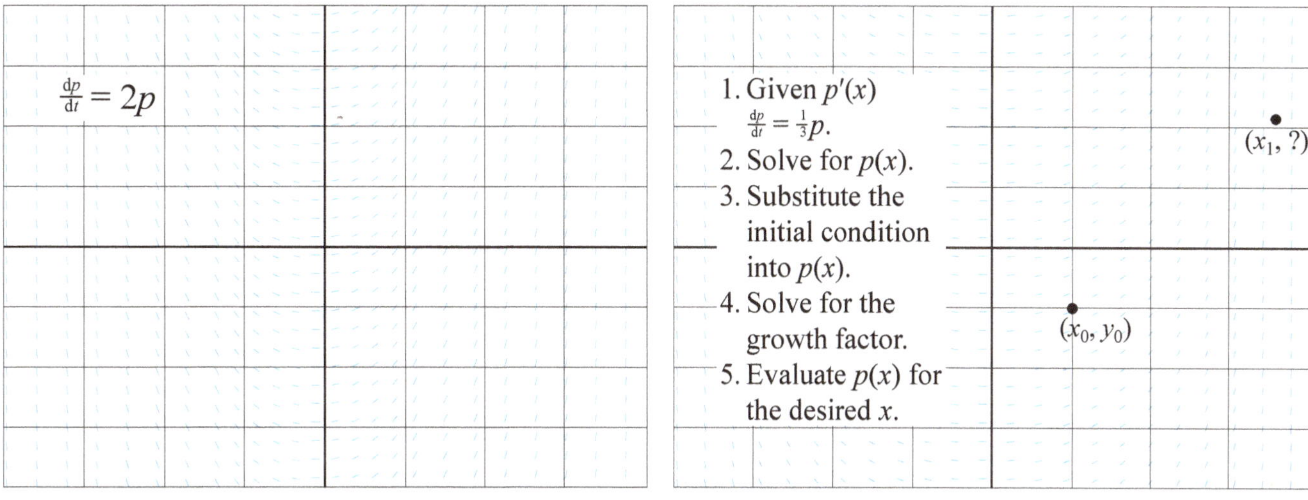

$\frac{dp}{dt} = 2p$
1. Given $p'(x)$ $\frac{dp}{dt} = \frac{1}{3}p$.
2. Solve for $p(x)$.
3. Substitute the initial condition into $p(x)$.
4. Solve for the growth factor.
5. Evaluate $p(x)$ for the desired x.
$(x_1, ?)$
(x_0, y_0)

Previously, we looked at a direct-variation problem that could be solved using algebra—$y = kx$, a linear function. When that happened, if you had information that the graph passed through a point, you could use the point's (x, y) value, substitute into $y = kx$, obtain k and then use k to back substitute into the given equation. With the now-known value of k, you could solve for y given any x value. In differential equations, you might be faced with an equation of the form $\frac{dp}{dt} = kp$ (or $p'(t) = kp$). Here you are given a direct variation where a rate of change, $\frac{dp}{dt}$, is directly related to something. You will need to solve the differential equation, $\frac{dp}{dt} = kp$, for p (or $p(t)$). If, at that time, you had information regarding a point (t, p) that the function $p(t)$ passed through you could solve for the *growth factor* (similar to solving for k in $y = kx$) and use that growth factor in the function $p(t)$ to solve for any t. Here you would need differential equations to solve the problem. If you understand that solving the equation $\frac{dp}{dt} = kp$ for t is analogous to and very similar to solving $y = kx$, for a given x, then solving $\frac{dp}{dt} = kp$ is not so intimidating and formidable.

Geometric Sequence ($ar^0, ar^1, ar^2, ar^3, \ldots, ar^{n-1}$)

Suppose a bank paid 5% interest to a depositor at the end of every year (annually). However, at the depositor's request the money was not paid outright but added to the money already in the account. Let's trace the value of $1,000 for 20 years through a table using the formula $I = prt$. So, for $t = 1$, $I = pr$.

Ending Year of Deposit	Value of Deposit
0	$1,000
1	$1,000 + $1,000(0.05) = $1,000 + $50 = $1,050 $p_0 + p_0r = p_1$
2	$1,050 + $1,050(0.05) = $1,050 + $52.50 = $1,102.50 $p_1 + p_1r = p_2$
3	$1,102.50 + $1,102.50(0.05) = $1,102.50 + $55.125 = $1,157.625 $p_2 + p_2r = p_3$
4	$1,157.625 + $1,157.625(0.05) = $1,157.625 + $57.88125 = $1,215.50625 $p_3 + p_3r = p_4$
⋮	⋮
20	$p_{19} + p_{19}r = p_{20}$

Just kidding about that 20-year trace. The arithmetic is getting painful. There is a pattern. Perhaps, instead of continuing with this brute-force approach, we should be studying the pattern to obtain some sort of generic approach. The value of the deposit (principal) of each subsequent year is equal to the previous year deposit plus 0.05 times the value of that deposit. $p_n = p_{n-1} + (p_{n-1} \times r)$ where $r = 0.05$. Clearly if we had p_{19} it would be really easy to obtain p_{20}. But we don't have p_{19}. Let's look again, but we'll use what we learn in the table below. $1,000 invested for 20 years at 5% interest is $1{,}000 \times (1 + 0.5)^{20}$ = $2,653.30. (Here, interest is compounded annually.)

Ending Year of Deposit	Value of Deposit
0	p_0
1	$p_0 + p_0r = p_0(1 + r) = p_1$
2	$p_1 + p_1r = p_1(1 + r)$ $= [p_0(1 + r)](1 + r)$ substitute $p_1 = p_0(1 + r)$ $= p_0(1 + r)^2 = p_2$
3	$p_2 + p_2r = p_2(1 + r)$ $= [p_0(1 + r)^2](1 + r)$ substitute $p_2 = p_0(1 + r)^2$ $= p_0(1 + r)^3 = p_3$
4	$p_3 + p_3r = p_3(1 + r)$ $= [p_0(1 + r)^3](1 + r)$ substitute $p_3 = p_0(1 + r)^3$ $= p_0(1 + r)^4 = p_4$
⋮	⋮
20	$p_{19} + p_{19}r = p_{19}(1 + r)$ $= [p_0(1 + r)^{19}](1 + r)$ substitute $p_{19} = p_0(1 + r)^{19}$ $= p_0(1 + r)^{20} = p_{20}$

Compare the following two formulas! The first formula finds a value for a specific term by incrementing the ***previous*** *term. The second formula finds a value for a specific term by multiplying the* ***first*** *term by a growth factor. The latter approach will be used in the next chapter. Be sure to understand this difference!*

$$p_n = p_{n-1} + (p_{n-1} \times r) \qquad p_n = p_0(1 + r)^n$$

The $(1 + r)^n$ in the second formula can be called a growth factor. The larger r (the geometric ratio) or n (the number of periods), the larger the growth factor.

According to Nellie's Red Book of Car Values, a car's value depreciates 10% a year. Assuming that it is still running, how much would a car purchased for \$20,000 be worth after 20 years? (10% of \$20,000 is \$2,000, so the car is worth \$18,000 at the end of year one.)

Ending Year of Ownership	Value of the Car at the End of the Year	
0	\$20,000	
1	\$20,000 – \$20,000(0.1) = \$20,000 – \$2,000 = \$18,000 $p_0 - p_0(r) = p_0(1-r) = p_1$	
2	\$18,000 – \$18,000(0.1) = \$18,000 – \$1,800 = \$16,200 $p_1 - p_1(r) = p_1(1-r) = p_0(1-r)^2 = p_2$	Substitute $p_1 = p_0(1-r)$
3	\$16,200 – \$16,200(0.1) = \$16,200 – \$1,620 = \$14,580 $p_2 - p_2(r) = p_2(1-r)$ $= \left[p_0(1-r)^2\right](1-r)$ $= p_0(1-r)^3 = p_3$	Substitute $p_2 = p_0(1-r)^2$
4	\$14,580 – \$14,580(0.1) = \$14,580 – \$1,458 = \$13,122 $p_3 - p_3(r) = p_3(1-r)$ $= \left[p_0(1-r)^3\right](1-r)$ $= p_0(1-r)^4 = p_4$	Substitute $p_3 = p_0(1-r)^3$
5	\$13,122 – \$13,122(0.1) = \$13,122 – \$1,312.20 = \$11,809.80 $p_4 - p_4(r) = p_4(1-r)$ $= \left[p_0(1-r)^4\right](1-r)$ $= p_0(1-r)^5 = p_5$	Substitute $p_4 = p_0(1-r)^4$
⋮	⋮	
20	$p_{19} - p_{19}(r) = p_{19}(1-r)$ $= \left[p_0(1-r)^{19}\right](1-r)$ $= p_0(1-r)^{20} = p_{20}$	Substitute $p_{19} = p_0(1-r)^{19}$

Under the assumptions stated above, the price of the car after 20 years is

$$p_n = \$20{,}000(1-r)^n$$
$$p_{20} = \$20{,}000(1-0.1)^{20}$$
$$p_{20} = \$20{,}000(0.9)^{20}$$
$$p_{20} = \$20{,}000 \times 0.12157665$$
$$p_{20} = \$2{,}432.$$

Mathematics is a great deal easier if you can see connecting principles and ideas that allow you to see that *what appear to be two* different problems are actually *two different versions of the same problem*! The original (principal) growth and the (value) decay problems were logically the same and solved using (basically) the same formula.

Take note of the r in the two formulas above involving geometric sequences: $p_n = p_0(1+r)^n$ and $p_n = p_0(1-r)^n$. In the compound interest problem the r was 5% a year so $(1+r)^n$ was $(1+0.05)^n = 1.05^n$ whereas in the price decay problem the r was 10% a year so $(1-r)^n$ was $(1-0.10)^n = 0.9^n$. Note that for growth problems the growth factor is greater than one, while for decay problems the growth factor is less than one. Having the numbers 1.05 and 0.9 as bases for exponentiation should be comfortable and familiar. When you go into the next chapter, you will find geometric ratios involving the number e, such as e^n. Do not let the appearance of e as a base or as a common ratio bother you; e is just a number. Oh, and also consider the following: In this chapter, the exponentiation was always an integer power. In the next chapter the exponentiation will not be limited to integer values and furthermore the exponent will be a direct variation of the domain. Combining together both of these points the reader is urged to anticipate the existence of a function that looks like $y = e^{rx}$. ***1) The base is e. 2) The x value is not restricted to the set of integers. 3)*** $y = e^{rx}$***, not*** $y = rx$ ***as in earlier mathematics. Finally, 4) assuming that*** $x > 0$***, if*** $r > 0$***, there will be growth; if*** $r < 0$***, there will be decay.***

Because the formula $p_n = p_0(1 + r)^n$ has four variables, you need to understand that it can be applied in four different situations:

1. What is the **principal** that results from compounding $1,000 annually for 20 years at 5%.

$$p_n = \$1{,}000(1 + 0.05)^{20}, \quad \text{solve for } p_n.$$

2. $5,000 was the result of compounding $1,000 annually for 20 years. What was the **rate of interest**?

$$\$5{,}000 = \$1{,}000(1 + r)^{20}, \quad \text{solve for } r.$$

3. How many **years** will it take for $1,000 principal compounded annually at 6% annual interest to grow to be $3,000?

$$\$3{,}000 = \$1{,}000(1 + 0.06)^n, \quad \text{solve for } n.$$

4. After 15 years, an amount of money compounded annually at 8% interest is worth $4,000. What was the **original amount** of the money?

$$\$4{,}000 = p_0(1 + 0.08)^{15}, \quad \text{solve for } p_0.$$

Discrete Versus Continuous Variables

Data from the recent compound interest problem is repeated here in the table at left and shown in the graph at right.

Year	Value of deposit
0	$1,000.00
1	$1,050.00
2	$1,102.50
3	$1,157.625
4	$1,215.50625
5	$1,276.28156

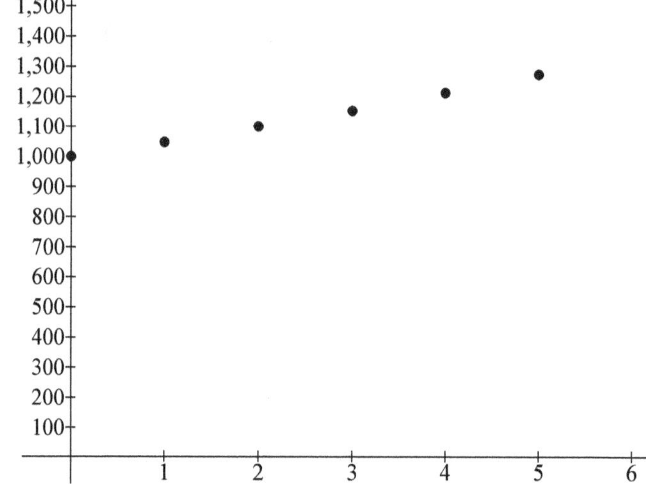

The disproportionate labeling of the x- and y-axes makes the curve at right look like a straight line, but you will notice the x increment from 0 to 1 has a corresponding y increment of $50 while the x increment from 1 to 2 has a corresponding y increment of $52.50. If a line were drawn through the points it would be curved.

Data from the car depreciation problem is repeated here in the table at left and shown in the graph at right.

Year	Value of deposit
0	$18,000
1	$16,200
2	$14,580
3	$13,122
4	$11,810
5	$10,629

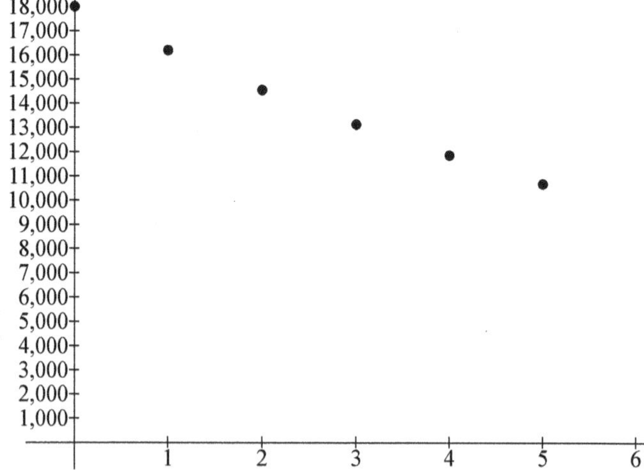

The disproportionate labeling of the x- and y-axes makes the points at right look like a straight line, but you will notice the x increment from 0 to 1 has a corresponding y decrement of $1,800 while the x increment from 1 to 2 has corresponding y decrement of $1,620. A line connecting the points would be curved.

In anticipation of the ideas presented in the following chapter, you need to notice that the tables and graphs above are examples of discrete data. That is because the x values in both examples were counting numbers by their context. Both represented yearly increases in the domain. *In the following chapter, functions will be continuous* in their domain and the ratio of the geometric sequence will be raised to an exponent to indicate a direct variation. Previously, in algebra, we saw direct variation as shown by the expression $y = kx$. Now, in differential calculus, we could see direct variation as shown by the expression $y = b^{kx}$. From algebra we are mostly familiar with exponentiation with a base of ten or some fraction and integral exponents, such as 10^3 or 0.05^2. In differential equations, you will frequently see base e with a product in the exponent, such as e^{kx}. For many people, it will be helpful to anticipate these ideas before seeing them and to realize that they are merely extensions of more primitive work done in algebra class.

Mathematics is easier, faster, and much more comfortable when you can connect new ideas with old ones.

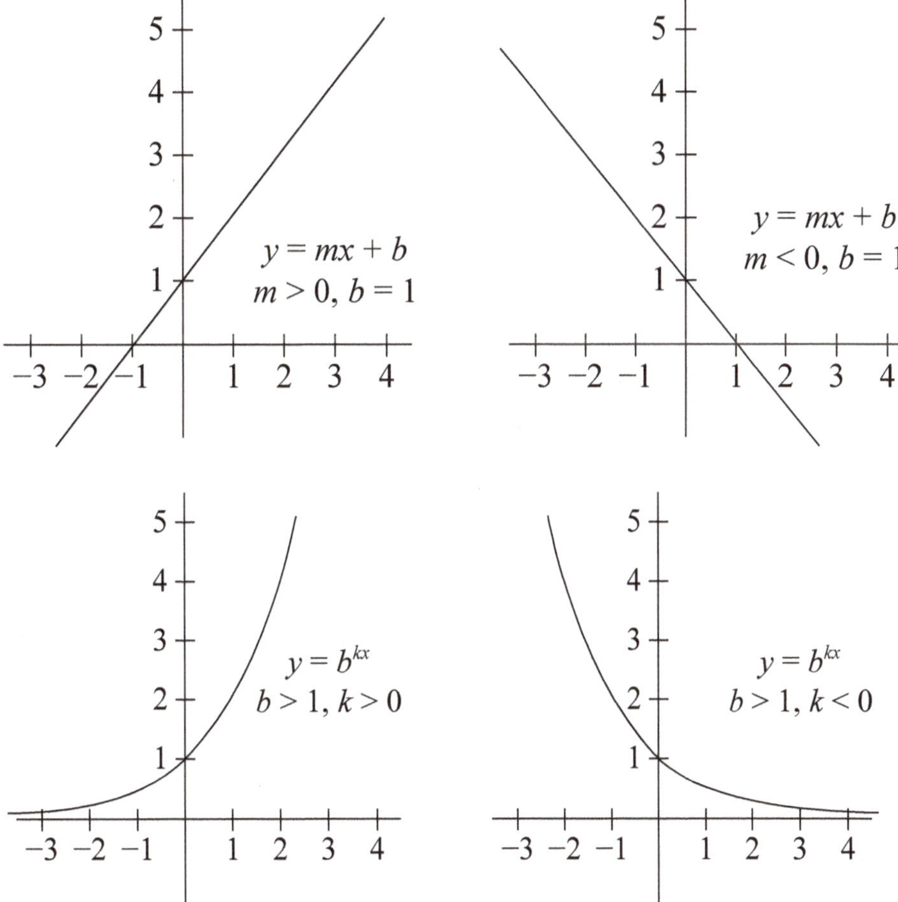

Integration, $\lim_{n\to\infty} \sum_{i=1}^{n} f(x_i)x$

In previous calculus classes, the term *integration* was always used synonymously with "area under a curve" in a specified domain. In differential calculus, you will hear the expression "take the integral of both sides." It is tempting to make the association with previous calculus classes and assume that you are finding area. It will also be tempting to assume that "taking the integral of both sides" is supported by a property such as if $a = b$, then $\int a = \int b$. This seems reasonable, but it is not the case.

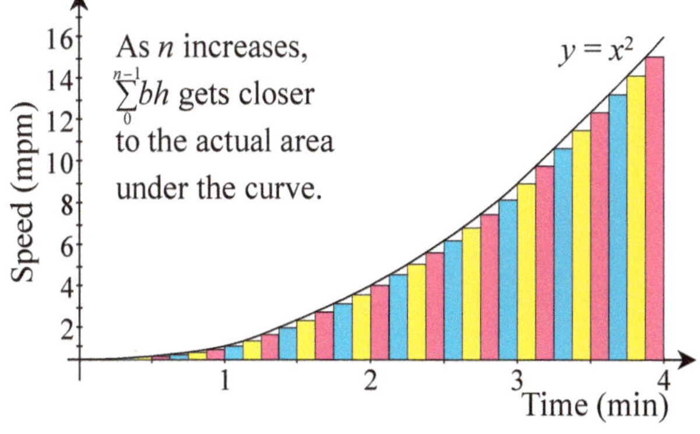

$$\frac{dy}{dx} = 2x$$

$$dy = 2x\,dx$$

"Take the integral of both sides."

$$\int dy = \int 2x\,dx$$

$$\int y^0\,dy = 2\int x^1\,dx$$

$$\frac{y^{0+1}}{0+1} + c_1 = 2\frac{x^{1+1}}{1+1} + c_2$$

$$y = 2\frac{x^2}{2} + c_2 - c_1$$

$$y = x^2 + c$$

$$f(x) = x^2 + c$$

In the work at left, it appears that we are multiplying both sides by the denominator of $\frac{dy}{dx}$ and then taking the integral of both sides using some sort of implied property of equality. such as if $a = b$, then $\int a = \int b$. That is actually what is happening, but a sharp-eyed student might be disturbed because we are integrating with respect to "y" on the left and with respect to "x" on the right. There is a subtle application of the "chain rule" going on here that justifies this step. If this bothers you, it may be best to think of "taking the integral of both sides of a differential equation" as "undoing the derivative of both sides."

Chapter 2 Review

In a perfect world, a math student will have been taught and have retained all necessary prerequisite knowledge to have success in learning new math ideas and skills. For students for whom that is not the case, Chapter 2 quickly reviewed four skills necessary to proceed and understand the material in Chapter 3.

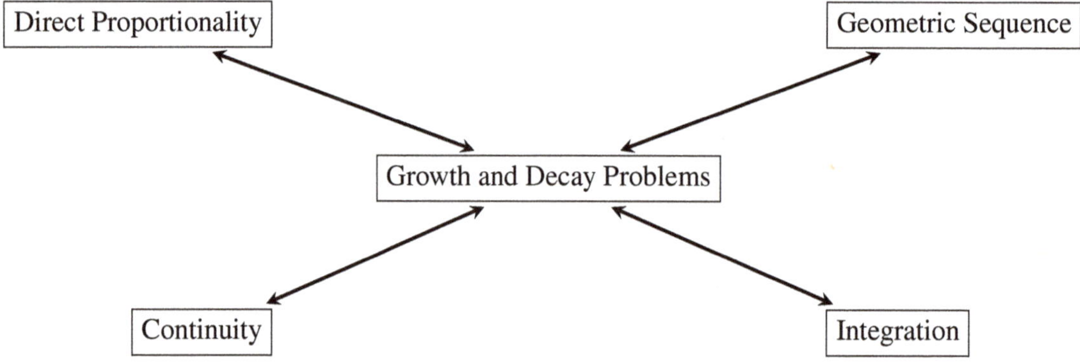

Chapter 3

Using Continuous Math to Solve Growth and Decay Problems; Introduction to Separable Differential Equations

Continuous Growth, Differential Equations in Finance

Building on the annual compounding from Chapter 2, how much would \$1,000 invested in a bank at 5% per year be worth after 20 years if the interest were compounded continuously? The formula for such a problem is

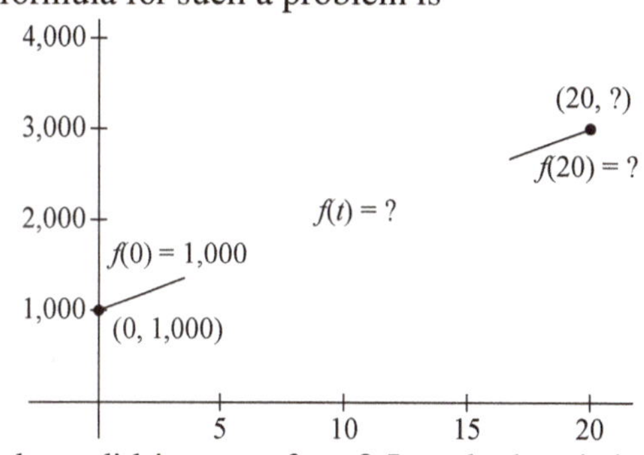

$p_f = p_0 \times e^{rt}$, where p_f is the final principal, p_0 original principal, r is the interest rate, and t is time in years

$$p_f = \$1,000 \times e^{0.05\times 20}$$
$$= \$1,000 \times e$$
$$= \$2,718.28$$

It is traditional to use this slick formula in precalculus math, but where did it come from? In calculus, it is traditional to derive the formula. Toward that end, the two tables that supported the two formulas learned in the previous chapter on discrete math are repeated on the next page to review and prepare the reader for their extension into continuous math.

End Year	Value of Deposit
0	\$1,000
1	\$1,000 + \$1,000(0.05) = \$1,050 $p_0 + p_0 \times r = p_1$
2	\$1,050 + \$1,050(0.05) = \$1,102.50 $p_1 + p_1 \times r = p_2$
3	\$1,102.50 + \$1,102.50(0.05) = \$1,157.63 $p_2 + p_2 \times r = p_3$
4	\$1,157.63 + \$1,157.63(0.05) = \$1,215.51 $p_3 + p_3 \times r = p_4$
⋮	⋮
20	$p_{19} + p_{19} \times r = p_{20}$
n	$p_{n-1} + p_{n-1} \times r = p_n$

End Year	Value of Deposit
0	p_0
1	$p_0 + p_0 \times r = p_0(1+r) = p_1$
2	$p_1 + p_1 \times r = p_1(1+r)$ $= [p_0(1+r)](1+r)$ $= p_0(1+r)^2 = p_2$
3	$p_2 + p_2 \times r = p_2(1+r)$ $= \left[p_0(1+r)^2\right](1+r)$ $= p_0(1+r)^3 = p_3$
4	$p_3 + p_3 \times r = p_3(1+r)$ $= \left[p_0(1+r)^3\right](1+r)$ $= p_0(1+r)^4 = p_4$
⋮	⋮
20	$p_{19} + p_{19} \times r = p_{19}(1+r)$ $= \left[p_0(1+r)^{19}\right](1+r)$ $= p_0(1+r)^{20} = p_{20}$
n	$p_{n-1} + p_{n-1} \times r = p_{n-1}(1+r)$ $= \left[p_0(1+r)^{n-1}\right](1+r)$ $= p_0(1+r)^n = p_n$

At left, the principal value in year n can be calculated by ***adding*** the interest on the principal in the previous year to that year's principal. At right, the principal for year n is calculated by ***multiplying*** the original term, p_0, by a growth factor. This formula can be made even more generic by stripping away the association with principal: $a_f = a_0(1+r)^t$. Here, a_f is the final amount (of whatever), a_0 is the original amount (of whatever), r is the geometric ratio of change, and t represents some unit of time. In this chapter, t is no longer restricted to the set of whole numbers.

Normally interest is not compounded yearly but for a smaller time interval, say **quarterly**. Instead of $p_f = p_0(1+r)^y$, the formula used would be

$$p_f = p_0\left[\left(1+\frac{r}{4}\right)^4\right]^y,$$

where p_f is the final principal after y years at r yearly interest rate and p_0 is the beginning principal. The money is compounded quarterly and depends on both r and y. $\left[\left(1+\frac{r}{4}\right)^4\right]^y$ can be called a growth factor. The inside part of the formula above, $\left(1+\frac{r}{4}\right)$, signifies the quarterly growth rate (based on the quarterly interest rate), and $\left(1+\frac{r}{4}\right)^4$ signifies the resulting annual growth rate.

Compare the two formulas below. The left formula comes from Chapter 2 while the right is introduced here in Chapter 3.

$$\boxed{p_f = p_0\mathbf{(1+r)^y}} \quad \longleftrightarrow \quad \boxed{p_f = p_0\left[\left(1+\frac{r}{4}\right)^4\right]^y}$$

On the left, the bold part, $\mathbf{(1+r)^y}$ (multiplied by p_0), is the growth factor much like the k in the algebraic direct-variation $y = kx$. Similarly, on the right, the bold part, $\left[\left(\mathbf{1}+\frac{r}{4}\right)^4\right]^y$ (multiplied by p_0), also acts like a growth factor like the k in the algebraic direct-variation $y = kx$. When compounding interest over other intervals, the formula would look like

$$p_f = p_0\left[\left(1+\frac{r}{n}\right)^n\right]^y,$$

where p_f is the final principal after y years at r yearly interest rate, p_0 is the beginning principal and n is the number of times per year that interest is compounded.

Observe the progression of detail in the following!!! That last expression did not just appear by magic. It results from a progression of thinking. If you understand that, then the final stage, $p_f = p_0 \times e^{rt}$, will not seem so magical.

1. $\boxed{p_f = p_0(1+r)^y}$

2. $\boxed{p_f = p_0\left[\left(1+\frac{r}{4}\right)^4\right]^y}$

3. $\boxed{p_f = p_0\left[\left(1+\frac{r}{n}\right)^n\right]^y}$

Let's do an experiment. In the subformula of $p_0\left[\left(1+\frac{r}{k}\right)^k\right]^y$—i.e., $\left(1+\frac{r}{k}\right)^k$—let $r = 1$ and let k get bigger and bigger. Watch what happens.

$$\left(1+\tfrac{1}{10}\right)^{10} = 2.593742460$$
$$\left(1+\tfrac{1}{100}\right)^{100} = 2.704813829$$
$$\left(1+\tfrac{1}{1{,}000}\right)^{1{,}000} = 2.716923932$$
$$\left(1+\tfrac{1}{10{,}000}\right)^{10{,}000} = 2.718145927$$
$$\left(1+\tfrac{1}{100{,}000}\right)^{100{,}000} = 2.718268237$$
$$\left(1+\tfrac{1}{1{,}000{,}000}\right)^{1{,}000{,}000} = 2.718280469$$

The fact that k is getting larger and larger in the formula at left means that our formula is compounding interest in more and more time intervals over one (1) year. In fact, as k approaches infinity we say that we are computing continuously compounded interest.

The number $\lim_{k\to\infty}\left(1+\frac{1}{k}\right)^k$ showing up here in a finance problem also appears all the time in science and engineering problems. In fact, it shows up so much that we give it a special symbol, e, so that we can shorten formulas using it. See Chapters 5 & 6 in my book *Explaining Logarithms*, available as a free pdf at www.mathlogarithms.com.

By a similar experiment, you can specify any value for r in $\left(1+\frac{r}{k}\right)^k$ and you will see as $k\to\infty$ that $\left(1+\frac{r}{k}\right)^k \to e^r$. So, substituting $\lim_{k\to\infty}\left(1+\frac{r}{k}\right)^k = e^r$ into $p_f = p_0\left[\left(1+\frac{r}{k}\right)^k\right]^y$, we get $p_f = p_0\,[e^r]^y$, which, in final form, would be $p_f = p_0e^{ry}$, where p_f is the final principal, p_0 is the original principal, r is the annual interest rate on the loan, and y is the number of years of the loan. Here, e^{ry} is dependent upon both r and y and can be called a growth factor for the initial value, in this case p_0. The initial value, p_0, times the growth factor equals the final value, p_f.

Behold!

1	2	3	4
$p_f = p_0(1+r)^y$	$p_f = p_0\left[\left(1+\frac{r}{4}\right)^4\right]^y$	$p_f = p_0\left[\left(1+\frac{r}{k}\right)^k\right]^y$	$p_f = p_0e^{ry}$

If you understand that the Stage 4 formula for p_f is an advanced abstract version of Stage 1 and that $(1+r)^y$ in **Stage 1**, $\left[\left(1+\frac{r}{4}\right)^4\right]^y$ in **Stage 2**, $\left[\left(1+\frac{r}{k}\right)^k\right]^y$ in **Stage 3**, and e^{ry} in **Stage 4** are all growth factors for the initial amount p_0, which is growing in each situation, then the final Stage 4 formula loses it magic and mystery. In all four stages—1, 2, 3, and 4—these growth factors are a result of raising a base to a power. The resulting value is used as a growth factor for an initial value. That e^{ry} in Stage 4 works very much like the product, kx, in the algebra direct-variation equation, $y = kx$. However, the result is a curvilinear function this time.

If the linear direct-variation equation $y = kx$ were taught as $y = 1\times kx$, the analogy would be complete: $y = 1\times kx$ is analogous to $p_f = p_0 \times e^{ry}$

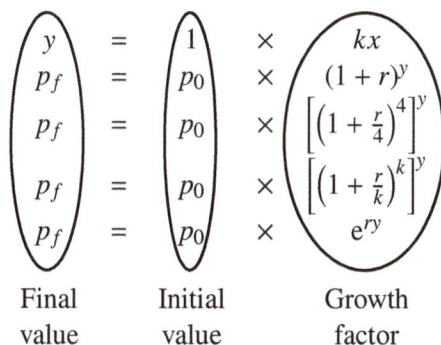

Final value		Initial value		Growth factor
y	=	1	×	kx
p_f	=	p_0	×	$(1+r)^y$
p_f	=	p_0	×	$\left[\left(1+\frac{r}{4}\right)^4\right]^y$
p_f	=	p_0	×	$\left[\left(1+\frac{r}{k}\right)^k\right]^y$
p_f	=	p_0	×	e^{ry}

Or as Monsieur Jourdain stated in Chapter 1, "Par ma foi! Il y a plus de quarante ans que je dis de la prose sans que j'en susse rien, et je vous suis le plus obligé du monde de m'avoir appris cela."

Using the formula $p_f = p_0e^{ry}$, it can be determined that \$1,000 invested at 5% interest per year compounded continuously for 20 years would be p_{20} = \$1,000 × $e^{0.05\times20}$ = \$1,000 × e^1 = **\$1,000 × 2.71828 = \$2,718.28.**

Differential equations show us a different way to derive this formula, $p_f = p_0 e^{ry}$. First, note that the change in principal over time is directly proportional to the interest rate and the principal: $\frac{dp}{dt} = r \times p$. *Here, $r = 0.05$, but, regardless of r, the slope field for such a differential equation would look like the figure below left because the slope of the function p(t) at each point in time would be $r \times p$. The slope of the unknown function p(t)—that is, $\frac{dp}{dt}$—varies directly with p.*

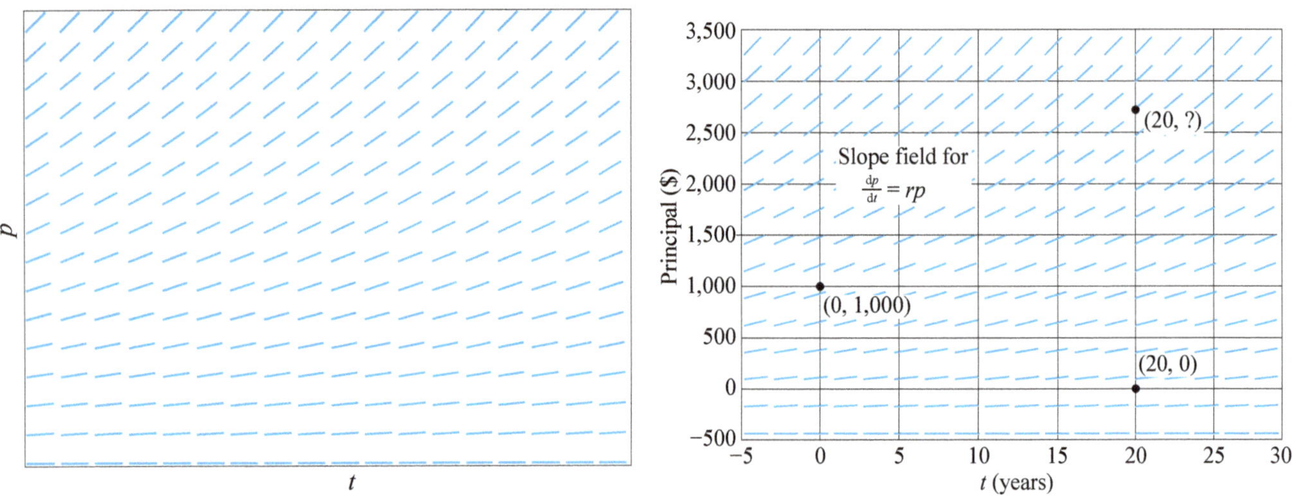

Now, there are a lot of slope lines indicated in this slope field with a slope of r (0.05 here), but only one of the curves they imply will pass through point (0, \$1,000)—\$1,000 invested at time 0. That curve will cross the vertical line at $t = 20$ years to give us the point we seek, (20, ?). If only, from the given slope ($\frac{dy}{dt}$, aka $f'(t)$), we could determine $p(t)$; it would be easy to determine the point (20, ?)—see figure above right—which would represent the amount of \$1,000 principal after 5% interest was calculated and compounded continuously for 20 years.

Below left is the Java code showing Euler's Method (see Chapter 1) for successively approximating a solution. On the right is its table output for $\Delta t = 0.001$.

```
public class EulerCompoundInterestApproximation
{
  public static void main(String args[])
  {
    double t = 0;
    double p = 1,000;
    double deltaT =.001; // run as deltax
    System.out.println(" deltaT = " + deltaT);
    System.out.println(" t p"); // column headers
    while (t < = 20)
    {
      System.out.println(t + ", " + p);
      p = p + (0.05 * p) * deltaT;
      t = t + deltaT;
    } // end while
  } // end main
} // end class
```

deltaT = 0.0010

t	p
0.000	\$1,000.000
0.001	\$1,000.050
0.002	\$1,000.100
0.003	\$1,000.150
⋮	⋮
19.996	\$2,717.670
19.997	\$2,717.806
19.998	\$2,717.942
19.999	\$2,718.078
20.000	\$2,718.214

The formula $p_f = p_0 e^{ry}$ was derived using differential calculus. **It can also be derived using differential equations.** Assume that the change in principal over time is directly proportional to the rate of interest and principal: $\frac{dp}{dt} = r \times p$.

$$\frac{dp}{dt} = r \times p$$
$$dp = (r \times p)\,dt$$
$$\frac{dp}{p} = \frac{(r \times p)\,dt}{p}$$
$$\frac{1}{p} \times dp = r\,dt$$
$$p^{-1}\,dp = r\,dt$$
$$\int p^{-1}\,dp = \int r\,dt$$
$$\int p^{-1}\,dp = r \int t^0\,dt$$
$$\ln p + c_1 = r \times t + c_2 \quad \text{because } p > 0$$
$$\ln p = r \times t + c_2 - c_1$$
$$\ln p = r \times t + C, \quad \text{where } C = c_2 - c_1$$
$$e^{\ln p} = e^{rt+C}$$
$$p = e^{rt} \times e^{C}$$
$$p = e^{C} \times e^{rt}$$

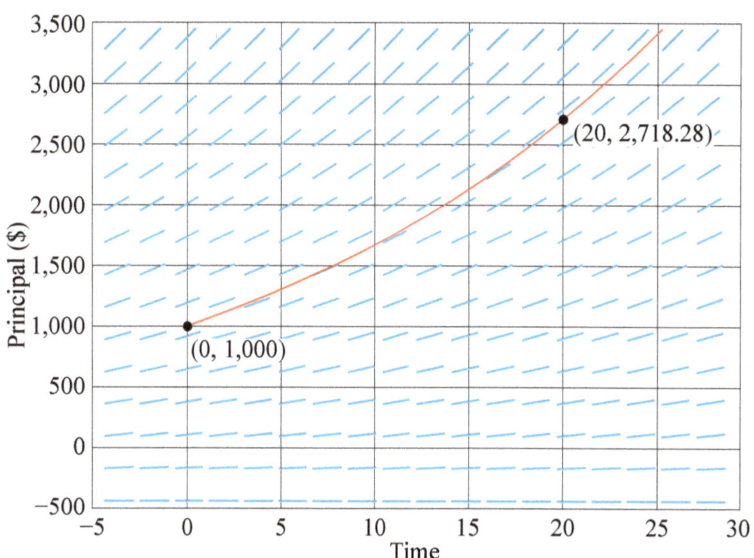

MATLAB script for the above graph is located in Appendix A.

Whatever C is, it is a constant; therefore, e^C *would be a constant, and* e^{rt} *can be called a growth factor for the term* e^C. *This is analogous to k in the algebra direct proportion* $y = kx$ *where k is a growth factor for x. In the above, p is the result of multiplying an initial value by a growth factor,* $p = e^c \times e^{rt}$. *We can start our growth at any place we wish. Choosing* p_0 *as the starting place for the growth of the principal, this equation can be rewritten as* $P_f = P_0 \times e^{rt}$. *This is how we can derive the formula for continuous compound interest using a differential equation.*

Substituting $r = 5\%$ *and* $t = 20$ *years, the original \$1,000 invested at 5% and compounded continuously is*

$$p = e^{C} \times e^{rt}$$
$$P_f = P_0 \times e^{rt}$$
$$P_{20} = \$1{,}000 \times e^{(0.05 \times 20)}$$
$$= \$1{,}000 \times e$$
$$= \$2{,}718.28$$

Fortunately, this answer using the formula $p_f = p_0 e^{rt}$ *matches well with the answer we got using Euler's Method above as well as the MATLAB graph. Otherwise, there would be egg on the author's face.*

Continuous Growth: Differential Equations in Game Biology

Game and wildlife workers have just introduced 125 deer to an incredibly large island in the land of Mathopia. Vegetation is plentiful, the deer are disease free, and there are no predators on the island. Based upon years and years of data, a wildlife scientist believes that the *change of the deer population over time will be directly proportional to the current number of deer.* Also, according to existing data, the reproductive rate of the deer is such that there will be more deer born each year than die. Hence, the population will grow. In fact it will double in 28 years. a.) Write a differential-equation model of these assumptions using P to represent the initial population of deer and t to represent time in years. Solve the equation for $P(t)$.

$$\frac{dP}{dt} = kP \quad \text{differential equation}$$

$$dP = k \times P\,dt \quad \text{separate the } dP \text{ and the } dt$$

$$\frac{1}{P} \times dP = \frac{1}{P} \times k \times P\,dt \quad \text{put } P \text{ on the same side as its differential}$$

$$\frac{1}{P} \times dP = k \times dt \quad \text{cancel } P\text{'s on the right and convert } \frac{1}{P} = P^{-1} \text{ to prepare to integrate}$$

$$\int P^{-1}\,dp = k \int t^0\,dt \quad \text{integrate both sides}$$

$$\ln P + c_1 = kt + c_2, \quad \text{actually, } \ln|P| + c_1 = kt + c_2 \text{ for purists, but } P > 0$$

$$\ln P = kt + c_2 - c_1 \quad \text{isolate the } P \text{ term}$$

$$\ln P = kt + c_3 \quad \text{consolidate the two constants of integration}$$

$$e^{\ln P} = e^{kt+c_3} \quad \text{if } x = y,\ b^x = b^y$$

$$P = e^{kt} \times e^{c_3} \quad \text{inverse operation on left, } b^{x+y} = b^x \times b^y \text{ on the right}$$

$$P = e^{kt} \times c_4 \quad \text{since both e and } c_3 \text{ are constant, } e^{c_3} \text{ would also be a constant value}$$

$$P = e^{kt} \times C, \quad \text{eliminate the subscript}$$

$$P_f = P_0 e^{kt} \quad \text{let } C = P_0\text{, the algebraic form of } \frac{dP}{dt} = kP$$

We have here four unknown values, two variables and two constants. Going back to the original problem, we see, at time $t = 0$ years, there are 125 deer and, at time $t = 28$ years, there are 250 deer. That information will allow us to solve for k.

$$P_f = P_0 \times e^{kt}$$

$$250 = 125 \times e^{k \times 28}$$

$$2 = e^{28k}$$

$$\ln 2 = \ln e^{28k}$$

$$0.6931471806 = 28k$$

$$k = 0.0247552564$$

How many deer will there be in 50 years? Substituting k into the original equation, $p_f = p_0 \times e^{kt}$ with $P = p_{50}$, $k = 0.0247552564$, and the initial population, $p_0 = 125$ (here, k is the growth factor)

$$p_{50} = 125e^{0.0247552564 \times 50}$$

$$p_{50} = 125e^{1.23776282}$$

$$p_{50} = 125 \times 3.447891277$$

$$p_{50} = 431 \text{ deer}$$

Year	Number deer
0	125
28	250
50	?
56	500

Is this a reasonable answer? Check out the table above. There were so many places in this process to go wrong. Do you see how easy it is to do a "reasonableness check"? With practice, you could do a check like this in your head.

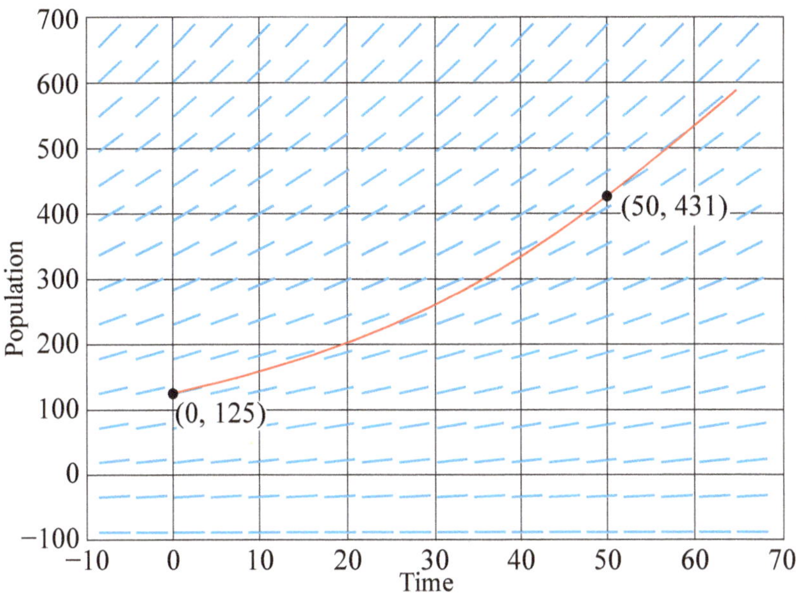

MATLAB script for the graph at left is located in Appendix A.

For those with obsessive compulsive disorder, we could further check our work if we wished with a quick modification to the previous Java code using the given conditions:

$t = 0, P = 125$ and $t = 28, P = 250$

$t = 0, P = 125$ and $t = 50, P = 431$

```
public class DeerEulers
{
  public static void main(String args[])
  {
    double t = 0;
    double p = 125; // start with 125 deer
    double deltaT = 0.0001; // run as deltax
    System.out.println(" deltaT = " + deltaT);
    System.out.println(" t x"); // column headers
    while (t < = 50)
    {
      System.out.println(t + ", " + p); // p(n) = p(n-1) + k * p(n-1)
      p = p + ((0.0247552564 * p) * deltaT);
      t = t + deltaT;
    } // end while
  } // end main
} // end class
```

deltaT = 0.001	
t	p
0.000	125.000
0.001	125.003
0.002	125.006
0.003	125.009
0.004	125.012
⋮	⋮
27.997	249.981
27.998	249.987
27.999	249.994
28.000	250.000

t	p
0.0000	125.0000
0.0001	125.0003
0.0002	125.0006
0.0003	125.0009
0.0004	125.0012
⋮	⋮
49.9996	430.9815
49.9997	430.9825
49.9998	430.9836
49.9999	430.9847
50.0000	431.0000

The rate of increase necessary to double the population of deer in 28 years is 0.0247552564. See work in text box.

$$2p = p \times e^{kt}$$
$$2 = e^{k\times 28}$$
$$\ln 2 = \ln e^{28\times k}$$
$$0.6931471806 = 28k$$
$$K = 0.0247552564$$

Continuous Decay of Carbon 14, Differential Equations Used to Date Artifacts

Photosynthesis is the chemical process that takes place in the green leaves of a plant. During photosynthesis, the plant takes in carbon dioxide (CO_2) and gives off oxygen (O_2). Some of the CO_2 that the plant takes in contains the carbon isotope called carbon 14. Carbon 14 results when sunlight strikes the CO_2 in the atmosphere in just the right way. When the plant dies, the carbon 14 in the plant slowly decays, changing into nitrogen 14. It takes approximately 5,730 years for the amount of carbon 14 originally in the plant to steadily decay to one half $\left(\frac{1}{2}\right)$ of its original amount. What expression would give the amount of carbon 14 many years after the tree died? Assume that the *change in the amount of carbon is directly proportional to the amount of carbon remaining:* $\frac{dc}{dt} = k \times c$. Previously, with continuous compounding of interest, we took the *direct-proportion differential equation,* $\frac{dp}{dt} = k \times p$, and developed the noncalculus equation $P_f = P_0 \times e^{rt}$. No need to reinvent the wheel. Since $p_f = p_0 \times e^{rt}$, we recognize that we can write $c_n = c_0 \times e^{kt}$, replacing the independent variable p (amount of principal) with c (amount of carbon). Let k be the common geometric ratio.

$$c_n = c_0 \times e^{kt} \quad \textit{This is a disguised differential equation!!}$$
$$50\% c_0 = c_0 \times e^{k \times 5{,}730}$$
$$50\% = e^{k \times 5{,}730} \quad \text{divide both sides by } c_0 \text{ resulting in one equation with one unknown}$$
$$0.50 = e^{5{,}730 \times k}$$
$$\ln(0.50) = \ln(e^{5{,}730 \times k})$$
$$-0.6931471806 = 5{,}730 \times k$$
$$k = \frac{-0.6931471806}{5{,}730}$$
$$k = -0.000120968094 \quad \text{here } k < 0 \text{ indicating that data will “decay,” not “grow”}$$
$$c_n = c_0 \times e^{-0.000120968094 \times t} \quad \textit{final amount of carbon 14, t years after the plant dies}$$

An artifact in a museum was displayed as being the masthead of a Viking ship dating back to the 1200s. An analysis of a sample of that artifact revealed that 90% of its carbon 14 remained. Could that artifact have dated back to the 1200s?

$$c_n = c_0 \times e^{-0.000120968094 \times t}$$
$$90\% c_0 = c_0 \times e^{-0.000120968094 \times t}$$
$$90\% = e^{-0.000120968094 \times t}$$
$$0.90 = e^{-0.000120968094 \times t}$$
$$\ln(0.90) = \ln(e^{-0.000120968094 \times t})$$
$$-0.1053605157 = -0.000120968094 \times t$$
$$t = \frac{-0.1053605157}{-0.000129680940}$$
$$t = 871 \text{ years}$$

Yes, the results of the carbon 14 testing support the idea that the artifact dates back to the 1200s!

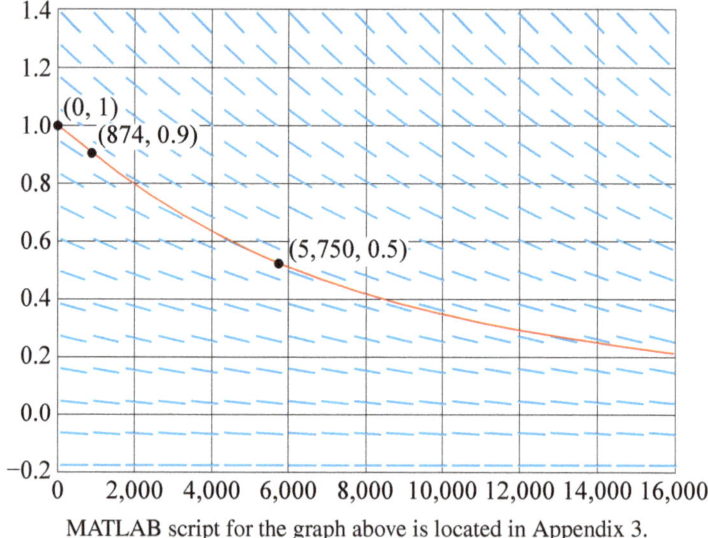

MATLAB script for the graph above is located in Appendix 3.

Just in case it slipped by you, the r in the formula $P_f = P_0 \times e^{rt}$ was greater than 0 ($r > 0$). Hence the values for P_f grew… $P_f > P_0$ (final p > original p). However, the k in the formula $c_n = c_0 \times e^{kt}$ was negative ($k < 0$). Hence the values for c_n decrease … $c_n < c_0$ (final c < original c). Also notice that the graph is “sinking” to zero. What is the significance of that?

Chapter 3 Review

The formula for exponential growth was reviewed at length and derived in two different ways: 1) using ideas about limits from earlier calculus and 2) using the new ideas (where limits are implied "under the hood" or "behind the scenes") in differential calculus. A distinction was made between the "differential equation" form of an equation and the "integrated algebraic form" of the same equation. Whenever possible, we wish to develop and use the integrated-algebraic form of an equation because it is more mechanical and requires much less mental effort. A differential equation, $\frac{dp}{dt} = r \times p$, in its integrated-algebraic form, $P_f = P_0 \times e^{rt}$ was used in three different scenarios: continuous compounding of interest, population growth, and radioactive decay. It is very helpful to see that the same math skills can be used in different situations.

Continuous Direct Proportion

Application	Differential Equation	Integrated-Algebraic Form
Increase in principal over time is *directly proportional* to the current amount of principal	$\frac{dp}{dt} = r \times p$ p is principal.	$p_f = p_0 \times e^{rt}$ p is principal.
Increase in deer population over time is *directly proportional* to the current deer population	$\frac{dP}{dt} = r \times P$ P is population.	$P_f = P_0 \times e^{rt}$ P is population.
Decrease in carbon 14 is *directly proportional* to the current amount of carbon 14	$\frac{dc}{dt} = k \times c$ c is carbon 14.	$c_n = c_0 \times e^{kt}$ c is carbon 14.

We studied a type of differential equation that can be solved using a skill called "solving separable differential equations." The three problems we just examined—1) continuous compounding of interest, 2) deer population growth, and 3) radioactive decay—are three different versions of the same problem. That is, if you understand the idea of continuous direct variation, then you don't have to learn how to solve three problems, only one, and then make small situation-dependent changes to get your answer. Also, if you understand that the integrated-algebraic form of a differential equation lets you skip worrying about many details of specific differential equations and work with the problem condition in a mechanical ("plug and chug") format, then your life becomes much easier. Taking that idea back another step further, if you understand linear direct variation, $y = kx$, then understanding curvilinear direct variation is much easier. The ideas at each level of abstraction are used in the next. Or, as Monsieur Jourdain stated in Chapter 1, "Par ma foi! Il y a plus de quarante ans que je dis de la prose sans que j'en susse rien, et je vous suis le plus obligé du monde de m'avoir appris cela."

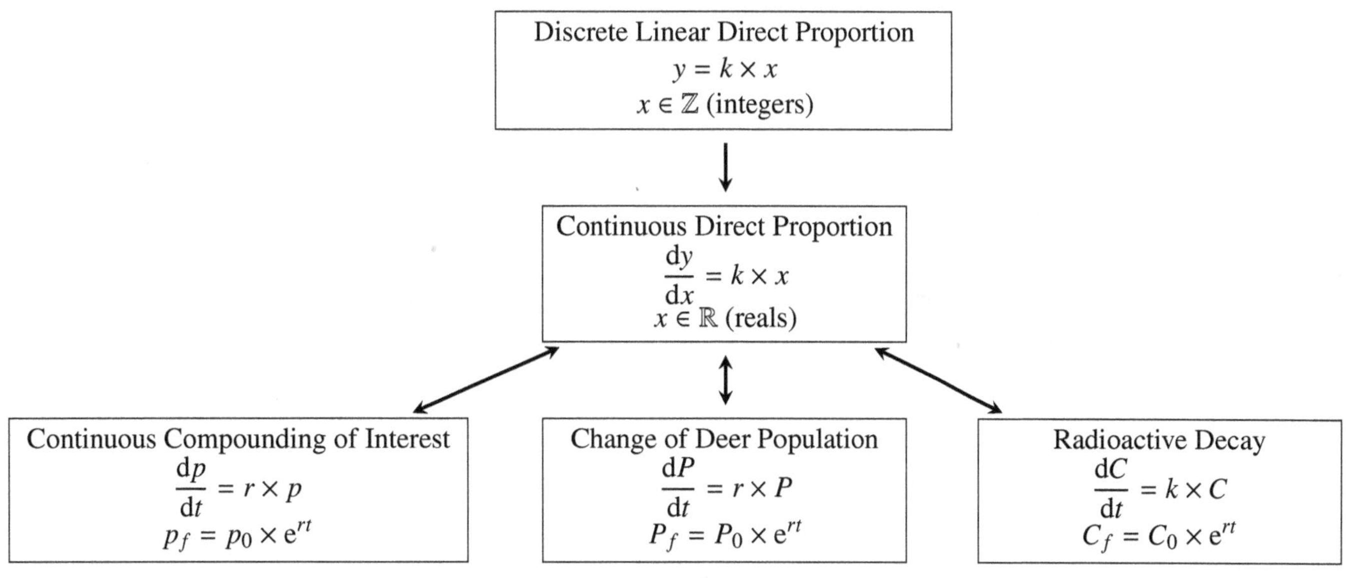

One last thing. Each of the differential equations presented,

1. $\frac{dp}{dt} = r \times p$, continuous growth of principal, new principal varies directly with current principal
2. $\frac{dP}{dt} = r \times P$, population growth, new population varies directly with current population
3. $\frac{dC}{dt} = k \times C$, carbon 14 decay, new carbon 14 amount varies directly with the current amount,

could be solved by the same technique. In each case, both sides of the equation were multiplied by the denominator differential of the derivative, and the variable associated with the numerator differential of the derivative was divided from the right. The result in each case was that all the variables of the numerator differential were together on one side of the equation and all the variables of the denominator differential were together on the other, separated from each other.

$$1.\ \frac{dp}{dt} = r \times p \qquad 2.\ \frac{dP}{dt} = r \times P \qquad 3.\ \frac{dC}{dt} = k \times C$$

$$\frac{1}{p}\,dp = r\,dt \qquad \frac{1}{P}\,dP = r\,dt \qquad \frac{1}{C}\,dC = k\,dt$$

Differential equations solved using this technique are called "separable differential equations." They are the easiest type of nontrivial differential equation to work with.

The two equivalent symbolisms shown in Chapter 0,

$$\boldsymbol{f'(x)} = \lim_{p \to x} \frac{f(x) - f(p)}{x - p} \text{ in derivative form (associated with Sir Isaac Newton) and}$$

$$\frac{\mathbf{dy}}{\mathbf{dx}} = \lim_{p \to x} \frac{f(x) - f(p)}{x - p} \text{ in differential form (associated with Gottfried Leibniz),}$$

show why much of the symbolism we use today came from Gottfried Leibniz instead of Sir Isaac Newton. The Newton symbolism, $f'(x)$, implies (but does not explicitly show) the ratio of two separately decreasing distances which Mr. Newton called fluxions. The Leibniz symbolism $\frac{dy}{dx}$ also implies the ratio of two separately decreasing distances but also has individual symbols dy and dx that actually represent those decreasing distances and can be manipulated independently, as though they were variables.

Chapter 4
Solving Separable Differential Equations

In Chapter 3, we studied differential equations that, through algebraic techniques, could be rewritten in such a way that the numerator differential and its respective variable were together on one side of the equation and the denominator differential and its respective variable were together on the other side. Then, both sides were integrated.

Continuous Compounding of Interest

$$\frac{dp}{dt} = r \times p$$
$$\frac{1}{p}\,dp = r\,dt$$
$$\int \frac{1}{p}\,dp = \int r\,dt$$

etc.

Continuous Population Growth

$$\frac{dP}{dt} = r \times P$$
$$\frac{1}{P}\,dP = r\,dt$$
$$\int \frac{1}{P}\,dP = \int r\,dt$$

etc.

Carbon 14 Decay

$$\frac{dC}{dt} = r \times C$$
$$\frac{1}{C}\,dC = r\,dt$$
$$\int \frac{1}{C}\,dC = \int r\,dt$$

etc.

Differential equations in which this technique can be used are called "separable differential equations." They are the easiest type of differential equations to work with. Moving from general to specific,

$$\frac{dy}{dx} = f(x) \times g(y),\ g(y) \neq 0$$ **standard form for a separable differential equation**

$$\frac{1}{g(y)}\,dy = f(x)\,dx$$
$$\int \frac{1}{g(y)}\,dy = \int f(x)\,dx$$

etc.,

we get the following examples.

Ex 1

$$\frac{dy}{dx} = 3$$
$$\frac{dy}{dx} = \left(3x^0\right) \times \left(1y^0\right)$$

Separable!

Ex 2

$$\frac{dy}{dx} = \sqrt{\frac{x}{y}}$$
$$\frac{dy}{dx} = \sqrt{x} \times \sqrt{y^{-1}}$$
$$\frac{dy}{dx} = \left(x^{\frac{1}{2}}\right) \times \left(y^{-\frac{1}{2}}\right)$$

Separable!

Ex 3

$$\frac{dy}{dx} = x^2 + 1$$
$$\frac{dy}{dx} = \left(x^2 + 1\right) \times \left(1y^0\right)$$

Separable!

Ex 4

$$\frac{dy}{dx} = \frac{x^2}{1 - y^2}$$
$$\frac{dy}{dx} = \left(x^2\right) \times \left(1 - y^2\right)^{-1}$$

Separable!

Ex 5

$$\frac{dy}{dx} = \frac{3x^2 + 4x + 2}{2(y - 1)}$$
$$\frac{dy}{dx} = \left(3x^2 + 4x + 2\right) \times \left[\frac{1}{2}(y - 1)^{-1}\right]$$

Separable!

Ex 6

$$\frac{dy}{dx} = \frac{y^2 + 1}{x - 1}$$
$$\frac{dy}{dx} = (x - 1)^{-1} \times \left(y^2 + 1\right)$$

Separable!

Ex 7

$$\frac{dy}{dx} = \frac{y \cos x}{1 + 2y^2}$$
$$\frac{dy}{dx} = (\cos x) \times \frac{y}{1 + 2y^2}$$

Separable!

The point that is being made here is that if a differential equation can be put into the form

$$\frac{\mathbf{dy}}{\mathbf{dx}} = f(x) \times g(y),\ g(y) \neq 0,$$

then it can be solved by the "separable technique" as shown above. As a general rule, it is easier to immediately strive for separation of the variables and differentials from the beginning rather than to put the equation into the generic form shown here and then to separate. The example above is just to create an understanding of how to categorize differential equations.

Your differential-equations teacher sometimes has a perspective, an understanding and a schema that give an intangible advantage over you when looking at problems. Here is an example.

"Solving a separable differential equation is the reverse of implicit differentiation."

So, what is a new skill to the student is just a backward version of what the differential equation teacher taught back in differential calculus. They blend seamlessly in the teacher's mind.

If you understood that last sentence go ahead and skip over this next part. Otherwise let's go back to differential calculus and review explicit and implicit differentiation and then revisit this idea. Find the slope of the tangent to the circle $x^2 + y^2 = 25$ at the point when $x = 4$. By substitution, then, $4^2 + y^2 = 25$ and $y = 3$. Using ideas taught in beginning calculus, then, the slope of the tangent would be equal to the derivative of the function evaluated at (4, 3).

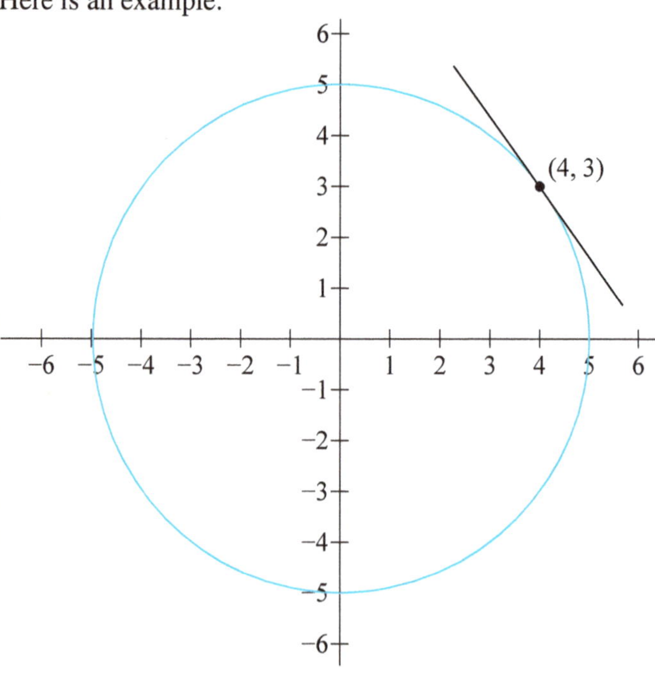

Explicit Differentiation

$$x^2 + y^2 = 25 \text{ (25 is } r^2\text{)}$$
$$y^2 = 25 - x^2$$
$$y = \pm\left(25 - x^2\right)^{\frac{1}{2}} \quad ***$$
$$\frac{dy}{dx} = \pm\frac{1}{2}\left(25 - x^2\right)^{-\frac{1}{2}}(-2x)$$
$$= \frac{-x}{\pm(25 - x^2)^{\frac{1}{2}}}$$
$$= \frac{-x}{y} \text{ (from *** above)}$$

Implicit Differentiation

$$x^2 + y^2 = 25$$
$$\frac{d}{dx}\left(x^2 + y^2\right) = \frac{d}{dx}(25)$$
$$2x + 2y\frac{dy}{dx} = 0$$
$$2y\frac{dy}{dx} = -2x$$
$$\frac{dy}{dx} = \frac{-x}{y}$$

Here, the derivative is in terms of both x and y.

Reverse Implicit Differentiation

$$\frac{dy}{dx} = \frac{-x}{y}$$
$$y\,dy = -x\,dx$$
$$\int y\,dy = -\int x\,dx$$
$$\frac{y^2}{2} + c_1 = -\frac{x^2}{2} + c_2$$
$$\frac{x^2}{2} + \frac{y^2}{2} = C$$
$$x^2 + y^2 = K^2$$

For $x = 4$ and $y = 3$, $K = 5$.

$$x^2 + y^2 = 25$$

"Solving a separable differential equation is the reverse of implicit differentiation."

Understanding connections makes mathematics more interesting. It also lets you see that new ideas are often connected to or related to old ones.

$$\frac{dy}{dx} = f(x) \times g(y), \; g(y) \neq 0$$
$$\frac{1}{g(y)}\,dy = f(x)\,dx$$
$$\int \frac{1}{g(y)}\,dy = \int f(x)\,dx$$

We'll spend the rest of this chapter reviewing those slope fields. Since this is otherwise a short chapter and slope fields will be used extensively in the next chapter, it might be good to talk more about them again. Slope fields are useful in anticipating solutions to differential equations and are a necessity when trying to get information about unsolvable differential equations. Today's software graphing packages really do all the work for you. However, like many topics in math, they obscure what really goes on. Just like everything else in postcalculator mathematics education, those graphing packages have potential for harm. For example, in my first book, *Explaining Logarithms*, I posited the scenario of a student claiming that $234 \times 4{,}192 = 8{,}219$ or $y = \log_{4.8} 714.6$, $y = 22.9$ because "the calculator said so."

All high school math teachers have heard statements such as these. As another example, when I was in college, the education majors would take $\frac{1}{2}$-inch thick SAS or SPSS printouts off the printer and proceed to make incredible conclusions based on their "data." When asked what a correlation or standard deviation was, they didn't know and they did not care that they didn't know.

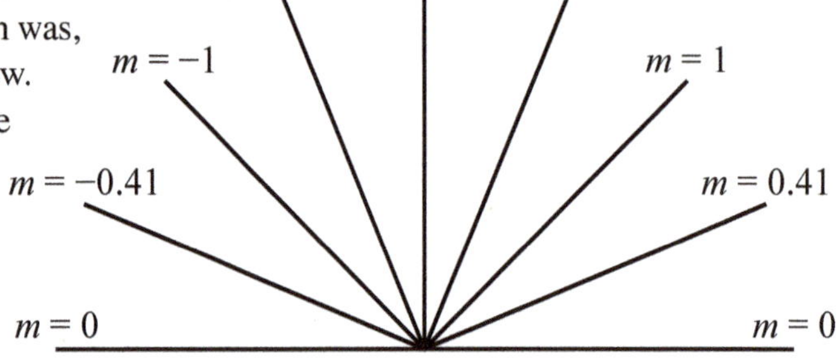

It is best for most people to see how slope fields arise by doing a few of them the old fashioned way. We'll start with a quick review from trig regarding selected slope values before proceeding. The following values are obtained by taking the tangents of 180, 157.5, 135, 112.5, 90, 67.5, 45, 22.5, and 0 degrees.

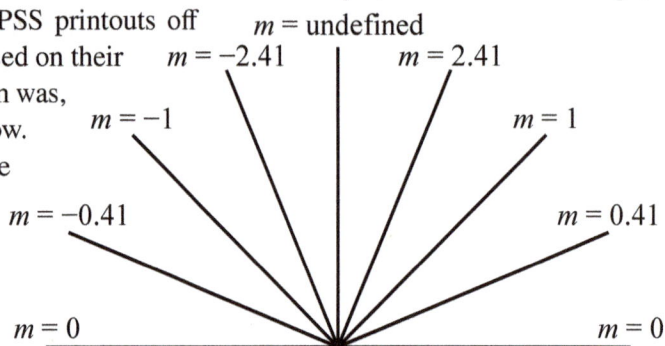

The following shows three different examples of slope fields being graphed "by hand" (without a computer).

1. Graph the slope field for $\frac{dy}{dx} = -\frac{x}{y}$

Step 1: Red slope segments for $x = 0$.

(x, y)	$\frac{dy}{dx} = -\frac{x}{y}$ (m)	Slope color
$(0, 3)$	0	red
$(0, 2)$	0	red
$(0, 1)$	0	red
$(0, 0)$	indet	red
$(0, -1)$	0	red
$(0, -2)$	0	red
$(0, -3)$	0	red

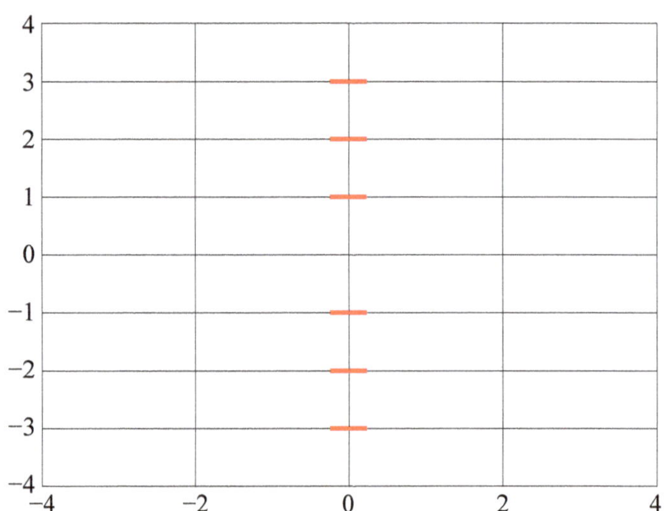

Step 2: Green slope segments for $x = 1$.

(x, y)	$\frac{dy}{dx} = -\frac{x}{y}$ (m)	Slope color
$(1, 3)$	$-\frac{1}{3}$	green
$(1, 2)$	$-\frac{1}{2}$	green
$(1, 1)$	-1	green
$(1, 0)$	undef	green
$(1, -1)$	1	green
$(1, -2)$	$\frac{1}{2}$	green
$(1, -3)$	$\frac{1}{3}$	green

Step 3: Blue slope segments for $x = -1$.

(x, y)	$\frac{dy}{dx} = -\frac{x}{y}$ (m)	Slope color
$(-1, 3)$	$\frac{1}{3}$	blue
$(-1, 2)$	$\frac{1}{2}$	blue
$(-1, 1)$	1	blue
$(-1, 0)$	undef	blue
$(-1, -1)$	-1	blue
$(-1, -2)$	$-\frac{1}{2}$	blue
$(-1, -3)$	$-\frac{1}{3}$	blue

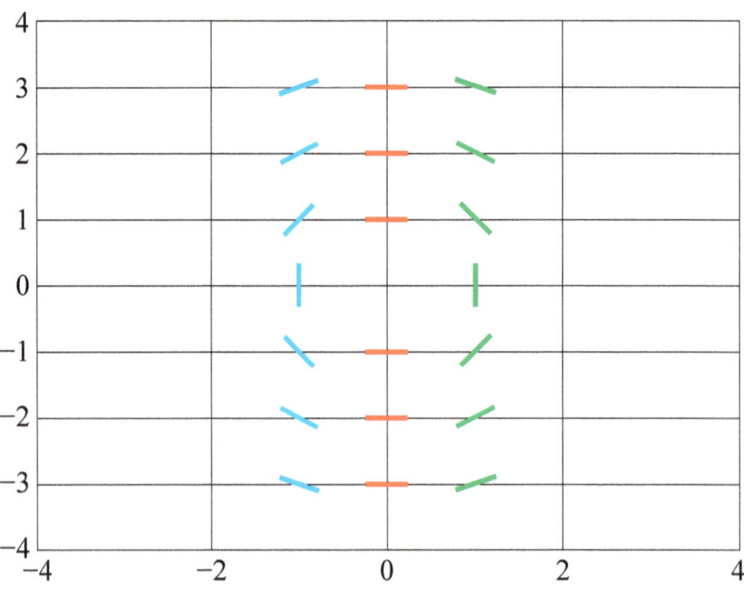

Step 4: Brown slope segments for $x = 2$.

(x, y)	$\frac{dy}{dx} = -\frac{x}{y}$ (m)	Slope color
$(2, 3)$	$-\frac{2}{3}$	brown
$(2, 2)$	-1	brown
$(2, 1)$	-2	brown
$(2, 0)$	undef	brown
$(2, -1)$	2	brown
$(2, -2)$	1	brown
$(2, -3)$	$\frac{2}{3}$	brown

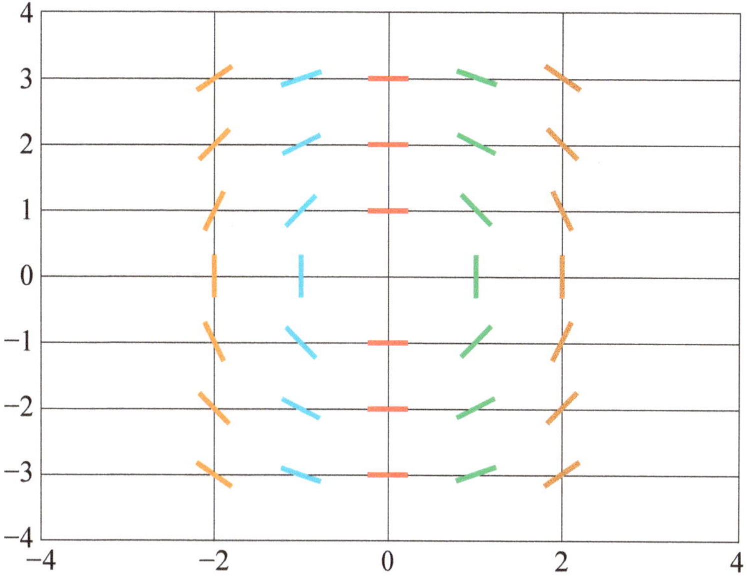

Step 5: Orange slope segments for $x = -2$.

(x, y)	$\frac{dy}{dx} = -\frac{x}{y}$ (m)	Slope color
$(-2, 3)$	$\frac{2}{3}$	orange
$(-2, 2)$	1	orange
$(-2, 1)$	2	orange
$(-2, 0)$	undef	orange
$(-2, -1)$	-2	orange
$(-2, -2)$	-1	orange
$(-2, -3)$	$-\frac{2}{3}$	orange

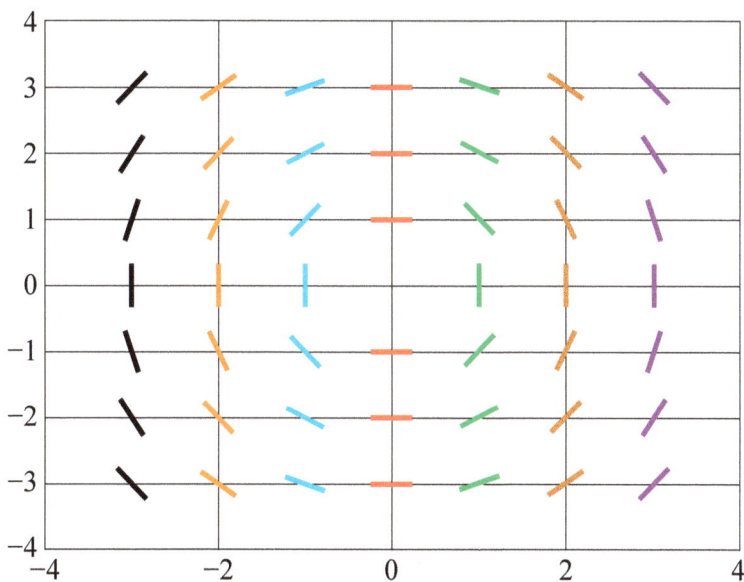

Step 6: Purple slope segments for $x = 3$.

(x, y)	$\frac{dy}{dx} = -\frac{x}{y}$ (m)	Slope color
$(3, 3)$	-1	purple
$(3, 2)$	$-\frac{3}{2}$	purple
$(3, 1)$	-3	purple
$(3, 0)$	undef	purple
$(3, -1)$	3	purple
$(3, -2)$	$\frac{3}{2}$	purple
$(3, -3)$	1	purple

Step 7: Black slope segments for $x = -3$.

(x, y)	$\frac{dy}{dx} = -\frac{x}{y}$ (m)	Slope color
$(-3, 3)$	1	black
$(-3, 2)$	$\frac{3}{2}$	black
$(-3, 1)$	3	black
$(-3, 0)$	undef	black
$(-3, -1)$	-3	black
$(-3, -2)$	$-\frac{3}{2}$	black
$(-3, -3)$	-1	black

What you need to do is to try to imagine many, many slope indicators each with diminished length until they reach the length of a point (points actually do not have length, think pixel). If you are successful in your imagination, you should be able to “see” (in your mind) the graph of the original function $f(x)$ that, when differentiated, resulted in $\frac{-x}{y}$: concentric circles.

2. Graph the slope field for $\frac{dy}{dx} = y - x$

Step 1: Red slope segments for $x = 0$.

(x, y)	$\frac{dy}{dx} = y - x$ (m)	Slope color
$(0, 3)$	3	red
$(0, 2)$	2	red
$(0, 1)$	1	red
$(0, 0)$	0	red
$(0, -1)$	-1	red
$(0, -2)$	-2	red
$(0, -3)$	-3	red

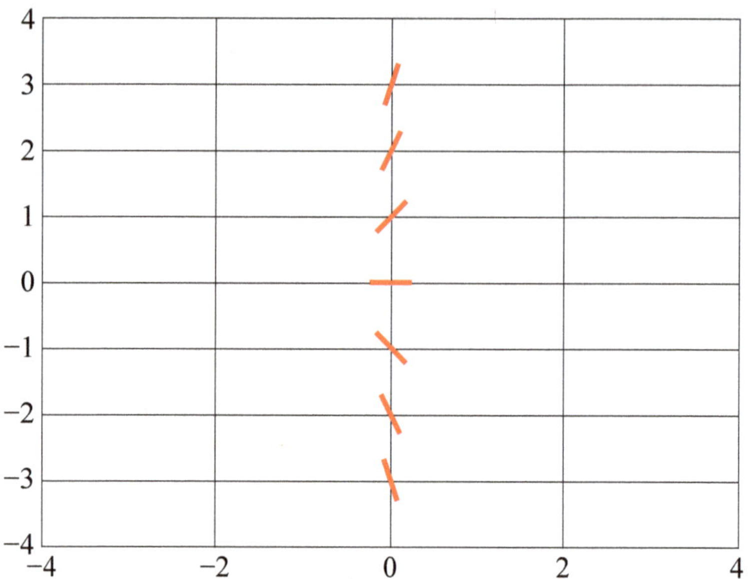

Step 2: Green slope segments for $x = 1$.

(x, y)	$\frac{dy}{dx} = y - x$ (m)	Slope color
$(1, 3)$	2	green
$(1, 2)$	1	green
$(1, 1)$	0	green
$(1, 0)$	-1	green
$(1, -1)$	-2	green
$(1, -2)$	-3	green
$(1, -3)$	-4	green

Step 3: Blue slope segments for $x = -1$.

(x, y)	$\frac{dy}{dx} = y - x$ (m)	Slope color
$(-1, 3)$	4	blue
$(-1, 2)$	3	blue
$(-1, 1)$	2	blue
$(-1, 0)$	1	blue
$(-1, -1)$	0	blue
$(-1, -2)$	-1	blue
$(-1, -3)$	-2	blue

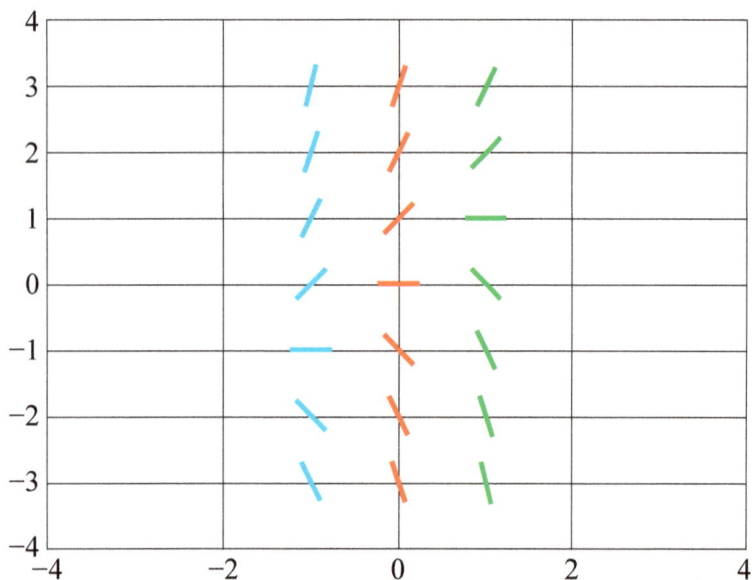

Step 4: Brown slope segments for $x = 2$.

(x, y)	$\frac{dy}{dx} = y - x$ (m)	Slope color
(2, 3)	1	brown
(2, 2)	0	brown
(2, 1)	−1	brown
(2, 0)	−2	brown
(2, −1)	−3	brown
(2, −2)	−4	brown
(2, −3)	−5	brown

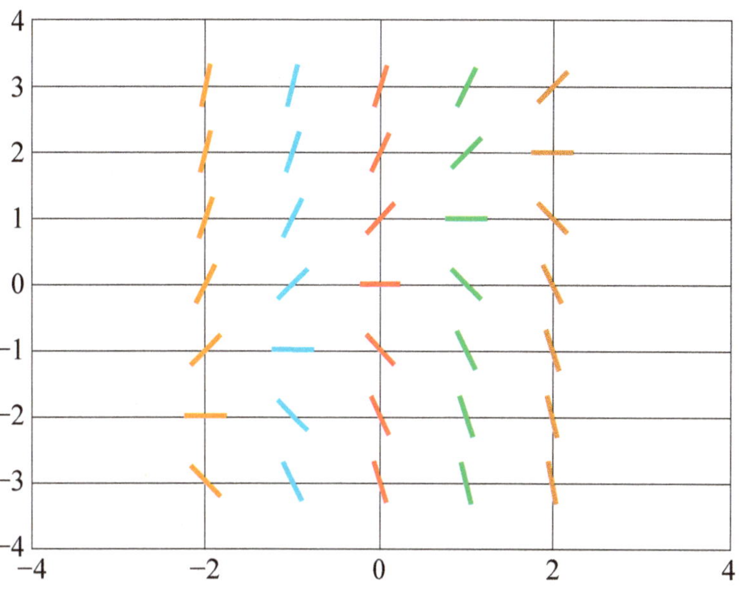

Step 5: Orange slope segments for $x = -2$.

(x, y)	$\frac{dy}{dx} = y - x$ (m)	Slope color
(−2, 3)	5	orange
(−2, 2)	4	orange
(−2, 1)	3	orange
(−2, 0)	2	orange
(−2, −1)	1	orange
(−2, −2)	0	orange
(−2, −3)	−1	orange

Step 6: Purple slope segments for $x = 3$.

(x, y)	$\frac{dy}{dx} = y - x$ (m)	Slope color
(3, 3)	0	purple
(3, 2)	−1	purple
(3, 1)	−2	purple
(3, 0)	−3	purple
(3, −1)	−4	purple
(3, −2)	−5	purple
(3, −3)	−6	purple

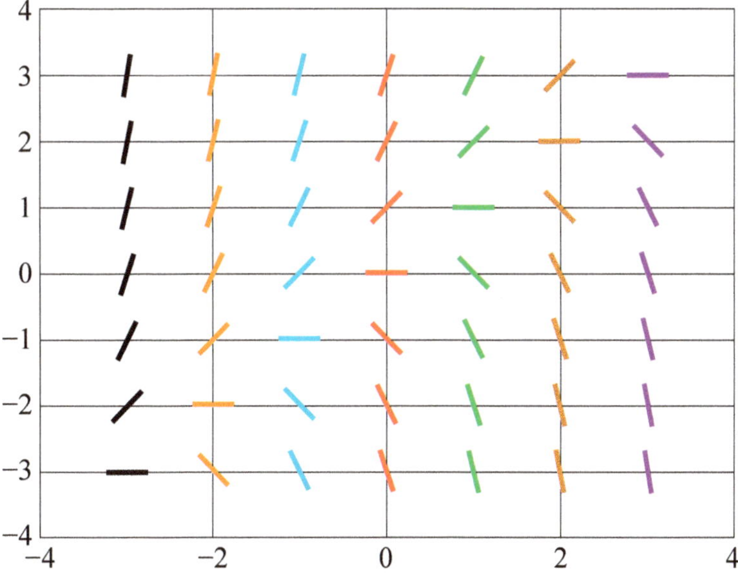

Step 7: Black slope segments for $x = -3$.

(x, y)	$\frac{dy}{dx} = y - x$ (m)	Slope color
(−3, 3)	6	black
(−3, 2)	5	black
(−3, 1)	4	black
(−3, 0)	3	black
(−3, −1)	2	black
(−3, −2)	1	black
(−3, −3)	0	black

Ta-Da!!

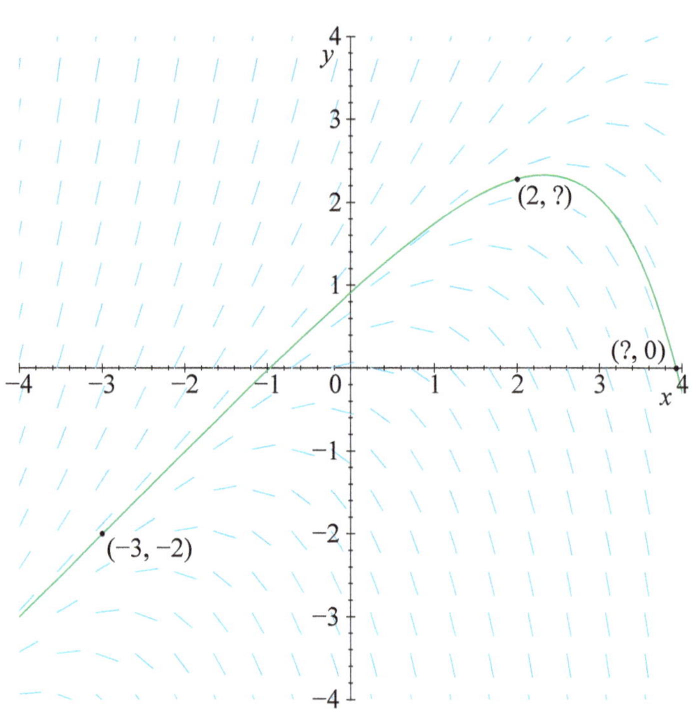

The slope field at right shows sample slopes of the differential equation $\frac{dy}{dx} = y - x$. As indicated in the graph, there are many functions that have a slope of $y - x$. If you are given that the graph of $f(x)$ that you want passes through a specific point, say $(-3, -2)$, that will help you to identify a specific function, $f(x)$, from the family of functions. This is shown in the figure at right. The progression of ideas is as follows: You are given a differential equation, $f'(x)$, or the problem you are working on dictates a differential equation. The solution to that differential equation would be an entire family of possibilities, $f(x) + c$, where c is unknown. You may be given (or the problem that you are working on will dictate) that the solution needed for your case will pass through a specified point. You are then to use the fact that the general solution, $f(x) + c$, of the given differential equation, f', passes through the given point in order to solve for c. From that $f(x) + c$ with known c, you could then find the value of $f(x)$ when $x = 2$ or perhaps you would be asked to find the x or y-intercept of the particular $f(x)$ passing through $(-3, -2)$ that is indicated by the differential equation $f'(x)$, also known as $\frac{dy}{dx}$. This is called an initial value problem (IVP) and is possible due to something called the "existence and uniqueness theorem."

3. Graph the slope field for $\frac{dy}{dx} = y^3 - y^2 - 6y$

This example is shown because it demonstrates a differential equation with no explicit x variables. There is a fancy word for such equations: autonomous. A differential equation with no explicit x values is called autonomous. This fact will have an impact on the hand-graphing process: If there is no explicit x value in the differential equation, all the y values will be the same for any given x. The following autonomous differential equation was chosen because the polynomial in y will factor, making for a nice clean example. $y^3 - y^2 - 6y = y(y^2 - y - 6) = y(y - 3)(y + 2)$. Regardless of the x value, the derivative (slope) will be zero when $y = -2$, 0, and 3.

Start with the zero-slope parts of the field.

All x values	y	$\frac{dy}{dx} = y^3 - y^2 - 6y$ $= y(y - 3)(y + 2)$	Slope color
	4		
	3	0	red
	2		
	1		
	0	0	red
	−1		
	−2	0	red
	−3		

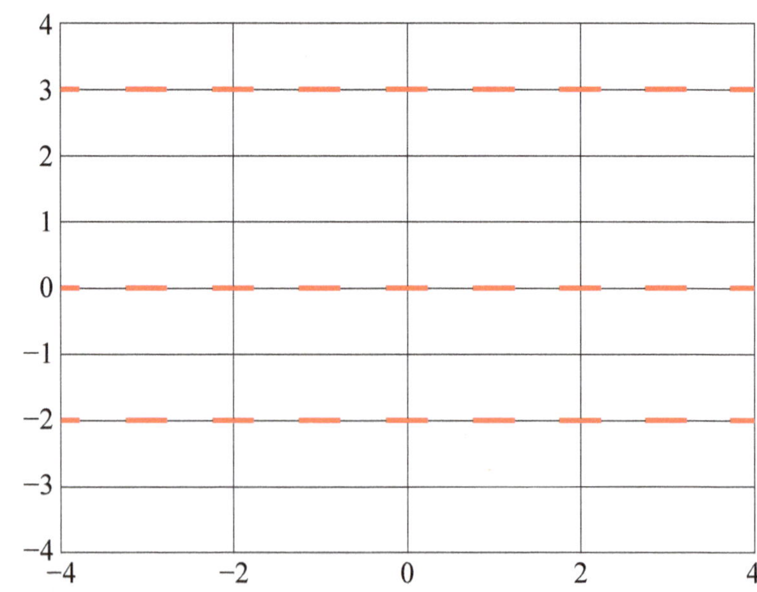

Now calculate several y-value slopes at once.

All x values	y	$\frac{dy}{dx} = y^3 - y^2 - 6y$ $= y(y-3)(y+2)$	Slope color
	4	$(4)(4-3)(4+2) = +24$	green
	3	0	red
	2	$(2)(2-3)(2+2) = -8$	blue
	1	$(1)(1-3)(1+2) = -6$	blue
	0	0	red
	−1	$(-1)(-1-3)(-1+2) = +4$	green
	−2	0	red
	−3	$(-3)(-3-3)(-3+2) = -18$	blue

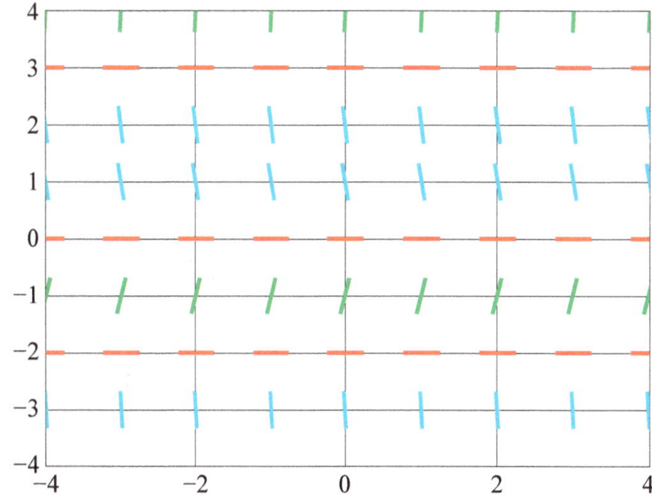

Finally, since slope lines are "parallel" (do not intersect each other), we can anticipate that the slope lines will approach the horizontal slope lines at $y = -2$, 0, and 3 asymptotically. Actually, since the curved slope lines cannot intersect each other either, each of those (curved) lines will approach each other asymptotically as well.

All x values	y	$\frac{dy}{dx} = y^3 - y^2 - 6y$ $= y(y-3)(y+2)$	Slope color
	4	$(4)(4-3)(4+2) = +24$	green
	3.1	$(3.1)(3.1-3)(3.1+2) =$ $(3.1)(0.1)(5.1) = 1.6$	green
	3	0	red
	2.9	$(2.9)(2.9-3)(2.9+2) =$ $(2.9)(-0.1)(4.9) = -1.4$	blue
	2	$(2)(2-3)(2+2) = -8$	blue
	1	$(1)(1-3)(1+2) = -6$	blue
	0.1	$(0.1)(0.1-3)(0.1+2) =$ $(0.1)(-2.9)(2.1) = -0.6$	blue
	0	0	red
	−0.1	$(-0.1)(-0.1-3)(-0.1+2) =$ $(-.1)(-3.1)(-1.9) = 0.6$	green
	−1	$(-1)(-1-3)(-1+2) = +4$	green
	−1.9	$(-1.9)(-1.9-3)(-1.9+2) =$ $(-1.9)(-4.9)(0.1) = 0.9$	green
	−2	0	red
	−2.1	$(-2.1)(-2.1-3)(-2.1+2) =$ $(-2.1)(-5.1)(-0.1) = -1.1$	blue
	−3	$(-3)(-3-3)(-3+2) = -18$	blue

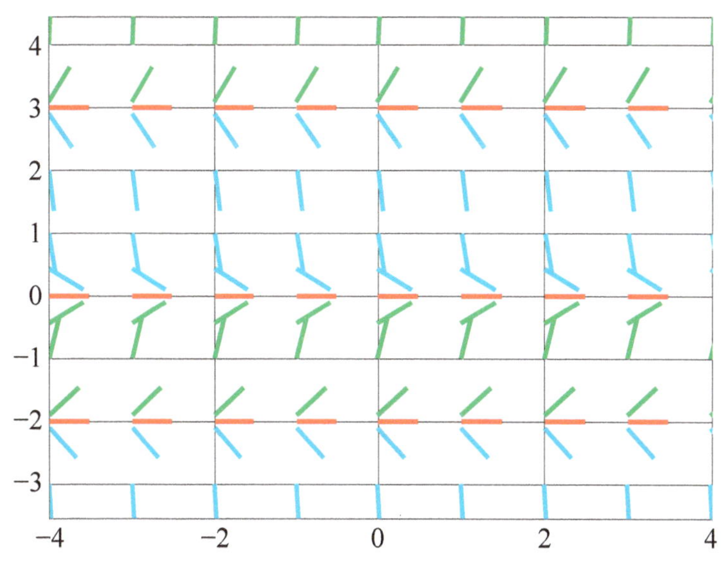

Except for instructional materials on beginning differential equations, there is very little need to graph a lot of slope fields by hand. In this third case, graphing the slope field for $\frac{dy}{dx} = y^3 - y^2 - 6y$, there is the benefit in that we get a visual on what the term *autonomous* means. Above, the term *autonomous differential equation* was defined as a differential equation that did not have any x terms. We can now add to that definition. An autonomous equation is a differential equation that, for each y, has the same slope, $\frac{dy}{dx}$, for all x values. This is also a good time to talk about sources and sinks. A *source asymptote* ($y = 3$, $y = -2$ at right) is a line from which the slope lines on each side are "leaving." A *sink asymptote* ($y = 0$) is a line that is being approached by slope lines on both sides. Note that approaching and leaving are taken from the point of view of x increasing.

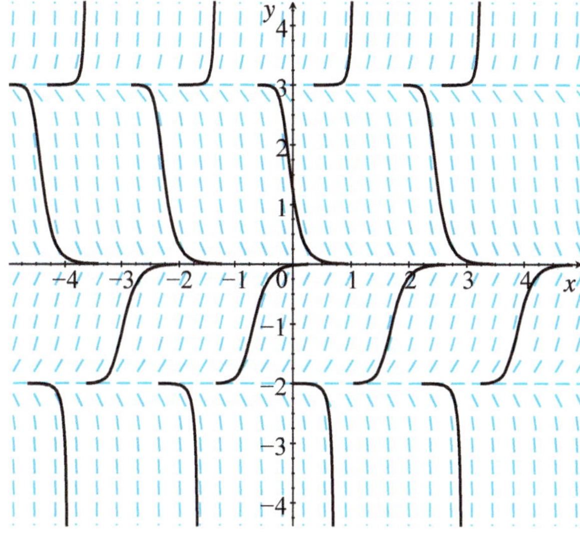

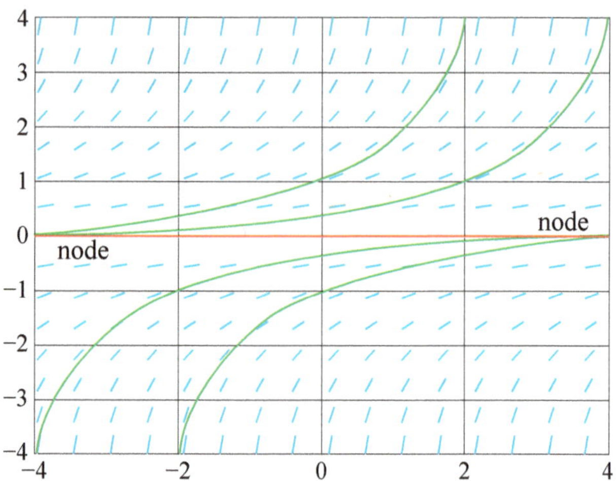

A *node asymptote* has slopes approaching from one side but leaving from the other as shown in this unrelated graph. Another benefit to hand graphing an autonomous differential equation (e.g., $\frac{dy}{dx} = y^3 - y^2 - 6y$) is to introduce the concept of *phase-plane analysis*. Above, $\frac{dy}{dx} = y^3 - y^2 - 6y$ was factored as $\frac{dy}{dx} = y(y - 3)(y + 2)$. That looks very much like the polynomials we studied in precalculus. We remember graphing cubic polynomials. Often they crossed the independent axis at three places. This cubic polynomial would obviously cross at −2, 0, and 3. That suggests the following. Graph $y' = y^3 - y^2 - 6y$ just as you were taught in precalculus except, since all the independent and dependent values have changed, **you would label the vertical and horizontal axes accordingly: the vertical axis as y' and the horizontal axis as y** (figure below left). Since the coefficient of the y^3 term is positive, the graph will increase from negative infinity, cross at $y = -2$, then ascend above the y-axis until it reaches a local maximum where it descends and crosses again at $y = 0$. It then continues descending below the y-axis until it reaches a local minimum where it again starts ascending until it crosses the y-axis at $y = 3$ and continues on to positive infinity. For each place where the graph of $y' = y^3 - y^2 - 6y$ crosses the y-axis, the graph of the solution to our differential equation, $\frac{dy}{dx} = y^3 - y^2 - 6y$, would have a slope value of zero at that y value. This is shown in the x–y graph at the right.

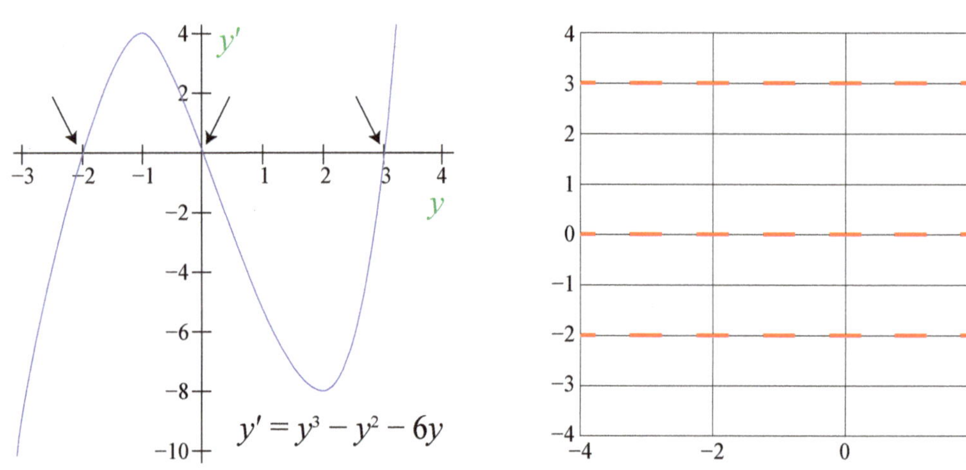

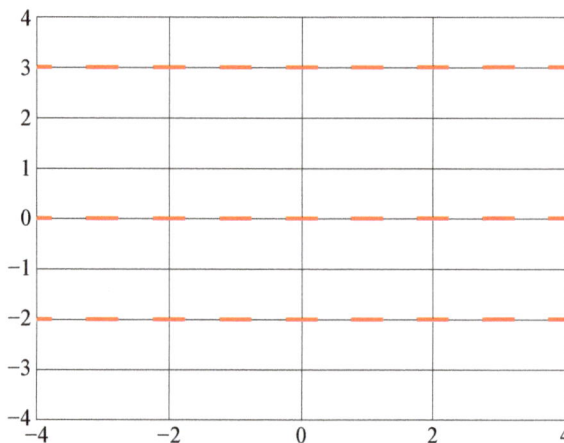

Well alrighty now! The y–y' graph in the phase-plane analysis indicates that the slopes are all negative and asymptotic to $y = -2$ in the interval $(-\infty, -2)$ and all positive and asymptotic to $y = 3$ in the interval $(3, \infty)$. All the graphs in both those intervals get steep very fast as they approach infinity. (See the middle figure below.) In the y–y' graph in the phase-plane analysis, slopes in the interval $(0, 3)$ are all negative and cannot pass the lines $y = 3$ and $y = 0$ and hence are asymptotic to them both. Similarly, this can be said for the slopes in the interval $(-2, 0)$ except that the slopes are positive in that interval. (See figure below right and compare it with the one at the bottom of page 51.

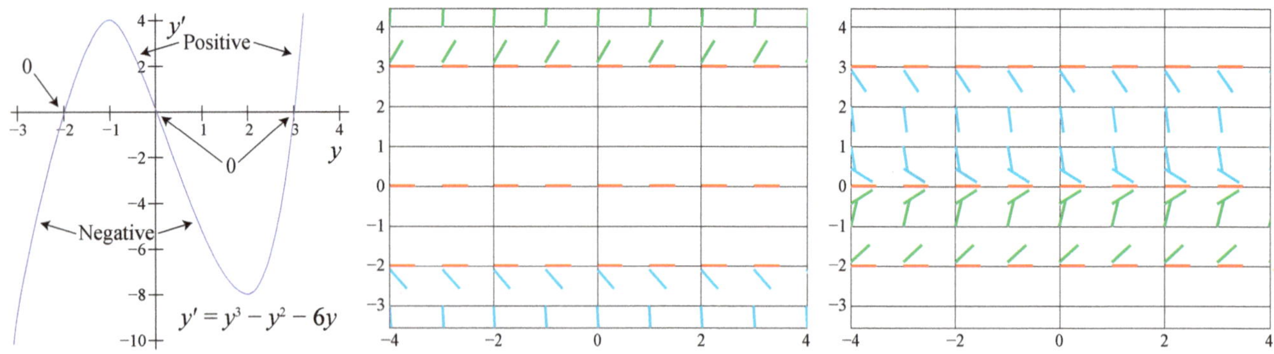

Chapter 4 Review

Chapter 4 showed how to solve a type or category of differential equation called separable differential equations. Here, the numerator and denominator of the derivative of the differential equation were separated by multiplying both sides of the differential equation by the denominator of the derivative. Additionally, the dependent and independent variables were moved to the appropriate sides of the equation using algebra. Then, the respective independent and dependent variables are obtained by integrating their respective differentials.

$$\frac{\mathrm{d}p}{\mathrm{d}t} = r \times p$$

$$\frac{1}{p}\,\mathrm{d}p = r\,\mathrm{d}t$$

$$\int \frac{1}{p}\,\mathrm{d}p = \int r\,\mathrm{d}t$$

We learned to identify separable differential equations from their generic form, which either looked like or could be transformed to look like $\frac{\mathrm{d}y}{\mathrm{d}x} = f(x) \times g(y),\ g(y) \neq 0$.

It was shown that solving a separable differential equation was the reverse of implicit differentiation. As a learning exercise, "slope fields" for three different differential equations were drawn in stages "by hand." In practice, these slope fields just magically appear in Mathematica or MATLAB, but it is helpful in the learning process to draw a few slope fields the old fashioned way to take the magic out of the process.

Several new vocabulary words were introduced:

1. Autonomous differential equations—a differential equation that does not have any x terms
2. Source asymptote—an asymptote from which a particular solution to a differential equation is diverging
3. Sink asymptote—an asymptote toward which a particular solution is converging
4. Node asymptote—an asymptote that has a particular solution converging on one side but diverging from the other
5. Phase-plane analysis—a technique for studying the behavior of a differential equation. With a differential equation the axis is labeled using the dependent and independent variables of the equation. With the phase-plane analysis the axis of the graph is labeled using the dependent variable (of the differential equation) as the horizontal axis and the derivative of the dependent variable (of the differential equation) as the vertical axis.

Online Application

Visit demonstrations.wolfram.com/DirectionFieldsForDifferentialEquations and demonstrations.wolfram.com/SlopeFields for two great little applications that will demonstrate some of the concepts from this chapter. "Direction Fields for Differential Equations" from the Wolfram Demonstrations Project. Contributed by Stephen Wilkerson and Charles E. Oelsner.

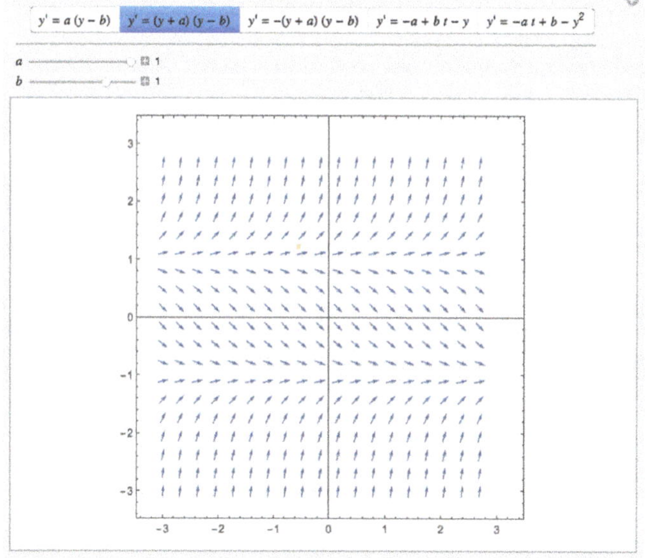

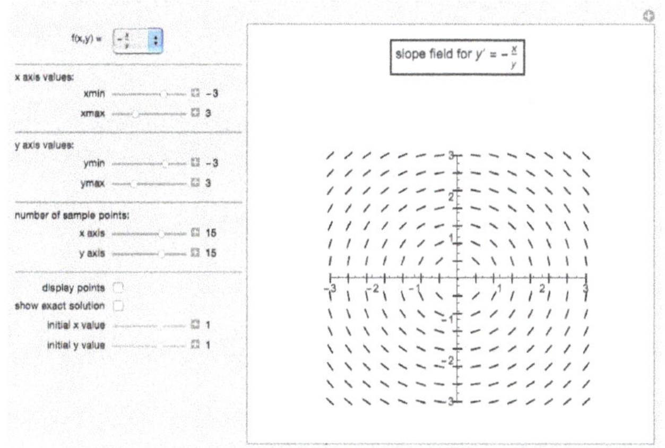

Chapter 5
More Applied Separable Differential Equations

Differential Equations in Chemistry

In a chemical reaction, 100 g of substance A are being dissolved at a rate that is directly proportional to the square of the undissolved amount remaining. After one hour, only 20 g of substance A remain undissolved. How much of substance A is present after two hours? How long before the amount of substance A decreases to 1 g?

Let A be the amount of substance A at any time t. From the data and assumption about the rate of dissolution, you can write the following differential equation: $\frac{\mathrm{d}A}{\mathrm{d}t} = kA^2$. Convert this differential equation to its integrated-function form to obtain the amount of substance after two hours.

$$\frac{\mathrm{d}A}{\mathrm{d}t} = kA^2$$

can be put into the following form,

$$\frac{\mathrm{d}A}{\mathrm{d}t} = t^0 \times kA^2, \; t^0 \neq 0,$$

which has the form

$$\frac{\mathrm{d}A}{\mathrm{d}t} = f(t) \times g(A).$$

Hence, this problem can be solved as a separable differential equation.

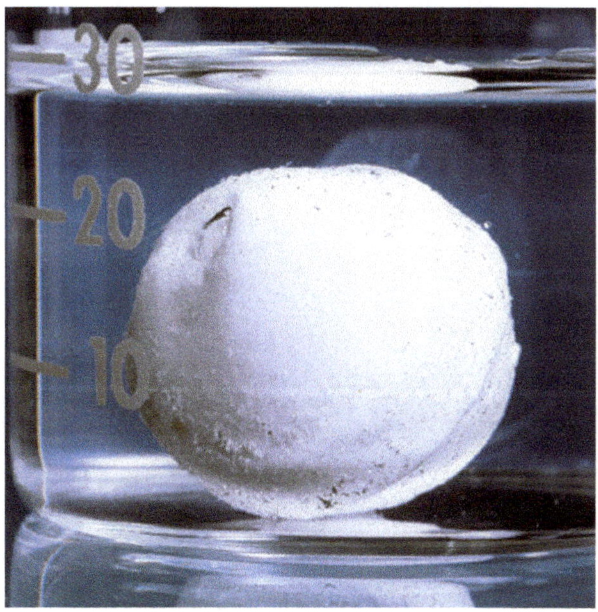

Fundamental Photographs

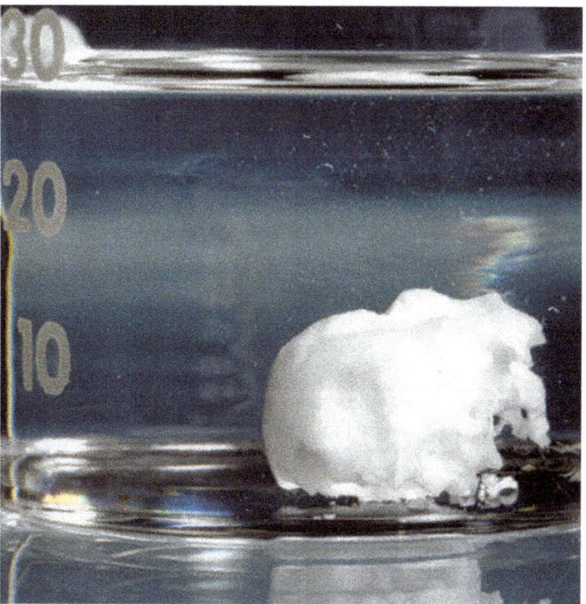

Fundamental Photographs

$$\frac{dA}{dt} = t^0 \times kA^2 \quad \text{has form } \frac{dA}{dt} = f(t) \times g(A)$$

$$dA = kA^2\,dt \quad \text{separate variables}$$

$$A^{-2}\,dA = k\,dt \quad \text{prepare to integrate}$$

$$\int A^{-2}\,dA = \int k\,dt \quad \text{integrate both sides}$$

$$\frac{A^{-1}}{-1} + c_1 = kt + c_2$$

$$\frac{-1}{A} = kt + c_2 - c_1$$

$$\frac{-1}{A} = kt + c_3 \quad \text{combine constants}$$

$$-1 = A(kt + c_3) \quad \text{eliminate the fraction}$$

$$\frac{-1}{kt + c_3} = A \quad \text{solve for } A$$

$$A = \frac{-1}{kt + c_3} \quad A \text{ to the left}$$

$$100 = \frac{-1}{k \times 0 + c_3} \quad \text{at } t = 0\text{, there was 100 g}$$

$$100 = \frac{-1}{c_3}$$

$$c_3 = -0.01$$

At one hour, only 20 g of the substance remained. Substituting $c_3 = -0.01$ into the integrated form of the function, we get

$$A = \frac{-1}{kt + c_3} \quad \text{we can solve for } k$$

$$20 = \frac{-1}{(k \times 1) + (-0.01)} \quad \text{substitution}$$

$$20 = \frac{-1}{k - 0.01}$$

$$20(k - 0.01) = -1 \quad \text{eliminate the fraction}$$

$$20k - 0.20 = -1 \quad \text{distributive property}$$

$$20k = -0.8 \quad \text{isolate the variable term}$$

$$k = -0.04 \quad \text{divide both sides by 20}$$

$$A = \frac{-1}{-0.04t + (-0.01)} \quad \text{substitute for } k$$

$$= \frac{-1}{-0.04t + (-0.01)} \times \frac{-100}{-100} \quad \text{simplify}$$

$$= \frac{100}{4t + 1} \quad \text{final integrated form}$$

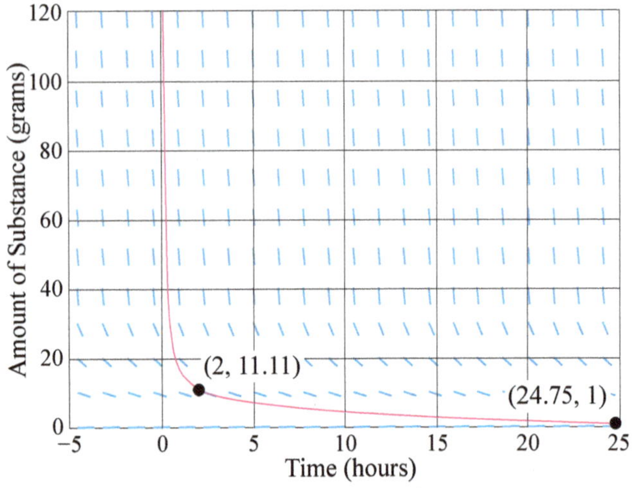

MATLAB script for the above graph is located in Appendix A.

How much of Substance A is present after two hours?

$$A = \frac{100}{4t + 1} = \frac{100}{4 \times 2 + 1} = 11.11\text{ g}$$

How long before the amount of the substance decreases to 1 g?

$$1 = \frac{100}{4t + 1}$$

$$4t + 1 = 100$$

$$4t = 99$$

$$t = 24.75\text{ h}$$

before the substance is reduced to one gram.

Differential Equations in Meteorology—Barometric Pressure

It seems counterintuitive, but air has weight. A column of air above your head is pushing down on you right now. At sea level the weight of that column of air would be more than if you were standing on top of Mt. Everest because there would be a shorter column of air over your head on top of Mt. Everest than there would be at sea level. For points on the earth's surface, a simplistic model of barometric pressure, p (in inches of mercury in a barometer), is that, with increasing altitude, pressure decreases in direct proportion to the current pressure: $\frac{dp}{dh} = -0.2p$ where $p = 29.92$ inches of mercury at sea level where $h = 0$ (miles). Find the barometric pressure at the top of Mt. Everest at 29,029 ft.

$$\frac{dp}{dh} = h^0 \times -0.2p \quad \text{form: } \frac{dp}{dh} = f(h) \times g(p)$$

$$\frac{dp}{p} = -0.2\,dh \quad \text{separate variables}$$

$$p^{-1}\,dp = -0.2\,dh \quad \frac{1}{p} = p^{-1}$$

$$\int p^{-1}\,dp = \int -0.2\,dh \quad \text{integrate both sides}$$

$$\ln|p| + c_1 = -0.2h + c_2 \quad \int p^{-1} = \ln|p|$$

$$\ln p = -0.2h + c_3 \quad \text{constants, } p > 0$$

$$e^{\ln p} = e^{-0.2h+c_3} \quad a = b \text{ so } e^a = e^b$$

$$p = e^{-0.2h} \times e^{c_3} \quad b^{m+n} = b^m \times b^n$$

$$p = e^{-0.2h} \times c_4 \quad \text{new constant ***}$$

$$29.92 = c_4 \times e^{-0.2\times0} \quad p = 29.92 \text{ at sea level}$$

$$29.92 = c_4 \times 1 \quad e^0 = 1$$

$$c_4 = 29.92$$

$$p = 29.92 \times e^{-0.2h} \quad \text{from *** above}$$

$$p = 29.92e^{-0.2\times\frac{29{,}029}{5{,}280}} \quad \text{Everest's } h \text{ in miles}$$

$$p = 29.92 \times e^{-0.2\times5.497916667}$$

$$p = 29.92 \times e^{-1.099583333}$$

$$p = 29.92 \times 0.3330098089$$

$$p = 9.96 \text{ inches of mercury}$$

MATLAB script for the graph at right is in Appendix A.

Author's note: $\frac{29{,}029 \text{ ft}}{5{,}280 \text{ ft/mi}} = 5.5$ miles.

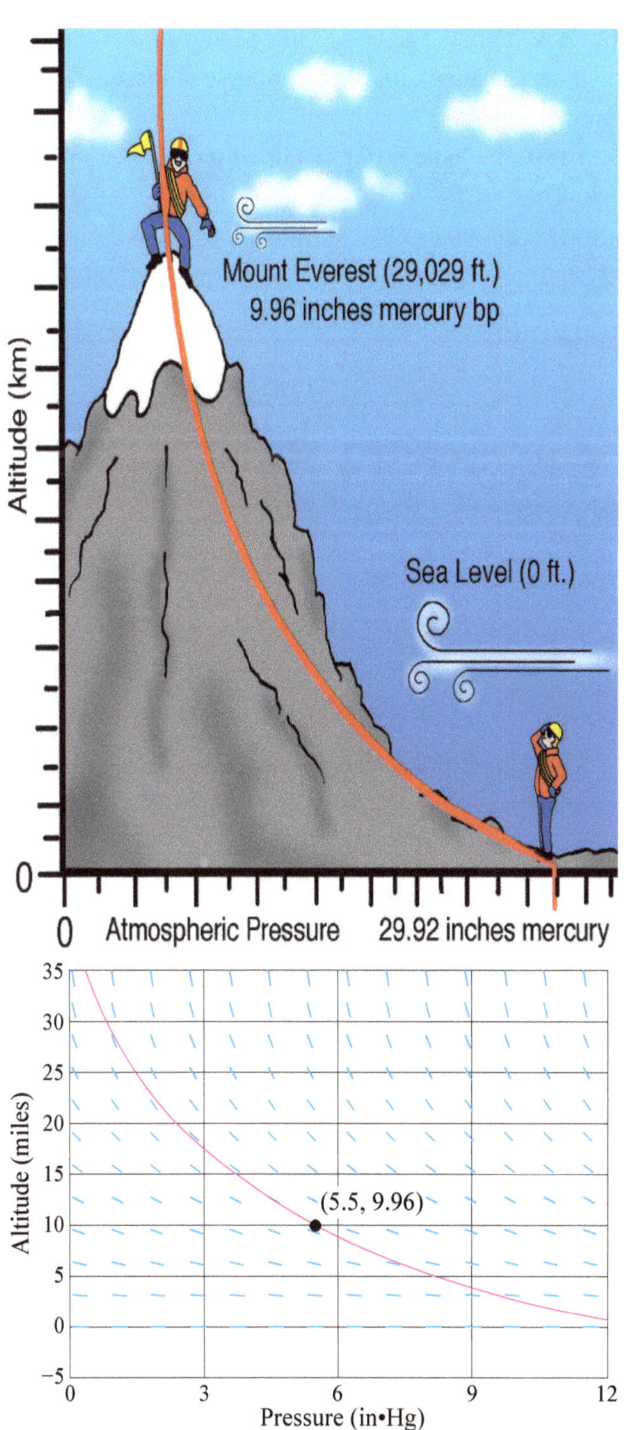

Mixing as a Differential Equation

After all his salt-water fish died, the owner of a large 1,000 L cylindrical fish tank has decided to replace them with freshwater fish. The owner could have just drained and cleaned the tank but instead he decided to drain the salt water from the tank at 10 L per minute while simultaneously replacing it with freshwater. The water had been kept at 3.5% (0.035) salt content meaning that in the 1,000 L tank there was 35 kg of salt.* The solution in the tank is kept mixed during the replacement process. How much salt will be in the tank after five hours? The owner has been told that the water must be less than 0.1% salt before he can put in his freshwater fish. How long will that take?

Let $A(t)$ = amount of salt in kilograms remaining after t minutes. Then, $A(0) = 35$ kg, the amount of salt at time 0: $0.035 \times 1{,}000$ L. $A(t)$ = ??, amount of salt at time t. $A(300)$ = amount of salt at time 5 h. $\frac{dA}{dt}$ = amount of salt coming in − amount of salt exiting. $\frac{dA}{dt} = \frac{0}{10}\frac{\text{kg}}{\text{L}} \times 10\frac{\text{L}}{\text{min}} - \frac{A}{1{,}000}\frac{\text{kg}}{\text{L}} \times 10\frac{\text{L}}{\text{min}}$. $\frac{dA}{dt} = 0\frac{\text{kg}}{\text{min}} - \frac{A}{100}\frac{\text{kg}}{\text{min}}$, the first term will drop out.

$$\frac{dA}{dt} = t^0 \times \frac{-A}{100}\frac{\text{kg}}{\text{min}} \quad \text{has form } \frac{dA}{dt} = f(t) \times g(A)$$
$$\frac{1}{A}\,dA = -0.01\,dt$$
$$A^{-1}\,dA = -0.01\,dt$$
$$\int A^{-1}\,dA = \int -0.01\,dt$$
$$\ln|A| + c_1 = -0.01t + c_2$$
$$\ln A = -0.01t + c_2 - c_1 \quad \text{where } A > 0$$
$$\ln A = -0.01t + c_3 \quad \text{where } c_3 = c_2 - c_1$$

$$e^{\ln A} = e^{-0.01t + c_3}$$
$$A = e^{-0.01\times t} \times e^{c_3}$$
$$A = e^{-0.01\times t} \times c_4$$
$$* \quad A = c_4 \times e^{-0.01\times t}$$
$$A_0 = 35 \text{ at } t_0$$
$$35 = c_4 \times e^{-0.01\times 0}$$
$$35 = c_4 \times e^0$$
$$35 = c_4$$

How much salt will be in the tank after five hours (300 min)? The owner has been told that the water must be less than 0.1% salt (less than one kilogram of salt in 1,000 L) for his new freshwater fish. How long will that take?

$$A = 35e^{-0.01\times t} \quad * \text{ above}$$
$$A = 35e^{-0.01\times 300} \quad 5\text{ h} = 300\text{ min}$$
$$A = 35e^{-3}$$
$$A = 1.74\text{ g salt}$$

$$1 = 35e^{-0.01\times t} \quad * \text{ above}$$
$$\frac{1}{35} = e^{-0.01\times t}$$
$$0.0285714286 = e^{-0.01\times t}$$
$$\ln(0.0285714286) = \ln\left(e^{-0.01\times t}\right)$$
$$-3.55534806 = -0.01t$$
$$t = 355.534806\text{ min}$$

y
Salt (kg)
Time (min)
x
(300, 1.74)
(355, 1)

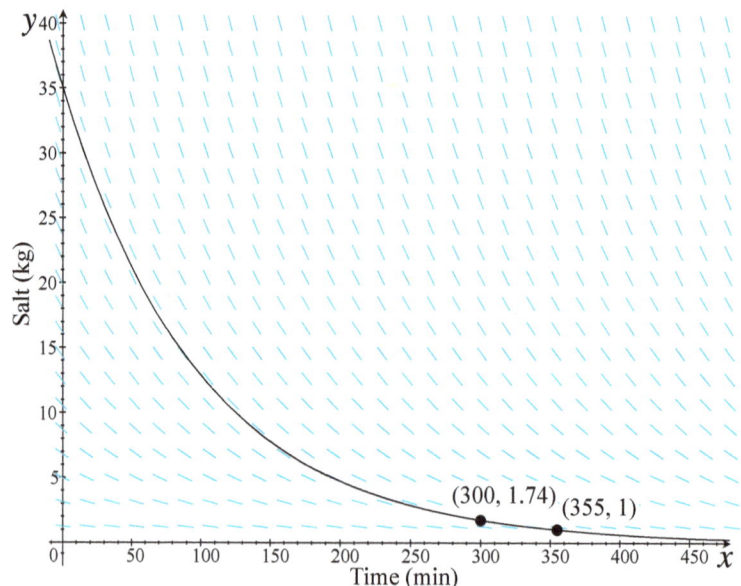

*A mililiter of water has a mass of one gram, so a liter (1,000 mL) has a mass of one kilogram. Then, 3.5% of a kilogram is 35 g, so 1,000 L must have 35 kg of salt.

Differential Equations in Electrical Circuits—Kirchhoff's Laws

In 1845 German physicist Gustav Kirchhoff (1824–1887) announced Kirchhoff's laws, which allowed calculation of the currents, voltages, and resistances of electrical networks. Kirchhoff's current law states that the current (flow of electrons) entering a junction is equal to the current leaving the junction. Kirchhoff's voltage law states that, in a circuit with only resistors and inductors, the sum of the voltage drops of the resistors and the inductors in a closed loop equals the total voltage gain of the source. That is, the sum of the voltages around a closed loop is 0, $\sum v_i = 0$. The resistors and the inductors in a circuit both work to reduce voltage; the difference is that a resistor works to impede or reduce voltage whereas the inductor only works to impede or reduce a change in current. Kirchhoff's voltage law is often written as $L\frac{dI}{dt} + RI = V(t)$ using Ohm's law ($V = IR$) as a measure of voltage drop in an inductor. Solving for I gives you an integrated formula to calculate current at a given time.

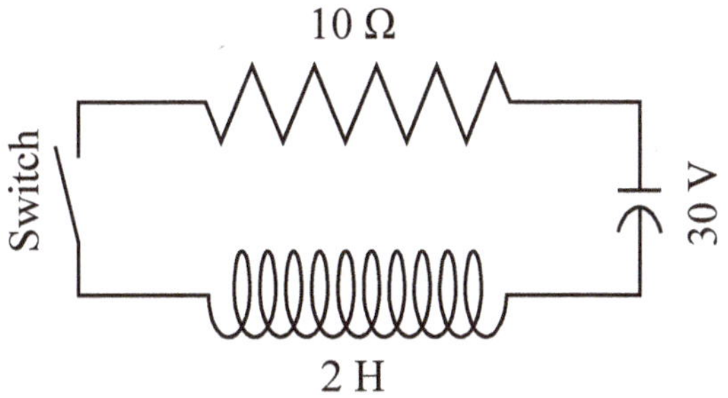

If a battery supplies a constant voltage of 30 V, the inductance is 2 H (henrys), the resistance is 10 Ω (ohms) and $I(0) = 0$, find $I(t)$ in general and the current after 0.3 seconds.

$$L\frac{dI}{dt} + RI = V(t)$$
$$2\frac{dI}{dt} + 10I = 30$$
$$\frac{dI}{dt} = 15 - 5I$$
$$dI = (15 - 5I)\,dt$$
$$\frac{dI}{15 - 5I} = dt$$
$$(15 - 5I)^{-1}\,dI = dt$$
$$-\frac{1}{5}\int (15 - 5I)^{-1}(-5)\,dI = \int dt$$

Note that $-\frac{1}{5}\int u^{-1}\,du = -\frac{1}{5}\ln|u|$

$$-\frac{1}{5}\ln|15 - 5I| + c_1 = t + c_2$$
$$-\frac{1}{5}\ln|15 - 5I| = t + c_2 - c_1$$
$$-\frac{1}{5}\ln|15 - 5I| = t + c_3$$
$$\ln|15 - 5I| = -5t - 5c_3$$
$$\ln|15 - 5I| = -5t + c_4$$
$$e^{\ln|15-5I|} = e^{-5t+c_4}$$
$$|15 - 5I| = e^{-5t} \times e^{c_4}$$
$$|15 - 5I| = c_5e^{-5t}$$
$$15 - 5I = c_5e^{-5t}$$
$$I = c_6e^{-5t} + 3 \quad ***$$

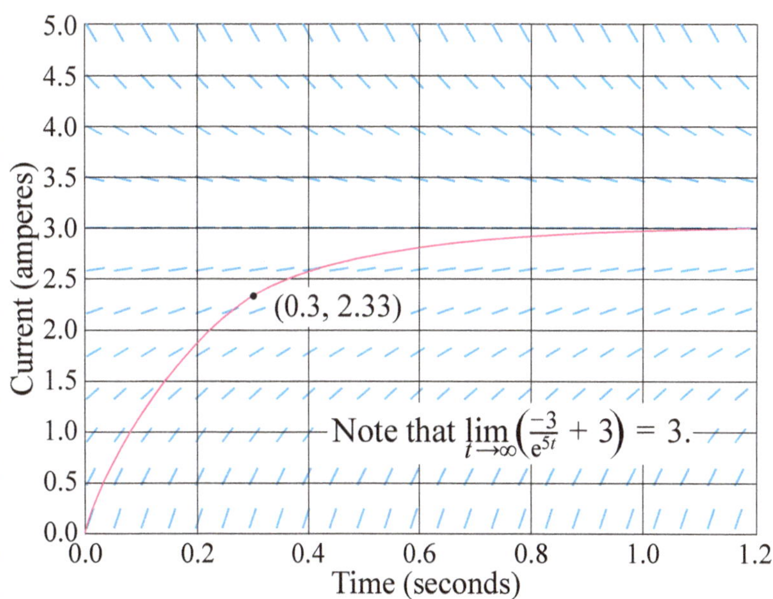

MATLAB script for the graph above in Appendix A.

$I(0) = 0$

$$0 = c_6e^{-5\times 0} + 3$$
$$0 = c_6 \times 1 + 3$$
$$c_6 = -3$$

Therefore, from ***,

$$I = -3e^{-5t} + 3.$$

Hence, at 0.3 seconds

$$I = -3e^{-5\times 0.3} + 3$$
$$I = -3e^{-1.5} + 3$$
$$I = -3 \times 0.2231301601 + 3$$
$$I = -0.6693904804 + 3$$
$$I = 2.33060952 \text{ A at } 0.3 \text{ s}$$

Newton's Law of Cooling

Sir Isaac Newton found that the rate at which an object cools is directly proportional to the difference between the object's temperature and the temperature of the surrounding medium (air, water). Hopefully your attention was caught by the following phrases: "rate of cooling (over time)" and "directly proportional." Déjà vu! Something seems familiar here. Follow the following progression of thought from words to symbols.

Rate of cooling (over time) is directly proportional to the difference between the object's temperature and the ambient temperature.

Change of temp (over time) = k × (object's temp − ambient temp)

$\frac{dT}{dt} = k \times (T - T_a)$ T = object's temperature, T_a = temperature of surrounding medium (air or H_2O)

$\frac{dT}{T - T_a} = k \times dt$ dT on the left is differential of temperature, dt on right is differential of time

$\int \frac{dT}{T - T_a} = k \int dt$

$\int (T - T_a)^{-1}\, dT = k \int t^0\, dt$

$\ln|T - T_a| + c_1 = kt + c_2$ the ts are all confusing. Remember that the capital Ts represent the object's and the ambient temperature, whereas the lowercase t represents time

$\ln|T - T_a| = kt + c_2 - c_1$

$\ln|T - T_a| = kt + c_3$

$e^{\ln|T - T_a|} = e^{kt + c_3}$ $b^m \times b^n = b^{m+n}$, therefore $e^{kt+c_3} = e^{kt}e^{c_3}$

$T - T_a = e^{kt}e^{c_3}$ object is cooling toward the ambient temp, therefore $T - T_a > 0$

$T = T_a + e^{kt}C$ $C = e^{c_3}$

$T_f = T_a + Ce^{kt}$ *** final temp = ambient temp + original temp × a decay factor. What a mess! There are lots of unknowns. We can simplify a bit by letting T_f be T_0 when $t = 0$.

$T_0 = T_a + Ce^{k \times 0}$

$T_0 = T_a + Ce^0$

$T_0 = T_a + C \times 1$

$C = T_0 - T_a$

$T_f = T_a + (T_0 - T_a) \times e^{kt}$ substitute $C = T_0 - T_a$ into the equation above ***

A person was murdered in a mansion. Police forensic specialists arrive at (arrival time, t_0) and note the temperature of the body to be (initial temperature upon arrival, T_0). t hours later, the temperature of the body was rechecked and found to be (temperature t hours later, T_f). The thermostat in the room had been set to maintain a constant temperature of (thermostat temperature, T_a). Assume the person's temperature was 98.6 °F while living. What was the time of death, t? Since the room was cooler than the body at the time of death, the ambient air would have acted to cool the body. Since "e" is a known constant and variables T_f, T_a, T_0, and t are all known, you could substitute all that known information into the Newton's Law of Cooling $T_f = T_a + (T_0 - T_a) \times e^{kt}$, and solve for k, the constant of variation for that equation.

Let's do that again. A person was murdered in a mansion. Police forensic specialists arrive at 10 am and note the temperature of the body to be 83 °F (T_0). After 2.5 (t) hours (12:30 pm), the temperature of the body was rechecked and found to be 72 °F (T_f). The thermostat in the room had been set to maintain a constant temperature of 67 °F (T_a). Assume the person's temperature was 98.6 °F when alive. Estimate the time of death. In Newton's Law, there are five unknowns. After substituting in for T_f, T_a, T_0, and t, we only have one unknown, k, for which we can solve.

With k now known, we can rework the problem with the newly found k in conjunction with a different T_f, 98.6 °F, to solve for t.

$$T_f = T_a + (T_0 - T_a) \times e^{kt}$$
$$72 = 67 + (83 - 67) \times e^{k \times 2.5}$$
$$5 = 16 \times e^{k \times 2.5}$$
$$\frac{5}{16} = e^{k \times 2.5}$$
$$0.3125 = e^{2.5k}$$
$$\ln 0.3125 = \ln e^{2.5k}$$
$$-1.16315081 = 2.5k$$
$$k = -0.4652603239$$

Moving to the right, the red line appears asymptotic to 67 °F. Why? Do you recall a sink asymptote from Chapter 4?

To determine the time (t) of death, substitute 98.6 °F into T_f (final temperature) and solve for t.

$$98.6 = 67 + (83 - 67) \times e^{-0.4652603239 \times t}$$
$$31.6 = 16 \times e^{-0.4652603239 \times t}$$
$$\frac{31.6}{16} = e^{-0.4652603239 \times t}$$
$$\ln 1.975 = \ln e^{-0.4652603239 \times t}$$
$$0.6805683983 = -0.4652603239 \times t$$
$$t = \frac{0.6805683964}{-0.4652603239}$$
$$= -1.46\,\text{h}$$
$$= -1{:}28$$

10 am − 1:28 = 8:32 am time of death

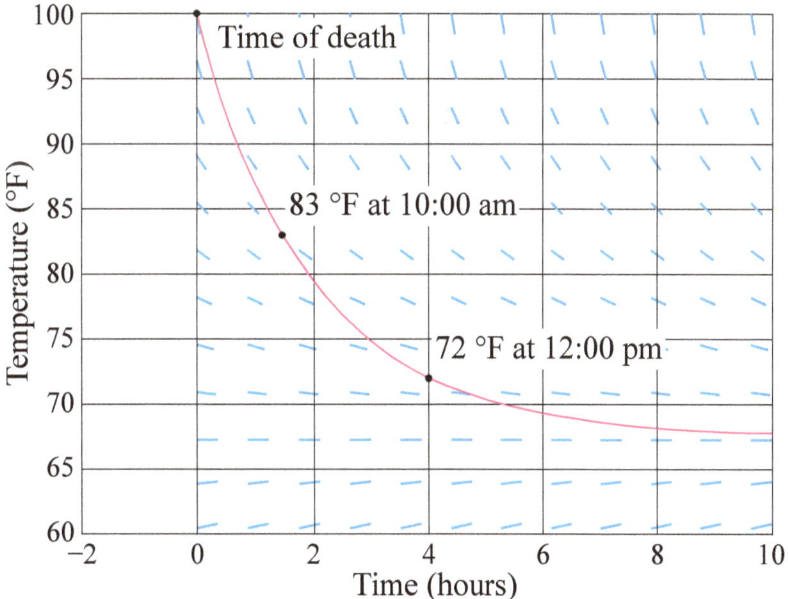

MATLAB script for graph above in Appendix A.

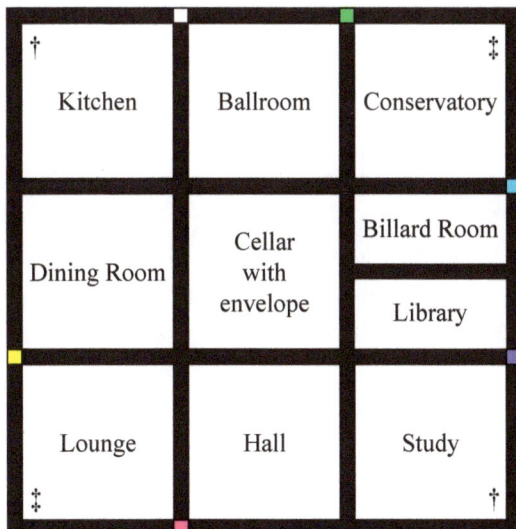

Kitchen
Ballroom
Conservatory
Dining Room
Cellar with envelope
Billard Room
Library
Lounge
Hall
Study

Differential Equations for Continuous Compound Interest—Revisited

At the risk of oversimplifying, investments are categorized as equities (stocks), with various levels of risk, and bonds with lower, but "guaranteed," fixed levels of income. In planning for retirement a retired person moved \$300,000 out of an equity account into income-producing bonds. If that account returned 5% a year compounded continuously, how much money could the retiree count on to spend from the interest from his account each year? Well, in Chapter 3, we studied the formula $p_f = p_0 \times e^{rt}$. For the conditions stated above, this would be $p_1 = \$300{,}000 \times e^{0.05\times1}$. $p = \$300{,}000 \times 1.051271096 = \$315{,}381.33$. Therefore, the retiree could withdraw \$15,381.33 a year without drawing down his principal. However, without a paycheck coming in anymore, the retiree needs more than \$15,381.33 a year to cover his living expenses. Given his family history, the retiree decides to plan on 20 years of additional life expectancy. In addition to the 5% annual growth of his money, could he draw down an additional \$23,000 a year and still not outlive his money at the end of that 20 years? We solved the previous continuous interest problem using the differential equation $\frac{dp}{dt} = 0.05p$, so it is tempting to solve this problem with $\frac{dp}{dt} = 0.05p - \$23{,}000$. There is a flaw with that approach, but we will start out that way and then discuss the results.

$$\frac{dp}{dt} = 0.05p - 23{,}000$$

$$dp = (0.05p - 23{,}000)\,dt$$

$$\frac{dp}{0.05p - 23{,}000} = dt$$

$$(0.05p - 23{,}000)^{-1}\,dp = dt$$

$$\int (0.05p - 23{,}000)^{-1}\,dp = \int dt$$

$$20\int (0.05p - 23{,}000)^{-1} \times 0.05\,dp = t + c_1$$

$$20\ln|0.05p - 23{,}000| + c_2 = t + c_1$$

$$20\ln|0.05p - 23{,}000| = t + c_1 - c_2$$

$$20\ln|0.05p - 23{,}000| = t + c_3$$

$$\ln|0.05p - 23{,}000| = \frac{1}{20}(t + c_3)$$

$$\ln|0.05p - 23{,}000| = \frac{t}{20} + \frac{c_3}{20}$$

$$\ln|0.05p - 23{,}000| = \frac{t}{20} + c_4$$

$$e^{\ln|0.05p - 23{,}000|} = e^{\frac{t}{20} + c_4}$$

$$|0.05p - 23{,}000| = e^{\frac{t}{20}} \times e^{c_4}$$

$$|0.05p - 23{,}000| = c_5 \times e^{\frac{t}{20}} \quad c_5 > 0$$

$$0.05p - 23{,}000 = \pm(c_5 \times e^{\frac{t}{20}})$$

$$0.05p - 23{,}000 = c_6 \times e^{\frac{t}{20}} \quad c_6 > 0$$

$$100(0.05p - 23{,}000) = 100 \times c_6 \times e^{\frac{t}{20}}$$

$$5p - 2{,}300{,}000 = c_7 \times e^{\frac{t}{20}}$$

$$5p = 2{,}300{,}000 + c_7 \times e^{\frac{t}{20}}$$

$$p = 460{,}000 + \frac{c_7 \times e^{\frac{t}{20}}}{5}$$

$$p = 460{,}000 + c_8 \times e^{\frac{t}{20}} \quad *** \quad t = 0,\ p = 300{,}000$$

$$300{,}000 = 460{,}000 + c_8 \times e^{\frac{0}{20}}$$

$$-160{,}000 = c_8 \times 1$$

$$c_8 = -160{,}000$$

Finally, substituting into *** above

$$p = 460{,}000 - 160{,}000 \times e^{\frac{t}{20}}$$

For $t = 20$ yrs

$$p_{20} = 460{,}000 - 160{,}000 \times e^{1}$$

$$p_{20} = 460{,}000 - 434{,}925.0926$$

$$p_{20} = 25{,}074.91$$

It appears that 20 years after drawing down his principal at \$23,000 a year, he will still have principal remaining: \$25,074.91. The flaw in that approach can best be shown by a visual. As was shown before, continuous compounding of principal results in a continuous accumulation of principal. However, the yearly drawdown of principal described in the problem above would be, for practical purposes, discrete: It happens all at once. If the yearly drawdown is greater than the amount of money earned during the preceding twelve months, then, over time, the original principal (\$300,000) will be depleted.

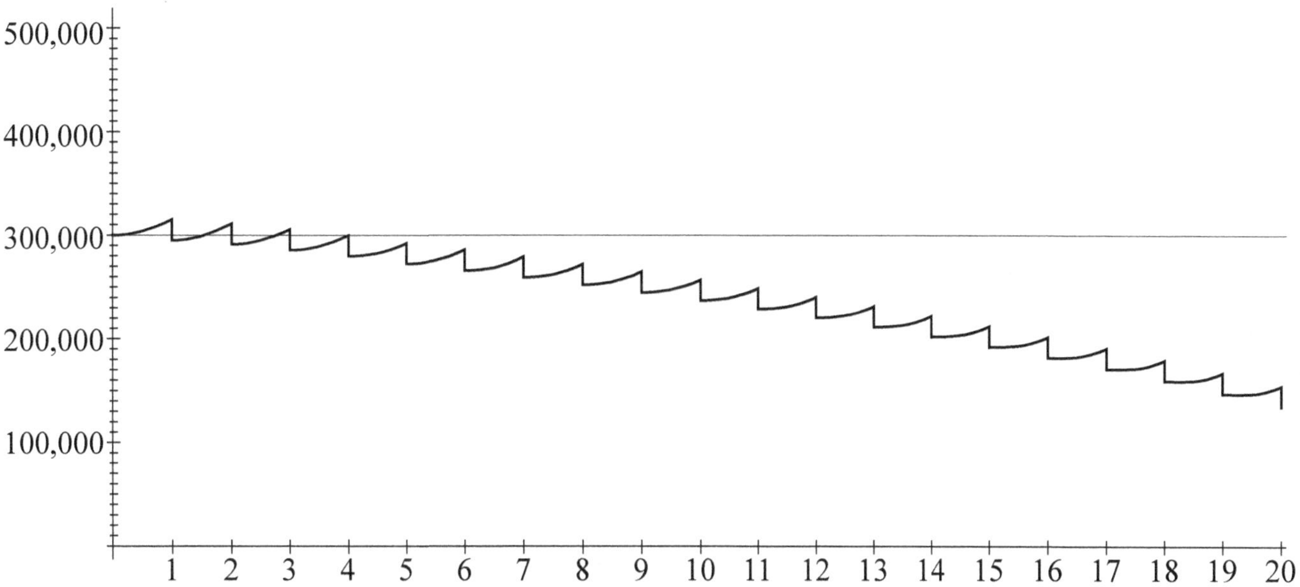

If the yearly drawdown is less than the amount of money earned during the preceding twelve months, then, over time, the original principal ($300,000) will increase, but at a much slower rate than in the earlier continuous accumulation problem.

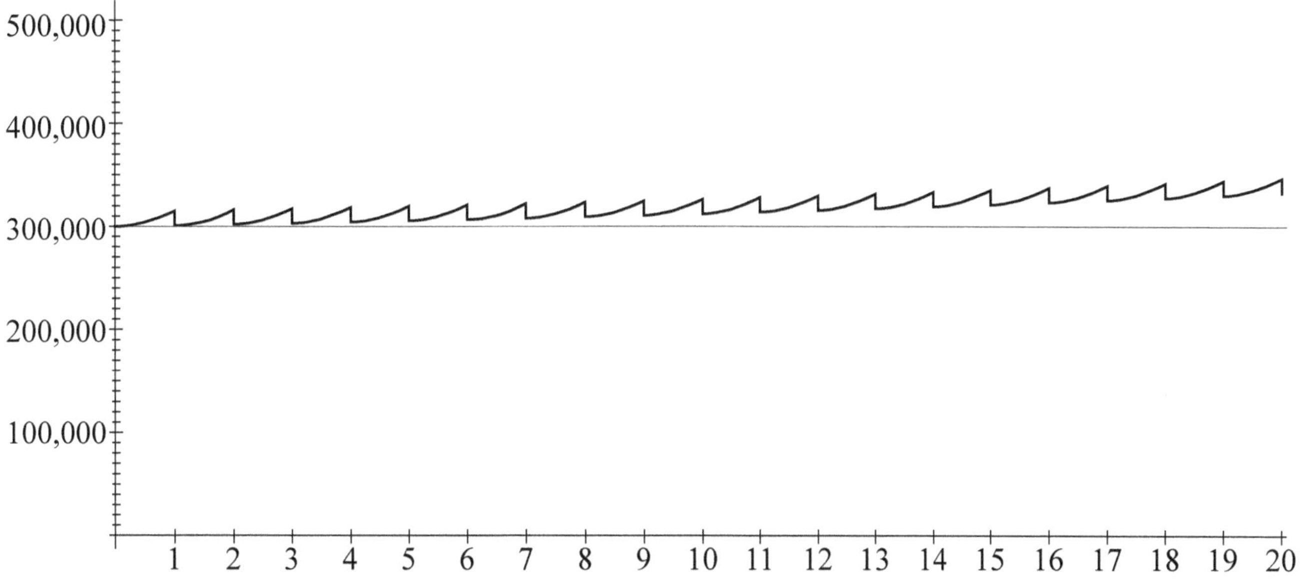

If you understand computer programming, you can use Euler's method to simulate both continuous compounding of interest and continuous withdrawals (combining both) to analyze the problem above (continuous compounding of interest with yearly lump sum drawdown) to any specified tolerance allowable in your computer.

The following program was run first showing Euler's method used with a Δt of 10^{-7} year to calculate compound interest accumulating on $300,000 at 5% for 20 years. The answer, as you know now, would be $p_f = p_0 e^{rt} = 300{,}000 \times e^{0.05\times 20} = \$300{,}000 \times e^1 = \$815{,}484.5485$. The answer obtained by the code was $815,484.57. The numbers differ because Euler's method gives discrete-math results whereas differential equations gives continuous-math results. The code that does the "continuous compounding" is combined with the code for "continuous withdrawal." Here the term "continuous" reflects the 10^{-7} Δt used by Euler's algorithm in the code. However, as pointed out above, "continuous" withdrawal is theoretically possible but not convenient in the real world. The yearly drawdown might be done in a lump sum at either the beginning or end of the year. This is done on the following page with the drawdown done at the beginning of the year. Notice that first of the year drawdown deprives the retiree of interest from $23,000 that year.

```
public class ContinuousGrowthOfPrincipal
{
  public static void main(String args[])
  {
    double t = 0;
    double r = 0.05;
    double p = 300000;
    double deltaT = 0.000001; // 10E-6
    double annualWithdrawal = 23000;
    boolean exactYear = true;
    System.out.println("Annual Withdrawal is " + AnnualWithdrawal);
    System.out.println("Delta t is " + deltaT);
    System.out.println(" t p"); // column headers
    System.out.println("\n " + (int)t + ", " + p); // initial
        // remove Java comment symbol for output at right
    p -= annualWithdrawal;      // starting annual drawdown
    while (t < 20)
    {
      // output for p shown top right
      p += ((r * p) * deltaT); // compound annual interest calculated
          // "continuously"
      t += deltaT;
      exactYear = (t - (int)t) < deltaT;
      if(exactYear)
      {
        System.out.println(" " + (int)t + ", " + p);
        // remove Java comment symbol for output right
        p -= annualWithdrawal; // annual drawdown
      } // end if
    } // end while
  } // end main
} // end class
```

Code check for "continuous" interest

$$p_f = p_0 e^{rt}$$
$$= 300{,}000 \times e^{0.5\times 20}$$
$$= 300{,}000 \times e$$
$$= 815{,}484.5485$$

```
Annual withdrawal is
$0
deltaT is 0.000001
```

t	p
0	300,000.00
1	315,381.33
2	331,551.29
3	348,550.27
⋮	⋮
18	737,880.95
19	775,712.92
20	815,484.57

```
Annual withdrawal is
$23,000
deltaT is 0.000001
```

t	p
0	300,000.00000000000
1	291,202.09511577286
2	281,953.10913769866
3	272,229.92028027700
⋮	⋮
18	49,538.19527624527
19	27,898.83764041088
20	5,150.00639080917

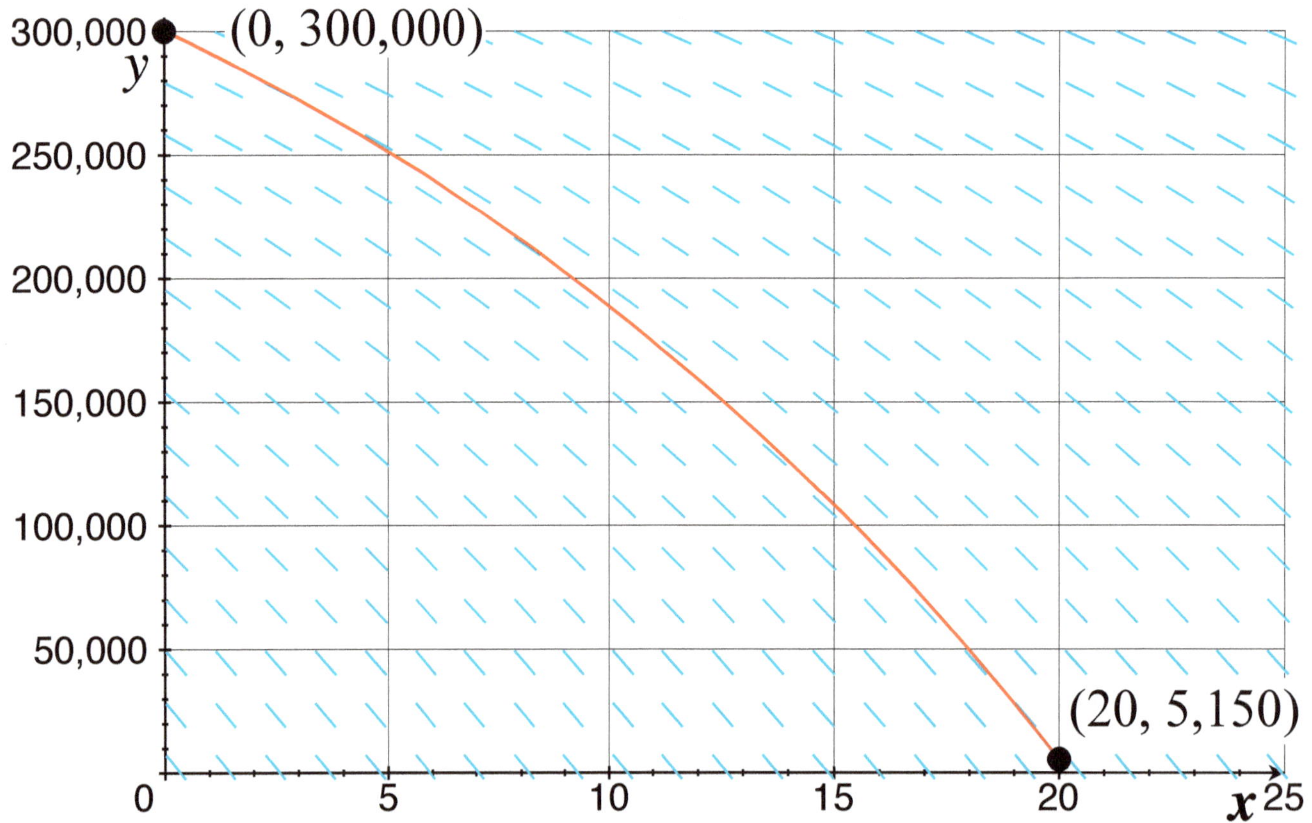

Checking Solutions to Differential Equations

$$\frac{2}{3}x - 2 = 4$$
$$3\left(\frac{2}{3}x - 2\right) = 3 \times 4$$
$$2x - 6 = 12$$
$$2x = 18$$
$$x = 9$$
check by back substituting
$$\frac{2}{3} \times 9 - 2 \stackrel{?}{=} 4$$
$$6 - 2 \stackrel{?}{=} 4$$
$$4 = 4 \text{ ck}$$

Since beginning algebra, you should have been encouraged to check your work by substituting the scalar answer back into the original algebraic equation.

Up to this point, all of our work has been checked using slope fields drawn in MATLAB and fitting a particular solution through given initial condition. The old way (before Mathematica and MATLAB) to check work done solving a differential equation was to take the derivative of the solution and see if it matches with the given differential equation. For the sake of nostalgia, that way to check work on a differential equation is shown below.

Is $x^2 + y^2 = Cy$ a solution of $\frac{dy}{dx} = \frac{2xy}{x^2-y^2}$? In other words, is $x^2 + y^2 = Cy$ a solution of $y' = \frac{2xy}{x^2-y^2}$? (See text box below.)

If $C = 5$, will the particular (specific) solution pass through the point $(2, 4)$? Substituting as appropriate into the original general solution, we get $x^2 + y^2 = Cy \to 2^2 + 4^2 \stackrel{?}{=} 5 \times 4 \to 4 + 16 \stackrel{?}{=} 20 \to 20 = 20$. Yes, the point $(2, 4)$ will lie on the particular curve that results when $C = 5$.

$$\frac{d}{dx}\left(x^2 + y^2\right) = \frac{d}{dx}(Cy)$$
$$2x + 2y\frac{dy}{dx} = C\frac{dy}{dx}$$
$$2y\frac{dy}{dx} - C\frac{dy}{dx} = -2x$$
$$\frac{dy}{dx}(2y - C) = -2x$$
$$\frac{dy}{dx} = \frac{-2x}{2y - C}$$
$$\frac{dy}{dx} = \frac{-2x}{2y - C} \times \left(\frac{y}{y}\right)$$
$$\frac{dy}{dx} = \frac{-2xy}{2y^2 - Cy}$$
$$\frac{dy}{dx} = \frac{-2xy}{2y^2 - (x^2 + y^2)}$$
$$\frac{dy}{dx} = \frac{-2xy}{y^2 - x^2}$$
$$\frac{dy}{dx} = \frac{-2xy}{y^2 - x^2} \times \frac{-1}{-1}$$
$$\frac{dy}{dx} = \frac{2xy}{x^2 - y^2} \text{ ck}$$

Yes, $x^2 + y^2 = Cy$ is a solution of $\frac{dy}{dx} = \frac{2xy}{x^2-y^2}$.

Will the point $(3, -1)$ lie on that particular curve when $C = 5$? $3^2 + (-1)^2 \stackrel{?}{=} 5 \times (-1) \to 9 + 1 \stackrel{?}{=} -5 \to 10 \stackrel{?}{=} -5$. No.

Chapter 5 Review

Following up on the three examples of separable differential equations in Chapter 4, several more examples of that type or category of differential equation were presented. The differential equations applied to

1. chemistry
2. meteorology
3. mixing problems
4. electrical circuits
5. Newton's Law of Cooling
6. continuous compound interest with withdrawals

Finishing out Chapter 5, an example of the pre-Mathematica, pre-MATLAB technique of checking solutions to differential equations was shown. Today many differential equation books start out showing that technique when introducing the reader to what a differential equation is. The book will have a function that it says it got somewhere and, by substitution, show that it satisfies a differential equation. The author then assumes that he/she has properly introduced what a differential equation is. Of course many beginning students will be fixated on where that "mystery" function came from and not be able to focus on the successful substitution and on the point that the author is trying to introduce.

Many, many math teachers and authors believe that math instruction should at all times be complete and rigorous and deal with all contingencies at the time of introduction. My philosophy is that sort of thinking discounts the importance of "student readiness" and is counterproductive for most beginners. What follows is a primitive chart to help beginning students (the target audience for this book) differentiate and mentally organize their thoughts to this point in the book. Bear in mind that it will have to be adjusted as more knowledge and information is covered.

Ordinary Differential Equations (ODE) by category

* Introduced
*** Covered

- Only a decimal approximation can be found
  - Euler's method (*** Chapter 1)
  - Runge–Kutta
- Closed form (analytical solution) and decimal approximation can be found
  - Separable (*** Chapters 1–5)
    - $y' = f(x) \times g(y),\ g(y) \neq 0$
    - $\frac{dy}{dx} = f(x) \times g(y),\ g(y) \neq 0$
    - Exponential growth
    - Exponential decay

Online Application

Visit demonstrations.wolfram.com/MixingSaltInWater InOneTank for a great little application that will demonstrate some of the concepts from this chapter. "Mixing Salt in Water in One Tank" from the Wolfram Demonstrations Project. Contributed by Stephen Wilkerson.

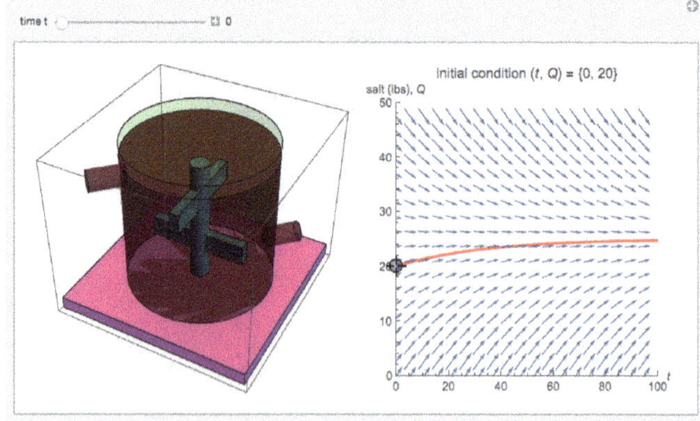

Chapter 6
Solving Separable Logistic Equations

Exponential Growth versus Logistic Growth

The term *exponential growth* is often used when talking about population growth. Assuming unlimited supplies of various resources, the growth rate is proportional to the existing population. In logistic growth, the growth rate is also proportional to the existing population, but the concept takes into account limited resources, disease, and predation. Consequently, logistic growth is bounded. That boundary is called the "carrying capacity of the environment." See the figures below.

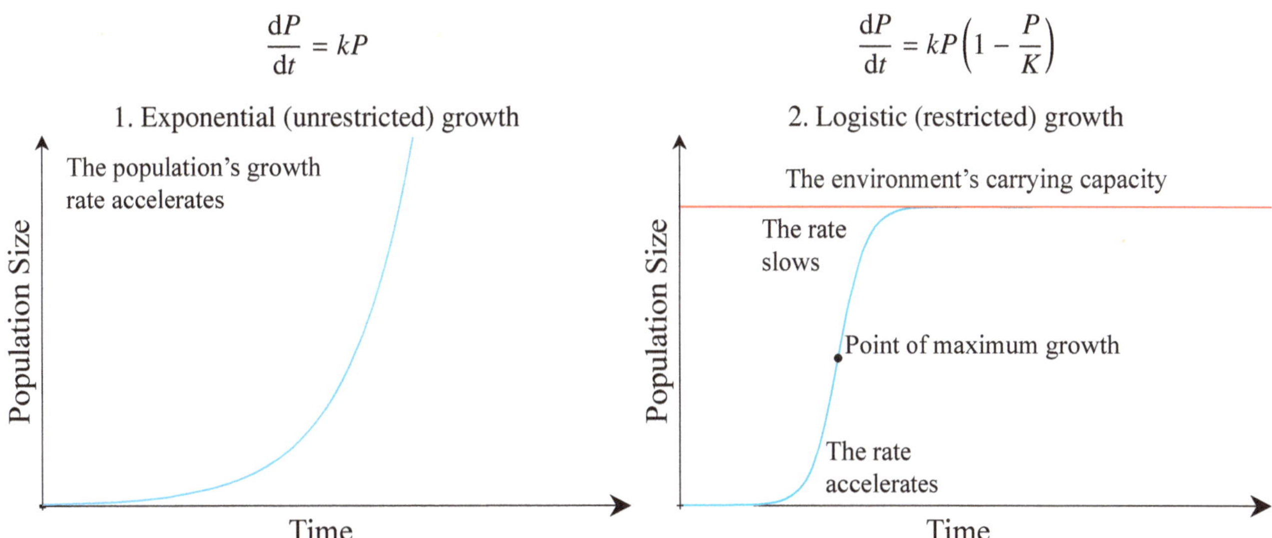

Thomas Malthus

In exponential growth, there is no upper limit, but growth is limited in logistic growth. Logistic growth is usually considered more realistic than exponential growth when studying populations. In 1798 a British economist, Thomas Malthus, became quite famous for writing that, while population growth might be exponential, shortages in food supply, war, disease and starvation create conditions which would slow and even limit that exponential growth. See the figures above.

When we studied continuous compound interest the exponential growth model was appropriate, but the logistic growth model is more appropriate when studying population growth. A better match between the problem being studied and the model used to study it generally gives more reliable study results.

The K or carrying capacity in the logistic differential equation sets a maximum bound on P.

$\frac{dP}{dt} = kP$

exponential growth

$k = 0.05$

t	P	deltaP
1	1,050.00	50.00
2	1,102.50	52.50
3	1,157.63	55.13
4	1,215.51	57.88
5	1,276.28	60.78
6	1,340.10	63.81
7	1,407.10	67.00
⋮	⋮	⋮
195	13,549,189.55	645,199.50
196	14,226,649.03	677,459.48
197	14,937,981.48	711,332.45
198	15,684,880.56	746,899.07
199	16,469,124.59	784,244.03
200	17,292,580.82	823,456.23

Growth here is unbounded

$\frac{dP}{dt} = kP\left(1 - \frac{P}{K}\right)$

logistic growth

$k = 0.05$, $K = 2{,}000$

t	P	deltaP
1	1,025.00	25.00
2	1,049.98	24.98
3	1,074.92	24.94
4	1,099.78	24.86
5	1,124.53	24.75
6	1,149.14	24.61
7	1,173.59	24.44
⋮	⋮	⋮
195	1,999.91	0.005
196	1,999.91	0.0046
197	1,999.92	0.0045
198	1,999.92	0.0042
199	1,999.923	0.0040
200	1,999.927	0.0038

Growth here is bounded

Unbounded Growth

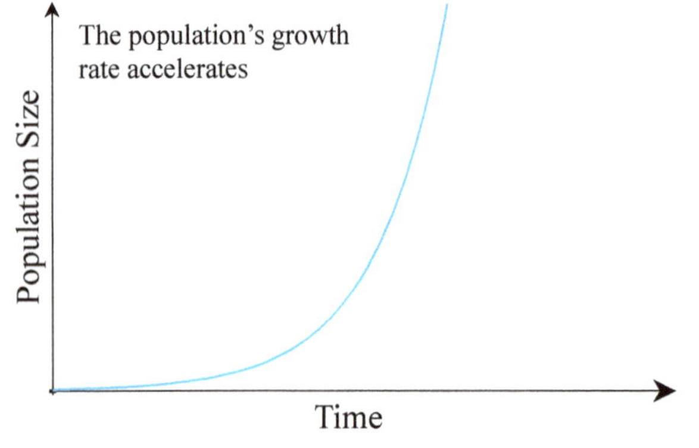

Bounded Growth

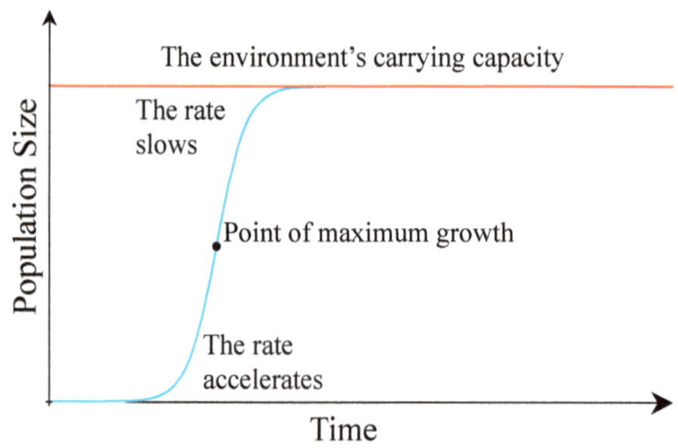


```
public class JavaTemplate
{
  public static void main(String args[])
  {
    double r = 0.05; // rate of growth
    int K = 2000; // carrying capacity
    double P = 1000; // population, principal, whatever
    System.out.println(" t P deltaP" + "\n");
    for (int t = 1; t < = 200; t++)
    {
      double oldP = P;
      P += r * P * (K - P)/K; // logistic growth
      // P += r * P; // exponential growth
      double deltaP = P - oldP;
      System.out.println(" " + t + " " + P + " " + deltaP);
    } // end loop
  } // end main
} // end class
```

Following are five examples of situations where there is a limit to how much something can grow. Each of these situations could be modeled using the logistic growth curve.

$$\frac{dP}{dt} = kP\left(1 - \frac{P}{K}\right)$$

1. **Differential Equations in Disease Control**

   After spring break, a college campus nurse identified 10 students with cooties. At Day 4, she identified 30 students. There are 800 students on campus. **Assuming the infection grows logistically,** how many students will be infected after five weeks? Disease spread is limited to 800 students.

2. **Differential Equations in Sociology—Measuring the Spread of Rumors**

   In a high school with 2,000 students, a couple breaks up. By the end of the day, 20 people know about it. By the end of the next day, 70 people know about it. **Assuming the gossip spreads logistically,** how many days will it take for half the school to know? Rumor spread is limited to 2,000 students.

3. **Differential Equations in Biology—Measuring Population Growth**

   A population of prairie dogs is estimated to be 800 prairie dogs. After four weeks there are 1,000 prairie dogs. Biologists estimate that there is only food for 1,800 prairie dogs. **Assuming the population grows logistically,** how many prairie dogs will there be after 10 weeks? After 20 weeks? The number of prairie dogs is limited to 1,800.

4. **Differential Equations in Genetics—Monitoring Inheritance of Characteristics**

   A geneticist is studying a population of mice to determine how quickly a physical trait will spread into the next generation. At the start of the study ($t = 0$), she has established that 30% of the population has the characteristic. After five generations ($t = 5$) she finds that 85% of the population has the characteristic. **Use the logistic function** to determine the percentage of mice that will have the studied trait after eight generations. The percentage of population is limited to 100%.

5. **Differential Equations in Advertising—Monitoring Effectiveness of an Advertising Campaign**

   A company making and selling widgets is engaged in a marketing campaign in a large metropolitan area of 1 million people. **Assuming logistic growth,** use the given IVP conditions to measure the effectiveness of the campaign. The maximum number of people who could hear about the product is 1 million people.

We'll look at each of these examples in more detail later, but first we should look a little more closely at the theory behind logistic equations and their relationship to separable differential equations. Logistic equations are a subset of separable differential equations, although this is not evident when comparing the two:

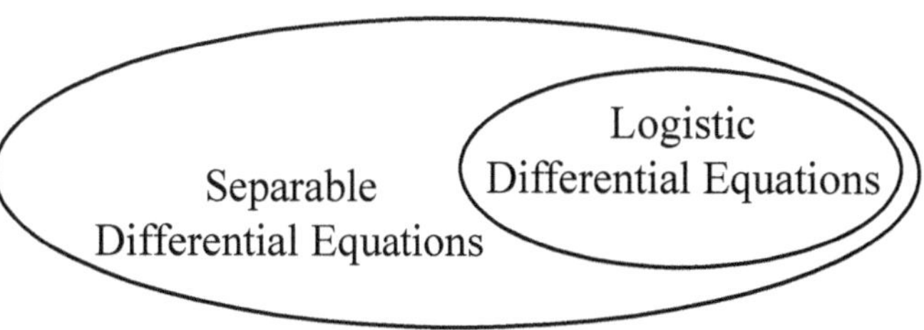

$$\frac{dp}{dt} = kP \quad \text{separable differential equation}$$

$$\frac{dP}{dt} = kP\left(1 - \frac{P}{K}\right) \quad \text{logistic differential equation}$$

Hold onto this idea as you proceed through your differential-equations class. There will be many occasions where more than one technique can be used to solve a problem.

If $\frac{dP}{dt} = kP\left(1 - \frac{P}{K}\right)$ is separable, then it should be possible to convert it to the form: $f(t) \times g(P)$. Let's try.

Attempt #1

$$\frac{dP}{dt} = kP\left(1 - \frac{P}{K}\right)$$
$$dP = kP\left(1 - \frac{P}{K}\right)dt$$
$$dP = \left(kP - \frac{P}{K}kP\right)dt$$
$$dP = kP\,dt - \frac{P}{K}kP\,dt???$$

Attempt #2

$$\frac{dP}{dt} = kP\left(1 - \frac{P}{K}\right)$$
$$dP = kP\left(1 - \frac{P}{K}\right)dt$$
$$\frac{dP}{P} = k\left(1 - \frac{P}{K}\right)dt$$

????????????

Attempt #3

$$\frac{dP}{dt} = kP\left(1 - \frac{P}{K}\right)$$
$$dP = kP\left(1 - \frac{P}{K}\right)dt$$
$$dP = \left(kP - \frac{kP^2}{K}\right)dt$$
$$\frac{dP}{kP - \frac{kP^2}{K}} = dt \text{ separated!}$$

The integration of the left side in #3 can be very difficult. It turns out that there is a way to avoid such difficult integrations. Chapter 6 teaches a clever way to do this. Be patient. What follows is a review from beginning algebra. It is often helpful to bridge into new ideas by connecting them to old ones.

Solve the following quadratic equation.

$x^2 - 2x - 15 = 0$ original quadratic

$(x + 3)(x - 5) = 0$ factor the trinomial into the product of two binomials

$x + 3 = 0$ or $x - 5 = 0$ zero product property, if $a \times b = 0$ then $a = 0$, $b = 0$, or both.

$x = -3$ or $x = 5$

But wait! *What could you do if the quadratic did not factor*? In that case you switched to another attack called "Completing the square."

Solve

$2x^2 - 9x - 4 = 0$ original quadratic

$x^2 - \frac{9}{2}x - \frac{4}{2} = 0$ divide both sides by 2 so that the x^2 term has a coefficient of 1

$x^2 - \frac{9}{2}x = \frac{4}{2}$ isolate the x terms

$x^2 - \frac{9}{2}x + \frac{81}{16} = 2 + \frac{81}{16}$ complete the binomial square by adding $\left(\frac{1}{2} \times \frac{9}{2}\right)^2$ to both sides

$\left(x - \frac{9}{4}\right)^2 = \frac{32}{16} + \frac{81}{16}$ change form: $a^2 - 2ab + b^2 = (a - b)^2$, prepare to add fractions

$\left(x - \frac{9}{4}\right)^2 = \frac{113}{16}$ add like fractions

$x - \frac{9}{4} = \pm\frac{\sqrt{113}}{4}$ take the square root of both sides

$x = \frac{9}{4} \pm \frac{\sqrt{113}}{4}$ isolate the x

$x = \frac{9 \pm \sqrt{113}}{4}$ combine like fractions

Oh, that was fun. Let's do it again! Solve the following quadratic equation.

$$3x^2 - 7x - 5 = 0$$

Just kidding!! We could solve this equation using the technique shown above. However, the process for solving this equation will be the same as the process for solving the first one above with only the coefficients being different. Perhaps we could do the same process with symbols (a, b, and c) for the constants, $ax^2 + bx + c = 0$; then we would have a generic template for all quadratics.

$$ax^2 + bx + c = 0$$
$$\frac{ax^2 + bx + c}{a} = \frac{0}{a}$$
$$x^2 + \frac{b}{a}x + \frac{c}{a} = 0$$
$$x^2 + \frac{b}{a}x = -\frac{c}{a}$$
$$x^2 + \frac{b}{a}x + \left(\frac{b}{2a}\right)^2 = -\frac{c}{a} + \left(\frac{b}{2a}\right)^2$$
$$\left(x + \frac{b}{2a}\right)^2 = -\frac{c}{a} + \frac{b^2}{4a^2}$$
$$\left(x + \frac{b}{2a}\right)^2 = \frac{b^2}{4a^2} - \frac{c}{a}$$
$$\left(x + \frac{b}{2a}\right)^2 = \frac{b^2}{4a^2} - \frac{c}{a} \times \frac{4a}{4a}$$
$$\left(x + \frac{b}{2a}\right)^2 = \frac{b^2}{4a^2} - \frac{4ac}{4a^2}$$
$$\left(x + \frac{b}{2a}\right)^2 = \frac{b^2 - 4ac}{4a^2}$$
$$\sqrt{\left(x + \frac{b}{2a}\right)^2} = \sqrt{\frac{b^2 - 4ac}{4a^2}}$$
$$\pm\left(x + \frac{b}{2a}\right) = \pm\frac{\sqrt{b^2 - 4ac}}{2a}$$
$$x = \frac{-b \pm \sqrt{b^2 - 4ac}}{2a}$$

a generic formula which will work for all quadratic equations!

The same sort of thing happens when working with "logistic equations." The differential equation form of a "logistic equation" is $\frac{dP}{dt} = kP\left(1 - \frac{P}{K}\right)$ where P is the population at time t, k is the constant of proportionality, and K is the "carrying capacity." From that point forward there is a **significant** amount of calculus and algebraic acrobatics needed to solve this equation for the specifics of the equation at hand, very much like all that effort shown above to solve an unfactorable quadratic equation by the process of completing the square. Just like finding the general-form quadratic equation, $ax^2 + bx + c = 0$, once to get a general-form answer, $x = \frac{-b\pm\sqrt{b^2-4ac}}{2a}$, it is best to take the time to solve the general-form logistic differential equation, $\frac{dP}{dt} = kP\left(1 - \frac{P}{K}\right)$, for its analytic form, $P = \frac{K}{(A\times e^{-kt}+1)}$, where $A = \frac{K-P_0}{P_0}$. From that point forward, we can substitute into the problem-specific analytic (integrated) form quickly, and relatively painlessly obtain the desired information.

If you wish to know how to convert from the general logistic differential equation form to its algebraic (differentiated) form please do access Appendix B. It is two pages long and rather involved.

Now let's return to those concrete examples! Following are five examples of situations where there is a limit on how much something can grow: cooties, a rumor, genetic transfer, population growth, and product awareness. Therefore, each of these situations could be modeled using the logistic growth curve, $\frac{dP}{dt} = kP(1-\frac{P}{K})$, or its equivalent integrated form.

Differential Equation Using the Logistic Model	Integrated-Equation Form Using the Logistic Model
$\frac{dP}{dt} = kP(1-\frac{P}{K})\ldots$ this is separable, but how???	$P = \frac{K}{(A \times e^{-kt}+1)}$, where $A = \frac{K-P_0}{P_0}$ (See Appendix B for a derivation of this formula.)

Ex 1. Differential Equations in Disease Control—Logistic Model

After spring break, a college campus nurse identified 10 students with cooties. At Day 4, she identified 30 infected students. There are 800 students on campus. Assuming the logistic model, how many students will be infected after two weeks? After five weeks?

$$\frac{dP}{dt} = kP\left(1-\frac{P}{K}\right) \quad \text{carrying capacity} = 800$$

$$\frac{dP}{dt} = kP\left(1-\frac{P}{800}\right) \quad \text{switch to integrated form}$$

$$P = \frac{K}{(A \times e^{-kt}+1)} \quad \text{where } A = \frac{K-P_0}{P_0}$$

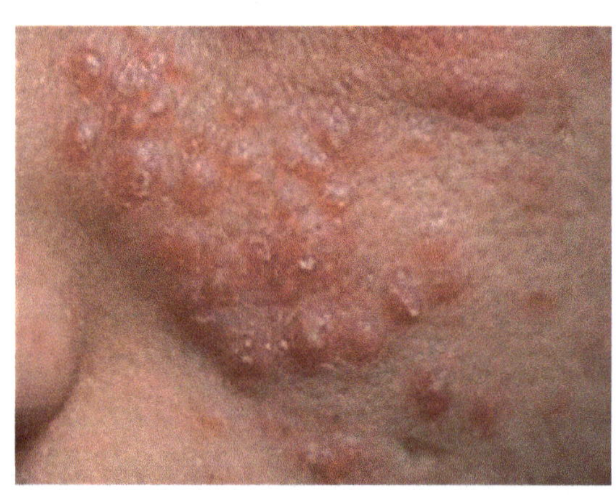

Step 1: Identify the carrying capacity and the initial population. Solve for A.

$$A = \frac{K-P_0}{P_0} = \frac{800-10}{10} = \frac{790}{10} = 79$$

Step 2: Use K and A to solve for k on Day 4 using the integrated-equation form

$$P = \frac{K}{(A \times e^{-kt}+1)}$$

$$30 = \frac{800}{(79 \times e^{-k\times 4}+1)} \quad \text{substitute into the integrated logistic growth model equation}$$

$$3 = \frac{80}{(79 \times e^{-4k}+1)} \quad \text{divide both sides by 10}$$

$$3 \times \left(79 \times e^{-4k}+1\right) = 80 \quad \text{multiply to eliminate the denominator}$$

$$\left(79 \times e^{-4k}+1\right) = \frac{80}{3} \quad \text{divide both sides by 3}$$

$$79 \times e^{-4k} = \frac{80}{3} - \frac{3}{3} \quad \text{subtract 1 from both sides}$$

$$79 \times e^{-4k} = \frac{77}{3} \quad \text{combine like fractions}$$

$$e^{-4k} = \frac{77}{3} \times \frac{1}{79} = \frac{77}{237} \quad \text{divide away the 79}$$

$$e^{-4k} = 0.3248945148 \quad \text{convert } \frac{77}{237} \text{ to a decimal}$$

$$\ln\left(e^{-4k}\right) = \ln(0.3248945148) \quad \text{take the natural log of both sides}$$

$$-4k = -1.124254719 \quad \ln e^u = u$$

$$k = 0.2810636798 \quad \text{finally, solve for } k$$

Step 3: Use A, K and k to determine logistic population numbers after two and five weeks.

$$P = \frac{K}{(A \times e^{-kt} + 1)}$$

After 2 weeks (14 days)

$$P = \frac{800}{79 \times e^{-0.2810636798 \times 14} + 1}$$

$$P = \frac{800}{79 \times e^{-3.934891517} + 1}$$

$$P = \frac{800}{1.544277766 + 1}$$

$$P = \frac{800}{2.544277766}$$

$$P = 314 \text{ infected}$$

After 5 weeks (35 days)

$$P = \frac{800}{79 \times e^{-0.2810636798 \times 35} + 1}$$

$$P = \frac{800}{79 \times e^{-9.837228793} + 1}$$

$$P = \frac{800}{0.004220588 + 1}$$

$$P = \frac{800}{1.0042205875}$$

$$P = 797 \text{ infected}$$

MATLAB script for graph below in Appendix A.

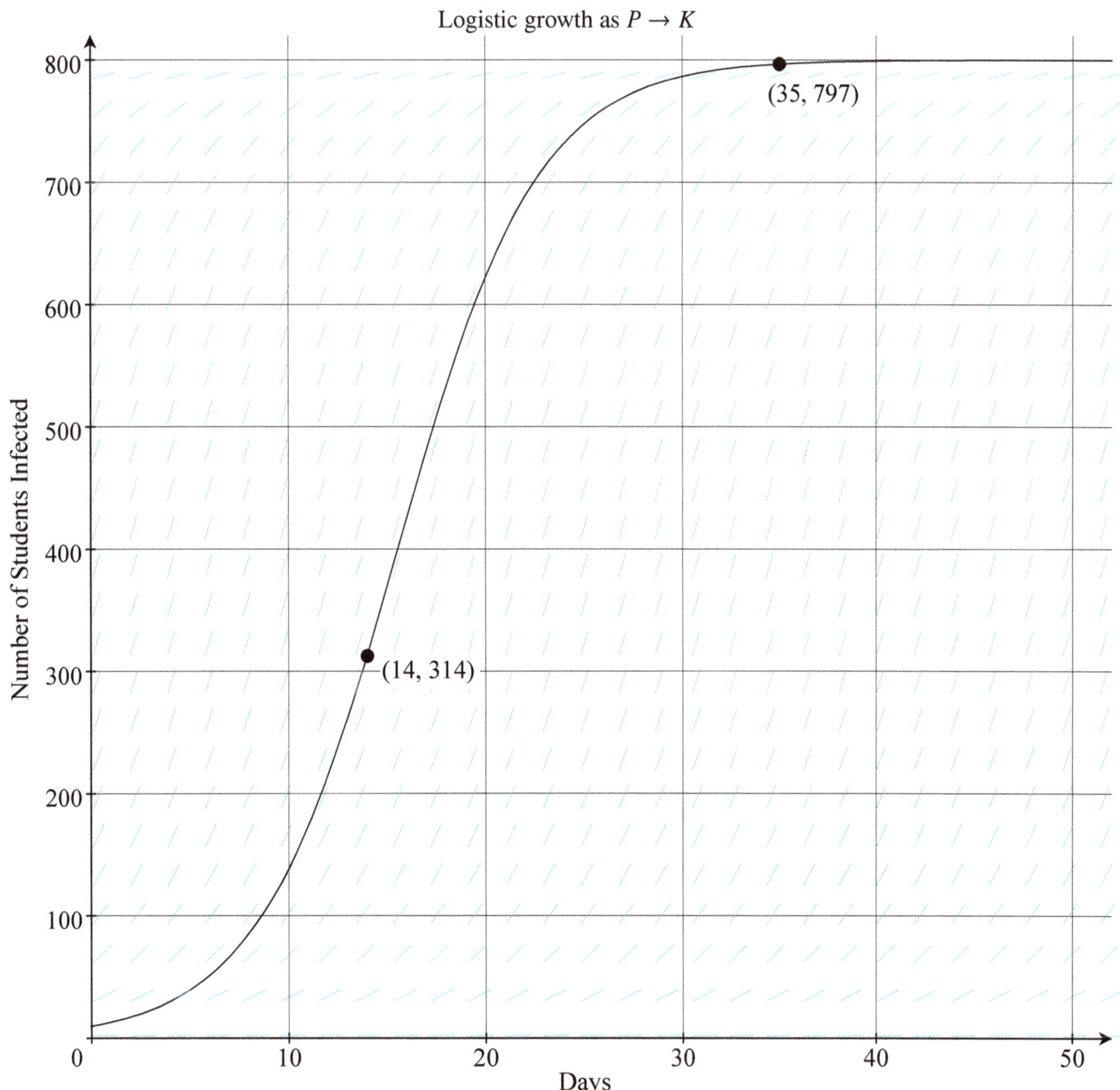

Ex 2. Differential Equations in Sociology—Measuring the Spread of Rumors

In a high school with 2,000 students, a couple breaks up. By the end of the day, 20 people know. By the end of the next day, 70 people know. If the logistic model holds, how many days will it take for before half the school knows?

$$\frac{dP}{dt} = kP\left(1 - \frac{P}{K}\right) \quad \text{carrying capacity} = 2{,}000$$

$$\frac{dP}{dt} = kP\left(1 - \frac{P}{2{,}000}\right) \quad \text{switch to integrated form}$$

$$P = \frac{K}{(A \times e^{-kt} + 1)} \quad \text{where } A = \frac{K - P_0}{P_0} \text{ and } P_o = 20$$

Bertha Elena

Step 1: Identify the carrying capacity and the initial population. Solve for A.

$$A = \frac{K - P_0}{P_0} = \frac{2{,}000 - 20}{20} = \frac{1{,}980}{20} = 99$$

Step 2: Use K and A to solve for k on Day 1 using the integrated-equation form (70 people know at the end of Day 1).

$$P = \frac{K}{(A \times e^{-kt} + 1)}$$

$$70 = \frac{2{,}000}{(99 \times e^{-k \times 1} + 1)} \quad \text{substitute into the integrated logistic growth model equation}$$

$$70 \times \left(99 \times e^{-k} + 1\right) = 2{,}000 \quad \text{multiply to eliminate the fractions}$$

$$\left(99 \times e^{-k} + 1\right) = 28.57142857 \quad \text{divide both sides by 70}$$

$$99 \times e^{-k} = 27.57142857 \quad \text{subtract 1 from both sides}$$

$$e^{-k} = 0.2784992785 \quad \text{divide away the 99}$$

$$\ln\left(e^{-k}\right) = \ln(0.2784992785) \quad \text{take the natural log of both sides}$$

$$-k = -1.27833981 \quad \ln e^{u} = u \text{ on left, ln value taken on right}$$

$$k = 1.27833981$$

Step 3: Use A, K and k to determine how many days before half the school knows.

$$P = \frac{K}{(A \times e^{-kt} + 1)} \quad \text{integrated form}$$

$$1{,}000 = \frac{2{,}000}{(99 \times e^{-1.27833981t} + 1)} \quad \text{substitute } \tfrac{1}{2} \times 2{,}000 = 1{,}000 \text{ students into the integrated logistic growth model}$$

$$\frac{1}{2} = \frac{1}{99 \times e^{-1.27833981t} + 1} \quad \text{divide both sides by 2,000}$$

$$\left(99 \times e^{-1.27833981t} + 1\right) \times 1 = 2 \times 1 \quad \text{cross multiply the proportion}$$

$$99 \times e^{-1.27833981t} + 1 = 2 \quad a \times 1 = a$$

$$99 \times e^{-1.27833981t} = 1 \quad \text{subtract 1 from both sides}$$

$$e^{-1.27833981t} = \frac{1}{99} \quad \text{divide away the 99}$$

$$e^{-1.27833981t} = 0.01010101 \quad \text{change } \frac{1}{99} \text{ to a decimal}$$

$$\ln(e^{-1.27833981t}) = \ln(0.010101) \quad a = b \text{ iff } \ln a = \ln b$$

$$-1.27833981t = -4.59511985 \quad \ln e^{u} = u$$

$$t = 3.59 \text{ days}$$

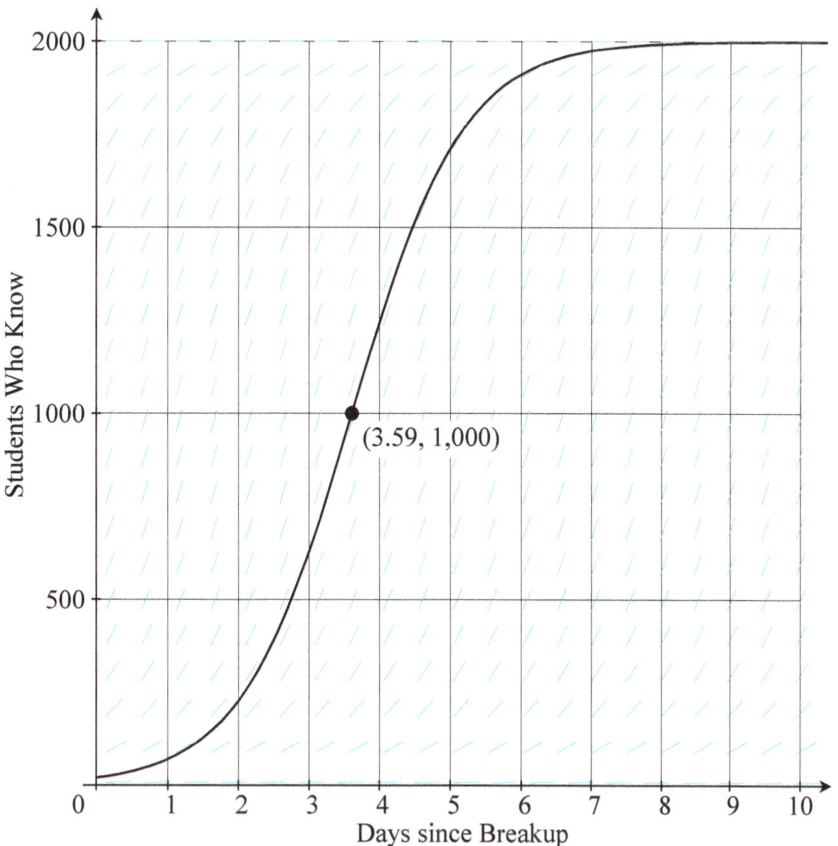

MATLAB script for the graph above in Appendix A.

Ex 3. Differential Equations in Game Management—Measuring Population Growth

A population of prairie dogs is estimated to be 800. After four weeks, there are 1,000 prairie dogs. Biologists estimate that there is only food for 1,800 prairie dogs. Assuming the logistic model holds, how many prairie dogs there will be after 10 weeks? 20 weeks? 30 weeks?

$$\frac{dP}{dt} = kP(1 - \frac{P}{K})$$

$$\frac{dP}{dt} = kP(1 - \frac{P}{1{,}800})$$ carrying capacity = 1,800

$$P = \frac{K}{(A \times e^{-kt} + 1)},$$ where $A = \frac{K - P_0}{P_0}$, integrated form of the logistic-model equation

Step 1: Solve for A.

$$A = \frac{1{,}800 - 800}{800} = \frac{1{,}000}{800} = 1.25$$

$$P = \frac{1{,}800}{(1.25 \times e^{-kt} + 1)}$$

Step 2: Use K and A to solve for k at time four weeks

$$P = \frac{K}{(A \times e^{-kt} + 1)}$$

$$1{,}000 = \frac{1{,}800}{(1.25 \times e^{-k \times 4} + 1)}$$ substitute for A, K, and t into the integrated form of logistic-model equation

$$1{,}000 \times (1.25 \times e^{-k \times 4} + 1) = 1{,}800$$ multiply to eliminate the fraction

$$(1.25 \times e^{-k \times 4} + 1) = 1.8$$ divide away the 1,000

$$1.25 \times e^{-k\times 4} = 0.8 \quad \text{subtract 1 from both sides}$$
$$e^{-k\times 4} = \frac{0.8}{1.25} \quad \text{divide away the 1.25}$$
$$e^{-k\times 4} = 0.64 \quad \text{convert fraction to a decimal}$$
$$\ln e^{-k\times 4} = \ln 0.64 \quad a = b \text{ iff } \ln a = \ln b$$
$$-k \times 4 = -0.4462871026 \quad \ln e^{u} = u \text{ on left, ln value on right}$$
$$k = 0.1115717757 \quad \text{solve for } k$$

Substitute A, K, and k into the integrated logistic-growth-model equation. **Solve for t = 10, 20, 30.**

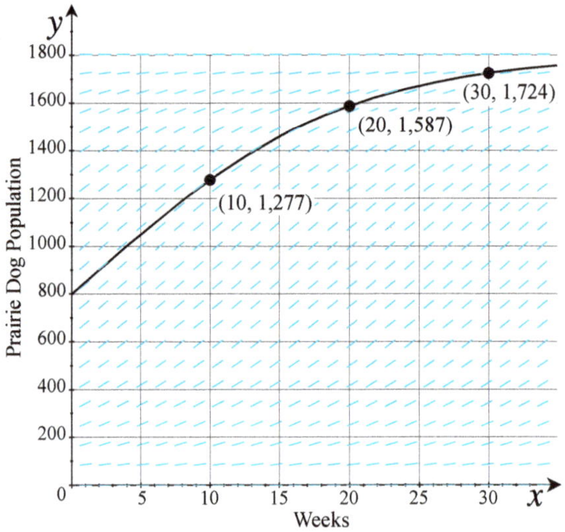

$$P = \frac{K}{(A \times e^{-kt} + 1)}, \quad \text{for } t = 10$$
$$P = \frac{1{,}800}{(1.25 \times e^{-0.1115717757\times 10} + 1)}$$
$$P = \frac{1{,}800}{(1.25 \times e^{-1.115717757} + 1)} \quad \text{multiply by 10 in the exponent of e}$$
$$P = \frac{1{,}800}{(1.25 \times 0.32768000 + 1)} \quad \text{calculate } e^{-1.115717757}$$
$$P = \frac{1{,}800}{(0.4096000 + 1)} \quad \text{calculate product}$$
$$P = \frac{1{,}800}{1.4096000} \quad \text{add 1 in the denominator}$$
$$P = 1{,}277$$

$$P = \frac{K}{(A \times e^{-kt} + 1)}, \quad \text{for } t = 20$$
$$P = \frac{1{,}800}{(1.25 \times e^{-0.1115717757\times 20} + 1)}$$
$$P = \frac{1{,}800}{(1.25 \times e^{-2.231435514} + 1)} \quad e^{-0.1115717757\times 20} = e^{-2.231435514}$$
$$P = \frac{1{,}800}{(1.25 \times 0.1073741823 + 1)} \quad \text{calculate } e^{-2.231435514}$$
$$P = \frac{1{,}800}{(0.1342177279 + 1)}$$
$$P = \frac{1{,}800}{1.134217728}$$
$$P = 1{,}587$$

$$P = \frac{K}{(A \times e^{-kt} + 1)}, \quad \text{for } t = 30$$
$$P = \frac{1{,}800}{(1.25 \times e^{-0.1115717757\times 30} + 1)}$$
$$P = \frac{1{,}800}{(1.25 \times e^{-3.3471532710} + 1)} \quad e^{-0.1115717757\times 30} = e^{-3.3471532710}$$
$$P = \frac{1{,}800}{(1.25 \times 0.0351843720 + 1)} \quad \text{calculate } e^{-3.3471532710}$$
$$P = \frac{1{,}800}{(0.0439804650 + 1)}$$
$$P = \frac{1{,}800}{1.0439804650}$$
$$P = 1{,}724$$

Ex 4. Differential Equations in Genetics—Monitoring Inheritance of Characteristics

A geneticist is studying mice to determine how quickly a physical trait will spread through a population from one generation to the next. At the start of the study ($t = 0$), she has established that 30% of the population has the characteristic. After five generations ($t = 5$), she finds that 85% of the population has the characteristic. Assume the logistic growth model to determine the percentage of mice that will have the studied trait after eight generations and after 10 generations.

$$\frac{dP}{dt} = kP\left(1 - \frac{P}{K}\right)$$

$$\frac{dP}{dt} = kP\left(1 - \frac{P}{100\%}\right) \quad \text{carrying capacity: 100\%}$$

$$P = \frac{K}{A \times e^{-kt} + 1}, \quad \text{logistic model integrated form}$$

Step 1: Solve for A.

$$A = \frac{K - P_0}{P_0} = \frac{100\% - 30\%}{30\%} = \frac{70\%}{30\%} = 2.333$$

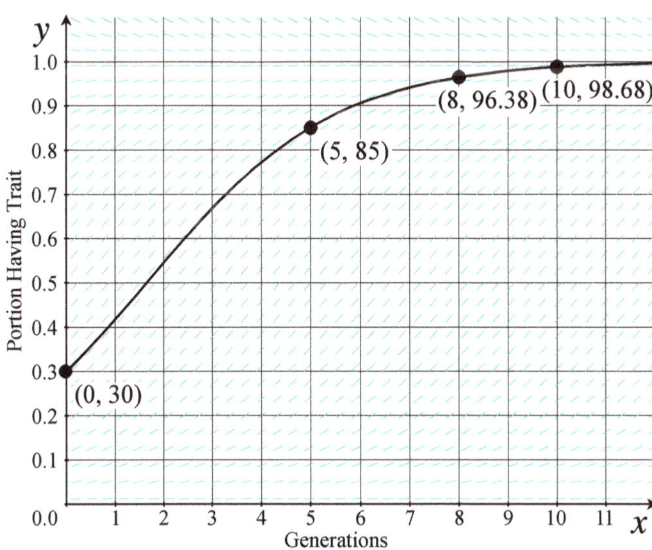

MATLAB script for graph above in Appendix A.

Generation	Have the trait
0	30.0%
1	
2	
3	
4	
5	85.0%
6	
7	
8	96.4%
9	
10	98.7%

Step 2: After five generations

$$P = \frac{K}{A \times e^{-kt} + 1}$$

$$85\% = \frac{100\%}{2.333e^{-k \times 5} + 1} \quad P = 85\% \text{ when } t = 5$$

$$0.85 \times \left(2.333e^{-5k} + 1\right) = 1$$

$$2.333e^{-5k} + 1 = 1.176$$

$$2.333e^{-5k} = 0.176$$

$$e^{-5k} = 0.076$$

$$\ln e^{-5k} = \ln(0.076) \quad a = b \text{ iff } \ln a = \ln b$$

$$-5k = -2.582 \quad \ln e^{u} = u$$

$$k = 0.516$$

After eight generations

$$P = \frac{100\%}{2.333e^{-0.516 \times 8} + 1} \quad \text{substitute } t = 8$$

$$P = \frac{1}{2.333e^{-4.131} + 1} \quad e^{-0.516 \times 8} = e^{-4.131}$$

$$P = \frac{1}{2.333 \times 0.0161 + 1} \quad e^{-4.131} = 0.0161$$

$$P = \frac{1}{0.037 + 1}$$

$$P = \frac{1}{1.037}$$

$$P = 0.9638$$

After ten generations

$$P = \frac{100\%}{2.333e^{-0.516 \times 10} + 1}$$

$$P = \frac{1}{2.333e^{-5.164} + 1}$$

$$P = \frac{1}{2.333 \times 0.0057 + 1}$$

$$P = \frac{1}{0.013 + 1}$$

$$P = \frac{1}{1.013}$$

$$P = 0.9868$$

Ex 5. Differential Equation Modeling Advertising Awareness

A company making and selling widgets is engaged in a marketing campaign in a large metropolitan area of 1 million people. At the beginning of an advertising campaign, only 10% (100,000) of the citizens had heard of the product. At the end of one year, 40% (400,000) of the citizens had heard of the product.. Assuming the amount of advertising stays the same and assuming that awareness of the product grows as in the logistic model, how many people will have heard of the product at the end of the second year? At the end of the third year?

$$\frac{dP}{dt} = kP\left(1 - \frac{P}{K}\right)$$

$$\frac{dP}{dt} = kP\left(1 - \frac{P}{1{,}000{,}000}\right) \quad \text{maximum awareness}$$

$$P = \frac{K}{A \times e^{-kt} + 1}, \quad \text{where } A = \frac{K - P_0}{P_0}$$

Step 1: Solve for A.

$$A = \frac{K - P_0}{P_0} = \frac{1{,}000{,}000 - 100{,}000}{100{,}000} = \frac{900{,}000}{100{,}000} = 9$$

Step 2: Solve for k at **Year 1**

$$400{,}000 = \frac{1{,}000{,}000}{9 \times e^{-k \times 1} + 1} \quad \text{substitute for } A, K, t$$

$$4 = \frac{10}{9e^{-k \times 1} + 1}$$

$$4 \times (9e^{-k \times 1} + 1) = 10$$

$$9e^{-k} + 1 = 2.5$$

$$9e^{-k} = 1.5$$

$$e^{-k} = 0.166666667 \quad \text{divide away the 9}$$

$$\ln(e^{-k}) = \ln(0.166666667) \quad a = b \text{ iff } \ln a = \ln b$$

$$-k = -1.791759469 \quad \ln e^u = u$$

$$k = 1.791759469$$

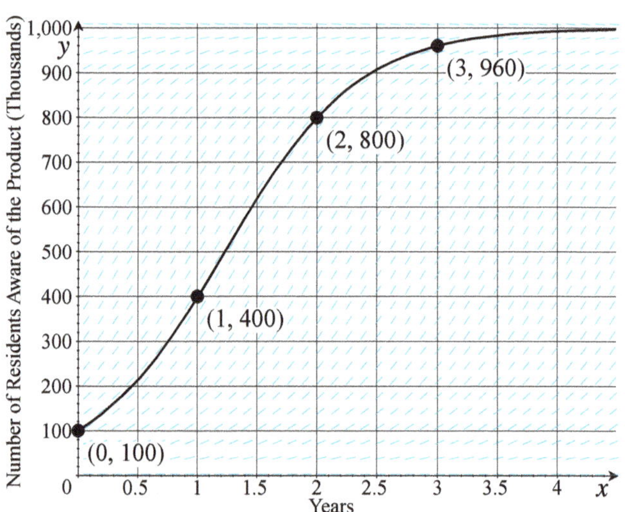

Substitute to find awareness after **Year 2**

$$P = \frac{1{,}000{,}000}{9e^{-1.791759473 \times 2} + 1}$$

$$P = \frac{1{,}000{,}000}{9e^{-3.583518938} + 1}$$

$$P = \frac{1{,}000{,}000}{9 \times 0.0277777778 + 1}$$

$$P = \frac{1{,}000{,}000}{0.25000000 + 1}$$

$$P = \frac{1{,}000{,}000}{1.25000000}$$

$P = 800{,}000$ people had heard of the product

Substitute to find awareness after **Year 3**

$$P = \frac{1{,}000{,}000}{9e^{-1.791759469 \times 3} + 1}$$

$$P = \frac{1{,}000{,}000}{9e^{-5.375278408} + 1}$$

$$P = \frac{1{,}000{,}000}{9 \times 0.0046296296 + 1}$$

$$P = \frac{1{,}000{,}000}{0.0416666667 + 1}$$

$$P = \frac{1{,}000{,}000}{1.0416666666}$$

$P = 960{,}000$ people had heard of the product

A better match between the problem being studied and the model used to study it generally provides more reliable study results. The logistic-growth model is only one of several hypothetical models that can be used to study phenomena. Following are a few others that we will not cover in this book.

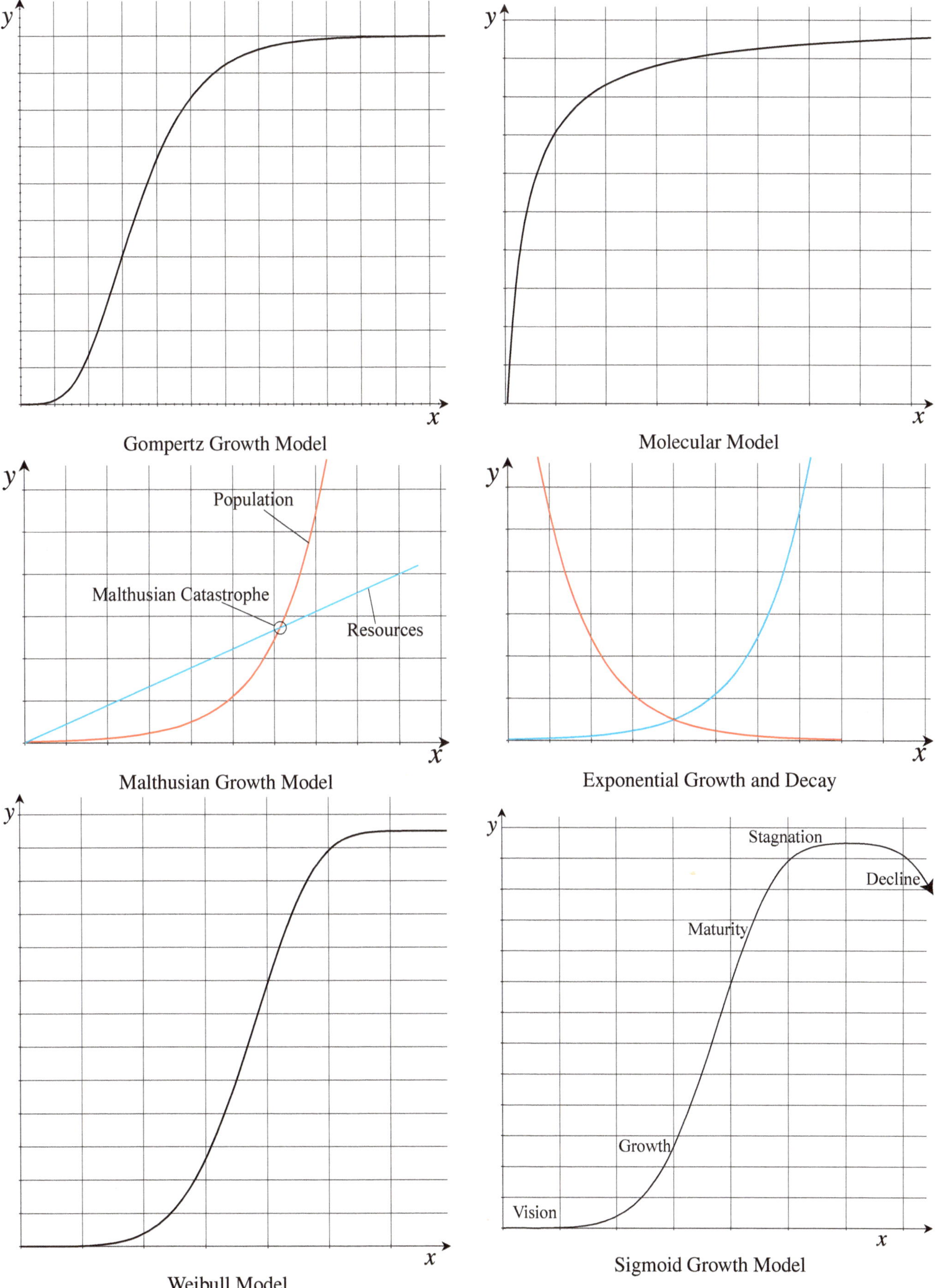

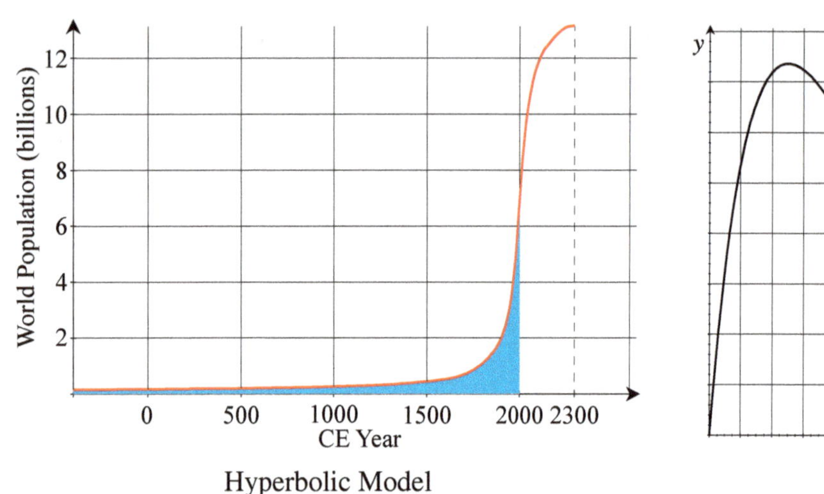

Hyperbolic Model

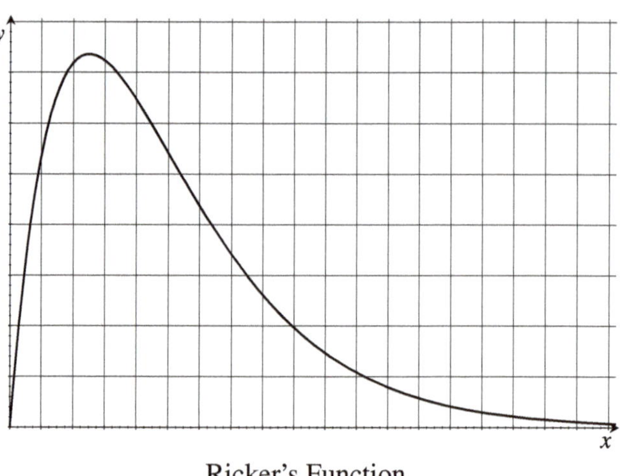

Ricker's Function

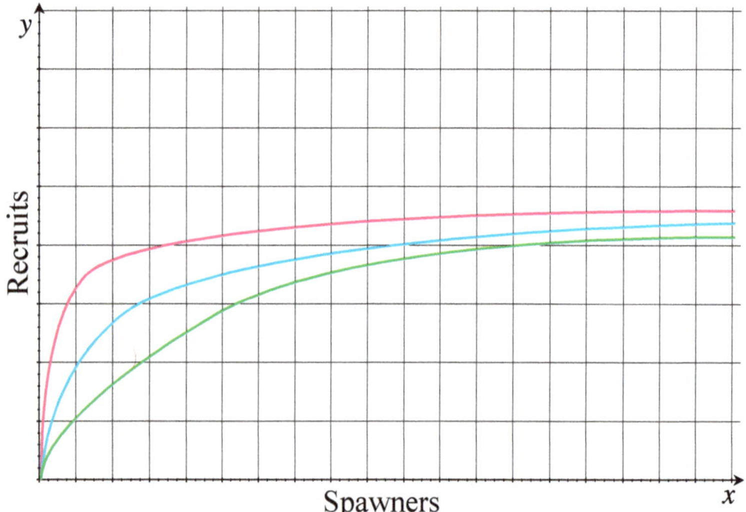

Beverton-Holt stock-recruitment curves

Chapter 6 Review

Chapter 6 introduced a new kind of differential equation, a logistic differential equation. The exponential-growth differential equation differs from logistic differential equation in that the latter had some sort of natural bounds placed upon the upper limit of growth of the data being observed. Five examples were given:

1. disease control
2. sociology
3. population growth bounded by available food
4. the spread of genetic traits
5. advertising awareness

(a) Exponential (unrestricted) growth

$$\frac{dp}{dt} = kP$$

Population Size

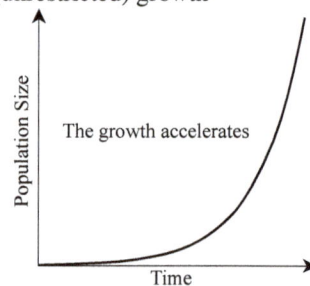

The growth accelerates

Time

(b) Logistic (restricted) growth

$$\frac{dP}{dt} = kP\left(1 - \frac{P}{K}\right)$$

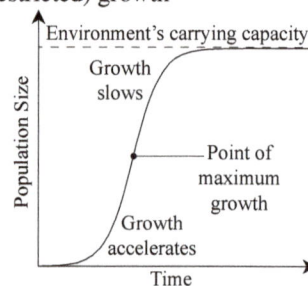

Environment's carrying capacity

Growth slows

Point of maximum growth

Growth accelerates

Population Size

Time

Although logistic equations were a type of separable differential equation, it was noted that their signature form and resulting curve differed.

It was observed that the algebra necessary to solve a logistic differential equation was significantly more difficult than solving a separable exponential-growth problem, and considerable time, effort, and page space (Appendix B) were spent on obtaining a closed algebraic formula

$$P = \frac{K}{(A \times e^{-kt} + 1)}, \text{ where } A = \frac{K - P_0}{P_0}$$

which would allow for a "plug-and-chug substitution solution" for all future logistic differential equations. Thus, all the tedious details involved in the development of the generic form were eliminated from all future logistic-problem situations.

Several other modeling curves were shown for the reader's consideration, and it was noted that a model chosen to more closely model a situation generally produces better the prediction results.

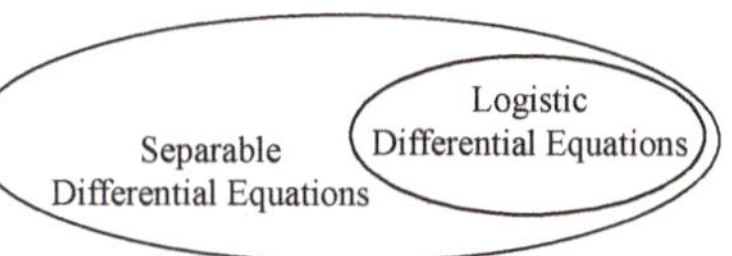

It was pointed out that logistic differential equations are a subset of separable differential equations, so it should be obvious that categorizing the different types of differential equations will become more complicated. The following chart should be considered a primitive and simplistic attempt to help beginning students (the target audience for this book) review, differentiate, and mentally organize their thoughts to this point in the book.

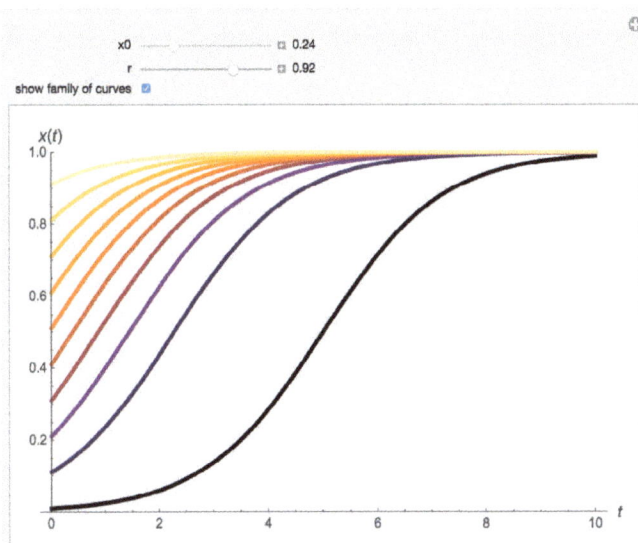

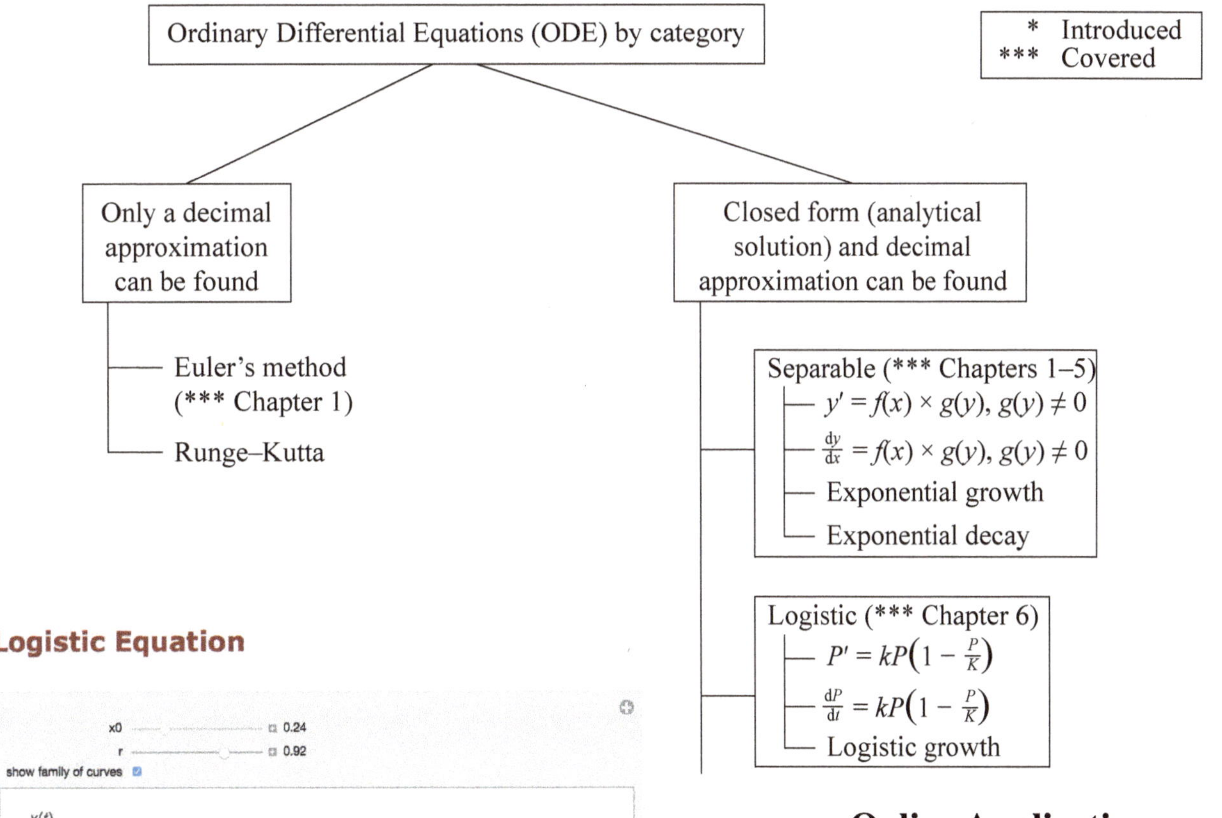

Online Application

Visit demonstrations.wolfram.com/LogisticEquation for a great little application that will demonstrate some of the concepts from this chapter. "Logistic Equation" from the Wolfram Demonstrations Project. Contributed by Jeff Bryant.

Chapter 7

Predator–Prey: Introduction to Systems of Differential Equations

In Chapter 3, we studied a "population growth" situation where a population of deer had unlimited food and no predators. The result was that, theoretically, if the natural birth rate of the deer was greater than their natural death rate the population could grow exponentially to infinity.

This would apply to rabbits as well and be represented in a differential equation as $\frac{dr}{dt} = kr$. However, modeling population growth is more complicated than that. Suppose a thriving population of rabbits were discovered by a group of foxes. With the introduction of foxes into the area, the population of rabbits would not grow exponentially. Their population would be reduced by a rate directly proportional to their encounters with the foxes: $\frac{dr}{dt} = k_1 r - k_2 r f$. The rabbit population would become cyclical. The foxes would eat them making their numbers decline. Then the fox population would decline due to the lack of food until the rabbit population would start growing again. The k_1 and k_2 in the differential equation above are obtained through field observations and data collection and allow you to identify a specific solution from the infinite family of solutions. Below is an example of one such solution.

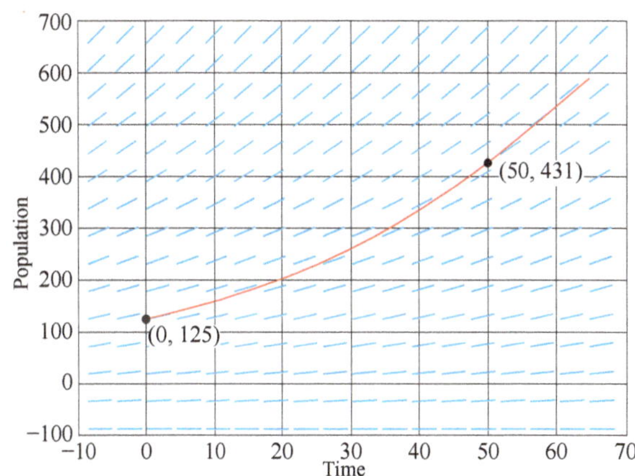

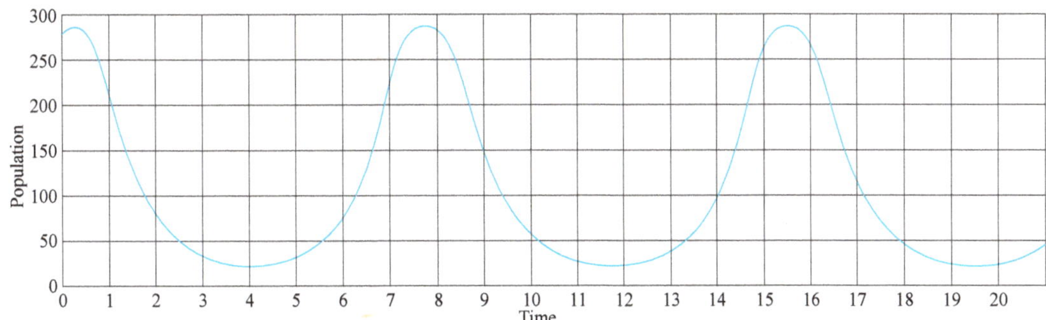

For simplicity, we model the fox population as $\frac{df}{dt} = -k_3 f + k_4 r f$. Here the fox population decreases if there are no rabbits to eat $(-k_3 f)$ and increases proportionally to the number of rabbit–fox encounters $(+k_4 r f)$. Here k_1, k_2, k_3, and k_4 are all *positive* constants of proportionality. All these values would have to be obtained through field observations and data collection. These equations are an example of a system of differential equations.

$$\frac{dr}{dt} = k_1 r - k_2 f r$$
$$\frac{df}{dt} = -k_3 f + k_4 f r$$

Modeling change is actually more complicated than shown above. The following is adapted from *Mathematical Modelling with Case Studies* by Belinda Barnes and Glenn Fulford, page 109, 2nd edition, CRC Press.

rate of change of rabbits = rabbit births − natural rabbit deaths − rabbits killed by predators

$$\frac{dr}{dt} = b_1 \times r(t) - a_1 \times r(t) - k_1 r(t) f(t)$$

Where b_1, a_1, and k_1 are all constants of proportionality.

$$\frac{dr}{dt} = (b_1 - a_1) \times r(t) - k_1 r(t) f(t)$$

$$\frac{dr}{dt} = \beta \times r(t) - k_1 r(t) f(t) \qquad \beta = b_1 - a_1 \text{ *** } (1)$$

Modeling the foxes' population growth might look like

rate of change of foxes = predator births − natural predator deaths

Predator births would be determined by natural fox fertility and also availability of food (rabbits), so

$$\text{rate of change of foxes} = [b_2 f(t) + k_2 r(t) f(t)] - a_2 f(t)$$

$$[b_2 f(t) - a_2 \times f(t)] + k_2 r(t) f(t)$$

$$[b_2 - a_2] \times f(t) + k_2 r(t) f(t)$$

$$\alpha f(t) + k_2 r(t) f(t) \qquad \alpha = b_2 - a_2 \text{ *** } (2)$$

$$\frac{dr}{dt} = \beta r - k_1 f r \quad (1)$$
$$\frac{df}{dt} = \alpha f + k_2 f r \quad (2)$$

This specific system is known as the Lotka–Volterra equations in honor of the mathematicians who first studied them. The system of equations shown at right is presented with an error in it. After a slight digression that error will be corrected. Rabbits are herbivores and foxes are carnivores. That means that, absent foxes, rabbits can acquire food and their numbers will grow independently. $\frac{dr}{dt} = \beta \times r - k_1 r f = \beta \times r$ when $f = 0$. However, absent rabbits ($r = 0$), the fox population should not grow: $\frac{df}{dt} = \alpha \times f + k_2 r f = \alpha \times f$. The way to correct this is to rewrite the equation modeling fox population growth as $\frac{df}{dt} = -\alpha \times f + k_2 r f$. Now without rabbits the fox population will decline. So the original system of equations is rewritten as shown at right.

$$\frac{dr}{dt} = \beta r - k_1 f r \quad (1)$$
$$\frac{df}{dt} = -\alpha f + k_2 f r \quad (2)$$

When plotting the population of foxes on the same graph as the population of rabbits, it is important here to recall the idea of parametric equations. There are solutions to two different equations at work here. Hence, for one domain (time) there are two ranges (populations of rabbits and foxes, respectively.)

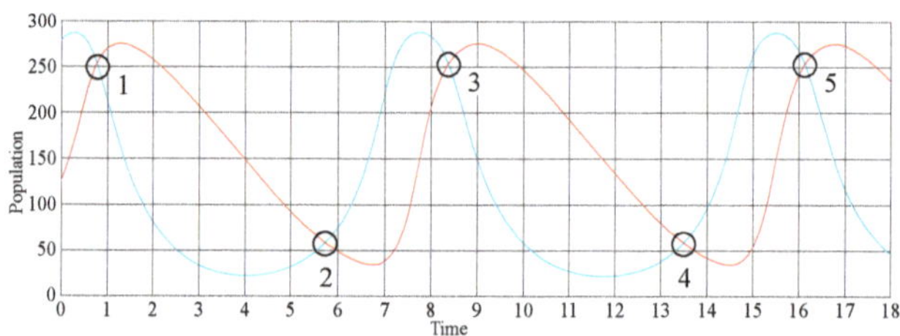

time	$r(t)$	$f(t)$
1	↓	↑
2	↑	↓
3	↓	↑
4	↑	↓
5	↓	↑

For constants of proportionality $\beta = 1$, $\alpha = 0.5$, $k_1 = 0.01$, and $k_2 = 0.005$ the graphs above and below are drawn using MATLAB. The trained eye sees a massive amount of information. The numbered points labeled above show the points at which the population of rabbits equals the population of foxes. The circled points below show the maximum and minimum number of rabbits and foxes and the fact that the predator–prey model shown here for the constants of proportion chosen above will cycle predictably and indefinitely.

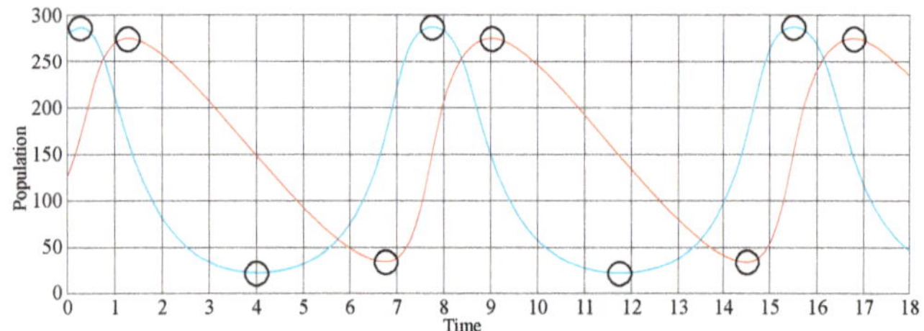

Below, Point 1 indicating rabbit population is located at a point of inflection (second derivative = 0, remember?) in the rabbit population. It is located at the same time point when the fox population is at a local maximum and indicates where the rabbit population decline is at its fastest.

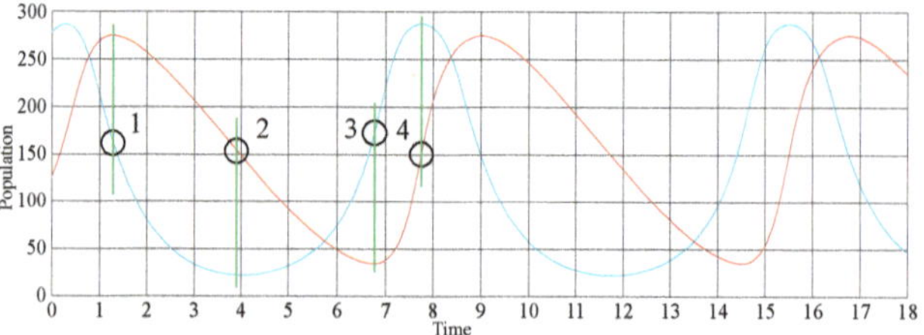

Point 2 indicates a point of inflection for the fox graph and it is no surprise that the steepest decline of the fox population happens at the time that the rabbit population is at a local minimum. What do Points 3 and 4 indicate?

Hopefully you recall studying systems of equations in algebra.

$$\begin{aligned} 2x + y &= 10 \\ x - y &= -1 \end{aligned}$$

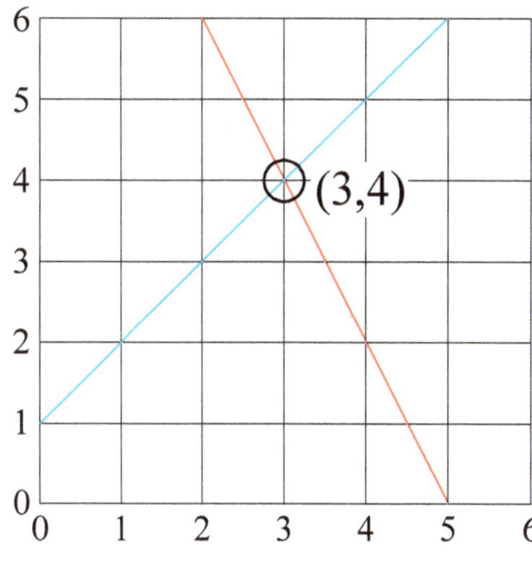

Several techniques were taught in algebra to solve them: 1) linear combinations, 2) substitution, 3) matrix algebra, 4) determinants using Cramer's Rule, and 5) graphical. The graphical technique of solving is shown at right. The (x, y) values $(3, 4)$ satisfy both equations in the given system. Those same skills (or variations of them) are used when studying systems of differential equations. Or as Monsieur Jourdain, in the play *Le Bourgeois Gentilhomme* (see Chapter 1), stated: "Par ma foi! Il y a plus de quarante ans que je dis de la prose sans que j'en susse rien, et je vous suis le plus obligé du monde de m'avoir appris cela."

In Chapter 6, when studying a population of prairie dogs, we took food supply into account and said that, in practice, the growth of a population would be restrained by the availability of food. The resulting growth curve was called "logistic" and looked like the graph below right.

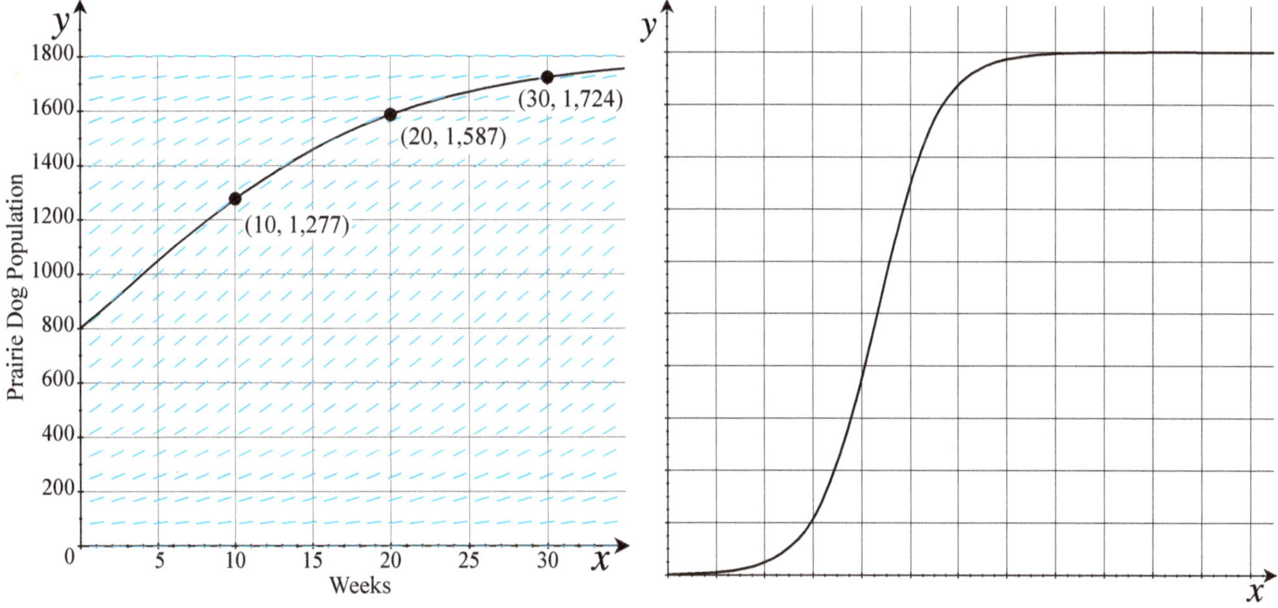

Incorporating the idea of logistic growth into our system of equations, the system of equations above is now modified as shown at right to constrain the rabbit population by the $\left(1 - \frac{r}{K}\right)$ term. The fox population is already constrained by the $-\alpha f$ term. α, β, k_1, and k_2 are all still constants of proportionality while K is the carrying capacity for the population of rabbits (1,800 for the graph above). Again all these values would have to be obtained through field observations and data collection. The resulting equations are examples of a system of differential equations. (This specific system is known as the Lotka–Volterra equations with bounded growth, again in honor of the mathematicians who first studied them.)

$$\begin{aligned} \frac{dr}{dt} &= \beta r - k_1 r f \\ \frac{df}{dt} &= -\alpha f + k_2 r f \end{aligned}$$

$$\begin{aligned} \frac{dr}{dt} &= \beta r \times \left(1 - \frac{r}{K}\right) - k_1 r f \\ \frac{df}{dt} &= -\alpha f + k_2 r f \end{aligned}$$

The MATAB script for rabbit exponential growth would have to be modified to reflect logistic growth.

```
function c_cp_predprey
  global beta alpha k1 k2;
  beta = 1.0; alpha = 0.5; k1 = 0.01; k2 = 0.005;
  tend = 20; % set the end time to run the simulation
  u0 = [200; 80]; % set initial conditions as a column vector
  [tsol, usol] = ode45(@rhs, [0, tend], u0);
  Xsol = usol(:, 1); Ysol = usol(:, 2);
  plot(tsol, Xsol, 'b'); hold on; plot(tsol, Ysol, 'r');
function udot = rhs(t, u)
  global beta alpha k1 k2;
  X = u(1); Y = u(2);
  Xdot = beta * X - k1 * X * Y;
  Ydot = -alpha * Y + k2 * X * Y;
  udot = [Xdot; Ydot];
```

← IVP data that can be modified to test various hypotheses

Mathematical Modelling with Case Studies by Belinda Barnes and Glenn Fulford, page 109, 2nd edition, CRC Press

By changing values for β, α, k_1, and k_2, a biologist can quickly test outcomes for different scenarios.

The following passage is taken from page 172 of *Calculus in Context,* by James Callahan, David Cox, Kenneth Hoffman, Donal O'Shea, Harriet Pollatsek, and Lester Senechal, W.H. Freeman and Company, 1995.

"Wine is made by yeast; yeast digests the sugars in grape juice and produces alcohol as a waste product. This process is called fermentation. The alcohol is toxic to the yeast, though, and the yeast is eventually killed by the alcohol. This stops fermentation, and the liquid has become wine with about 8–12% alcohol."

Analogy time!

Predator is to prey as fox is to rabbit as yeast is to sugar. It should be anticipated that systems of equations have applications far beyond the fox–rabbit, predator–prey system shown in this chapter.

Chapter 7 Review

Just as we studied "two equations with two unknowns" in algebra,

$$\begin{aligned} 2x + y &= 10 \\ x - y &= -1 \end{aligned}$$

we now study "two equations with two unknowns" in differential equations.

$$\begin{aligned} \frac{dr}{dt} &= \beta r - k_1 r f \\ \frac{df}{dt} &= -\alpha f + k_2 r f \end{aligned}$$

Here, α, β, k_1 and k_2 are all constants determined by field observations and used to identify particular solutions.

In algebra, we studied systems of equations primarily to find a point of intersection. In differential equations, we also find points of intersection. However, as shown in Chapters 7 and 16, we use a system of differential equations to study how changes in one equation can impact another.

Alfred J. Lotka

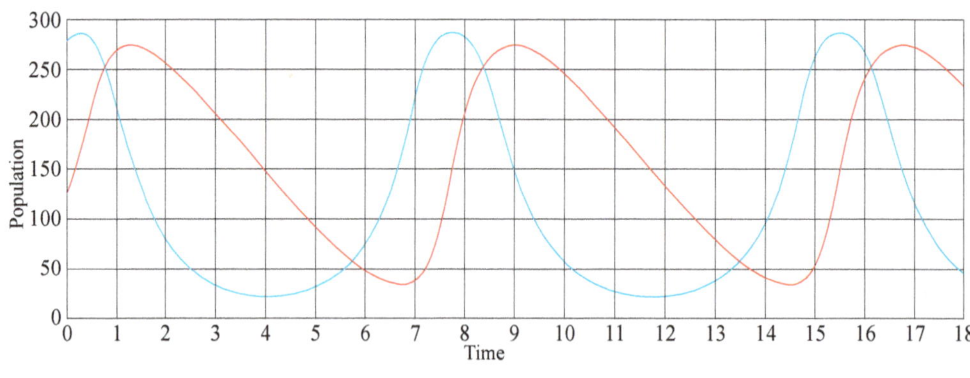

The following chart should be considered a primitive and simplistic attempt to help beginning students (the target audience for this book) review, differentiate, and mentally organize their thoughts to this point in the book. Bear in mind that it will have to be adjusted as more knowledge and information is gained.

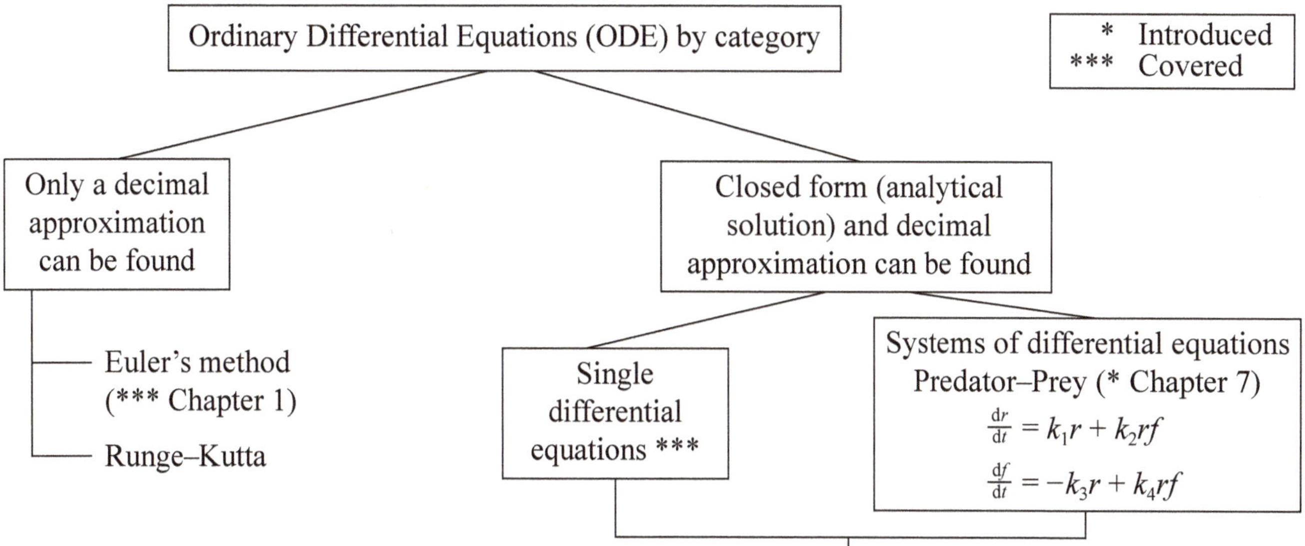

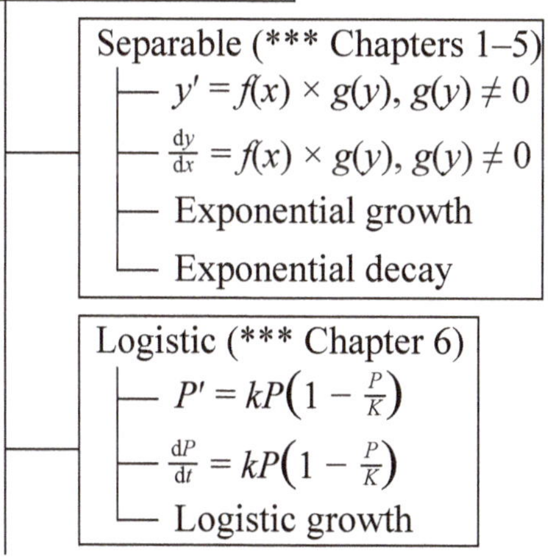

Ordinary Differential Equations (ODE) by category
* Introduced
*** Covered
Only a decimal approximation can be found
Euler's method (*** Chapter 1)
Runge–Kutta
Closed form (analytical solution) and decimal approximation can be found
Single differential equations ***
Systems of differential equations
Predator–Prey (* Chapter 7)
$\frac{dr}{dt} = k_1 r + k_2 rf$
$\frac{df}{dt} = -k_3 r + k_4 rf$

Separable (*** Chapters 1–5)
$y' = f(x) \times g(y),\ g(y) \neq 0$
$\frac{dy}{dx} = f(x) \times g(y),\ g(y) \neq 0$
Exponential growth
Exponential decay
Logistic (*** Chapter 6)
$P' = kP\left(1 - \frac{P}{K}\right)$
$\frac{dP}{dt} = kP\left(1 - \frac{P}{K}\right)$
Logistic growth

Online Application

Visit demonstrations.wolfram.com/PredatorPreyModel and demonstrations.wolfram.com/PredatorPreyEquations for two great little applications that will demonstrate some of the concepts from this chapter. "Predator–Prey Model" from the Wolfram Demonstrations Project. Contributed by Stephen Wilkerson. "Predator–Prey Equations" from the Wolfram Demonstrations Project. Contributed by Eric W. Weisstein.

Predator–Prey Equations

Predator–Prey Model

parameters
end time
rabbits x
foxes y

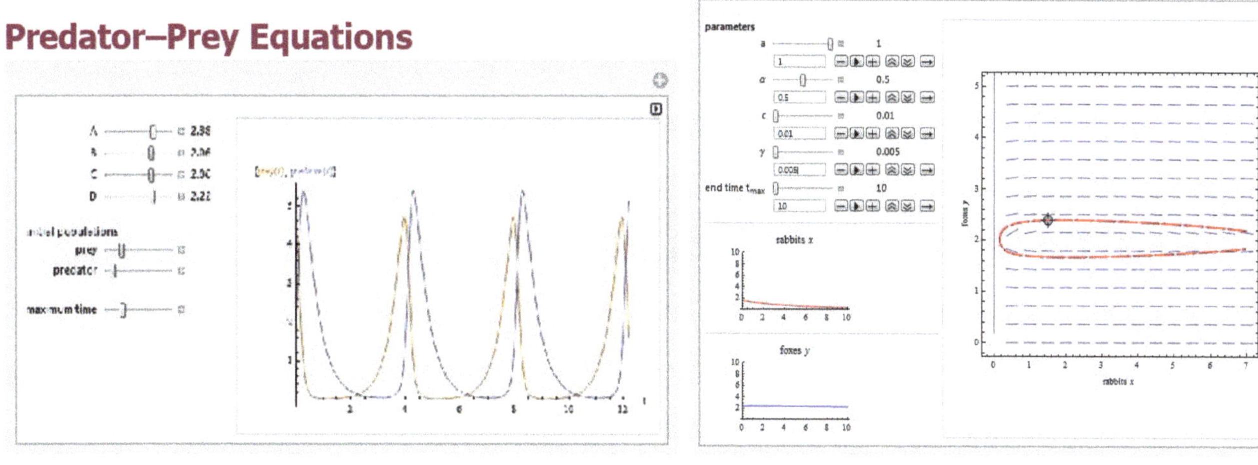

Chapter 8
Solving "Linear" Differential Equations Using an Integrating Factor

In Chapters 3 through 5 and before we looked at general separable differential equations. In Chapter 6, we looked at a special kind of separable differential equations called logistic equations. Then, in Chapter 7, we examined systems of differential equations (not just single equations). Now that we've covered a fair bit of the separable-differential-equation universe, we'll turn our attention to a group of differential equations that are not generally separable in the sense we have been discussing, but they do allow for a closed-form solution by using a neat trick called the *integrating factor*. We'll start with a motivating example.

A chemical has spilled upstream from a small pond. The pond is estimated to contain 1,000,000 gallons of water. Its outlet releases 20,000 gallons of water per day. The stream carrying the chemical flows into the pond at 21,000 gallons per day. This means that the pond's volume is increasing by 1,000 gallons a day (during the rainy season). When the chemical spill enters the pond, it will have a concentration of four grams of the spilled chemical per gallon of stream water. Find and solve a differential equation that will model the pollution of this pond and determine the concentration of pollutant after 10 days.

As with the aquarium problem in Chapter 5, we start with the general equation description:
Change in pollutant = pollutant coming in − pollutant leaving. Note that g_{pol} is grams of pollutant.

$$\text{Pollutant coming in: } \frac{g_{pol}}{d} = \frac{gal}{d} \times \frac{g_{pol}}{gal}$$

$$\frac{21{,}000\text{ gal}}{1\text{ d}} \times \frac{4\text{ }g_{pol}}{1\text{ gal}} = \frac{84{,}000\text{ }g_{pol}}{1\text{ d}}$$

Assume the pollutant gets diluted uniformly as it enters the pond. Let p be the number of gallons in the pond.

SACRE BLEU!
AY, CARAMBA!
AIEEE!
THESE VARIABLES AND THEIR DIFFERENTIALS CANNOT BE SEPARATED FOR RESPECTIVE INTEGRATION AS BEFORE!
WHAT TO DO?
Dr. Richard Detmer

$$\text{Pollutant leaving: } \frac{g_{pol}}{d} = \frac{gal}{d} \times \frac{g_{pol}}{gal} \times p$$

$$\frac{20{,}000\text{ gal}}{1\text{ d}} \times \frac{p g_{pol}}{(1{,}000{,}000 + 1{,}000t)\text{gal}} = \frac{20{,}000 p g_{pol}}{(1{,}000{,}000 + 1000t)\text{d}}$$

$$\frac{dp}{dt} = 84{,}000\frac{g_{pol}}{d} - \frac{20{,}000p}{1{,}000{,}000 + 1{,}000t}\frac{g_{pol}}{d} \quad \text{simplify the fraction}$$

$$\boxed{\frac{dp}{dt} = 84{,}000 - \frac{20p}{1{,}000 + t}}$$

Multiply by dt to separate the differentials.

$$dp = \left(84{,}000 - \frac{20p}{1{,}000 + t}\right) dt \text{ ???}$$

$$dp = 84{,}000\, dt - \frac{20p}{1{,}000 + t} dt \text{ ???}$$

$$dp(1{,}000 + t) = 84{,}000(1{,}000 + t)\, dt - 20p\, dt$$

?????????????

Back in the Foreword, I shared a story about a 25-hour differential-equations video series that began by spending five minutes of that set of lectures as shown in the text box below left. Following a hunch, we juxtapose the equation we are looking to solve $\frac{dp}{dt} = 84{,}000 - \frac{20p}{1{,}000+t}$ with the generic equation in the text box. Is that what the smart man was trying to tell me? That some differential equations are not separable and there is a process to solve such equations?

How to Solve Linear (Differential) Equations $y'(x) + P(x) \times y(x) = Q(x)$	$\frac{dp}{dt} = 84{,}000 - \frac{20p}{1{,}000+t}$
1. Calculate the integrating factor, $I(x) = e^{\int P(x)\,dx}$, and multiply that by both sides.	Changing form $p'(t) + \frac{20}{1000+t} \times p(t) = 84{,}000$
2. This makes the left-hand side into $e^{\int P(x)\,dx}y' + P(x)e^{\int P(x)\,dx}y = Iy' + I'y = (Iy)'$.	Linear equation from video $y'(x) + P(x) \times y(x) = Q(x)$
3. We can then integrate both sides.	

Separable differential equations are equations which can be put into form $\frac{dy}{dx} = f(x) \times g(y)$, $g(y) \neq 0$, and can be solved using the techniques shown in previous chapters. *Linear* differential equations are equations which can be put into form $y'(x) + P(x) \times y(x) = Q(x)$, or $\frac{dy}{dx} + P(x) \times y(x) = Q(x)$. Linear differential equations can be solved with the techniques shown in this chapter. Many people find it helpful to think of new ideas as modifications, extensions, or variations of old ideas. The next section is a little detour with those people in mind. The purpose of the next section is to smooth out the learning curve and prepare you for a critical phase of the process of solving a linear equation. Following is a side-by-side comparison of topics learned in algebra and topics in differential equations. They are not directly related by function; rather, they are related by analogy. Glove is to hand as shoe is to foot. If you don't need this help (or don't like this teaching technique) just follow the logic on the right side of the following text box.

Multiplication of Binomials	*Derivative of a Product of Functions Rule*
$(x+5)(x+5) = x^2 + 5x + 5x + 25$ $= x^2 + 10x + 25$	$\frac{d}{dx}[f(x) \times g(x)] = f(x) \times \frac{d}{dx}g(x) + g(x) \times \frac{d}{dx}f(x)$ or $[f(x) \times g(x)]' = f(x) \times g'(x) + g(x) \times f'(x)$
The symmetric property states that if a = b then b = a, so $(x+5)(x+5) = x^2 + 10x + 25$ *is equivalent to* $x^2 + 10x + 25 = (x+5)(x+5)$ *or* $(a+b)^2 = (a+b) \times (a+b) = a^2 + 2ab + b^2$ *is equivalent to* $a^2 + 2ab + b^2 = (a+b)^2$	*The symmetric property states that if a = b then b = a, so* $\frac{d}{dx}[f(x) \times g(x)] = f(x) \times \frac{d}{dx}g(x) + g(x) \times \frac{d}{dx}f(x)$ *is equivalent to* $f(x) \times \frac{d}{dx}g(x) + g(x) \times \frac{d}{dx}f(x) = \frac{d}{dx}[f(x) \times g(x)]$ *or* $[f(x) \times g(x)]' = f(x) \times g'(x) + g(x) \times f'(x)$ *is equivalent to* $f(x) \times g'(x) + g(x) \times f'(x) = [f(x) \times g(x)]'$
Solving a Quadratic Equation by Factoring and Using the Zero-Product Property Solve $x^2 + x - 12 = 0$ $(x+4) \times (x-3) = 0$ $x + 4 = 0$ and $x - 3 = 0$	*Solving a Differential Equation by Separating Variables and Associated Differentials and Integrating Both Sides* $\frac{dy}{dx} = 2$ $dy = 2\,dx$ $\int dy = \int 2\,dx$ $y = 2x + C$

Sometimes the quadratic in the equation will not factor. When that happens, it is possible to create a perfect binomial square by adding (on both sides) a value that will result in a perfect square trinomial.

$2x^2 + 16x + 10 = 0$

$\frac{2x^2 + 16x + 10}{2} = \frac{0}{2}$ must have a 1 as coefficient of x^2

$$x^2 + 8x + 5 = 0 \qquad (1)$$

$x^2 + 8x = -5$ separating x from the scalar.

Notice that the coefficient of x is 8:

$$\left(\frac{1}{2} \times 8\right)^2 = 16$$

Add 16 to both sides of the equation above.

$$x^2 + 8x + 16 = -5 + 16$$

****

$$x^2 + 8x + 16 = 11 \qquad (2)$$

****

The result of adding 16 to both sides of the equation is that the trinomial on the left side of the equation will be the perfect square of a binomial

ck: $x^2 + 8x + 16 = (x + 4)(x + 4) = (x + 4)^2$

******* continuing after the check *******

Substituting into (2) above

$$(x + 4)^2 = 11$$

Sometimes the variables and differentials in a differential equation will not separate. When that happens, if the equation is in the form

$$y'(x) + P(x) \times y(x) = Q(x),$$

multiplying both sides of the equation by a specially chosen factor will create a differential expression on one side of the equation that is the derivative of a product of functions. This factor is called an "integrating factor." If the coefficient of the derivative is not already 1 you would have to divide to make it 1.

$$p'(t) + \frac{20}{1000 + t} \times p(t) = 84{,}000 \qquad (1)$$

Note that the polynomial coefficient of $p(t)$ is $\frac{20}{1{,}000+t}$. Let $I(x) = e^{\int \frac{20}{1{,}000+t} dt} = e^{20 \int (1{,}000+t)^{-1} dt} = e^{20 \ln(1{,}000+t)+c_1}$ (the c_1 can be ignored because there are initial conditions, $p = 0$ when $t = 0$) by step #1 of the smart man's notes from above. $I(t)$ is called an *integrating factor*. Multiply both sides of equation #1 above by the integrating factor.

$$I(t)\left[p'(t) + \frac{20}{1{,}000 + t} \times p(t)\right] = I(t) \times 84{,}000$$

*************** Distribute $I(t) = e^{20 \ln(1000+t)}$ ***************

$$e^{20 \ln(1{,}000+t)} p'(t) + e^{20 \ln(1{,}000+t)} \times \frac{20}{1{,}000 + t} \times p(t)$$
$$= e^{20 \ln(1{,}000+t)} \times 84{,}000 \qquad (2)$$

****

The result of multiplying both sides of the differential equation by the "integrating factor" is that the resulting expression will be the derivative of a product of functions.

ck: Let $e^{20 \ln(1{,}000+t)}$ be the first function, and let $p(t)$ be the second function.

$$\left[e^{20 \ln(1{,}000+t)} \times p(t)\right]' = \left[e^{20 \ln(1{,}000+t)} \times p'(t) + p(t) \times \left(e^{20 \ln(1{,}000+t)}\right) \times \left(\frac{20}{1{,}000 + t}\right)\right]$$

**************** continuing after the check *****************

Substituting into (2) above

$$\left(e^{20 \ln(1{,}000+t)} \times p(t)\right)' = e^{20 \ln(1{,}000+t)} \times 84{,}000$$

Take the square root of both sides	Integrate both sides
$$\sqrt{(x+4)^2} = \sqrt{11}$$	$$\int \left(e^{\ln(1{,}000+t)^{20}} p\right)' \, dt = \int \left(e^{\ln(1{,}000+t)^{20}} \times 84{,}000 \, dt\right)$$
Recall that square and square root are inverse functions, $\sqrt{u^2} = \pm u$, *leaving the path clear to solve for x.*	**For brevity, $p(t)$ is now represented as p.** Recall that taking the derivative and integrating are inverse functions, $\int u' \, du = u + c$, *leaving the path clear to solve for p.*
$$\pm(x+4) = \sqrt{11}$$ $$x+4 = \pm\sqrt{11}$$ $$x = -4 \pm \sqrt{11}$$ We have discussed how to choose a value to add on both sides of the algebraic equation to achieve a perfect square, which can then be eliminated by taking the square root of both sides. This removes the variable from the radical, making it clear how to proceed. In an analogous process, when solving a differential equation of the form $y'(x) + P(x) \times y(x) = Q(x)$, we multiply both sides by an integrating factor, $I(x) = e^{\int P(x)\,dx}$. This results in an expression that is the derivative of a product, which can be removed by integration, making it clear how to proceed.	$$e^{\ln(1{,}000+t)^{20}} p + c_1 = 84{,}000 \int e^{\ln(1{,}000+t)^{20}} \, dt$$ $$(1{,}000+t)^{20} p + c_1 = 84{,}000 \int (1{,}000+t)^{20} \, dt$$ $$(1{,}000+t)^{20} p + c_1 = 84{,}000 \frac{(1{,}000+t)^{21}}{21} + c_2$$ $$(1{,}000+t)^{20} p = 84{,}000 \frac{(1{,}000+t)^{21}}{21} + c_2 - c_1$$ $$(1{,}000+t)^{20} p = 4{,}000(1{,}000+t)^{21} + c_3$$ $$\frac{\cancel{(1{,}000+t)^{20}} p}{\cancel{(1{,}000+t)^{20}}} = \frac{4{,}000(1{,}000+t)^{21} + c_3}{(1{,}000+t)^{20}}$$ $$p = 4{,}000(1{,}000+t) + \frac{c_3}{(1{,}000+t)^{20}} \qquad (3)$$ In the beginning ($t = 0$), there's no pollution ($p = 0$). Solve for c_3. $$0 = 4{,}000(1{,}000+0) + \frac{c_3}{(1{,}000+0)^{20}}$$ $$0 = 4{,}000(1{,}000) + \frac{c_3}{(1{,}000)^{20}}$$ $$0 = 4{,}000{,}000 + \frac{c_3}{(1{,}000)^{20}}$$ $$-4{,}000{,}000 = \frac{c_3}{(1{,}000)^{20}}$$ $$-4{,}000{,}000 \times 1{,}000^{20} = c_3$$ $$-4 \times 10^6 \times {10^3}^{20} = c_3$$ $$-4 \times 10^6 \times 10^{60} = c_3$$ $$c_3 = -4 \times 10^{66}$$ Substituting c_3 into (3). $$p = 4{,}000(1{,}000+t) + \frac{-4 \times 10^{66}}{(1{,}000+t)^{20}}$$ when $t = 10$ days $$p = 4{,}000(1{,}000+10) + \frac{-4 \times 10^{66}}{(1{,}000+10)^{20}}$$ $$p = 4{,}000(1{,}010) + \frac{-4 \times 10^{66}}{1{,}010^{20}}$$ $$p = 4{,}040{,}000 - 3{,}278{,}178$$ $$p = 761{,}822 \text{ g}_{\text{pol}}$$

MATLAB script for graph at right is located in Appendix A

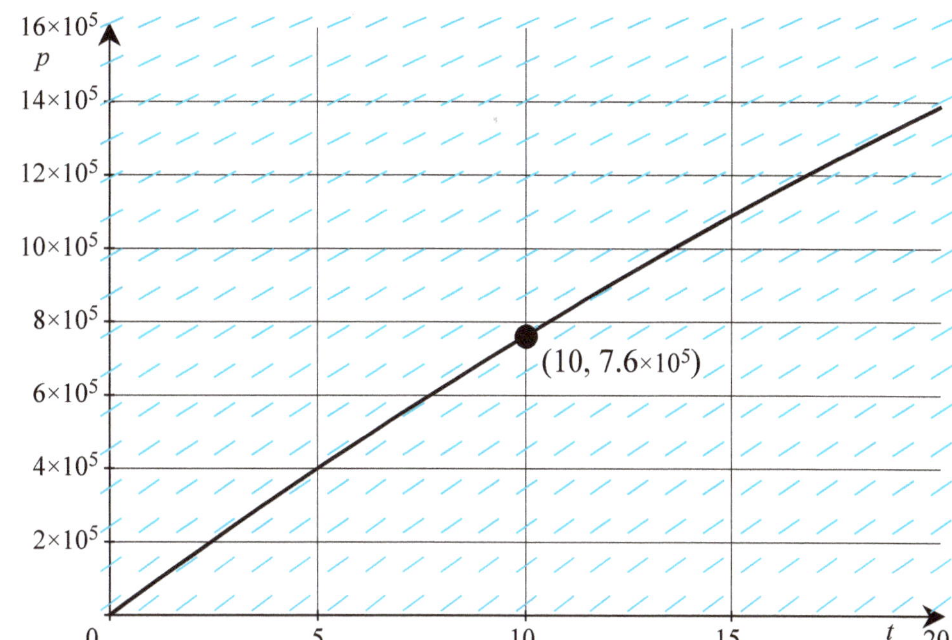

How do you pick that integrating factor to convert the left side of a differential equation into a derivative of a product of functions? Well, in brief, as the smart man said in his video, if your differential equation is in the form of $y'(x) + P(x) \times y(x) = Q(x)$, the integrating factor will be $I(x) = e^{\int P(x)\,dx}$. If you multiply both sides of the equation by $I(x)$—$I(x)[y'(x) + P(x) \times y(x)] = I(x) \times Q(x)$—the left side is the derivative of a product of functions. Integration will remove y from the polynomial expression, allowing you to solve for y. If that bothers you, detour to Appendix C where it is explained. If that doesn't bother you, continue on in the text.

Detour to Appendix C for a proof that, for a first-order differential equation of the form $y'(x)+P(x)\times y(x) = Q(x)$, multiplying both sides by an "integrating factor" $I(x) = e^{\int P(x)\,dx}$,

$$e^{\int P(x)\,dx}y' + P(x)e^{\int P(x)\,dx} \times y = e^{\int P(x)\,dx} \times Q(x)$$

$$I \times y' + I' \times y = e^{\int P(x)\,dx} \times Q(x)$$

$$(Iy)' = e^{\int P(x)\,dx} \times Q(x)$$

$$\int (Iy)' = \int e^{\int P(x)\,dx} \times Q(x)$$

Etc.,

results in a derivative on the left side which can be easily integrated and solved.

Review the pond problem below before proceeding.

A chemical has spilled upstream from a small pond. The pond is estimated to contain 1,000,000 gallons. It has an outlet that releases 20,000 gal of water per day. The stream carrying the chemical flows into the pond at 21,000 gallons per day. This means that the pond's volume is increasing by 1,000 gallons a day (during the rainy season). When the chemical spill enters the pond, it will have a concentration of four grams of the spilled chemical per gallon of stream water. Find and solve a differential equation that will model the pollution of this pond and determine the concentration of pollutant after 10 days.

Suppose there was a magical land called "Mathlandia" with 1,000,000 citizens. 20,000 citizens a year emigrated (left) from Mathlandia each year. At the same time, 21,000 citizens a year immigrated (moved in) to Mathlandia. Four percent of those immigrating citizens were from a neighboring war-torn country. Some of the new immigrants continued their movement to another country at the same rate as the native citizens. Find and solve a differential equation that will model how the population of Mathlandia was affected by the circumstances described here. How many citizens from the war-torn country will be there after 10 years?

A company is valued to be worth $1,000,000. The company has sales of $20,000 a month and has expenses of $19,000 a month. It decides to change one of the products in its product line for a new item, which will account for 4% of its sales each month. After 10 months, how much of the company's value can be attributed to this new product?

A person with 5 liters of blood undergoes an operation. He loses 1 liter of blood per hour during the operation, which is replaced with the same amount of blood over the same time period. The replacement blood has a blood additive that allows doctors to trace the blood for diagnostic purposes. That additive was 4% of the blood added. Over a 10-hour operation, how much of the additive built up in the person's body?

There are 1,000,000 Gucci purses in a wealthy country. Wealthy women buy 20,000 new Gucci purses a year. At the same time 20,000 purses a year are randomly stolen, thrown away, or lost. Starting in year X, it is found that 4% of the purses are fake. After 10 years, how many fake purses are there?

What can you conclude from the five examples shown above? (Hint: You are supposed to say, "They are all versions of the same problem.")

(Author's note: My thanks to Dr. Larry Green of Lake Tahoe JC for his suggestions on this chapter. Any errors are my own.)

In Chapter 5, we solved a differential equation involving Kirchhoff's voltage law. Let's review that problem.

If a battery supplies constant 30 V, and if the circuit's inductance is 2 H (henrys) and resistance is 10 Ω (ohms) and $I(0) = 0$, find $I(t)$, the current, after 0.3 seconds.

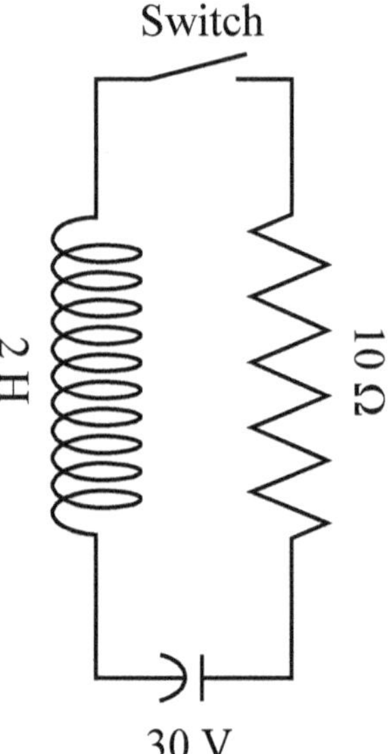

$2\frac{dI}{dt} + 10I = 30$ Solving this differential equation for I will give an integrated equation to determine current at time t.

$\frac{dI}{dt} + 5I = 15$ divide to make the $\frac{dI}{dt}$ term have a coefficient of 1

Hence, by the smart man's rule #1, the general equation $y'(x) + P(x) \times y(x) = Q(x)$ for us would be $\frac{dI}{dt} + 5I = 15$.

1. Calculate the integrating factor. For us, that would be $e^{\int 5\,dt} = e^{5t}$.

2. Multiply both sides of the original equation by the integrating factor, e^{5t}.

$$\frac{dI}{dt} + 5I = 15$$

$e^{5t}\left[\frac{dI}{dt} + 5I\right] = e^{5t} \times (15)$ multiply both sides of the differential equation by the integrating factor e^{5t}

$e^{5t} \times \frac{dI}{dt} + e^{5t} \times 5I = e^{5t} \times (15)$ distributing the integrating factor through. Here, you are multiplying both sides of the equation by an integrating factor, $e^{\int 5\,dt}$. The result on the left side of the equation will be the derivative of a product of functions. In the following steps, we will be able to integrate this derivative of a product of functions, $\int \frac{dy}{dx} y(x) = y(x)$, isolating the desired function!!!

$$e^{5t} \times \frac{dI}{dt} + e^{5t} \times 5I = e^{5t} \times (15)$$ copied from the previous page

$$\left[e^{5t} \times \frac{dI}{dt}\right] + \left[5e^{5t} \times I\right] = e^{5t}(15)$$ left, commute the 5 away from the $5I$

$$\left[e^{5t} \times \frac{dI}{dt}\right] + \left[5e^{5t} \times I\right] = 15e^{5t}$$

$$\left[e^{5t} \times \frac{dI}{dt}\right] + \left[I \times 5e^{5t}\right] = 15e^{5t}$$ left side is $(f(x) \times g(x))'$

Now, remembering the arguing Zax at the top of page 290,

$$f(x) \times g'(x) + g(x) \times f'(x) = [f(x) \times g(x)]'$$ so

$$\left[e^{5t} \times I\right]' = 15e^{5t}$$

$$\int \left[e^{5t} \times I\right]' \, dt = \int 15e^{5t} \, dt$$

$$e^{5t} \times I + c_1 = 3 \int e^{5t} 5 \, dt$$

$$e^{5t} \times I + c_1 = 3e^{5t} + c_2$$

$$e^{5t} \times I = 3e^{5t} + c_2 - c_1$$

$$I = \frac{3e^{5t}}{e^{5t}} + \frac{C}{e^{5t}}$$

$$I = 3 + Ce^{-5t}$$

Some people benefit from the teaching technique called "compare and contrast." For those of you in that camp, the preceding problem is now solved side by side by two different techniques: 1) separation and 2) integrating factor. If you do not benefit from this presentation then do skip on ahead.

Separation—Chapter 5	Linear Equation with Integrating Factor—Chapter 8
$2\frac{dI}{dt} + 10I = 30$ I denotes current	$2\frac{dI}{dt} + 10I = 30$ I represents current
$\frac{dI}{dt} + 5I = 15$ divide so that leading coefficient is 1	$\frac{dI}{dt} + 5I = 15$ divide so that leading coefficient is 1
$\frac{dI}{dt} = 15 - 5I$ isolate the differential $dI = (15 - 5I)\,dt$ separated! $\frac{dI}{15 - 5I} = dt$ separated and grouped	Multiply both sides of the differential equation by the integrating factor, $e^{\int 5\,dt} = e^{5t}$. $e^{5t}\left[\frac{dI}{dt} + 5I\right] = e^{5t}(15)$
The differentials and their respective variable are now grouped. $(15 - 5I)^{-1}\,dI = dt$ preparing to integrate	$e^{5t}\frac{dI}{dt} + e^{5t}5I = e^{5t}(15)$ distribute the integrating factor
$-\frac{1}{5}\int (15 - 5I)^{-1}(-5\,dI) = \int dt$	The expression on the left is a derivative of a product.
$-\frac{1}{5}\ln\lvert 15 - 5I\rvert + c_1 = t + c_2$	$e^{5t}\frac{dI}{dt} + e^{5t}5I = \left(e^{5t}I\right)'$

$-\frac{1}{5}\ln\lvert 15-5I\rvert = t + c_2 - c_1$ $-\frac{1}{5}\ln\lvert 15-5I\rvert = t + c_3$ $-\frac{1}{5}\ln\lvert 15-5I\rvert = t - c_4$ $\ln\lvert 15-5I\rvert = -5(t-c_4)$ $\ln\lvert 15-5I\rvert = -5t+5c_4$ $e^{\ln\lvert 15-5I\rvert} = e^{-5t+5c_4}$ $\lvert 15-5I\rvert = e^{-5t}e^{5c_4}$ $\lvert 15-5I\rvert = c_5e^{-5t}$ $15-5I = \pm\left(c_5e^{-5t}\right)$ $15-5I = c_6\times e^{-5t}$ $\quad c_6=\pm c_5$ or 0 $-5I = c_6\times e^{-5t} - 15$ $I = -\frac{1}{5}\left(c_6e^{-5t}-15\right)$ $I = c_7e^{-5t}+3$ $I = Ce^{-5t}+3$ *Here, we have an equation that describes current in the circuit at time t.*	$\left(e^{5t}I\right)' = e^{5t}(15)$ substitute it into the expression above $\int\left(e^{5t}I\right)'\,dt = \int e^{5t}(15)\,dt$ $e^{5t}I = 15\int e^{5t}\,dt$ $I = \frac{15\int e^{5t}\,dt}{e^{5t}}$ $\quad$ I (current) is finally isolated $I = \frac{15\left(\frac{1}{5}\right)\int e^{5t}5\,dt}{e^{5t}}$ $\quad$ preparing to integrate $e^u\,du$ $I = \frac{3e^{5t}+C}{e^{5t}}$ $I = 3+\frac{C}{e^{5t}}$ $I = 3+Ce^{-5t}$ $I = Ce^{-5t}+3$

More practice using the integrating factor to solve a linear ODE. ODEs with trigonometry.

Solve the following linear ODE $\frac{dy}{d\theta} + (\tan\theta)y = \cos\theta$ This is in form $\frac{dy}{dx}+P(x)y=Q(x)$. Hence, it is linear and can be solved using the integrating-factor technique. $I(x) = e^{\int P(x)\,dx} = e^{\int\tan\theta\,d\theta} = e^{\ln\lvert\sec\theta\rvert} = \sec\theta$ for $-\frac{\pi}{2}<\theta<\frac{\pi}{2}$. Multiply both sides by the integrating factor. $\sec\theta\left[\frac{dy}{d\theta}+\tan\theta\times y\right] = \sec\theta\cos\theta$ $\quad \sec\theta=\frac{1}{\cos\theta}$ $\sec\theta\frac{dy}{d\theta}+\sec\theta\tan\theta\times y = \frac{\cos\theta}{\cos\theta}$ $\quad (\sec\theta)'=\sec\theta\tan\theta$ $[\sec\theta\times y]' = 1$ $\int[\sec\theta\times y]'\,d\theta = \int 1d\theta$ $\sec\theta\times y = 1\int d\theta$ $\sec\theta\times y = \theta + c_1$ $y = \frac{\theta}{\sec\theta}+\frac{c_1}{\sec\theta}$ $\quad \frac{1}{\sec\theta}=\cos\theta$ $y = \theta\cos\theta + c_1\cos\theta$	Solve the following linear ODE $x\frac{dy}{dx}+3y = \frac{\sin x}{x^2}, \quad x\neq 0$ Since $x\neq 0$, $\frac{dy}{dx}+\frac{3}{x}y = \frac{\sin x}{x^3}, \quad x\neq 0.$ Now $\frac{dy}{dx}$ has a coefficient of 1. This is in form $\frac{dy}{dx}+P(x)y=Q(x)$. Hence, it is linear and can be solved using the integrating-factor technique. $I(x) = e^{\int\frac{3}{x}dx} = e^{3\int x^{-1}dx} = e^{3\ln\lvert x\rvert} = e^{\ln\lvert x^3\rvert} = x^3 \quad x>0$ Multiply both sides by the integrating factor. $x^3\left[\frac{dy}{dx}+\frac{3}{x}y\right] = x^3\frac{\sin x}{x^3}$ $x^3\frac{dy}{dx}+3x^2y = \sin x$ $\left[x^3y\right]' = \sin x$ $\int\left[x^3y\right]'\,dx = \int\sin x\,dx$ $x^3y = -\cos x + c$ $y = \frac{-\cos x + c}{x^3}$

Chapter 8 Review

Linear just means that the dependent variable in an equation appears only with a power of one. In algebra a general linear equation was in the form of $y^1 = mx + b$. The dependent variable, **y**, has a power of one. A linear differential equation is a differential equation in which the dependent variable, **y**, has a power of one. For example.

$$\frac{dy}{dx} + P(x) \times \mathbf{y}^1 = Q(x) \quad \text{(first-order } \textbf{linear differential equation}\text{).}$$

A nonlinear differential equation is one that is, well, not linear. A nonlinear function of the dependent variable or its derivatives cannot appear in a linear differential equation nor can the coefficient of the dependent variable or its derivatives depend upon y. Here are three examples of nonlinear differential equations.

1. $\frac{dy}{dx} + \cos y = 5$. The dependent variable is y. $\cos y$ is not linear.

2. $\frac{dy}{dx} + y^3 = 3x$. The dependent variable is y. y^3 is not linear.

3. $(2 + y)\frac{dy}{dx} + y = 0$. The dependent variable is y. The coefficient of $\frac{dy}{dx}$ depends on y.

In Chapter 8, we found that some first-order linear differential equations could not be separated. If they were in (or could be transformed into) the form $\frac{dy}{dx} + P(x)y = Q(x)$ such equations could be solved using the integrating-factor technique. This technique is based on the remarkable fact that, by multiplying both side of the equation by $I(x)$, the integrating factor, the left-hand side of this differential equation can be transformed into the form of the derivative of a product.

$$\frac{dy}{dx} + P(x)y = Q(x)$$

$$\boldsymbol{I(x)}\left[\frac{dy}{dx} + P(x)y\right] = \boldsymbol{I(x)} \times Q(x)$$

$$\boldsymbol{I(x)} \times \frac{dy}{dx} + \boldsymbol{I(x)} \times P(x) \times y = \boldsymbol{I(x)} \times Q(x)$$

$$[\boldsymbol{I(x)} \times y]' = \boldsymbol{I(x)} \times Q(x)$$

$$\int [\boldsymbol{I(x)} \times y]'\, dx = \int [\boldsymbol{I(x)} \times Q(x)]\, dx$$

Then, after integrating both sides, the path will be clear to solve for y.

$$[\boldsymbol{I(x)} \times y] = \int [\boldsymbol{I(x)} \times Q(x)]\, dx$$

$$y = \frac{\int [\boldsymbol{I(x)} \times Q(x)]\, dx}{\boldsymbol{I(x)}}$$

It was noted in passing that applications for linear differential equations are often seen in different forms: the pond problem, the immigration problem, the IV-drip problem, and the Gucci-purse problem.

We learned that some ODEs can be solved by more than one technique. For example the electrical circuit involving Kirchhoff's Law was solved by both the separable technique and also the integrating-factor technique for linear equations.

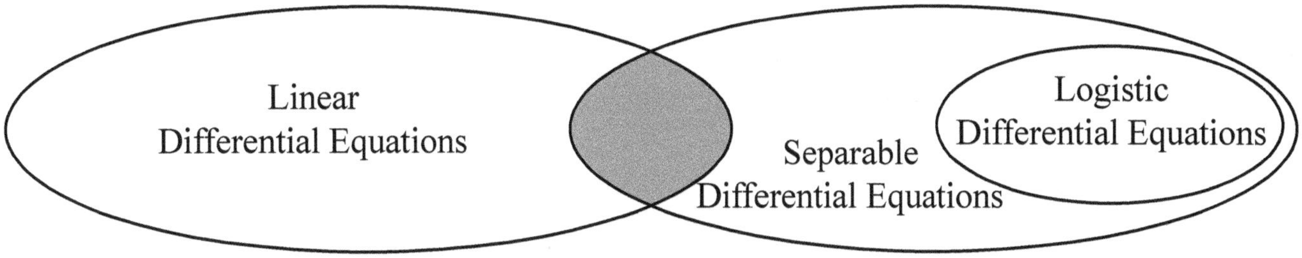

Linear Differential Equations

Separable Differential Equations

Logistic Differential Equations

Separable (*** Chapters 1–5)
$y' = f(x) \times g(y), \quad g(y) \neq 0$
$\frac{dy}{dx} = f(x) \times g(y), \quad g(y) \neq 0$
Exponential growth
Exponential decay
Separate the functions and their respective differentials.
$\frac{1}{g(y)}\,dy = f(x)\,dx$
Proceed to integrate each side using the integration formulas
$\int y^{-1}\,dy = \ln y + C$
and $\int x^n\,dx = \frac{x^{n+1}}{n+1}$.
$\int \frac{1}{g(y)}\,dy = \int f(x)\,dx$
Logistic (*** Chapters 6)
$P' = kP\left(1 - \frac{P}{K}\right)$
$\frac{dP}{dt} = kP\left(1 - \frac{P}{K}\right)$
Logistic equations are separable, but it is best to solve them by converting them to their integrated form.
$P(x) = \frac{K}{A \times e^{-kt} + 1}$
where $A = \frac{K - P_0}{P_0}$
Integrating factor technique for linear equations (*** Chapters 8)
$y' + P(x) \times y = Q(x)$
$I(x) = e^{\int P(x)\,dx}$
Multiply both sides of the linear differential equation by an integrating factor. Proceed to integrate both sides using integration to undo the resulting derivative of a product on the left side of the equation.
$\int (y(x) \times I(x))'\,dx = y(x) \times I(x)$
Then use whatever integration technique is appropriate on the right.

The following chart is a primitive and simplistic attempt to help beginning students (the target audience for this book) review, differentiate, and mentally organize their thoughts to this point in the book. Bear in mind that, as implied by the Venn diagram on the previous page, this chart is primitive and simplistic.

* Introduced
*** Covered

Ordinary Differential Equations (ODE) by category

- Only a decimal approximation can be found
  - Euler's method (*** Chapter 1)
  - Runge–Kutta
- Closed form (analytical solution) and decimal approximation can be found
  - Single differential equations ***
  - Systems of differential equations
    Predator–Prey (* Chapter 7)
    $\frac{dr}{dt} = k_1 r + k_2 rf$
    $\frac{df}{dt} = -k_3 r + k_4 rf$
  - Nonlinear Differential Equations (* Chapter 8)
    - Separable (*** Chapters 1–5)
      - $y' = f(x) \times g(y), g(y) \neq 0$
      - $\frac{dy}{dx} = f(x) \times g(y), g(y) \neq 0$
      - Exponential growth
      - Exponential decay
    - Logistic (*** Chapter 6)
      - $P' = kP\left(1 - \frac{P}{K}\right)$
      - $\frac{dP}{dt} = kP\left(1 - \frac{P}{K}\right)$
  - Linear Differential Equations $y' + P(x) \times y = Q(x)$ (* Chapter 8)
    - Separable (*** Chapters 1–5)
      - $y' = f(x) \times g(y), g(y) \neq 0$
      - $\frac{dy}{dx} = f(x) \times g(y), g(y) \neq 0$
      - Exponential growth
      - Exponential decay
    - Integrating Factor Technique (*** Chapter 8)
      $y' + P(x) \times y = Q(x)$
      $I(x) = e^{\int P(x)\,dx}$

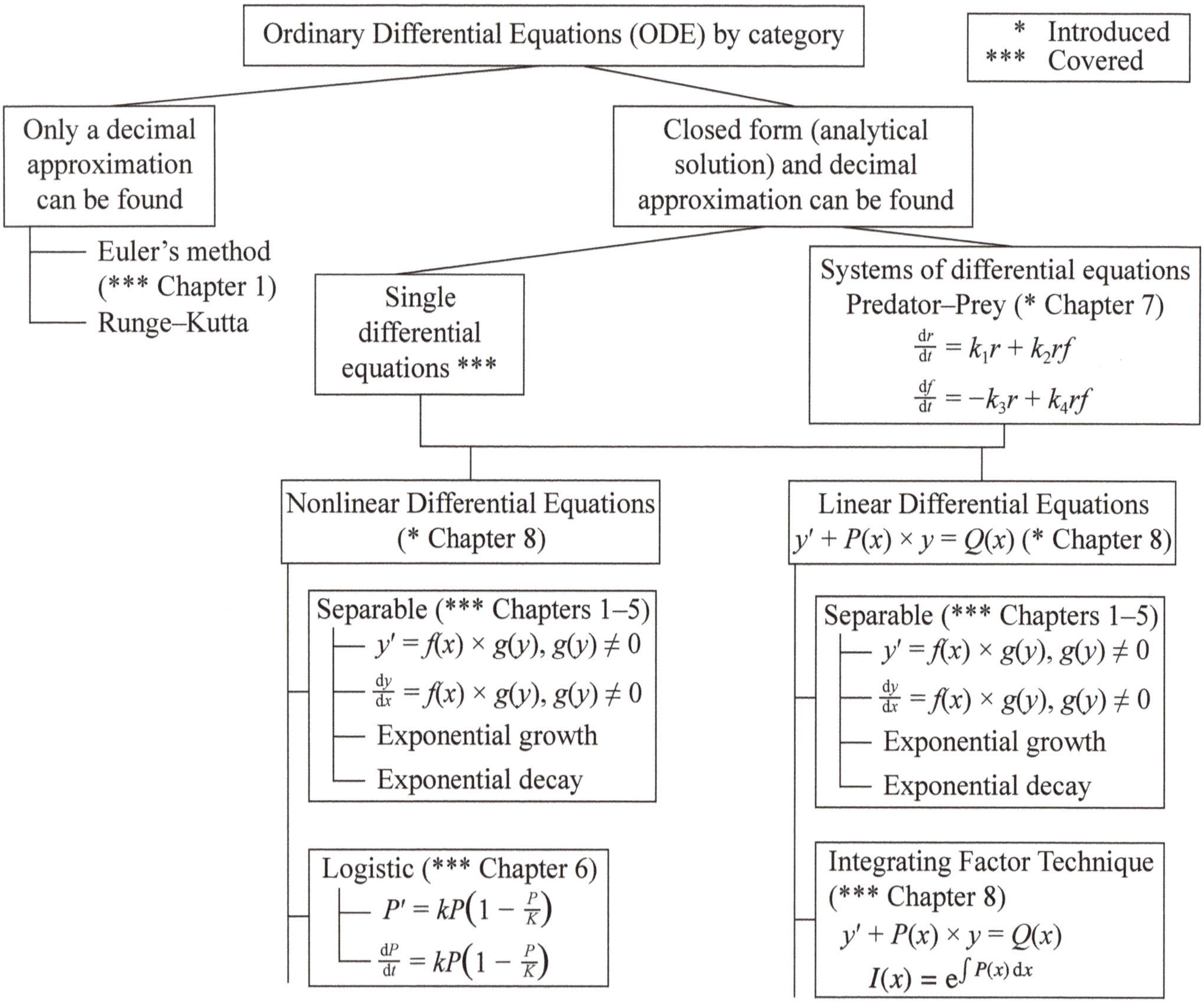

Using an Integrating Factor to Solve a Separable Equation

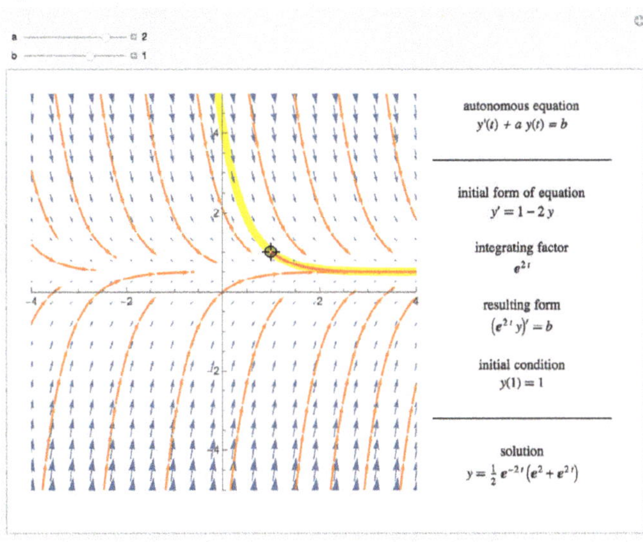

Online Application

Visit demonstrations.wolfram.com/UsingAnIntegratingFactorToSolveASeparableEquation for a great little application that will demonstrate some of the concepts from this chapter. "Using an Integrating Factor to Solve a Separable Equation" from the Wolfram Demonstrations Project. Contributed by Stephen Wilkerson.

Chapter 9
Differential Equations in Physics

Just about every beginning physics class introduces the formula, $p = -4.9t^2 + v_0t + p_0$, describing the position of a ball thrown with initial velocity v_0 from an arbitrary height p_0 above the Earth's surface. After t seconds, the ball's position would be equal to the sum of $\frac{1}{2}$ the acceleration due to gravity times t^2, the initial velocity times time, and the initial position of the ball. That formula just sort of appears, deus ex machina, like Athena jumping out of the head of Zeus, in those physics classes. Actually, in a different form, it jumped out of the head of the likes of Sir Isaac Newton. How did they come up with that formula?

A smart scientist named Galileo is among the first credited with determining the acceleration due to gravity. Since he did not have the measuring capabilities of today's scientists, he got at his information indirectly—by observing the movement of a marble down inclined ramps. By reducing the angle of the decline he reduced the speed of the marble so that he could take measurements using crude instruments. However, regardless of the angle of the ramp, he noticed the same pattern. At any time t (water clocks? pendulum clocks? pulse rate?), the distance that the marble traveled was the sum of the odd numbers, in sequence, to add up to the square of the time associated with that distance. As the time increased, the distance traveled was directly proportional to the square of the time, $d = kt^2$, where k is a constant of proportionality. This, for $k = 1$, is shown in the chart at right. As Galileo changed the angle of the ramp, the distance traveled in that initial second changed, but the pattern remained.

t	Diagonal distance covered	d[a]
0	0	0
1	1	1
2	1 + 3	4
3	1 + 3 + 5	9
4	1 + 3 + 5 + 7	16

[a]Total distance covered in units of the distance traveled during the initial second

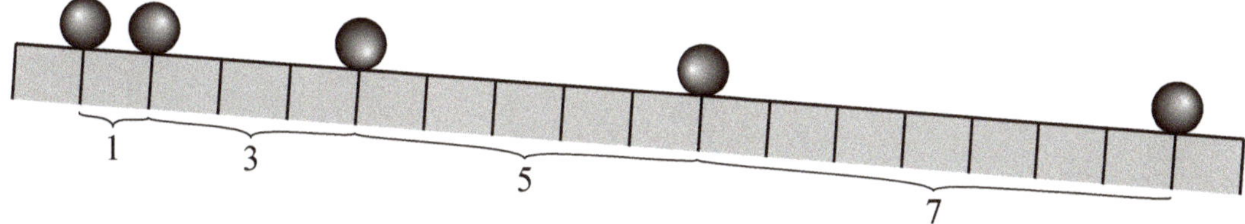

In the example above, 1 unit of diagonal distance was defined to be the distance the marbled traveled during the initial second after it was released. If the units shown above on the inclined plane corresponded to our familiar measuring units, we would be finished. However, we measure our lengths in terms of feet or meters. To compensate for the fact that our measuring units are not the same as shown above, we adjust a constant of proportionality, k. This direct proportion holds regardless of the angle of incline, regardless of the units being used, but k is different for each angle of incline. Today, thanks to Sir Isaac Newton, we recognize this constant of proportionality to be due to the effect of earth's gravitational pull, and we would write $d = kt^2$ as $d = \frac{1}{2}gt^2$ for a marble falling straight down. That is, as the angle of the incline approaches 90°, $k \to \frac{1}{2}g$. Through experiment and observation of data, one could then, for any desired linear unit, determine initial conditions which would allow you to solve for g. Today, for instructional purposes, that would be either $g = -9.8\,\frac{\text{m}}{\text{s}^2}$ or $g = -32\,\frac{\text{ft}}{\text{s}^2}$. Whether Sir Isaac Newton used Galileo's work or had his own method of measuring g, and the fact that the metric system did not exist in Newton's time, are not really relevant to these materials. The fact is, when Mr. Newton did his work, a value for g was known—$g = -9.8\,\frac{\text{m}}{\text{s}^2}$ in today's metric units.

pbslearningmedia.org/resource/phy03.sci.phys.mfw.galileoplane/galileos-inclined-plane

mcm.edu/academic/galileo/ars/arshtml/mathofmotion1.html

en.wikipedia.org/wiki/History_of_experiments

Differential equations can be used to derive $p_f = -4.9t^2 + v_0 t + p_0$. By definition, the velocity increases every second by $-9.8\,\frac{\text{m}}{\text{s}}$. That is the meaning of $g = -9.8\,\frac{\text{m}}{\text{s}^2}$. That is, the velocity is downward (hence the negative sign) and getting larger.

Definition: Acceleration is the change in velocity over time.

$$\frac{dv}{dt} = -9.8\,\frac{\text{m}}{\text{s}^2} \quad \text{a separable differential equation!}$$

$$dv = -9.8\,dt$$

$$\int dv = \int -9.8\,dt$$

$$v + c_1 = -9.8t + c_2$$

$$v = -9.8t + c_2 - c_1$$

$$v_f = -9.8t + c_3$$

At time $t = 0$, the velocity was v_0. Hence, $v_0 = -9.8 \times 0 + c_3$. $v_0 = 0 + c_3$, so $c_3 = v_0$. Finally, $v_f = -9.8t + v_0$.

t (time)	$v_f = -9.8t + v_0$, $v_f = \int g\,dt$	acceleration due to gravity (g)
0	0.0m/s	-9.8m/s^2
1	-9.8m/s	-9.8m/s^2
2	-19.6m/s	-9.8m/s^2
3	-29.4m/s	-9.8m/s^2
4	-39.2m/s	-9.8m/s^2
5	-49.0m/s	-9.8m/s^2

Definition: Velocity is change in position over time.

$$\frac{dp}{dt} = v_f = -9.8t + v_0$$

$$\frac{dp}{dt} = -9.8t + v_0 \quad \text{a separable differential equation!}$$

$$dp = (-9.8t + v_0)\,dt$$

$$\int dp = \int (-9.8t + v_0)\,dt$$

$$p + c_1 = -9.8\frac{t^2}{2} + v_0 t + c_2$$

$$p = -4.9t^2 + v_0 t + c_2 - c_1$$

$$p = -4.9t^2 + v_0 t + c_3 \quad c_3 = c_2 - c_1$$

$$p_f = -4.9t^2 + v_0 t + c_3$$

$$p_0 = -4.9 \times 0^2 + v_0 0 + c_3 \quad \text{when } t = 0\ p_f = p_0$$

$$c_3 = p_0$$

Finally, $p_f = -4.9t^2 + v_0 t + p_0$.

t (time)	p (position) at time t relative to starting position $p = -4.9t^2 + v_0 t + p_0$, $p = \int v\,dt$	$v_f = -9.8t + v_0$ $v_f = \int g\,dt$	Acceleration (g)
0	0.0m	0.0m/s	-9.8m/s^2
1	-4.9m	-9.8m/s	-9.8m/s^2
2	-19.6m	-19.6m/s	-9.8m/s^2
3	-44.1m	-29.4m/s	-9.8m/s^2
4	-78.4m	-39.2m/s	-9.8m/s^2
5	-122.5m	-49.0m/s	-9.8m/s^2

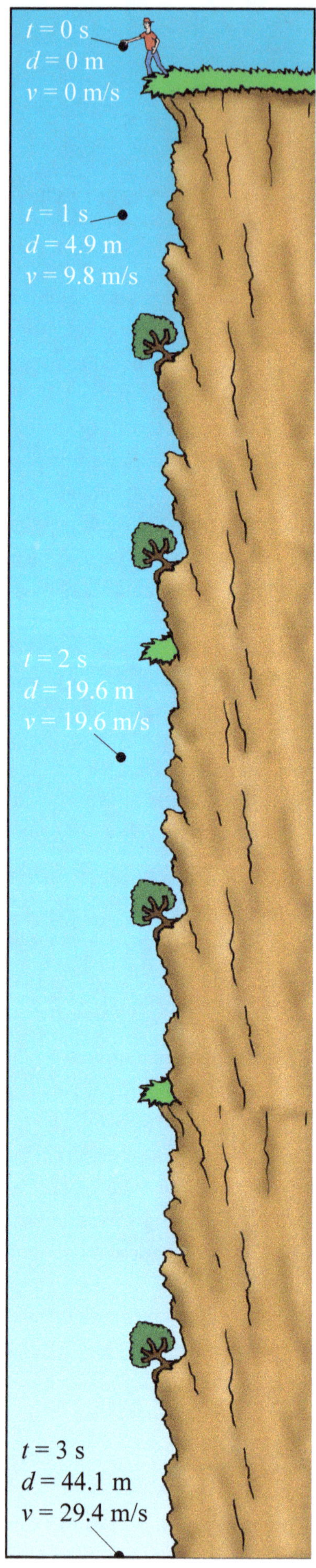

In physics, $p = -4.9t^2 + v_0t + p_0$ is taught as shown at right.

t (time) $\rightarrow\rightarrow\rightarrow$	p (position) at time t $p = -4.9t^2 + v_0t + p_0$ $\rightarrow\rightarrow\rightarrow\rightarrow\rightarrow$	v (velocity) $v_f = -9.8t + v_0$ $v_f = \frac{dp}{dt}$ $\rightarrow\rightarrow\rightarrow$	a (acceleration) $a = -9.8\ \frac{m}{s^2}$ $a = \frac{dv}{dt} = \frac{d}{dt}\left(\frac{dp}{dt}\right)$

Above: Differentiate left to right.

Below: Integrate right to left.

t (time)	p (position) at time t $p = -4.9t^2 + v_0t + p_0$ $p = \int v\,dt$ $\leftarrow\leftarrow\leftarrow\leftarrow\leftarrow$	v (velocity) $v_f = -9.8t + v_0$ $v_f = \int g\,dt$ $\leftarrow\leftarrow\leftarrow$	Acceleration (g) $a = -9.8\ \frac{m}{s^2}$ $a = \frac{dv}{dt} = \frac{d}{dt}\left(\frac{dp}{dt}\right)$ $\leftarrow\leftarrow\leftarrow$

However, it is possible that the formulas $p = -4.9t^2 + v_0t + p_0$ and $v_f = -9.8t + v_0$ at right were obtained in the reverse order from the way they are traditionally taught, by solving differential equations obtained from the empirically derived data: The acceleration due to gravity at the Earth's surface is $g = -9.8\ \frac{m}{s^2}$. Think about the logic of that statement.

There is a well-documented story about the Italian scientist Galileo dropping two cannon balls (greatly differing in weight) from the top of the Leaning Tower of Pisa. According to a famous philosopher named Aristotle, the object should fall at a rate in ratio with their weights. A 10-pound weight should fall 10 times as fast as a one-pound weight. If Galileo had known the formula above for distance, he could have saved himself the walk up the 296 steps (55 m).

Follow this logic. If we neglect air resistance,

$$p_f = -4.9t^2 + v_0t + p_0.$$

Larger cannon ball	Smaller cannon ball
$p = -4.9t^2 + v_0t + p_0$	$p = -4.9t^2 + v_0t + p_0$
$-55 = -4.9t^2 + 0 \times t + 55$	$-55 = -4.9t^2 + 0 \times t + 55$

These formulas are the same! There is no variable in the formula to account for mass. The distances traveled by the two canon balls are the same. The times will be the same.

Newton's Second Law

Newton's original Latin:

"Lex II: Mutationem motus proportionalem esse vi motrici impressae, et fieri secundum lineam rectam qua vis illa imprimitur."

Newton's second law as we know it:

The relationship between an object's mass m, its acceleration a, and the applied force F is $F = ma$.

Newton's law as it was originally stated:

Newton's second law, $F = ma$, was originally formulated in slightly different, but equivalent, terms: The original version states that the net force acting upon an object is equal to the rate at which its momentum changes. (Did you pick up on the word *rate* here, rate of momentum change? I smell a differential equation!) Momentum is defined in mechanics as "a quantity expressing the motion of a body or system equal to the product of the mass of a body and its velocity."

$F = \frac{d}{dt}(mv)$ force is equal to the rate of change of the product of the mass of a body and its velocity)

$F = \frac{d}{dt}(m \times v)$ arrived at through definitions

$F = m \times \frac{dv}{dt} + v \times \frac{dm}{dt}$ derivative of a product rule

$F = m \times \frac{dv}{dt} + v \times 0$ an object's mass does not change over time

$F = m \times \frac{dv}{dt}$ m is constant

$F = ma$ definition of acceleration $\left(\frac{dv}{dt} = a\right)$

Newton's Second Law	
Differential Equation Form	Algebraic Form
$F = \frac{d}{dt}(mv)$	$F = ma$

It is helpful to see that, since there are three variables in this equation, there are three immediate applications of Mr. Newton's law: 1) solving for F if given m and a, 2) solving for m if given F and a, and 3) solving for a if given F and m. It makes your life easier if you see that, despite differences in specifics, the process for each of the problems listed above is identical. For the sake of compare and contrast and for the sake of review of how units are canceled in physics, an example of each is shown below.

(1) $F = ma$	**(2) $F = ma$**	**(3) $F = ma$**
Let $F = ?$ $m = 10\,\text{kg}$ $a = -9.8\,\frac{\text{m}}{\text{s}^2}$ $F = m \times a$ Solve for F. $F = ma$ Substitute for m and a. $F = 10\,\text{kg}\left(-9.8\,\frac{\text{m}}{\text{s}^2}\right)$ $F = -98\,\frac{\text{kg m}}{\text{s}^2}$ Since $1\,\text{N} = 1\,\frac{\text{kg m}}{\text{s}^2}$, $F = -98\,\text{N}$ (Newtons)	Let $F = -98\,\text{N}$ $m = ?$ $a = -9.8\,\frac{\text{m}}{\text{s}^2}$ $F = m \times a$ Solve for m. $m = \frac{F}{a}$ Substitute for F and a. $m = \frac{-98\,\text{N}}{-9.8\,\frac{\text{m}}{\text{s}^2}}$ Since $1\,\text{N} = 1\,\frac{\text{kg m}}{\text{s}^2}$, $m = \frac{-98\,\frac{\text{kg m}}{\text{s}^2}}{-9.8\,\frac{\text{m}}{\text{s}^2}} = 10\,\text{kg}$	Let $F = -98\,\text{N}$ $m = 10\,\text{kg}$ $a = ?$ $F = m \times a$ Solve for a. $a = \frac{F}{m}$ Substitute for F and m. $a = \frac{-98\,\text{N}}{10\,\text{kg}}$ Since $1\,\text{N} = 1\,\frac{\text{kg m}}{\text{s}^2}$, $a = \frac{-98\,\frac{\text{kg m}}{\text{s}^2}}{10\,\text{kg}} = -9.8\,\frac{\text{m}}{\text{s}^2}$

Note that gravity pulls down. Using the convention that up is positive, g must be negative, as is the force exerted by the Earth on any other object.

Three different variations of the same problem, F = ma!!!

F = ma is an integrated algebraic form of a differential equation!!!

Let's go back to ideas shown a few pages back where we differentiated the polynomial $p = -4.9t^2 + v_0 t + p_0$.

t (time) →→→	p (position) at time t $p = -4.9t^2 + v_0 t + p_0$ →→→→→	v (velocity) $v_f = -9.8t + v_0$ $v_f = \frac{dp}{dt}$ →→→	a (acceleration) $a = 9.8\,\frac{\text{m}}{\text{s}^2}$ $a = \frac{dv}{dt} = \frac{d}{dt}\left(\frac{dp}{dt}\right)$

Here we see a polynomial $p = -4.9t^2 + v_0 t + p_0$, where the highest degree of the independent variable is two. This is said to be a polynomial of degree two. After two differentiations, we got $a = p''(t)$, or $a = \frac{d^2p}{dt^2} = g$, the constant for the acceleration due to gravity at the surface of the earth in this case.

Next we take a constant, $a = 9.8\,\frac{\text{m}}{\text{s}^2} = p''$, and do two integrations and get a *second degree polynomial*,

t (time)	p (position) at time t $p = -4.9t^2 + v_0 t + p_0$ $P = \int v\,dt$ ←←←←←	v (velocity) $v_f = -9.8t + v_0$ $v_f = \int g\,dt$ ←←←	a (acceleration) $a = -9.8\,\frac{\text{m}}{\text{s}^2}$ $a = \frac{dv}{dt} = \frac{d}{dt}\left(\frac{dp}{dt}\right)$ ←←←

If you do two integrations in order to obtain your desired function, then the differential equation you are working with is said to be a *second-order differential equation.*

Study the examples in the following table. Be able to distinguish the following terms: first-, second-, third-, and fourth-order differential equations.

Degree of Polynomial	Polynomial examples	Order of differential equation	Differential equation examples
1	$y = 5x$	1	$\frac{dy}{dx} = 7x$ or $y'(x) = 7x$
2	$y = 8x^2 + 3x - 2$	2	$\frac{d^2y}{dx^2} + 2\frac{dy}{dx} = 15x - 7$ or $y''(x) + 2y(x) = 15x - 7$
3	$y = 4x^3 - 2x + 6$	3	$4\frac{d^3y}{dx^3} + 5\frac{d^2y}{dy^2} - \frac{dy}{dx} = x^2 + 1$ or $4y'''(x) + 5y''(x) - y(x) = x^2 + 1$
4	$y = -2x^4 + 3x^2 - 5$	4	$\frac{d^4y}{dx^4} + 6\frac{dy}{dx} - y = 0$ or $y'''' + 6y' - y = 0$

Chapter 9 Review

The acceleration due to gravity and how Galileo obtained its value were discussed. That value was integrated twice and the result is the well-known physics formula for position of an object thrown from position p_0 with an initial velocity of v_0 after t seconds in Earth's gravity, $p = -4.9t^2 + v_0t + p_0$. The "degree of polynomial" was discussed and used to compare and contrast with the concept "order of a differential equation."

Chapter 10
Bernoulli Equations

"$y' + P(x)y = Q(x)y^n$, for $n \geq 2$, $n \in \mathbb{N}$, is called a **Bernoulli equation** and is named after Jacob Bernoulli, who discussed it and its solution in 1695. Bernoulli equations are special because they are nonlinear differential equations with known exact solutions" (Wikipedia).

Take a moment to compare and contrast a Bernoulli equation with a first-order linear equation.

$y' + P(x)y = Q(x)y^n$ Bernoulli Equation, Chapter 10

$y' + P(x)y = Q(x)y^0$ First-Order Linear Equation, Chapter 8

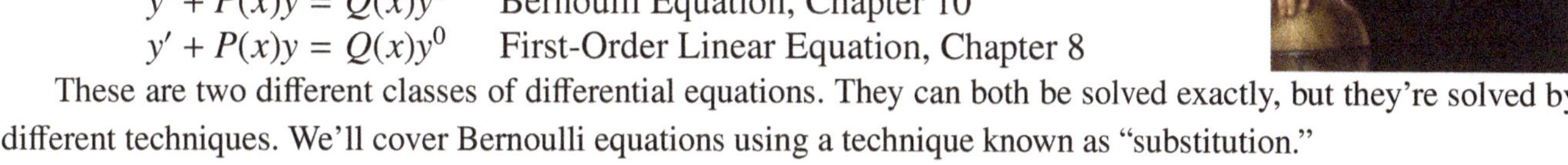

These are two different classes of differential equations. They can both be solved exactly, but they're solved by different techniques. We'll cover Bernoulli equations using a technique known as "substitution."

As has often been the case throughout this book, a new technique will be taught by comparing and contrasting it with a skill from beginning algebra. If this annoys you, by all means skip the left side of the following table and concentrate only on the right side.

Beginning algebra technique for solving polynomial equations in the form $P(x) = ax^4 + 0x^3 + cx^2 + 0x + f = 0$ A fourth-degree polynomial equation with missing third- and first-degree terms	Bernoulli equation technique for solving differential equations in the form $y' + P(x)y = Q(x)y^n$ When $n = 0$, this is a "First-Order Linear Equation" (Chapter 8) $y' + P(x) \times y = Q(x)y^0$. When $n = 1$, the $P(x)$ and $Q(x)$ terms can be combined to yield another first-order linear equation. Therefore, we won't concern ourselves with $n < 2$.
Consider $x^4 - 3x^2 - 10 = 0$. Note the similarity of this equation to be solved to the quadratic equation $x^2 - 3x - 10 = 0$. By cleverly substituting $v = x^2$ we can obtain a quadratic equation in v.	Consider $y' + xy = xy^2$, a Bernoulli equation with $n = 2$. Note the similarity of this equation to a first-order linear equation. $y' + xy = xy^2$ equation to be solved $y' + P(x)y = Q(x)y^0$ first-order linear equation By cleverly substituting $v = y^{1-n} = y^{1-2} = y^{-1} = \frac{1}{y}$, $y \neq 0$, we can reduce this equation to the form of a first-order linear equation in v.

<table>
<tr>
<td>

$$v^2 - 3v - 10 = 0$$

This is a quadratic equation in v. Solving it will not solve the original equation for x, but will be a major step toward that goal. This is a quadratic equation in v.

</td>
<td>

$$v = \frac{1}{y} \text{ so } y = \frac{1}{v} = v^{-1} \text{ making } y' = -v^{-2}\frac{dv}{dx}$$

Now convert the original equation to a linear equation in v by substituting the v terms above into that equation.

$$y' + xy = xy^2 \quad \text{Bernoulli equation to be solved}$$

$$-v^{-2}\frac{dv}{dx} + xv^{-1} = xv^{-2} \quad \text{divide both sides by } -v^{-2}$$

$$\frac{-v^{-2}\frac{dv}{dx} + xv^{-1}}{-v^{-2}} = \frac{xv^{-2}}{-v^{-2}}$$

$$\frac{dv}{dx} + (-xv^{-1-(-2)}) = -x$$

$$\frac{dv}{dx} + (-xv^{-1+2}) = -x$$

$$\frac{dv}{dx} + (-xv^{1}) = -xv^{0}$$

This is a first-order linear equation in v and can be solved using the techniques in Chapter 8.

</td>
</tr>
<tr>
<td>

Solve by factoring.

$$v^2 - 3v - 10 = 0$$

$$(v-5)(v+2) = 0$$

$$v = 5, -2$$

Having solved for v, we must now back substitue for x.

</td>
<td>

$$\frac{dv}{dx} + (-xv^{1}) = -xv^{0} \quad \text{solve this equation for } v \text{ using an integrating factor}$$

$$I(x) = e^{\int -x\,dx} = e^{\frac{-x^2}{2}}$$

$$I(x)\left[\frac{dv}{dx} + \left(-xv^{1}\right)\right] = I(x)(-x)$$

$$e^{\frac{-x^2}{2}}\left[\frac{dv}{dx} + \left(-xv^{1}\right)\right] = e^{\frac{-x^2}{2}}(-x)$$

$$e^{\frac{-x^2}{2}}\frac{dv}{dx} + e^{\frac{-x^2}{2}}\left(-xv^{1}\right) = e^{\frac{-x^2}{2}}(-x)$$

$$\left[e^{\frac{-x^2}{2}} \times v\right]' = e^{\frac{-x^2}{2}}(-x) \quad \text{derivative of a product rule}$$

$$\int \left[e^{\frac{-x^2}{2}} \times v\right]' dx = \int \left(e^{\frac{-x^2}{2}}\right)(-x)\,dx \quad \text{let } u = \frac{-x^2}{2}, \text{ so this is } \int e^{u}\,du$$

$$e^{\frac{-x^2}{2}} \times v + c_1 = e^{\frac{-x^2}{2}} + c_2$$

$$e^{\frac{-x^2}{2}} \times v = e^{\frac{-x^2}{2}} + c_2 - c_1$$

$$e^{\frac{-x^2}{2}} \times v = e^{\frac{-x^2}{2}} + c_3$$

Solve for v

$$\frac{e^{\frac{-x^2}{2}} \times v}{e^{\frac{-x^2}{2}}} = \frac{e^{\frac{-x^2}{2}} + c_3}{e^{\frac{-x^2}{2}}}$$

$$v = 1 + c_3 e^{\frac{x^2}{2}}$$

Having solved for v, we must now back substitute to solve for y.

</td>
</tr>
<tr>
<td>

Since $v = x^2$,

$$5 = x^2 \rightarrow x = \pm\sqrt{5}$$

$$-2 = x^2 \rightarrow x = \pm i\sqrt{2}$$

</td>
<td>

We have previously stated that $v = \frac{1}{y}$, so

$$y = \frac{1}{v} = \frac{1}{1 + c_3 e^{\frac{x^2}{2}}}$$

</td>
</tr>
</table>

Notice the technique used here. For both problems (left column and right column), a substitution was made transforming the original equation into a form we were familiar with from previous work. Then, after that phase, the answer was obtained by back substitution into the original equation. The differential equation $y' + xy = xy^2$ has the generic solution

$$y = \frac{1}{1 + c_3 e^{\frac{x^2}{2}}}.$$

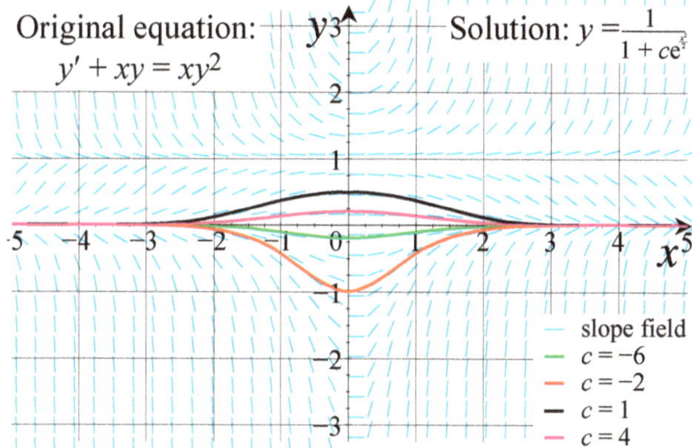

Particular solutions for arbitrary values of c_3 (-6, -2, 1, & 4) are shown at right. Note that, in this equation, $y \neq 0$ as noted on the first page of this chapter. We already know that $y = 0$ is a trivial solution to the original differential equation.

Bernoulli equations come about in projectile motion problems where the projectile is moving very fast. For a slow-moving object, the air resistance is approximately proportional to the velocity: $F_{\text{drag}} = -k_1 v$, but for a high-speed object like a bullet, the air resistance is more likely to be proportional to a higher power of the velocity or a combination of drag considerations, such as $F_{\text{drag}} = -k_1 v - k_2 v^2$. The $-k_2 v^2$ here is called the "quadratic turbulent forces term." At higher speeds, forces develop from the interaction of the projectile's "skin" (the material that is exposed to all the air friction) that are not there at low speeds. Here, the constants of proportionality above (k_1 and k_2) reflect physical qualities of the object's interaction with the medium it is moving in. It is easy to understand that the drag on a bullet passing through water would be greater than the drag on a bullet shot in the air. The Wikipedia chart at right shows that the drag of an object is impacted by the object's shape and size as well as its material.

Shape and flow	Form drag	Skin friction
	0%	100%
	~10%	~90%
	~90%	~10%
	100%	0%

Newton's second law states $F = ma$ or $F = m \times \frac{dv}{dt}$. Ignoring gravitational force (to achieve a more manageable beginner's differential equation) and applying the transitive property to F in the paragraph above, we get $m \times \frac{dv}{dt} = F = -k_1 v - k_2 v^2$.

$$m \times \frac{dv}{dt} = -k_1 v - k_2 v^2 \quad \text{transitive property } F_{\text{drag}} = -k_1 v - k_2 v^2,\ F = m \times \frac{dv}{dt}$$

$$m \times \frac{dv}{dt} + k_1 v = -k_2 v^2 \quad \text{isolating the } v^2 \text{ term}$$

$$\frac{dv}{dt} + \frac{k_1}{m} \times v = \frac{-k_2}{m} v^2 \quad \text{dividing by } m \text{ to make the leading coefficient 1} \qquad (1)$$

$$y' + P(x) \times y = Q(x) y^n \quad \text{standard form of the Bernoulli equation for comparison}$$

$$\frac{dv}{dt} + \frac{k_1}{m} \times v = \frac{-k_2}{m} v^2 \quad \text{Yes! We recognize a Bernoulli equation.}$$

Let $z = v^{1-n}$ so $z = v^{1-2} = v^{-1}$, so $v = \frac{1}{z} = z^{-1}$. Therefore, $\frac{dz}{dt} = -v^{-2} \frac{dv}{dt}$, so $\frac{dv}{dt} = -z^{-2} \frac{dz}{dt}$.

$$-z^{-2} \frac{dz}{dt} + \frac{k_1}{m} \times z^{-1} = \frac{-k_2}{m} \times z^{-2} \quad \text{substitute into (1) for } v,\ v^2 \text{ and } \frac{dv}{dt}$$

$$-z^2 \left[-z^{-2} \frac{dz}{dt} + \frac{k_1}{m} \times z^{-1} \right] = \left[\frac{-k_2}{m} \times z^{-2} \right] \times (-z^2) \quad \text{eliminate some } z \text{ terms by multiplying by } -z^2$$

$$\left[-z^2 \times -z^{-2} \frac{dz}{dt} \right] + \left[-z^2 \times \frac{k_1}{m} \times z^{-1} \right] = \frac{-k_2}{m} \times z^{-2} \times \left(-z^2 \right)$$

$$z^0 \frac{dz}{dt} - \frac{k_1}{m} z^1 = -\frac{k_2}{m} \times (-1) z^0$$

$$\frac{dz}{dt} = \frac{k_1}{m} z + \frac{k_2}{m} \quad \text{It is not clear, but this equation is a separable equation!}$$

$$\frac{dz}{dt} = \frac{k_1 z + k_2}{m} \quad \text{prepare to separate}$$

Continuing from the previous page.

$$\frac{dz}{dt} = (k_1 z + k_2)\left(\frac{1}{m}\right) \quad \text{ditto}$$

$$(k_1 z + k_2)^{-1}\frac{dz}{dt} = \frac{1}{m} \quad \text{ditto}$$

$$(k_1 z + k_2)^{-1}\,dz = \frac{1}{m}\,dt \quad z \text{ variables all separated to the left away from } dt \text{ differential.}$$

$$\int (k_1 z + k_2)^{-1}\,dz = \frac{1}{m}\int dt \quad \text{integrate both sides}$$

$$\frac{1}{k_1}\int (k_1 z + k_2)^{-1}k_1\,dz = \frac{1}{m}t + c_1 \quad \text{preparing to integrate left side as integral of } \frac{1}{u}$$

$$\frac{1}{k_1}\ln(k_1 z + k_2) + c_2 = \frac{1}{m}t + c_1 \quad \int u^{-1}\,du = \ln u \text{ (assume } k_1 z + k_2 > 0)$$

$$\frac{1}{k_1}\ln(k_1 z + k_2) = \frac{1}{m}t + c_1 - c_2$$

$$\frac{1}{k_1}\ln(k_1 z + k_2) = \frac{1}{m}t + c_3 \quad \text{combine constants of integration}$$

$$\ln(k_1 z + k_2) = k_1\left(\frac{1}{m}t + c_3\right)$$

$$\ln(k_1 z + k_2) = \frac{k_1}{m}t + k_1 c_3$$

$$\ln(k_1 z + k_2) = \frac{k_1}{m}t + c_4 \quad \text{let } k_1 c_3 = c_4$$

$$e^{\ln(k_1 z + k_2)} = e^{\left(\frac{k_1}{m}t + c_4\right)} \quad a = b \text{ iff } e^a = e^b$$

$$k_1 z + k_2 = e^{\frac{k_1}{m}t} \times e^{c_4} \quad e^{\ln u} = u\text{, also } b^{(m+n)} = b^m \times b^n$$

$$k_1 z + k_2 = c_5 e^{\frac{k_1}{m}t} \quad \text{let } c_5 = e^{c_4}$$

$$k_1 z = c_5 e^{\frac{k_1}{m}t} - k_2$$

$$\frac{k_1 z}{k_1} = \frac{c_5 e^{\frac{k_1}{m}t} - k_2}{k_1} \quad \text{divide to solve for } z$$

$$z(t) = \frac{c_5 e^{\frac{k_1}{m}t}}{k_1} - \frac{k_2}{k_1}$$

$z = z(t)$. It's actually been that way from the start, but it would have been "in the way" during all that work.

$$z(t) = \frac{c_5}{k_1} \times e^{\frac{k_1}{m}t} - \frac{k_2}{k_1}$$

$$z(t) = c_6 \times e^{\frac{k_1}{m}t} - \frac{k_2}{k_1}$$

substitute $\frac{c_5}{k_1} = c_6$ to get the solution for z. We still need the solution for v!

$$v(t) = \frac{1}{c_6 \times e^{\frac{k_1}{m}t} - \frac{k_2}{k_1}} \quad z = \frac{1}{v}\text{, so } v = \frac{1}{z}\text{; \textbf{general form for } } v(t) \tag{2}$$

$$v_0 = \frac{1}{c_6 \times e^{\frac{k_1}{m}\times 0} - \frac{k_2}{k_1}} \quad \text{at } t = 0,\ v = v_0$$

$$v_0 = \frac{1}{c_6 \times e^0 - \frac{k_2}{k_1}}$$

$$v_0 = \frac{1}{c_6 - \frac{k_2}{k_1}}$$

$$v_0 = \frac{1}{\frac{c_6 k_1}{k_1} - \frac{k_2}{k_1}}$$

Continuing from the previous page.

$$v_0 = \frac{1}{\frac{c_6 k_1 - k_2}{k_1}}$$

$$v_0 = \frac{k_1}{c_6 k_1 - k_2}$$

$$v_0(c_6 k_1 - k_2) = k_1$$

$$v_0 c_6 k_1 - v_0 k_2 = k_1$$

$$v_0 c_6 k_1 = k_1 + v_0 k_2$$

$$c_6 = \frac{k_1 + v_0 k_2}{v_0 k_1}$$

$$c_6 = \frac{1}{v_0} + \frac{k_2}{k_1}$$

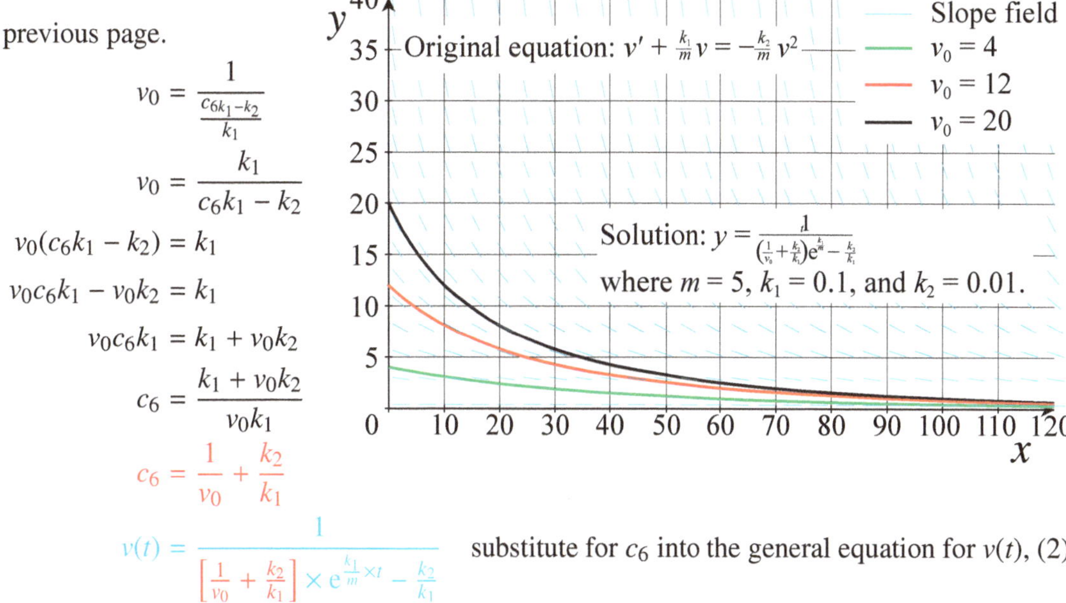

$$v(t) = \frac{1}{\left[\frac{1}{v_0} + \frac{k_2}{k_1}\right] \times e^{\frac{k_1}{m} \times t} - \frac{k_2}{k_1}}$$ substitute for c_6 into the general equation for $v(t)$, (2)

Generic solutions to the differential equation $m \times \frac{dv}{dt} = -k_1 v - k_2 v^2$ are shown above right as well as particular solutions for arbitrarily chosen parameters.

Chapter 10 Review

	Ordinary differential equations			
	Chapters 1–5	Chapter 6	Chapter 8	Chapter 10
	Separable	Logistic growth	Linear	Bernoulli
Prime notation	$y' = f(x) \times g(y)$	$P' = kP\left(1 - \frac{P}{K}\right)$	$y' + P(x) \times y = Q(x)$	$y' + P(x)y = Q(x)y^n$
Differential notation	$\frac{dy}{dx} = f(x) \times g(y)$	$\frac{dP}{dt} = kP\left(1 - \frac{P}{K}\right)$	$\frac{dy}{dx} + P(x) \times y = Q(x)$	$\frac{dy}{dx} + P(x)y = Q(x)y^n$
Integrated (algebraic) form	$y = Ce^{\ln x}$	$P(x) = \frac{K}{(Ae^{-kt}+1)}$, where $A = \frac{K-P_0}{P_0}$	$y = \frac{\int Q(x)e^{\int P(x)\,dx + C}\,dx}{e^{\int P(x)\,dx}}$	Let $v = y^{1-n}$. Solve for v. Back substitute for y.

Chapter 10 reviewed how substituting $v = x^2$ into the algebraic equation $x^4 - 3x^2 - 10 = 0$ resulted in the quadratic equation $v^2 - 3v - 10 = 0$, which could easily be solved for the *variable* v. The solution for v could then be used to solved for the original variable x. Similarly, the Bernoulli equation $y' + P(x)y = Q(x)y^n$ (or $\frac{dy}{dx} + P(x)y = Q(x)y^n$) could, by substituting $v = y^{1-n}$, be transformed to a linear equation in v which, after solving for v by the technique shown in Chapter 8, could then be used to solve for the original variable (y).

The following chart is a primitive and simplistic attempt to help beginning students (the target audience for this book) review, differentiate, and mentally organize their thoughts to this point in the book. This chart is simplistic because many equations can be solved by more than one technique.

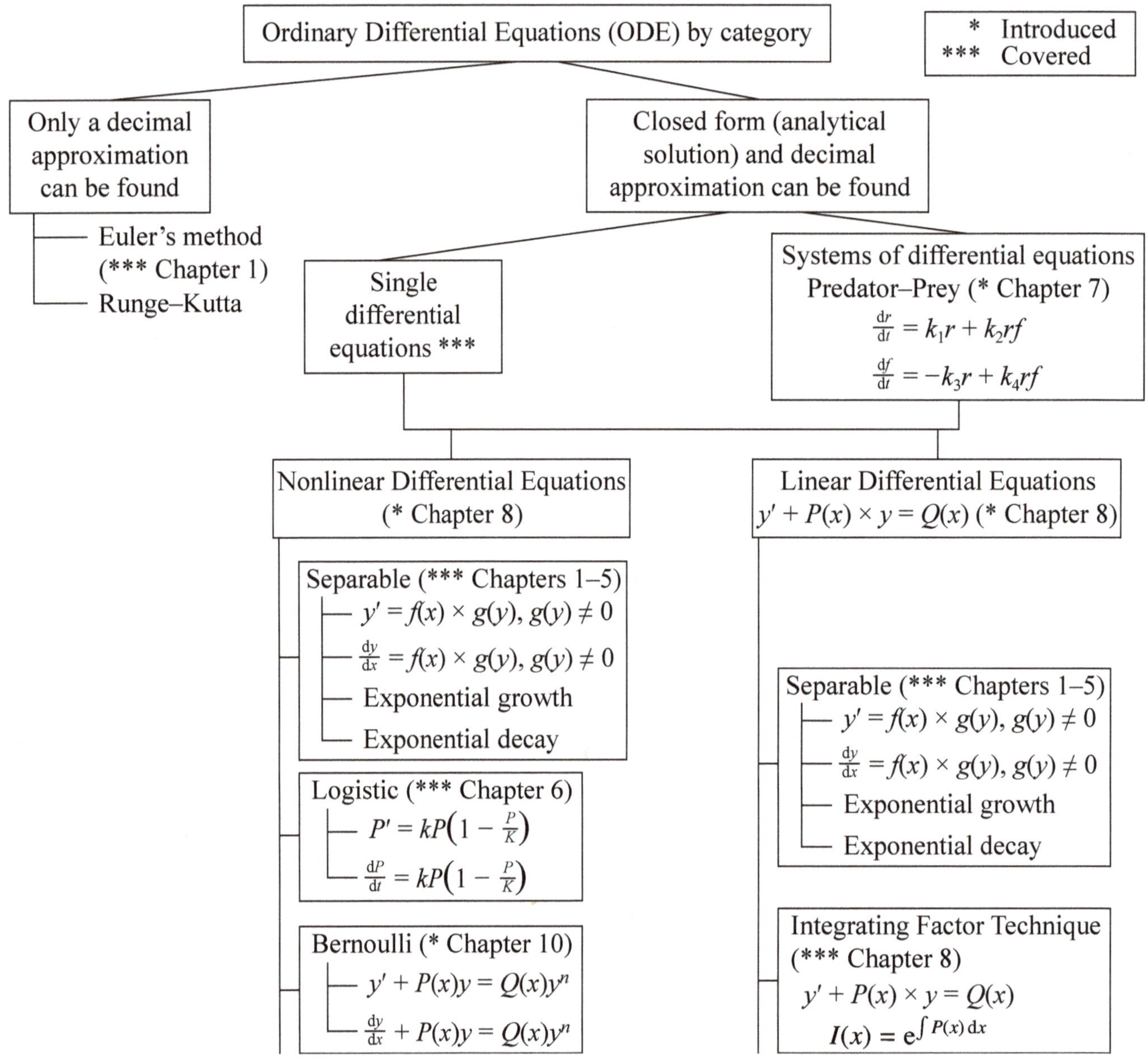

* Introduced
*** Covered

- Ordinary Differential Equations (ODE) by category
  - Only a decimal approximation can be found
    - Euler's method (*** Chapter 1)
    - Runge–Kutta
  - Closed form (analytical solution) and decimal approximation can be found
    - Single differential equations ***
    - Systems of differential equations Predator–Prey (* Chapter 7)
      $\frac{dr}{dt} = k_1 r + k_2 rf$
      $\frac{df}{dt} = -k_3 r + k_4 rf$
      - Nonlinear Differential Equations (* Chapter 8)
        - Separable (*** Chapters 1–5)
          - $y' = f(x) \times g(y),\ g(y) \neq 0$
          - $\frac{dy}{dx} = f(x) \times g(y),\ g(y) \neq 0$
          - Exponential growth
          - Exponential decay
        - Logistic (*** Chapter 6)
          - $P' = kP\left(1 - \frac{P}{K}\right)$
          - $\frac{dP}{dt} = kP\left(1 - \frac{P}{K}\right)$
        - Bernoulli (* Chapter 10)
          - $y' + P(x)y = Q(x)y^n$
          - $\frac{dy}{dx} + P(x)y = Q(x)y^n$
      - Linear Differential Equations $y' + P(x) \times y = Q(x)$ (* Chapter 8)
        - Separable (*** Chapters 1–5)
          - $y' = f(x) \times g(y),\ g(y) \neq 0$
          - $\frac{dy}{dx} = f(x) \times g(y),\ g(y) \neq 0$
          - Exponential growth
          - Exponential decay
        - Integrating Factor Technique (*** Chapter 8)
          $y' + P(x) \times y = Q(x)$
          $I(x) = e^{\int P(x)\,dx}$

Bernoulli's Differential Equation

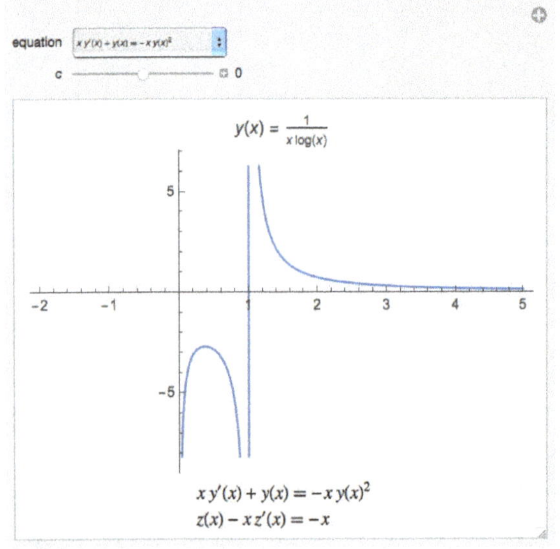

Online Application

Visit demonstrations.wolfram.com/BernoullisDifferentialEquation for a great little application that will demonstrate some of the concepts from this chapter. "Bernoulli's Differential Equation" from the Wolfram Demonstrations Project. Contributed by Izidor Hafner.

Chapter 11

Review of Partial Derivatives and Integrals: Getting Ready for Exact Differential Equations

Exact differential equations, another category of solvable differential equations, are also traditionally taught in a beginning differential-equations class. If a differential equation can be put into the *exact differential* form, and if certain conditions hold, then a time-tested technique can be applied to solving that differential equation by converting it to an integrated form.

The technique of solving an *exact differential equation* presumes that the student is familiar with *partial derivatives*. Finding a partial derivative is a skill and topic taught in differential calculus. However, based on the premise that some students never understood what was going on at that time or the possibility that some students may have forgotten about *partial derivatives,* this book takes a few pages to go back over that topic before taking up the topic of *exact differential equations*. If you understand *partial derivatives,* then you should skip on ahead to the next chapter.

A *partial derivative* is a type of derivative. To distinguish between a derivative and a partial derivative, we use different symbols. The symbol $\frac{dy}{dx}$ is used to indicate a derivative in two dimensions, whereas the symbol $\frac{\partial y}{\partial x}$ is used to indicate a partial derivative—a derivative of the two-dimensional cross section of a function in three or more dimensions.

In the graph of the three-dimensional surface at right, you will see the point (0, 0, 9) exaggerated for instructional purposes. Finding a derivative in differential calculus meant finding the slope of a tangent to a two-dimensional curve at a specific point. In three-dimensional space, we can talk about a tangent plane to a three-dimensional surface at a specific point but not, in general, about a tangent line to that point as there would be an infinity of such lines.

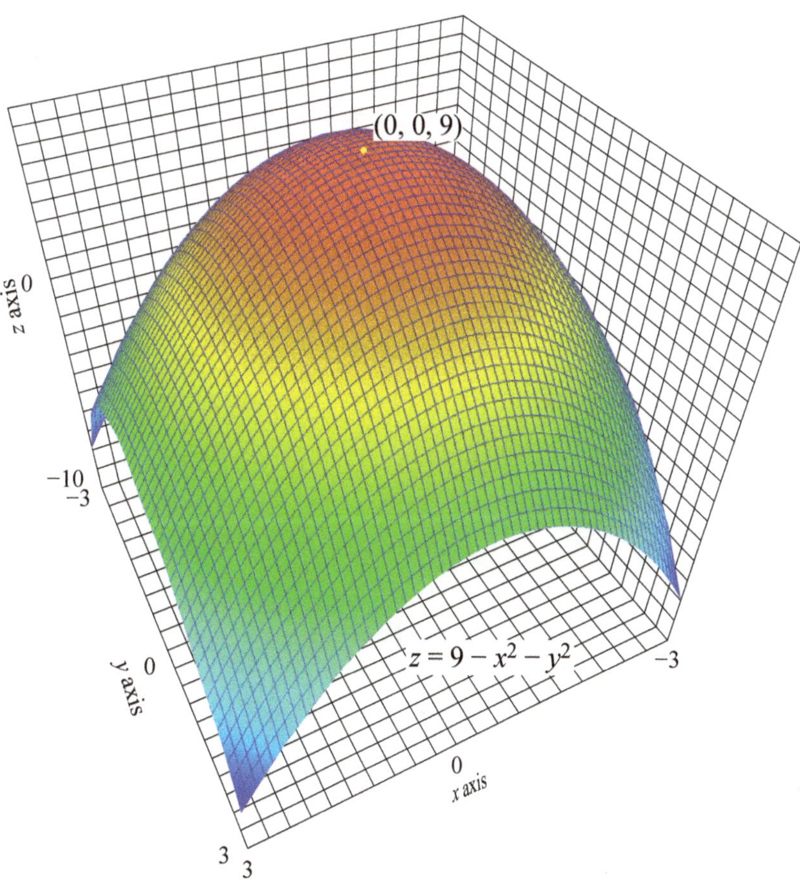

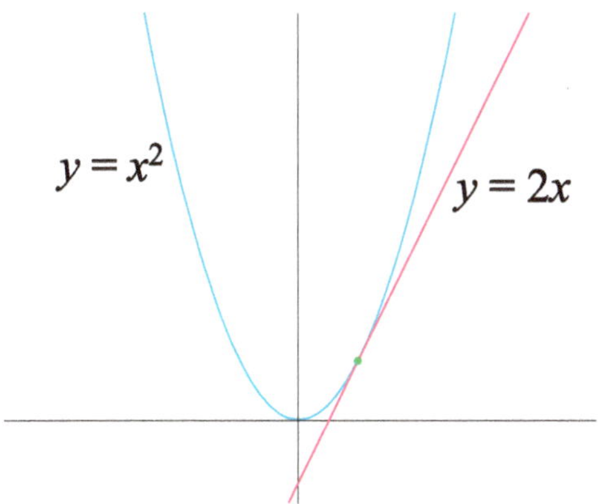

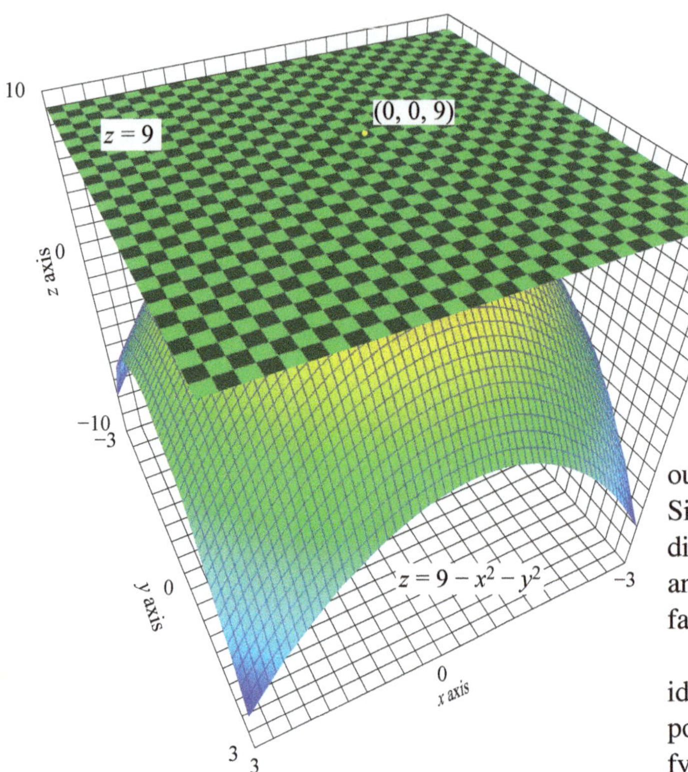

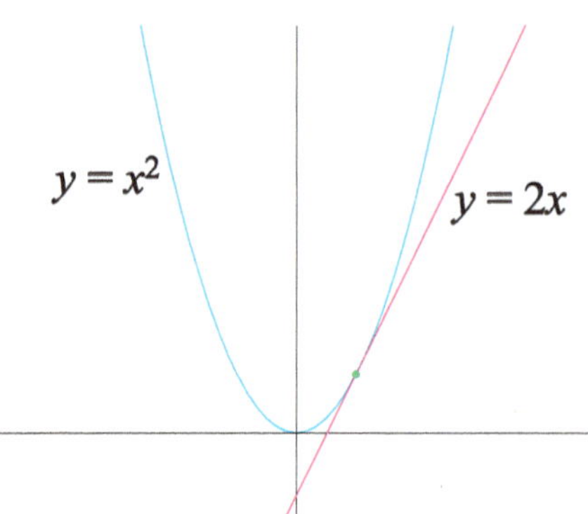

There can be only one line tangent to a continuous curve at a specified point in two-dimensional space. Similarly, there is only one tangent plane to a three-dimensional surface at a specific point. However, there is an infinity of lines tangent to the three-dimensional surface at a specific point.

You may recall from Chapter 1 that we discussed the idea that, out of an infinite family of functions, it was possible to select or indicate a specific function by specifying a point that lay on the curve. Keep that idea in mind because it is similar to what happens when taking a partial derivative. A partial derivative is a directional derivative specially chosen from all the infinite linear derivatives (slopes) that could be drawn to a three-dimensional surface at a specific point.

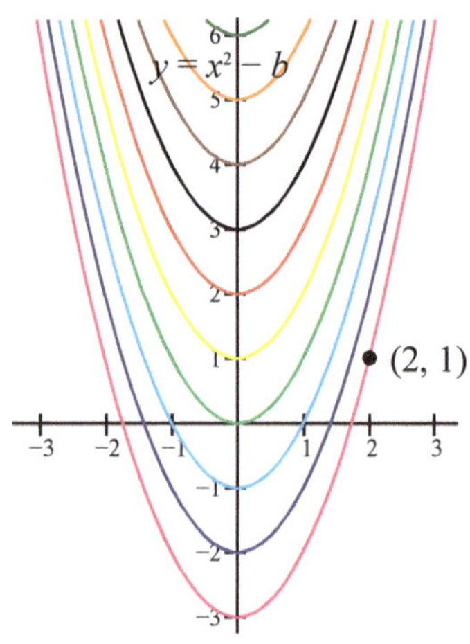

An infinity of directional derivatives can be drawn to each of the points on the curved surface shown below. However, of those infinite directional derivatives, some are very special and need to be discussed. Notice the white parabola near the middle of the three-dimensional surface below. That white parabola is drawn completely in a plane parallel to the y–z plane. That means the x value of every point on the white parabola is the same, or is *fixed* or constant. The parabola is concave down and is described by the function $f(y) = -500y^2$. The slope of every tangent to the three-dimensional surface in this plane is $f'(y) = -1{,}000y$. The white sine curve shown below is drawn completely in a plane parallel to the x–z plane. That means the y value of every point on the white sine curve is the same or is *fixed* or constant. The sine curve can be described by $f(x) = 2{,}000\sin(x)$. The slope of every tangent to the three-dimensional surface in this plane would be $f'(x) = 2{,}000\cos(x)$. When you take a partial derivative of a function dependent on two variables, you are just trying to get the slope of a tangent to a y–z, x–z, or x–y cross section of the three-dimensional function. For the three-dimensional surface shown at left, $f(x, y) = 2{,}000\sin(x) - 500y^2$, the partial derivative $\frac{\partial f}{\partial x}$ is $2{,}000\cos(x)$ (the y is treated as a constant), while the partial derivative $\frac{\partial f}{\partial y}$ is $-1{,}000y$ (the x is treated as a constant).

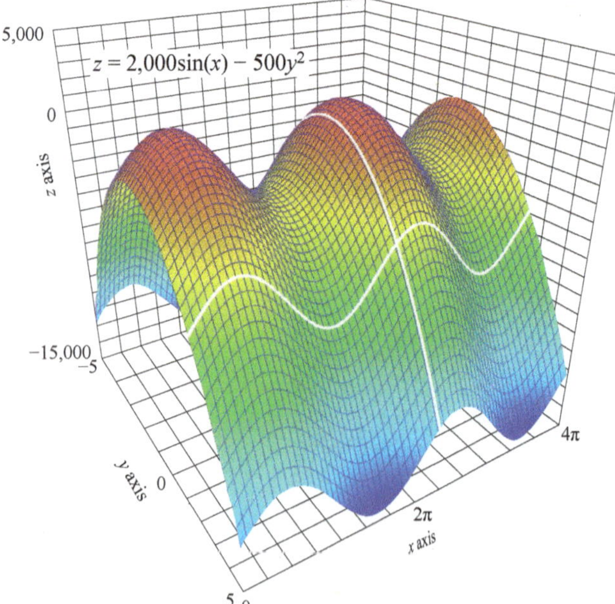

Without a visual to help, one can think of partial derivatives as taking the derivative of a "part" of the original equation in two variables—$z = f(x, y)$. This is done by treating a variable in the fixed plane as a constant.

For $f(x, y) = 2{,}000\sin(x) - 500y^2$,

$$\frac{\partial}{\partial x}f(x,y) = \frac{\partial}{\partial x}[2{,}000\sin(x)] - \frac{\partial}{\partial x}\left[500y^2\right] \quad \text{differentiating parallel to the } x\text{-axis, } y \text{ stays constant}$$
$$= 2{,}000\cos(x) - 0 = 2{,}000\cos(x)$$

and

$$\frac{\partial}{\partial y}f(x,y) = \frac{\partial}{\partial y}[2{,}000\sin(x)] - \frac{\partial}{\partial y}\left[500y^2\right] \quad \text{differentiating parallel to the } y\text{-axis, } x \text{ stays constant}$$
$$= 0 - 1{,}000y = -1{,}000y$$

Summarizing, for $f(x, y) = 2{,}000\sin(x) - 500y^2$, $\frac{\partial f}{\partial x} = 2{,}000\cos(x)$, while $\frac{\partial f}{\partial y} = -1{,}000y$.

Drill and practice on taking partial derivatives

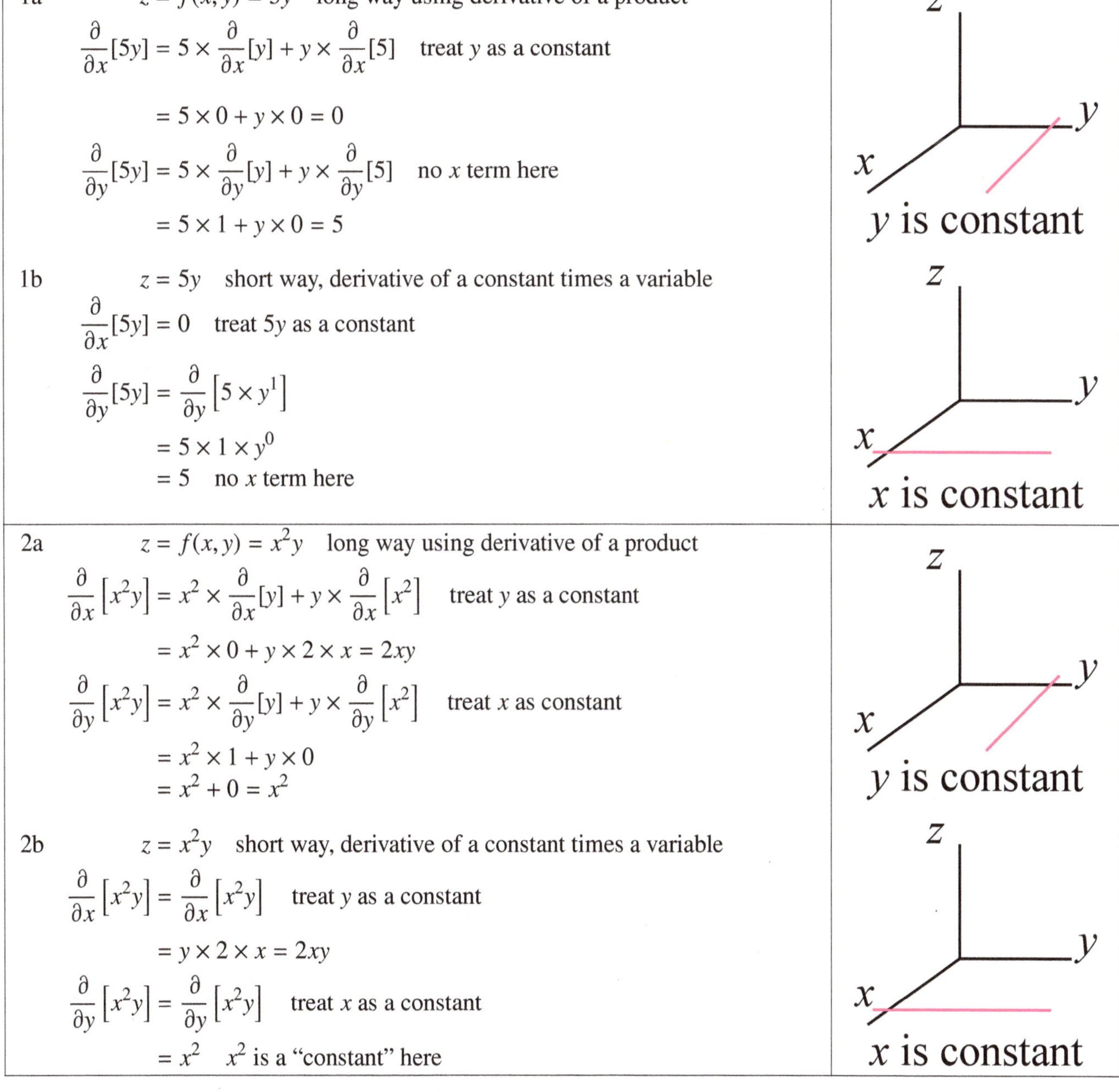

1a $z = f(x, y) = 5y$ long way using derivative of a product

$$\frac{\partial}{\partial x}[5y] = 5 \times \frac{\partial}{\partial x}[y] + y \times \frac{\partial}{\partial x}[5] \quad \text{treat } y \text{ as a constant}$$
$$= 5 \times 0 + y \times 0 = 0$$
$$\frac{\partial}{\partial y}[5y] = 5 \times \frac{\partial}{\partial y}[y] + y \times \frac{\partial}{\partial y}[5] \quad \text{no } x \text{ term here}$$
$$= 5 \times 1 + y \times 0 = 5$$

1b $z = 5y$ short way, derivative of a constant times a variable

$$\frac{\partial}{\partial x}[5y] = 0 \quad \text{treat } 5y \text{ as a constant}$$
$$\frac{\partial}{\partial y}[5y] = \frac{\partial}{\partial y}\left[5 \times y^1\right]$$
$$= 5 \times 1 \times y^0$$
$$= 5 \quad \text{no } x \text{ term here}$$

2a $z = f(x, y) = x^2y$ long way using derivative of a product

$$\frac{\partial}{\partial x}\left[x^2y\right] = x^2 \times \frac{\partial}{\partial x}[y] + y \times \frac{\partial}{\partial x}\left[x^2\right] \quad \text{treat } y \text{ as a constant}$$
$$= x^2 \times 0 + y \times 2 \times x = 2xy$$
$$\frac{\partial}{\partial y}\left[x^2y\right] = x^2 \times \frac{\partial}{\partial y}[y] + y \times \frac{\partial}{\partial y}\left[x^2\right] \quad \text{treat } x \text{ as constant}$$
$$= x^2 \times 1 + y \times 0$$
$$= x^2 + 0 = x^2$$

2b $z = x^2y$ short way, derivative of a constant times a variable

$$\frac{\partial}{\partial x}\left[x^2y\right] = \frac{\partial}{\partial x}\left[x^2y\right] \quad \text{treat } y \text{ as a constant}$$
$$= y \times 2 \times x = 2xy$$
$$\frac{\partial}{\partial y}\left[x^2y\right] = \frac{\partial}{\partial y}\left[x^2y\right] \quad \text{treat } x \text{ as a constant}$$
$$= x^2 \quad x^2 \text{ is a “constant” here}$$

3 $$z = e^{xy}$$ $$\frac{\partial}{\partial x}[e^{xy}] = e^{xy} \times \frac{\partial}{\partial x}[xy] = e^{xy} \times \frac{\partial}{\partial x}[yx] \quad \text{treat } y \text{ as constant}$$ $$= e^{xy} \times yx^0$$ $$= ye^{xy}$$ $$\frac{\partial}{\partial y}[e^{xy}] = e^{xy} \times \frac{\partial}{\partial y}[xy] \quad \text{treat } x \text{ as constant}$$ $$= e^{xy} \times xy^0$$ $$= xe^{xy}$$	z y x y is constant z y x x is constant
4 $$z = 3x^2e^{2y}$$ $$\frac{\partial}{\partial x}\left[3x^2e^{2y}\right] = 3e^{2y} \times \frac{\partial}{\partial x}\left(x^2\right) \quad \text{treat } 3e^{2y} \text{ as constant}$$ $$= 3e^{2y} \times 2x$$ $$= 6xe^{2y}$$ $$\frac{\partial}{\partial y}\left[3x^2e^{2y}\right] = 3x^2 \times e^{2y} \times \frac{\partial}{\partial y}[2y] \quad \text{treat } 3x^2 \text{ as constant}$$ $$= 3x^2e^{2y} \times 2$$ $$= 6x^2e^{2y}$$	z y x y is constant z y x x is constant
5 $$z = e^x \sin y$$ $$\frac{\partial}{\partial x}[e^x \sin y] = \frac{\partial}{\partial x}[\sin y \times e^x] \quad \text{treat } \sin y \text{ as constant}$$ $$= \sin y \times e^x \times \frac{\partial}{\partial x}[x]$$ $$= \sin y \times e^x \times 1$$ $$= e^x \sin y$$ $$\frac{\partial}{\partial y}[e^x \sin y] = e^x \times \cos(y)\frac{\partial}{\partial y}[y] \quad \text{treat } e^x \text{ as constant}$$ $$= e^x \times \cos(y) \times 1$$ $$= e^x \cos(y)$$	z y x y is constant z y x x is constant
6 $$z = \sqrt{x+y}$$ $$\frac{\partial}{\partial x}\left[(x+y)^{\frac{1}{2}}\right] = \tfrac{1}{2}(x+y)^{-\frac{1}{2}} \times \frac{\partial}{\partial x}[x+y]$$ $$= \tfrac{1}{2}(x+y)^{-\frac{1}{2}} \times \left(\frac{\partial}{\partial x}[x] + \frac{\partial}{\partial x}[y]\right) \quad \text{treat } y \text{ as constant}$$ $$= \tfrac{1}{2}(x+y)^{-\frac{1}{2}} \times (1+0)$$ $$= \tfrac{1}{2}(x+y)^{-\frac{1}{2}}$$ $$\frac{\partial}{\partial y}\left[(x+y)^{\frac{1}{2}}\right] = \tfrac{1}{2}(x+y)^{-\frac{1}{2}} \times \frac{\partial}{\partial y}[x+y]$$ $$= \tfrac{1}{2}(x+y)^{-\frac{1}{2}} \times \left(\frac{\partial}{\partial y}[x] + \frac{\partial}{\partial y}[y]\right) \quad \text{treat } x \text{ as constant}$$ $$= \tfrac{1}{2}(x+y)^{-\frac{1}{2}} \times (0+1)$$ $$= \tfrac{1}{2}(x+y)^{-\frac{1}{2}}$$	z y x y is constant z y x x is constant

7	$z = e^{3x} + 5e^{3x}y$ $\frac{\partial}{\partial x}\left[e^{3x} + 5e^{3x}y\right] = \frac{\partial}{\partial x}\left[e^{3x}\right] + \frac{\partial}{\partial x}\left[5e^{3x}y\right]$ $= e^{3x} \times \frac{\partial}{\partial x}[3x] + \frac{\partial}{\partial x}\left[5ye^{3x}\right]$ 5y is constant $= 3e^{3x} + 5y\frac{\partial}{\partial x}\left[e^{3x}\right]$ $= 3e^{3x} + 5ye^{3x}\frac{\partial}{\partial x}[3x]$ $= 3e^{3x} + 5ye^{3x} \times 3$ $= 3e^{3x} + 15ye^{3x}$ $\frac{\partial}{\partial y}\left[e^{3x} + 5e^{3x}y\right] = \frac{\partial}{\partial y}\left[e^{3x}\right] + \frac{\partial}{\partial y}\left[5e^{3x}y\right]$ e^{3x} is constant $= 0 + \frac{\partial}{\partial y}\left[5e^{3x}y\right]$ 5e^{3x} is constant $= 0 + 5e^{3x} \times 1$ $= 5e^{3x}$	z y x y is constant z y x x is constant
8	$z = x^2y + 5x^3 - \sqrt{x + 2y}$ $\frac{\partial}{\partial x}\left[x^2y + 5x^3 - \sqrt{x+2y}\right] = \frac{\partial}{\partial x}\left[x^2y\right] + \frac{\partial}{\partial x}\left[5x^3\right] - \frac{\partial}{\partial x}\left[\sqrt{x+2y}\right]$ y constant $= \frac{\partial}{\partial x}\left[yx^2\right] + \frac{\partial}{\partial x}\left[5x^3\right] - \frac{\partial}{\partial x}\left[(x+2y)^{\frac{1}{2}}\right]$ $= y \times 2x + 15x^2 - \frac{1}{2}(x+2y)^{-\frac{1}{2}} \times \frac{\partial}{\partial x}[x + 2y]$ $= 2xy + 15x^2 - \frac{1}{2}(x+2y)^{-\frac{1}{2}} \times 1$ $= 2xy + 15x^2 - \frac{1}{2}(x+2y)^{-\frac{1}{2}}$ $\frac{\partial}{\partial y}\left[x^2y + 5x^3 - \sqrt{x+2y}\right] = \frac{\partial}{\partial y}\left[x^2y\right] + \frac{\partial}{\partial y}\left[5x^3\right] - \frac{\partial}{\partial y}\left[\sqrt{x+2y}\right]$ x constant $= x^2(1) + 0 - \frac{\partial}{\partial y}\left[(x+2y)^{\frac{1}{2}}\right]$ $= x^2 - \frac{1}{2}(x+2y)^{-\frac{1}{2}}\frac{\partial}{\partial y}[x + 2y]$ $= x^2 - \frac{1}{2}(x+2y)^{-\frac{1}{2}} \times 2$ $= x^2 - (x+2y)^{-\frac{1}{2}}$	z y x y is constant z y x x is constant

Since we have been discussing the idea of "partial derivatives," it would be good to touch on the idea of "partial integration" before leaving this chapter. We have already discussed that there are infinite tangents to a 3D graph so that when differentiating one must specify, somehow, which tangent is desired. That is most often done by specifying the tangent to the surface that is parallel to the x-axis, the y-axis, or the z-axis. Finding the area under a surface curve in three space is not a problem for students who have taken multivariable calculus . . . simply do a double integral. However, finding the area under a cross-sectional curve gives us the same problem as finding the tangent to a surface curve. Which cross-sectional curve is to be used? Again, as with the tangent/derivative discussion, the cross-sectional curve to be integrated will be parallel to the x-axis, y-axis, or z-axis.

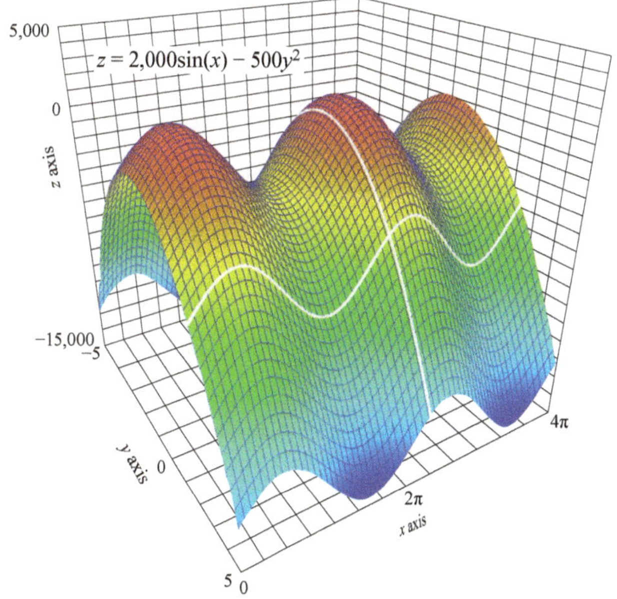

Following are two examples of partial integration.

Partial integration with respect to y	Partial integration with respect to x
$\int (2x + y^2)\,dy = \int 2x\,dy + \int y^2\,dy$	$\int (2x + y^2)\,dx = \int 2x\,dx + \int y^2\,dx$
Since x does not change, treat it as a constant.	Since y does not change, treat it as a constant.
$2x \int dy + \frac{y^3}{3} + f_1(x) \overset{?}{=}$ What is $f_1(x)$??? $2xy + f_2(x) + \frac{y^3}{3} + f_1(x) \overset{?}{=}$ What is $f_2(x)$??? $2xy + \frac{y^3}{3} + f(x) \overset{?}{=}$ Apparently, $f_2(x) + f_1(x) = f(x)$???	$2\int x\,dx + y^2 \int dx = \frac{2x^2}{2} + xy^2 + g(y)$ $= x^2 + xy^2 + g(y)$??? What is $g(y)$???

Traditionally, when we integrate, we indicate that there is more than one possible solution by adding in a "constant of integration."

$$\frac{dy}{dx} = 2x$$

$$dy = 2x\,dx$$

$$\int dy = \int 2x\,dx$$

$$y + c_1 = 2\int x\,dx$$

$$y + c_1 = 2\frac{x^2}{2} + c_2$$

$$y = x^2 + c_2 - c_1$$

$$y = x^2 + c$$

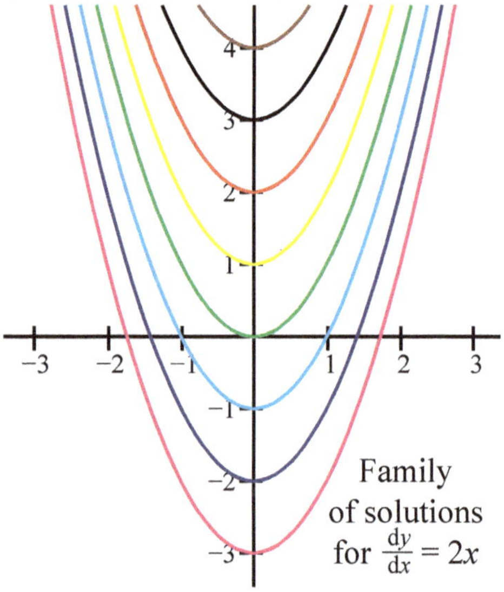

Family of solutions for $\frac{dy}{dx} = 2x$

Here, the constant of integration indicates that there is more than one function of the form $y = x^2$ whose derivative is $2x$. When doing partial integration, we get those curious "functions of integration." What is the significance of those "functions of integration"?

The last page or so of Chapter 2 discussed that while integration in differential calculus meant to find the area under a curve, it is better in differential equations to think of integration as "undoing" a derivative.

$$\frac{\partial}{\partial y}\left(2x + y^2\right) = \frac{\partial}{\partial y}(2x) + \frac{\partial}{\partial y}\left(y^2\right) = 0 + 2y = 2y$$

Reversing that process (undoing the partial derivative)

$$\int 2y\,dy = 2\int y\,dy = 2\int y^1\,dy = 2\frac{y^2}{2} = y^2??? \quad \text{Wrong!!! The answer should be } 2x + y^2!!!$$

In partial integrations such as this, we indicate that $\int 2y\,dy = y^2 + g(x)$. This answer indicates that the original function could have been $3x + y^2$, $4x + y^2$, $5x + y^2$, $x^2 + y^2$, $5x^3 + y^2$, etc., etc.

See this for yourself by partially differentiating and then partially integrating any of the examples shown above.

$$\int \frac{\partial}{\partial y}\left(5x^3 + y^2\right)dy = \int\left[\frac{\partial}{\partial y}\left(5x^3\right) + \frac{\partial}{\partial y}\left(y^2\right)\right]dy = \int[0+2y]dy = \int 2y\,dy = y^2 + f(x), \text{ where, in this case, } f(x) = 5x^3$$

Since the function $f(x)$ above is analogous to the constant of integration when doing full integration in two space, it could be called a "function of integration" when doing partial integration in three space. **Had the original partial derivative been $\frac{\partial}{\partial x}(5x^3 + y^2) = \frac{\partial}{\partial x}(5x^3) + \frac{\partial}{\partial x}(y^2) = 15x^2 + 0 = 15x^2$, then the partial integration would have been $\int 15x^2\,dx = 15\int x^2\,dx = 15\frac{x^3}{3} = 5x^3 + g(y)$.**

The graph at right demonstrates five functions, three explicit and two implicit.

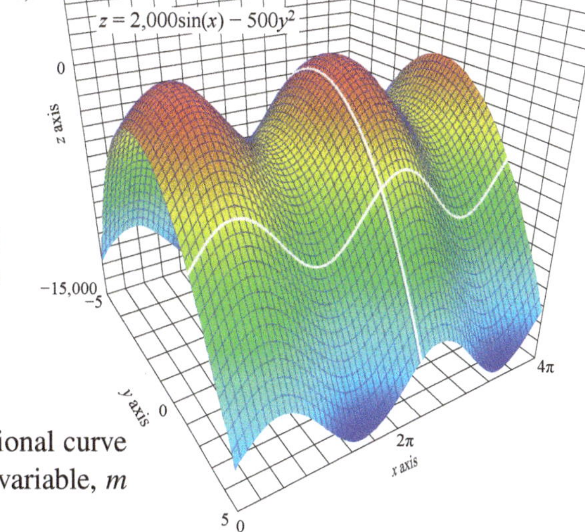

1. $z = 2{,}000\sin(x) - 500y^2$: This function is the 3D surface shown at right. It has one dependent variable (z) and two independent variables (x and y).

2. For $z = 2{,}000\sin(x) - 500y^2$,

$$\frac{\partial z}{\partial x} = \frac{\partial}{\partial x}[2{,}000\sin(x)] - \frac{\partial}{\partial x}\left[500y^2\right]$$ differentiate parallel to the x-axis, y stays constant

$$= 2{,}000\cos(x) - 0$$

$$= 2{,}000\cos(x)$$

This function is the slope of the tangent to any cross-sectional curve above that is parallel to the x-axis. It has one dependent variable, m (z', the slope), and one independent variable, x.

3. For $z = 2{,}000\sin(x) - 500y^2$,

$$\frac{\partial z}{\partial y} = \frac{\partial}{\partial y}[2{,}000\sin(x)] - \frac{\partial}{\partial y}\left[500y^2\right]$$ differentiate parallel to the y-axis, x stays constant

$$= 0 - 1{,}000y$$

$$= -1{,}000y$$

This function is the slope of the tangent to any cross-sectional curve above that is parallel to the y-axis. It has one dependent variable, m (z', the slope), and one independent variable, y.

4. $$\int \frac{\partial z}{\partial x}\,\mathrm{d}x = \int 2{,}000\cos(x)\,\mathrm{d}x$$

$$= 2{,}000\int \cos(x)\,\mathrm{d}x$$

$$= 2{,}000\sin(x) + g(y)$$ $g(y)$ indicates a specific x–z planar cross section

Here, $g(y)$ indicates which sine curve from the family of sine curves. This function is the cross section of the three dimensional surface above that is parallel to the x–z plane. It has one dependent variable, z, and one independent variable, x.

5. $$\int \frac{\partial z}{\partial y}\,\mathrm{d}y = \int (-1{,}000y)\,\mathrm{d}y$$

$$= -1{,}000\int y\,\mathrm{d}y$$

$$= -1{,}000\frac{y^2}{2} + f(x)$$

$$= -500y^2 + f(x)$$ $f(x)$ indicates a specific y–z planar cross section

Here, $f(x)$ indicates which concave down parabola from the family of concave down parabolas. This function is the cross section of the 3D surface above that is parallel to the y–z plane. It has one dependent variable, z, and one independent variable, y.

Notice that by summing the two integrals $[2{,}000\sin(x) + g(y)]$ and $\left[-500y^2 + f(x)\right]$ with $f(x) = 0$ and $g(y) = 0$, one gets $2{,}000\sin(x) - 500y^2$, the original surface function $f(x, y) = 2{,}000\sin(x) - 500y^2$. In other words, at the beginning of this chapter, we took the 3D function $z = 2{,}000\sin(x) - 500y^2$ and, by taking partial derivatives, obtained two functions: $2{,}000\cos(x)$ and $-1{,}000y$. Now we are taking the integrals of those functions, combining them and getting the original 3D function. **This example strongly suggests that, given two partial derivatives meeting certain preconditions, it may be possible to reconstruct the 3D function from which the partial derivatives were derived by summing their integrals. Chapter 12 shows that this will not always be the case.**

Chapter 11 Review

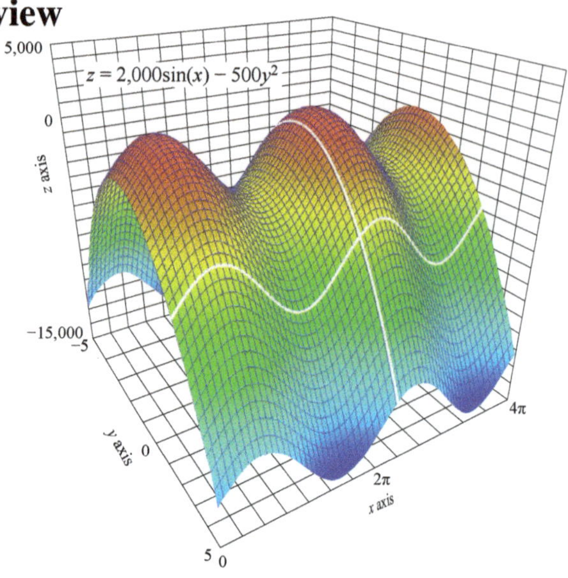

The skills of taking both partial derivatives and partial integrals, being necessary to solving "exact differential equations," were reviewed at length prior to their application in Chapter 12. In brief, taking a partial derivative can be thought of as finding the value of a "directional derivative" in three space, a derivative of a 3D surface restricted to a planar (x, y, or z) cross section. Similarly, finding a partial integral can be thought of as finding the area under a cross section of the surface that is perpendicular to one of the three axes.

It makes no sense to find the slope of a tangent to the surface of a three-dimensional surface as there are infinitely many of such tangents. However, by restricting ourselves to tangents that are parallel to either the x-axis, y-axis, or z-axis, we can find tangents to such a surface.

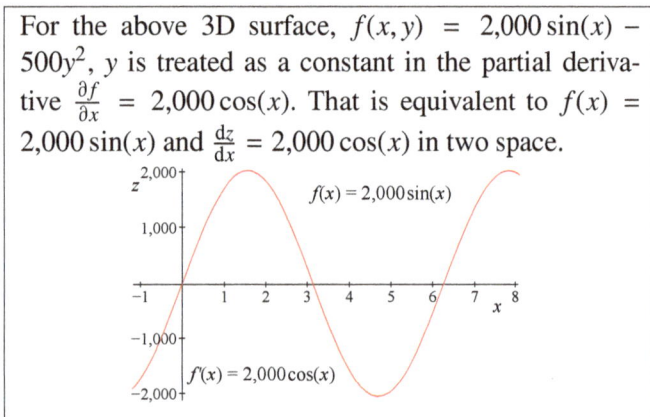

For the above 3D surface, $f(x,y) = 2{,}000\sin(x) - 500y^2$, y is treated as a constant in the partial derivative $\frac{\partial f}{\partial x} = 2{,}000\cos(x)$. That is equivalent to $f(x) = 2{,}000\sin(x)$ and $\frac{dz}{dx} = 2{,}000\cos(x)$ in two space.

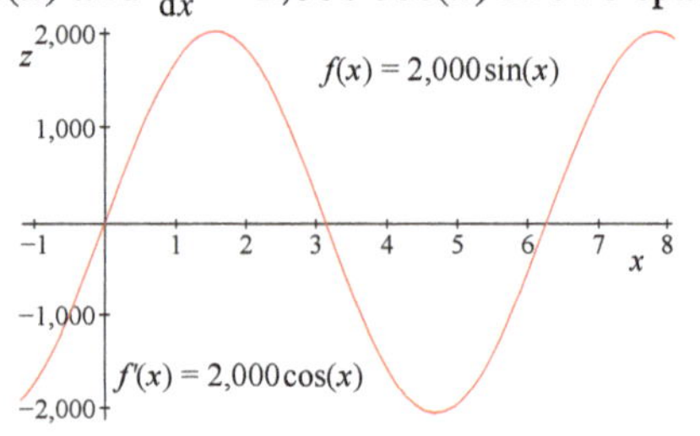

For the above 3D surface, $f(x,y) = 2{,}000\sin(x) - 500y^2$, x is treated as a constant in the partial derivative $\frac{\partial f}{\partial y} = -1{,}000cos(x)$. That would be equivalent to $f(x) = -500y^2$ and $\frac{dz}{dy} = -1{,}000y$ in two space.

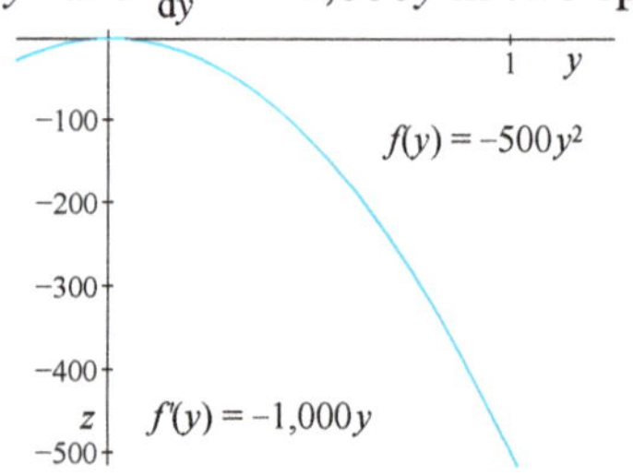

Then you are basically just finding the tangent to a curve in two space which is familiar stuff. This is done by finding partial derivatives. Finding the partial integral of those partial derivatives results in two-space functions. The example shown in Chapter 11 strongly suggests that, given two partial derivatives meeting certain preconditions, it will be possible to reconstruct the 3D function from which the partial derivatives were derived by summing their partial integrals.

$$\int \frac{\partial}{\partial x} f(x,y)\, dx + \int \frac{\partial}{\partial y} f(x,y)\, dy = f(x,y)$$

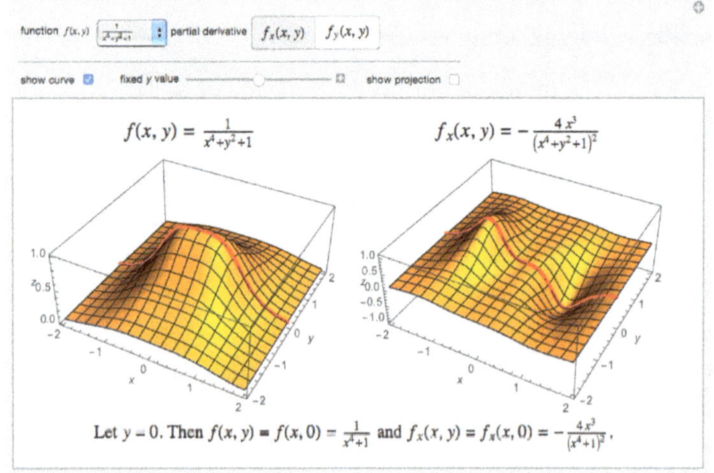

Online Application

Visit demonstrations.wolfram.com/PartialDerivativeFunctionsAndTheirPlots for a great little application that will demonstrate some of the concepts from this chapter. "Partial Derivative Functions and Their Plots" from the Wolfram Demonstrations Project. Contributed by Marc Brodie.

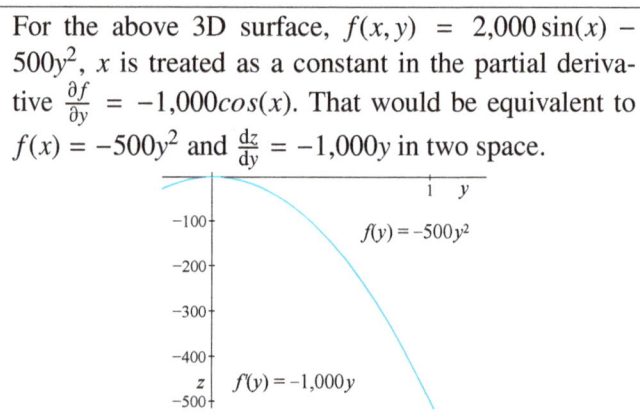

Chapter 12
Exact Differential Equations

(Author's note: These materials are being written to provide encouragement to mathophobes under the belief that if a foundation can be built, then such students will be prepared to study and understand traditional, rigorous, and complete treatments of the topics covered. Writing this chapter has been uniquely challenging in balancing the conflict between correctness, completeness, and rigor on the one hand and clarity and encouragement on the other. If the philosophy stated here bothers you, perhaps you should seek out any of a number of traditional sources. *Elementary Differential Equations* by Boyce and DiPrima is a reputable one. Better yet, if the presentation decisions I have made make you angry, why don't you take that anger and do better? The audience I am writing for is in great need and greatly underserved by the mathematics community!)

The MATLAB graph at right shows five functions, three explicit and two implicit.

$$f(x,y) = 2{,}000\sin(x) - 500y^2 \quad \text{is explicit.}$$

$$\frac{\partial}{\partial x}\left[2{,}000\sin(x) - 500y^2\right] = 2{,}000\cos(x) \quad \text{and}$$

$$\frac{\partial}{\partial y}\left(2{,}000\sin(x) - 500y^2\right) = -1{,}000y \quad \text{are both implicit.}$$

$$\int 2{,}000\cos(x)\,\mathrm{d}x = 2{,}000\sin(x) \quad \text{and}$$

$$\int -1{,}000y\,\mathrm{d}y = -500y^2 \quad \text{are both explicit.}$$

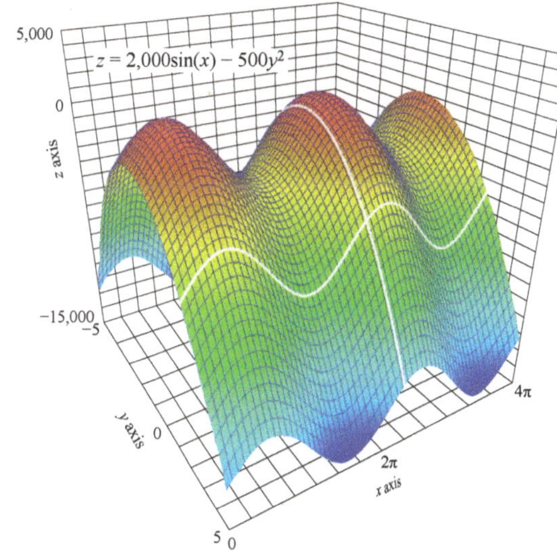

Here, we started with the three-dimensional function $f(x,y) = 2{,}000\sin(x) - 500y^2$ and found partial derivatives, $\frac{\partial f(x,y)}{\partial x}$ and $\frac{\partial f(x,y)}{\partial y}$. By integrating the two partial derivatives, $\int \frac{\partial f(x,y)}{\partial x}$ and $\int \frac{\partial f(x,y)}{\partial y}$, we find components allowing us to reconstruct the original f(x,y). If you see this, perhaps you can see the possibility of starting with the two partial derivatives of an unknown function, $\frac{\partial f(x,y)}{\partial x}$ and $\frac{\partial f(x,y)}{\partial y}$, partially integrating them, combining the results, and ending up with the original $f(x,y)$.

If $f(x,y) = c$ and
$\frac{\partial}{\partial x} f(x,y) = M(x,y)$ and
$\frac{\partial}{\partial y} f(x,y) = N(x,y)$
and $M(x,y)\,\mathrm{d}x + N(x,y)\,\mathrm{d}y = 0$,
it may be possible, under special preconditions, to partially integrate the partial derivatives $M(x,y)$ and $N(x,y)$, reversing them, and somehow combine the results to obtain the original $f(x,y) = c$.

Author's note: This example is simplified to introduce a difficult idea. The two integrals shown above would involve functions of integration, $g(x)$ and $g(y)$, that were covered in Chapter 11. Those functions of integration complicate things. *A more thorough treatment will follow.*

What is Mathematics? A Digression

Mathematics is a mosaic of different component parts:

1. **Common notions (Geometric Postulates):** "In a plane through a point not on a line only one line can be drawn through the point and parallel to the given line." (This is Euclid's famous fifth postulate.)

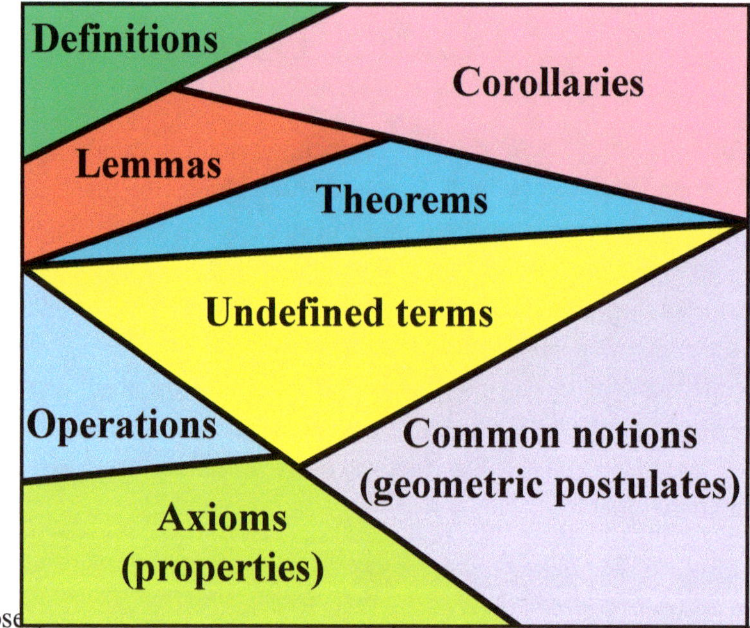

2. **Field axioms (Algebraic Properties):** $ab = ba$, $a(b + c) = ab + ac$, etc.

3. **Theorems:** if $y = x^n$ then $\frac{\mathrm{d}y}{\mathrm{d}x} = nx^{n-1}$.

4. **Lemmas:** Minor theorems used to develop major proofs. In Calculus, the lemma $\lim\limits_{\Delta x \to 0} \dfrac{\sin \Delta x}{\Delta x} = 1$ is used to prove the theorem $\frac{\mathrm{d}}{\mathrm{d}x}(\sin x) = \cos x$.

5. **Undefined terms:** In geometry point, line, plane, and space are all undefined terms.

6. **Corollaries:** A theorem that follows so clos

7. **Definitions:** $f'(x) = \lim\limits_{x \to p} \dfrac{f(x) - f(p)}{x - p}$, definition of derivative. **"Exact Differential Equation" is a definition.**

8. **Operations:** $+$, $-$, $\times$, and $\div$ are all operations.

> *Definitions: Exact Differential and Exact Differential Equation*
>
> A differential expression $M(x, y)\,\mathrm{d}x + N(x, y)\,\mathrm{d}y$ is an exact differential if it corresponds to the differential of some function $f(x, y)$ and if $M(x, y)$ and $N(x, y)$ are continuous and have continuous first partial derivatives in a rectangular region $a < x < b$, $c < y < d$ and $\frac{\partial M}{\partial y} = \frac{\partial N}{\partial x}$. A first-order differential equation of the form $M(x, y)\,\mathrm{d}x + N(x, y)\,\mathrm{d}y = 0$ is said to be an exact differential equation if the expression on the left-hand side is an exact differential.

For clarity an "exact" differential equation has the form

$$(\text{function of } x \,\&\, y)\,\mathrm{d}x + (\text{function of } x \,\&\, y)\,\mathrm{d}y = 0$$

$$M(x, y)\,\mathrm{d}x + N(x, y)\,\mathrm{d}y = 0 \quad M \text{ and } N \text{ could be the same function, but that rarely happens}$$

If $\frac{\partial M}{\partial y} = \frac{\partial N}{\partial x}$, then there are specific techniques for solving this form of differential equation that will always work.

Is the equation $-\frac{y\,\mathrm{d}x - x^3\,\mathrm{d}y}{x^2} = 0$ in exact differential equation form with $\frac{\partial M}{\partial y} = \frac{\partial N}{\partial x}$? If so, one can use a standard attack to solve it. Clearly the example equation is not in the exact-differential-equation form shown above. Could it be transformed into exact differential equation form? That is, is $-\frac{y\,\mathrm{d}x - x^3\,\mathrm{d}y}{x^2} = 0$ a disguised exact differential equation?

Let's see.

$$\begin{aligned} -\tfrac{y\,\mathrm{d}x - x^3\,\mathrm{d}y}{x^2} &= 0 \\ -x^{-2}y\,\mathrm{d}x \;+\; xy^0\,\mathrm{d}y &= 0 \\ \text{Compares with} & \\ M(x, y)\,\mathrm{d}x \;+\; N(x, y)\,\mathrm{d}y &= 0 \end{aligned}$$

but $\frac{\partial M}{\partial y} = \frac{\partial}{\partial y}\left(-x^{-2}y\right) = -x^{-2}$, while $\frac{\partial N}{\partial x} = \frac{\partial}{\partial x}\left(xy^0\right) = 1$.

Since $\frac{\partial M}{\partial y} = -x^{-2} \neq 1 = \frac{\partial N}{\partial x}$, $-\frac{y\,\mathrm{d}x - x\,\mathrm{d}y}{x^2} = 0$ is not an exact differential equation.

Is the equation $\left(xy^2 + x\right)\,\mathrm{d}x + \left(x^2y\right)\,\mathrm{d}y = 0$ an exact differential equation?

$$\begin{aligned} \left(xy^2 + x\right)\mathrm{d}x &+ \left(x^2y\right)\mathrm{d}y &= 0 \quad &\text{As shown below, this is already in the generic form.} \\ M(x,y)\,\mathrm{d}x &+ N(x,y)\,\mathrm{d}y &= 0 \quad &\text{Does } \tfrac{\partial M}{\partial y} = \tfrac{\partial N}{\partial x}? \end{aligned}$$

$$\frac{\partial M}{\partial y} = \frac{\partial}{\partial y}\left[xy^2 + x\right] = \frac{\partial}{\partial y}\left[xy^2\right] + \frac{\partial}{\partial y}[x] = x \times 2y + 0 = 2xy \quad \text{treat } x \text{ as constant}$$

$$\frac{\partial N}{\partial x} = \frac{\partial}{\partial x}\left[x^2y\right] = y(2x) = 2xy \quad \text{treat } y \text{ as constant}$$

$$\frac{\partial M}{\partial y} = 2xy = \frac{\partial N}{\partial x}.$$

Yes, $\left(xy^2 + x\right)\,\mathrm{d}x + \left(x^2y\right)\,\mathrm{d}y = 0$ is an exact differential equation. It satisfies the conditions set in the definition! One can use a standard attack to solve this equation.

Notice the inequalities in the definition on the previous page: "... and if $M(x,y)$ and $N(x,y)$ are continuous and have continuous first partial derivatives in a rectangular region $a < x < b$, $c < y < d$ ...." A picture can help understand what is being required here.

Let's review a concept from algebra. In two space on an x–y-axis, for $x_2 > x_1$, the function $y = 1$ can be drawn from any x_1 to any x_2. However for function $y(x) = \frac{x}{x}$ the function can't be graphed at $x = 0$ without lifting the pencil. There is a hole in the one dimensional x domain at $x = 0$. The function $y(x) = \frac{x}{x}$ cannot be graphed when the domain, x, includes zero. **There is a hole in the one-dimensional x domain at $x = 0$ because, with a value of zero in the denominator, the division is undefined.**

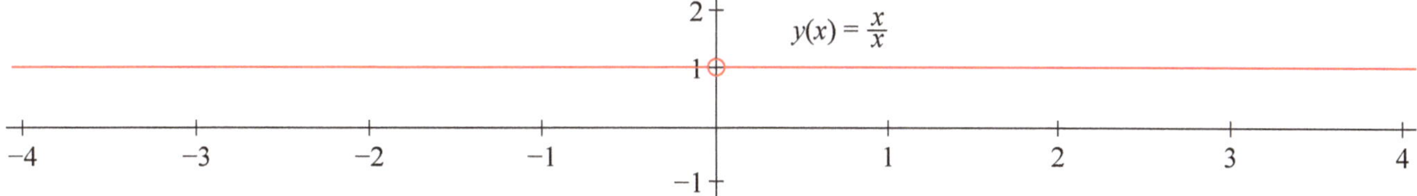

If a differential equation is exact, that means that $f(x,y)$ can be graphed for every point in its two-dimensional (x,y) domain. If you can draw the four sides of a rectangle (each side parallel to the x- or y-axis) and all sides and all area of the rectangle are in the domain of $f(x,y)$, then the function $z = f(x,y)$ can be graphed for all (x,y) pairs inside the rectangle.* The figures below give two examples where the test fails.

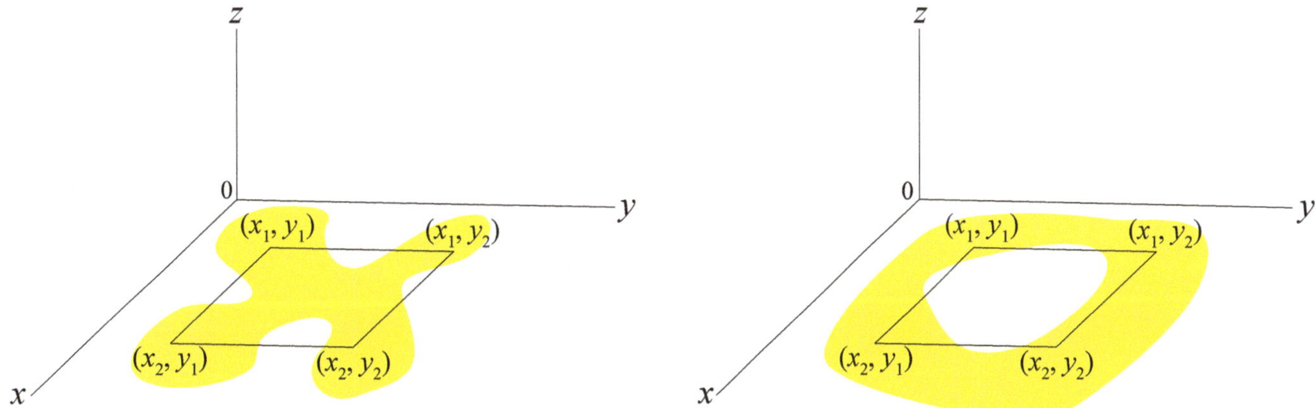

The test for exactness (above, "If you can draw ...") is a three-space version of the test shown previously ($x \neq 0$) for continuity in two space ("The function $y(x) = \frac{x}{x}$ cannot be graphed at $x = 0$ without lifting your pencil.") The significance of such a successful test is that, for any two-dimensional domain, (x,y), there will be no missing points inside a rectangle laid over that domain. In both graphs above, some x–y points are missing in the (two-dimensional) domain. Perhaps it is better to say that every point inside the rectangular domain must be defined. The pictures above

*_Ordinary Differential Equations,_ Morris Tenenbaum and Harry Pollard, Dover Publications, 1963, page 71

show that you could not partially integrate along either the x- or y-axis as the region has (x, y) points for which the function is not defined.

If $\frac{\partial M}{\partial y} = \frac{\partial N}{\partial y}$, then, for all (x, y) points in the domain of $f(x, y)$, there will be a solution to the differential equation. The differential equation $f(x, y)$ is not solvable for those points that are not in the domain. Take for example, the differential equation $\frac{x\,dy - y\,dx}{x^2+y^2} = 0$. This function is clearly not defined for the point $(0, 0)$ as that would result in a division by zero. The solution for defined (x, y) points according to Morris Tennenbum and Harry Pollard in their book *Ordinary Differential Equations*, 1963, page 81, is found to be $\arctan\left(\frac{y}{x}\right) = c$. Watch what happens in the graph at point $(0, 0)$ when MATLAB tries to graph this solution there!!

The graph at right is a 3D analog of the graph below of $y = \frac{1}{x}$, which is undefined where $x = 0$. Or, as Monsieur Jourdain said in Chapter 1 "Par ma foi! Il y a plus de quarante ans que je dis de la prose sans que j'en susse rien, et je vous suis le plus obligé du monde de m'avoir appris cela."

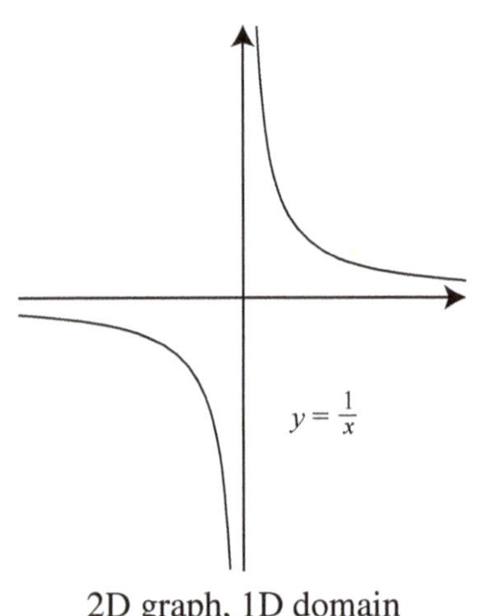
$y = \frac{1}{x}$

2D graph, 1D domain

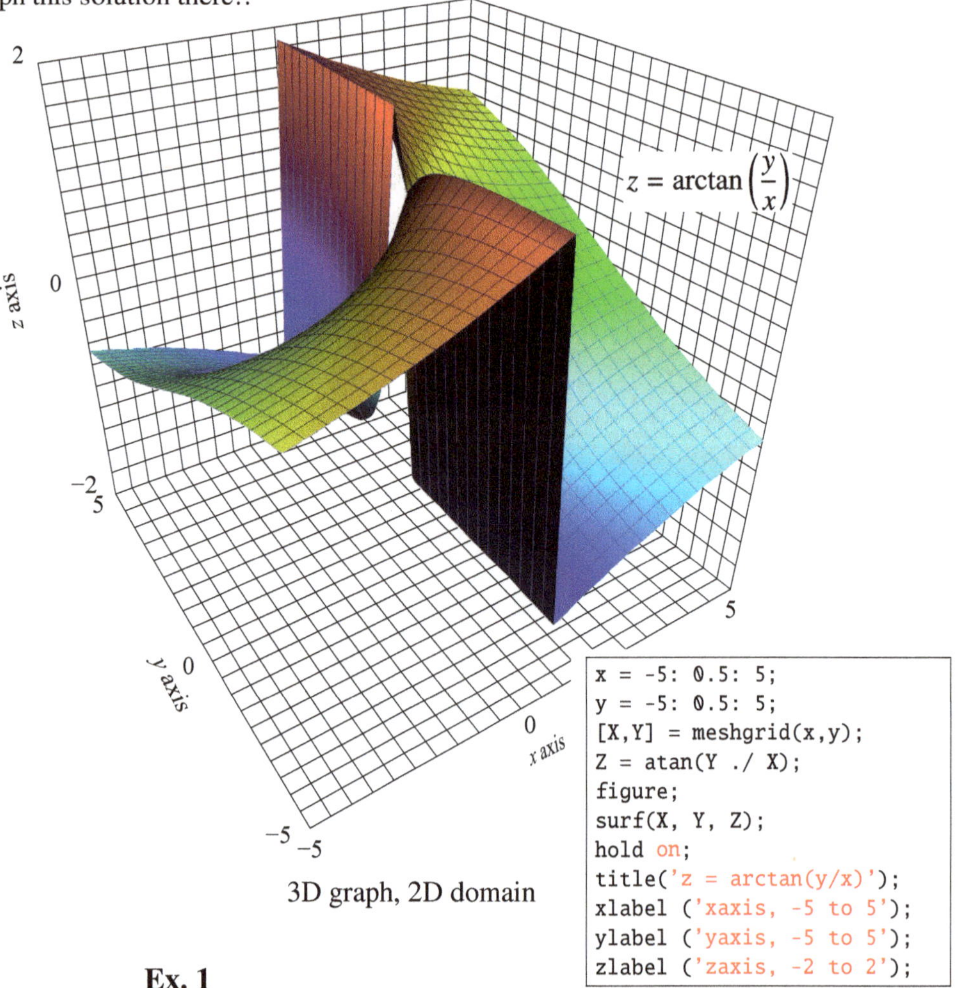
$z = \arctan\left(\frac{y}{x}\right)$
z axis
y axis
x axis
2
0
−2
5
0
−5
−5
0
5
x = -5: 0.5: 5;
y = -5: 0.5: 5;
[X,Y] = meshgrid(x,y);
Z = atan(Y ./ X);
figure;
surf(X, Y, Z);
hold on;
title('z = arctan(y/x)');
xlabel ('xaxis, -5 to 5');
ylabel ('yaxis, -5 to 5');
zlabel ('zaxis, -2 to 2');

3D graph, 2D domain

Ex. 1

Assume that there is a mysterious, unknown three-dimensional function: $f(x, y) = c$. We do not know what $f(x, y)$ is, but we are given a differential equation. $\left(xy^2 + x\right) dx + \left(x^2y\right) dy = 0$, which, under certain conditions, will allow us to find two partial derivatives of the unknown three-dimentional function $f(x, y) = c$.

Step 1: Find $f(x, y)$ for the differential equation $\left(xy^2 + x\right) dx + \left(x^2y\right) dy = 0$.

$\left(xy^2 + x\right) dx + \left(x^2y\right) dy = 0$ specific differential equation, compare with the exact differential equation form below

$M(x, y)\,dx + N(x, y)\,dy = 0$ generic exact differential equation form

Step 2: Partially differentiate each addend with respect to its "opposite differential" and see if they are equal.

$$\frac{\partial M}{\partial y} = \frac{\partial}{\partial y}\left[xy^2 + x\right] = \frac{\partial}{\partial y}\left[xy^2\right] + \frac{\partial}{\partial y}[x] = x \times 2y + 0 = 2xy \quad \text{treat } x \text{ as constant}$$

$$\frac{\partial N}{\partial x} = \frac{\partial}{\partial x}\left[x^2y\right] = y(2x) = 2xy \quad \text{treat } y \text{ as constant}$$

$$\frac{\partial M}{\partial y} = 2xy = \frac{\partial N}{\partial x}, \quad \text{exact differential equation!!}$$

Step 3: Since $\frac{\partial M}{\partial y} = \frac{\partial N}{\partial x}$, it can be proved that $f(x,y)$ exists such that $M(x,y)\,\mathrm{d}x + N(x,y)\,\mathrm{d}y = \frac{\partial f(x,y)}{\partial x}\,\mathrm{d}x + \frac{\partial f(x,y)}{\partial y}\,\mathrm{d}y$. Therefore, $M(x,y) = \frac{\partial f(x,y)}{\partial x}$, $N(x,y) = \frac{\partial f(x,y)}{\partial y}$, and $\frac{\partial M}{\partial y} = \frac{\partial}{\partial y}\left(\frac{\partial f(x,y)}{\partial x}\right) = \frac{\partial^2 f(x,y)}{\partial y\,\partial x} = \frac{\partial}{\partial x}\left(\frac{\partial f(x,y)}{\partial y}\right) = \frac{\partial N}{\partial x}$.

Step 4: **(1)** Assume that there is an unknown function, $f(x,y)$, whose partial derivatives are known. Use those partial derivatives to solve for $f(x,y)$.

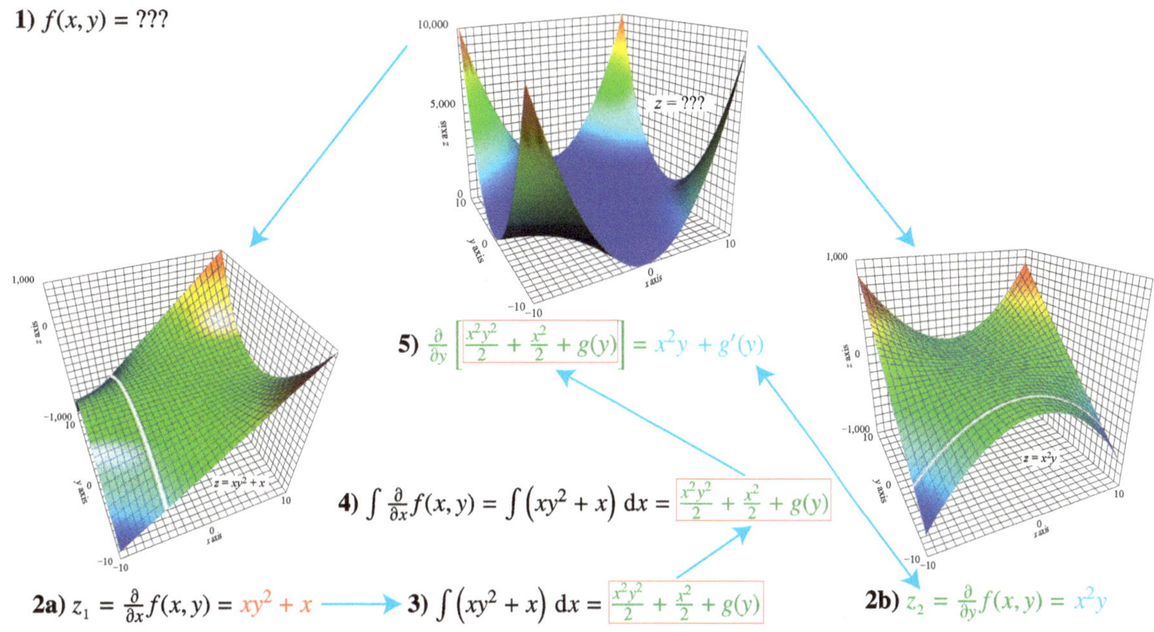

Since (on the left) the unknown $f(x,y)$ **(1)** partially differentiated with respect to x yielded $\frac{\partial}{\partial x} f(x,y) = xy^2 + x$ **(2a)**, it seems reasonable that by doing a partial integration on both sides of this equation with ***respect to x***, one would get the original, but unknown, $f(x,y)$: $\int \frac{\partial}{\partial x} f(x,y) = \int (xy^2 + x)\,\mathrm{d}x$ **(3)**. Then $f(x,y) = \int xy^2\,\mathrm{d}x + \int x\,\mathrm{d}x = y^2 \int x\,\mathrm{d}x + \frac{x^2}{2} = \frac{x^2y^2}{2} + \frac{x^2}{2} + g(y)$ **(4)**. Now, if we differentiate with ***respect to y***, the $f(x,y)$ just obtained in parts **3** and **4**: $\frac{\partial}{\partial y}\left(\frac{x^2y^2}{2} + \frac{x^2}{2} + g(y)\right) = \frac{x^2}{2}\left(\frac{\partial}{\partial y} y^2\right) + \frac{\partial}{\partial y}\left(\frac{x^2}{2}\right) + \frac{\partial}{\partial y} g(y) = \frac{x^2}{2}(2y) + 0 + g'(y) = x^2y + g'(y)$ **(5)**. But since the unknown $f(x,y)$ was also partially differentiated with respect to y to yield $\frac{\partial}{\partial y} f(x,y) = \left(x^2y\right)$ **(2b)**, we can infer that **5** equals **2b**: $x^2y + g'(y) = x^2y$. Therefore, it can be concluded that $g'(y) = 0$. If $g'(y) = 0$ then $g(y) = m$, some constant.

Having already determined in Step **4** that the unknown $f(x,y) = \frac{x^2y^2}{2} + \frac{x^2}{2} + g(y)$ after substituting $g(y) = m$, we conclude that the original mysterious, unknown differential equation whose partials were known would be $f(x,y) = \frac{x^2y^2}{2} + \frac{x^2}{2} + m$ or $x^2y^2 + x^2 = c$.

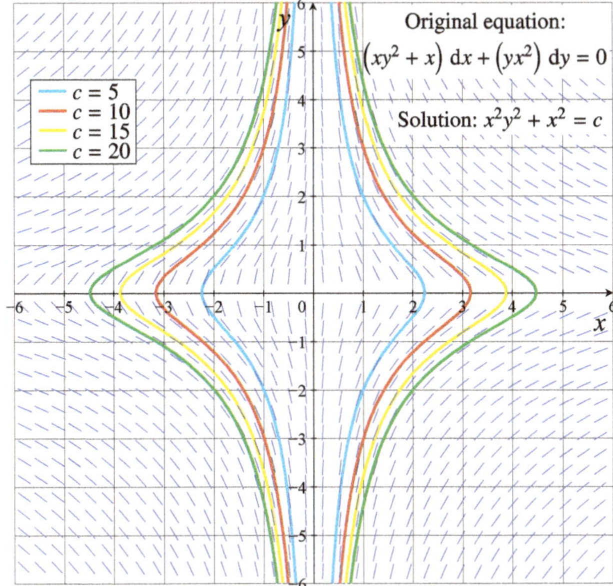

$$\left(xy^2 + x\right)\,\mathrm{d}x + \left(x^2y\right)\,\mathrm{d}y = 0$$

$$\left(x^2y\right)\,\mathrm{d}y = -\left(xy^2 + x\right)\,\mathrm{d}x$$

$$\frac{\mathrm{d}y}{\mathrm{d}x} = \frac{-\left(xy^2 + x\right)}{x^2y}$$

$$\frac{\mathrm{d}y}{\mathrm{d}x} = \frac{-x\left(y^2 + 1\right)}{x \times xy} = -\frac{y^2 + 1}{xy}$$

$$\text{As } x \to 0, \frac{\mathrm{d}y}{\mathrm{d}x} \to \infty$$

As x approaches zero, particular solutions will approach the y-axis asymptotically.

$$\text{As } x \to \infty, \frac{\mathrm{d}y}{\mathrm{d}x} \to 0,$$

but particular solutions will not result in asymptotes to the x-axis due to the effect of c, the constant of integration.

There would be no solution at $x = 0$ because the derivative goes to infinity: The slope fields approach vertical.

All of the steps shown above (1, 2a, 2b, 3, 4, and 5) can be organized cleanly and succinctly in a table. Doing so will 1) review and reinforce the concepts, 2) make more clear the necessity and importance of the requirement that $\frac{\partial M}{\partial y} = \frac{\partial N}{\partial x}$, and 3) allow for a quick lesson on the symmetry or mirroring of the technique being taught. $\left(xy^2 + x\right)\,\mathrm{d}x + \left(x^2 y\right)\,\mathrm{d}y = 0$: specific differential equation, compare with the exact differential-equation form below.

	Symmetric (mirror) solution technique
Given $M(x,y)\,\mathrm{d}x + N(x,y)\,\mathrm{d}y = 0$ with $\frac{\partial M}{\partial y} = \frac{\partial N}{\partial x}$ $\left(xy^2 + x\right)\mathrm{d}x + \left(x^2 y\right)\mathrm{d}y = 0$ with $\frac{\partial M}{\partial y} = 2xy = \frac{\partial N}{\partial x}$ Hence $\frac{\partial}{\partial x} f(x,y) = M(x,y) = (xy^2 + x)\,\mathrm{d}x$	Given $M(x,y)\,\mathrm{d}x + N(x,y)\,\mathrm{d}y = 0$ with $\frac{\partial M}{\partial y} = \frac{\partial N}{\partial x}$ $\left(xy^2 + x\right)\mathrm{d}x + \left(x^2 y\right)\mathrm{d}y = 0$ with $\frac{\partial M}{\partial y} = 2xy = \frac{\partial N}{\partial x}$ Hence $\frac{\partial}{\partial y} f(x,y) = N(x,y) = (x^2 y)\,\mathrm{d}y$
$\int \frac{\partial}{\partial x} f(x,y)\,\mathrm{d}x = \int M(x,y)\,\mathrm{d}x$ $f(x,y) = \int \left(xy^2 + x\right)\mathrm{d}x = \frac{x^2y^2}{2} + \frac{x^2}{2} + g(y)$ *** $\frac{\partial}{\partial y}[f(x,y)] = \frac{\partial}{\partial y}\left[\frac{x^2y^2}{2} + \frac{x^2}{2} + g(y)\right]$ $= \frac{\partial}{\partial y}\left[\frac{x^2y^2}{2}\right] + \frac{\partial}{\partial y}\left[\frac{x^2}{2}\right] + \frac{\partial}{\partial y}[g(y)]$ $= \frac{x^2}{2}\frac{\partial}{\partial y}(y^2) + 0 + g'(y)$ $= x^2 y + g'(y)$	$\int \frac{\partial}{\partial y} f(x,y)\,\mathrm{d}y = \int N(x,y)\,\mathrm{d}y$ $f(x,y) = \int \left(x^2 y\right)\mathrm{d}y = \frac{x^2y^2}{2} + g(x)$ ***** $\frac{\partial}{\partial x}[f(x,y)] = \frac{\partial}{\partial x}\left[\frac{x^2y^2}{2} + g(x)\right]$ $= \frac{y^2}{2}\frac{\partial}{\partial x}\left[x^2\right] + \frac{\partial}{\partial y}[g(x)]$ $= xy^2 + g'(x)$
But since $\frac{\partial}{\partial y} f(x,y) = N(x,y) = x^2 y$, it follows that $x^2 y + g'(y) = x^2 y$. Therefore, $g'(y) = 0$ and $g(y) = C$. So, from above (***), $f(x,y) = \int \left(xy^2 + x\right)\mathrm{d}x = \frac{x^2y^2}{2} + \frac{x^2}{2} + g(y)$ $= \frac{x^2y^2}{2} + \frac{x^2}{2} + C$ Or $x^2y^2 + x^2 = C$	But since $\frac{\partial}{\partial x} f(x,y) = M(x,y) = xy^2 + x$, it follows that $xy^2 + g'(x) = xy^2 + x$. Therefore, $g'(x) = x$ and $g(x) = \frac{x^2}{2} + C$. So, from above (*****), $f(x,y) = \int \left(x^2 y\right)\mathrm{d}y = \frac{x^2y^2}{2} + g(x)$ $= \frac{x^2y^2}{2} + \frac{x^2}{2} + C$ Or $x^2y^2 + x^2 = C$

This algorithm works because $\frac{\partial M}{\partial y} = \frac{\partial N}{\partial x}$! It always requires two integrations, but the second is simplified and may be trivial. Since you can start with either $\int M(x,y)\,\mathrm{d}x$ or $\int N(x,y)\,\mathrm{d}y$ as the first integration, you are free to pick the simpler integration to start with.

Here is a little visual that can help you remember this process … a sort of diamond. If the two partials at the bottom are equal, then somehow combining their antiderivatives will yield $f(x,y)$.

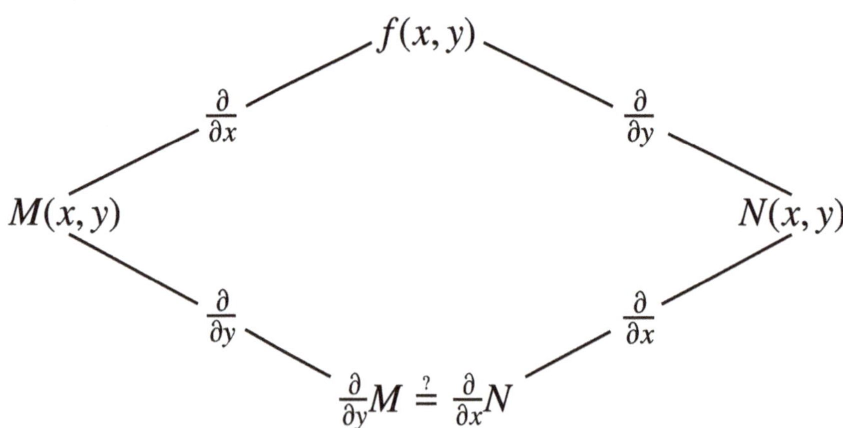

On page 249 of the very formidable book *Mathematical Handbook for Scientists and Engineers* by Granino and

Theresa Korn, Dover Publications, 2000, we find the closed-form formula

$$\boxed{f(x,y) = \int M(x,y)\,\mathrm{d}x + \int \left[N(x,y) - \frac{\partial}{\partial y}\int M(x,y)\,\mathrm{d}x\right]\mathrm{d}y = C}$$

for solving a differential equation that has passed the criteria for being an exact differential equation: $M(x,y)\,\mathrm{d}x + N(x,y)\,\mathrm{d}y = 0$ or $M(x,y) + N(x,y)\frac{\mathrm{d}y}{\mathrm{d}x} = 0$ or $M(x,y) + N(x,y)y'(x) = 0$ and $\frac{\partial M}{\partial y} = \frac{\partial N}{\partial x}$.

The downside of using closed-form formulas to solve equations is that, without derivation, they are always mysterious. The upside is that you can get a correct answer with little (or at least reduced) mental effort.

Let try the formula on Example 1, which, it has been established, is an exact differential equation.

$$0 = \left(xy^2 + x\right)\mathrm{d}x + \left(x^2y\right)\mathrm{d}y$$

$$0 = M(x,y)\,\mathrm{d}x + N(x,y)\,\mathrm{d}y \text{ and } \frac{\partial M(x,y)}{\partial y} = \frac{\partial N(x,y)}{\partial x}$$

$$f(x,y) = \int M(x,y)\,\mathrm{d}x + \int\left[N(x,y) - \frac{\partial}{\partial y}\int M(x,y)\,\mathrm{d}x\right]\mathrm{d}y = C$$

$$f(x,y) = \int \left(xy^2 + x\right)\mathrm{d}x + \int\left[x^2y - \frac{\partial}{\partial y}\int\left(xy^2 + x\right)\mathrm{d}x\right]\mathrm{d}y = C$$ applying formula from the *Mathematical Handbook*

Since the term $\int\left(xy^2 + x\right)\mathrm{d}x$ is used twice in this scary formula it seems reasonable to solve it first and then to substitute the result into the original $f(x,y)$ formula for $\int\left(xy^2 + x\right)\mathrm{d}x$. (To make things easier, ignore the integration constants for now. They can all be combined later.)

$$\int\left(xy^2 + x\right)\mathrm{d}x = \int xy^2\,\mathrm{d}x + \int x\,\mathrm{d}x = y^2\int x\,\mathrm{d}x + \frac{x^2}{2} = y^2\left(\frac{x^2}{2}\right) + \frac{x^2}{2} = \frac{x^2y^2}{2} + \frac{x^2}{2}$$

Substituting $\int\left(xy^2 + x\right)\mathrm{d}x = \frac{x^2y^2}{2} + \frac{x^2}{2}$ into the original equation,

$$\begin{aligned}
f(x,y) &= \int\left(xy^2 + x\right)\mathrm{d}x + \int\left[x^2y - \frac{\partial}{\partial y}\int\left(xy^2 + x\right)\mathrm{d}x\right]\mathrm{d}y = C\\
&= \frac{x^2y^2}{2} + \frac{x^2}{2} + \int\left[x^2y - \frac{\partial}{\partial y}\left(\frac{x^2y^2}{2} + \frac{x^2}{2}\right)\right]\mathrm{d}y = C\\
&= \frac{x^2y^2}{2} + \frac{x^2}{2} + \int\left[x^2y - \frac{\partial}{\partial y}\left(\frac{x^2y^2}{2}\right) - \frac{\partial}{\partial y}\frac{x^2}{2}\right]\mathrm{d}y = C\\
&= \frac{x^2y^2}{2} + \frac{x^2}{2} + \int\left[x^2y - \frac{x^2}{2}\frac{\partial}{\partial y}\left(y^2\right) - 0\right]\mathrm{d}y = C\\
&= \frac{x^2y^2}{2} + \frac{x^2}{2} + \int\left[x^2y - \frac{x^2}{2}\times 2y\right]\mathrm{d}y = C\\
&= \frac{x^2y^2}{2} + \frac{x^2}{2} + \int\left[x^2y - x^2y\right]\mathrm{d}y = C
\end{aligned}$$
Can you see the duplicate terms being canceled here???

$$\begin{aligned}
&= \frac{x^2y^2}{2} + \frac{x^2}{2} + \int 0\,\mathrm{d}y = C\\
&= \frac{x^2y^2}{2} + \frac{x^2}{2} + 0 = C
\end{aligned}$$

So $x^2y^2 + x^2 = C$. Fortunately, this matches with the previous approach or there would be egg on the author's face. Although this is more work than the trick shown above, one can feel better about the answer because the formula used to obtain it will always work and can be proven. (Not in this book!!!)

Ex. 2

Test the following initial value differential equation to see if it is in exact differential equation form and if so solve using a technique suited to this form.

Step 1:

$$\left(\cos x - x\sin x + y^2\right) \mathrm{d}x + 2xy\,\mathrm{d}y = 0, \quad y = 1 \text{ when } x = \pi, \text{ specific IVP differential equation}$$
$$M(x,y)\,\mathrm{d}x + N(x,y)\,\mathrm{d}y = 0 \quad \text{generic exact differential equation form}$$

Step 2: Partially differentiate both addends with respect to their opposite differential and see if the resulting partial derivatives are equal.

$$\frac{\partial M}{\partial y} = \frac{\partial}{\partial y}\left[\cos x - x\sin x + y^2\right] = \frac{\partial}{\partial y}[\cos x] - \frac{\partial}{\partial y}[x\sin x] + \frac{\partial}{\partial y}\left[y^2\right] = 0 - 0 + 2y = 2y \quad \text{treat } x \text{ as constant}$$
$$\frac{\partial N}{\partial x} = \frac{\partial}{\partial x}[2xy] = 2y\frac{\partial}{\partial x}x = 2y \times 1 = 2y \quad \text{treat } y \text{ as constant}$$
$$\frac{\partial M}{\partial y} = 2y = \frac{\partial N}{\partial x} \quad \text{exact differential equation!!}$$

Step 3: Since $\frac{\partial M}{\partial y} = \frac{\partial N}{\partial x}$, it can be proved that there exists a function $f(x,y)$ such that $M(x,y)\,\mathrm{d}x + N(x,y)\,\mathrm{d}y = \frac{\partial f(x,y)}{\partial x}\mathrm{d}x + \frac{\partial f(x,y)}{\partial y}\mathrm{d}y$. Therefore, $M(x,y) = \frac{\partial f(x,y)}{\partial x}$, $N(x,y) = \frac{\partial f(x,y)}{\partial y}$ and $\frac{\partial M}{\partial y} = \frac{\partial}{\partial y}\left(\frac{\partial f(x,y)}{\partial x}\right) = \frac{\partial^2 f(x,y)}{\partial y \partial x} = \frac{\partial}{\partial x}\left(\frac{\partial f(x,y)}{\partial y}\right) = \frac{\partial N}{\partial x}$.

Step 4: Assume that there is an unknown differential equation, $f(x,y)$, whose partial derivatives are known. Use these partial derivatives to solve for $f(x,y)$. Follow along in the figure below.

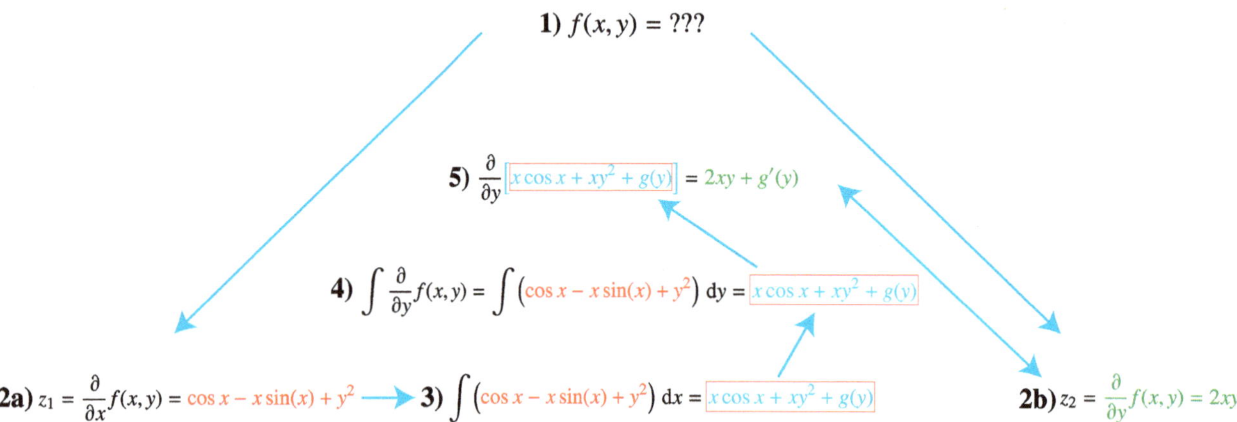

Since (on the left) the unknown $f(x,y)$ (**1**) was partially differentiated with respect to x to yield $\cos x - x\sin x + y^2$ $\left(\frac{\partial}{\partial x}f(x,y) = \cos x - x\sin x + y^2\right)$ (**2a**), it seems reasonable that by doing a partial integration on both sides of this equation ***with respect to x***, one would get the original, but unknown, $f(x,y)$ (**3**). So, $\int \frac{\partial}{\partial x}f(x,y) = \int\left[\cos x - x\sin x + y^2\right]\mathrm{d}x = \int(\cos x)\,\mathrm{d}x - \int(x\sin x)\,\mathrm{d}x + y^2\int x^0\,\mathrm{d}x = \sin x - (\sin x - x\cos x) + xy^2 = x\cos x + xy^2 + g(y)$ (**4**). Note that $\int(x\sin x)\,\mathrm{d}x = \sin x - x\cos x$ from a table of integrals. Now, if one differentiates, ***with respect to y,*** the $f(x,y)$ just obtained in **4**, that would give $\frac{\partial}{\partial y}\left[x\cos x + xy^2 + g(y)\right] = \frac{\partial}{\partial y}(x\cos x) + x\frac{\partial}{\partial y}y^2 + \frac{\partial}{\partial y}[g(y)] = 0 + x(2y) + g'(y) = 2xy + g'(y)$ (**5**). But, since the unknown $f(x,y)$ was also partially differentiated with respect to y to yield **2b**, $\frac{\partial}{\partial y}f(x,y) = 2xy$, $2xy + g'(y) = 2xy$ (**5 = 2b**). It can be concluded that $g'(y) = 0$. If $g'(y) = 0$, then $g(y) = C$.

Having already determined in **4** that the unknown $f(x,y) = x\cos x + xy^2 + g(y)$ after substituting $g(y) = C$, we conclude that the original mysterious, unknown differential equation whose partials were known is

$$f(x,y) = x\cos(x) + xy^2 + C \text{ or } x\cos(x) + xy^2 = C.$$

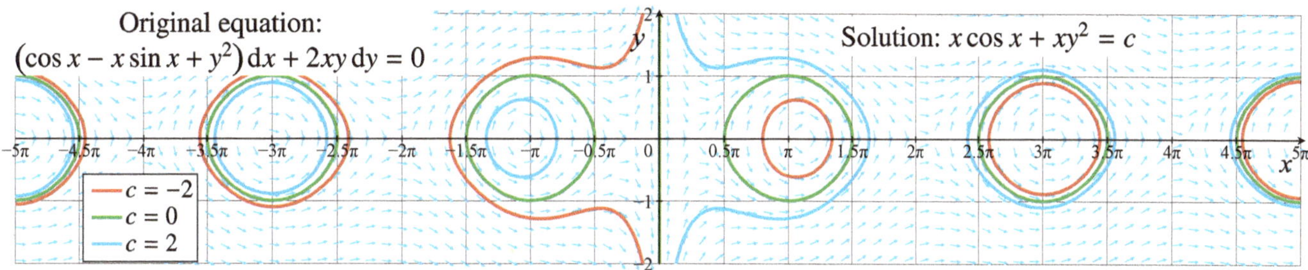

Again using the *Mathematical Handbook* formula on Example 2,

$$f(x,y) = \int M(x,y)\,dx + \int\left[N(x,y) - \frac{\partial}{\partial y}\int M(x,y)\,dx\right]dy = C,$$

which, it has been established, is an exact differential equation,

$$\left(\cos x - x\sin x + y^2\right)\,dx + 2xy\,dy = 0$$

$$M(x,y)\,dx + N(x,y)\,dy = 0 \text{ with } \frac{\partial M}{\partial y} = \frac{\partial N}{\partial x}$$

we get $f(x,y) = \int\left(\cos x - x\sin x + y^2\right)dx + \int\left[2xy - \frac{\partial}{\partial y}\int\left(\cos x - x\sin x + y^2\right)dx\right]dy = C$

Since the term $\int\left(\cos x - x\sin x + y^2\right)$ is used twice in this scary formula, it seems reasonable to solve it first and then to substitute the result into the original $f(x,y)$ formula. (To make things easier, ignore the integration constants for now. They can all be combined later.) $\int\left(\cos x - x\sin x + y^2\right)dx = \int\cos x\,dx - \int x\sin x\,dx + \int y^2\,dx = \sin x - (\sin x - x\cos x) + y^2\int dx$ (because $\int x\sin x\,dx = \sin x - x\cos x$). So $\int(\cos x - x\sin x + y^2)\,dx = x\cos x + y^2x$.

Substituting $\int\left(\cos x - x\sin x + y^2\right)dx = x\cos x + y^2x$ into the original equation

$$\begin{aligned}
f(x,y) &= \int\left(\cos x - x\sin x + y^2\right)dx + \int\left[2xy - \frac{\partial}{\partial y}\int\left(\cos x - x\sin x + y^2\right)dx\right]dy = C\\
&= x\cos x + y^2x + \int\left[2xy - \frac{\partial}{\partial y}\left(x\cos x + y^2x\right)\right]dy = C\\
&= x\cos x + xy^2 + \int\left[2xy - \frac{\partial}{\partial y}(x\cos x) - \frac{\partial}{\partial y}y^2x\right]dy = C\\
&= x\cos x + xy^2 + \int\left[2xy - 0 - x\frac{\partial}{\partial y}y^2\right]dy = C\\
&= x\cos x + xy^2 + \int[2xy - 2xy]\,dy = C \quad \text{here duplicate terms will be eliminated}\\
&= x\cos x + xy^2 + \int[0]\,dy = C\\
&= x\cos x + xy^2 = C
\end{aligned}$$

Author's note: It is part of my teaching philosophy that understanding how a skill is used makes it easier for many people to learn. Therefore, I have endeavored to give scenarios or applications for each type of differential equation that has been introduced. However, most applications of exact ODEs are in electricity and magnetism and would require a substantial amount of physics to understand: gradients, vectors, potential, flows, closed systems, particles, etc. Hence, my approach to exact ODEs will be limited to only working math equations and I will not have any applications for you. Below is a set of exact differential equations that I found on pages 80 & 81 of the book *Ordinary Differential Equations*, by Morris Tenenbaum and Harry Pollard, Dover Publications, 1963. They are included so that a motivated student can drill, drill, and drill on solving exact differential equations. **Remember, differential equations cannot be solved at points that are not defined for the equation.**

Differential Equation	Integrated/Algebraic Form
1) $y\,dx + x\,dy = 0$ $(x^0 y)\,dx + (xy^0)\,dy = 0$ convert to M, N form $M(x,y)\,dx + N(x,y)\,dy = 0$ for comparison $\frac{\partial M}{\partial y} = \frac{\partial}{\partial y}[y] = 1$ $\frac{\partial N}{\partial x} = \frac{\partial}{\partial x}[x] = 1$ $\frac{\partial M}{\partial y} = 1 = \frac{\partial N}{\partial x}$, hence, exact	1) Answer: $xy = c$.
2) $2xy\,dx + x^2\,dy = 0$ $2xy\,dx + x^2y^0\,dy = 0$ convert to M, N form $M(x,y)\,dx + N(x,y)\,dy = 0$ for comparison $\frac{\partial M}{\partial y} = \frac{\partial}{\partial y}[2xy] = 2x$ treat $2x$ as constant $\frac{\partial N}{\partial x} = \frac{\partial}{\partial x}\left[x^2\right] = 2x$ $\frac{\partial M}{\partial y} = 2x = \frac{\partial N}{\partial x}$, hence, exact	2) Answer: $x^2y = c$.
3) $y^2\,dx + 2xy\,dy = 0$ $M(x,y)\,dx + N(x,y)\,dy = 0$ for comparison $\frac{\partial M}{\partial y} = \frac{\partial}{\partial y}\left[y^2\right] = 2y$ $\frac{\partial N}{\partial x} = \frac{\partial}{\partial x}[2yx] = 2y$ treat $2y$ as constant $\frac{\partial M}{\partial y} = 2y = \frac{\partial N}{\partial x}$, hence, exact	3) Answer: $xy^2 = c$.
4) $2xy^2\,dx + 2x^2y\,dy = 0$ $M(x,y)\,dx + N(x,y)\,dy = 0$ for comparison $\frac{\partial M}{\partial y} = \frac{\partial}{\partial y}\left[2xy^2\right] = 4xy$ treat $2x$ as constant $\frac{\partial N}{\partial x} = \frac{\partial}{\partial x}\left[2x^2y\right] = 4xy$ treat $2y$ as constant $\frac{\partial M}{\partial y} = 4xy = \frac{\partial N}{\partial x}$, hence, exact	4) Answer: $x^2y^2 = c$.
5) $3x^2y^3\,dx + 3x^3y^2\,dy = 0$ $M(x,y)\,dx + N(x,y)\,dy = 0$ for comparison $\frac{\partial M}{\partial y} = \frac{\partial}{\partial y}\left[3x^2y^3\right] = 9x^2y^2$ treat $3x^2$ as constant $\frac{\partial N}{\partial x} = \frac{\partial}{\partial x}\left[3x^3y^2\right] = 9x^2y^2$ treat $3y^2$ as constant $\frac{\partial M}{\partial y} = 9x^2y^2 = \frac{\partial N}{\partial x}$, hence, exact	5) Answer: $x^3y^3 = c$.

Differential Equation	Integrated/Algebraic Form
6) $3x^2y\,dx + x^3\,dy = 0$ $M(x,y)\,dx + N(x,y)\,dy = 0$ for comparison $\frac{\partial M}{\partial y} = \frac{\partial}{\partial y}\left[3x^2y\right] = 3x^2$ treat $3x^2$ as constant $\frac{\partial N}{\partial x} = \frac{\partial}{\partial x}\left[x^3\right] = 3x^2$ $\frac{\partial M}{\partial y} = 3x^2 = \frac{\partial N}{\partial x}$, hence, exact	6) Answer: $x^3y = c$.
7) $y\cos x\,dx + \sin x\,dy = 0$ $M(x,y)\,dx + N(x,y)\,dy = 0$ for comparison $\frac{\partial M}{\partial y} = \frac{\partial}{\partial y}[y\cos x] = \cos x$ treat $\cos x$ as constant $\frac{\partial N}{\partial x} = \frac{\partial}{\partial x}[\sin x] = \cos x$ $\frac{\partial M}{\partial y} = \cos x = \frac{\partial N}{\partial x}$, hence, exact	7) $y\cos x\,dx + \sin x\,dy = 0$ $\frac{\partial}{\partial x} f(x,y) = y\cos x\,dx$ $\int \frac{\partial}{\partial x} f(x,y) = y \int \cos x\,dx$ $f(x,y) = y\sin x + g(y)$ *** $\frac{\partial}{\partial y} f(x,y) = \frac{\partial}{\partial y}[y\sin x + g(y)]$ $= \sin(x) + g'(y)$ But, since $\frac{\partial}{\partial y} f(x,y) = N(x,y) = \sin x$, it follows that $\sin x + g'(y) = \sin x$. Therefore, $g'(y) = 0$ and $g(y) = c$. So, from ***, $f(x,y) = y\sin x + c$.
8) $\sin y\,dx + x\cos y\,dy = 0$ $M(x,y)\,dx + N(x,y)\,dy = 0$ for comparison $\frac{\partial M}{\partial y} = \frac{\partial}{\partial y}[\sin y] = \cos y$ $\frac{\partial N}{\partial x} = \frac{\partial}{\partial x}[x\cos y] = \cos y$ treat $\cos y$ as constant $\frac{\partial M}{\partial y} = \cos y = \frac{\partial N}{\partial x}$, hence, exact	8) Answer: $x\sin y = c$.
9) $\frac{dx}{x} + \frac{dy}{y} = 0$ $M(x,y)\,dx + N(x,y)\,dy = 0$ for comparison $\frac{\partial M}{\partial y} = \frac{\partial}{\partial y}\left[x^{-1}\right] = 0$ treat x^{-1} as constant $\frac{\partial N}{\partial x} = \frac{\partial}{\partial x}\left[y^{-1}\right] = 0$ treat y^{-1} as constant $\frac{\partial M}{\partial y} = 0 = \frac{\partial N}{\partial x}$, hence, exact	9) Answer: $\log(xy) = c$.

10) $ye^{xy}\,dx + xe^{xy}\,dy = 0$

$$
\begin{aligned}
&M(x,y)\,dx + N(x,y)\,dy = 0 \quad \text{for comparison}\\
&\frac{\partial M}{\partial y} = \frac{\partial}{\partial y}[ye^{xy}] = ye^{xy}\frac{\partial}{\partial y}[xy] + e^{xy}\times 1 \quad x \text{ constant}\\
&\qquad = y\times e^{xy}\times[(x\times 1)+(y\times 0)] + e^{xy}\\
&\qquad = y\times e^{xy}\times x + e^{xy}\\
&\qquad = xye^{xy} + e^{xy}\\
&\frac{\partial N}{\partial x} = \frac{\partial}{\partial x}[xe^{xy}] = x\frac{\partial}{\partial x}[e^{xy}] + \left(e^{xy}\frac{\partial}{\partial x}[x]\right) \quad y \text{ constant}\\
&\qquad = x\times e^{xy}\frac{\partial}{\partial x}[xy] + e^{xy}\times 1\\
&\qquad = xe^{xy}\times x\times\frac{\partial}{\partial x}[y] + y\frac{\partial}{\partial x}[x] + e^{xy}\\
&\qquad = xe^{xy}\times(x\times 0 + y\times 1) + e^{xy}\\
&\qquad = xe^{xy}\times(y) + e^{xy}\\
&\qquad = xye^{xy} + e^{xy}\\
&\frac{\partial M}{\partial y} = xye^{xy} + e^{xy} = \frac{\partial N}{\partial x}, \quad \text{hence, exact}
\end{aligned}
$$

10) Answer: $e^{xy} = c$.

11)

$$
\begin{aligned}
&\frac{y\,dx - x\,dy}{y^2} = 0 \quad \text{divide numerator by } y^2\\
&x^0y^{-1}\,dx - xy^{-2}\,dy = 0 \quad \text{convert to } M, N \text{ form}\\
&x^0y^{-1}\,dx + \left(-xy^{-2}\right)dy = 0\\
&M(x,y)\,dx + N(x,y)\,dy = 0 \quad \text{for comparison}\\
&\frac{\partial M}{\partial y} = \frac{\partial}{\partial y}\left[y^{-1}\right] = -y^{-2}\\
&\frac{\partial N}{\partial x} = \frac{\partial}{\partial x}\left[-xy^{-2}\right] = -y^{-2} \quad \text{treat } -y^{-2} \text{ as constant}\\
&\frac{\partial M}{\partial y} = -y^{-2} = \frac{\partial N}{\partial x}, \quad \text{hence, exact}
\end{aligned}
$$

11) Answer: $\frac{x}{y} = c$.

12)

$$
\begin{aligned}
&-\frac{y\,dx - x\,dy}{x^2} = 0 \quad \text{divide numerator by } x^2\\
&-yx^{-2}\,dx + x^{-1}y^0\,dy = 0 \quad \text{convert to } M, N \text{ form}\\
&M(x,y)\,dx + N(x,y)\,dy = 0 \quad \text{for comparison}\\
&\frac{\partial M}{\partial y} = \frac{\partial}{\partial y}\left[-yx^{-2}\right] = -x^{-2} \quad \text{treat } -x^{-2} \text{ as constant}\\
&\frac{\partial N}{\partial x} = \frac{\partial}{\partial x}\left[x^{-1}\right] = -x^{-2}\\
&\frac{\partial M}{\partial y} = -x^{-2} = \frac{\partial N}{\partial x}, \quad \text{hence, exact}
\end{aligned}
$$

12) Answer: $\frac{y}{x} = c$.

Work	Answer
13) $\dfrac{2xy\,\mathrm{d}y - y^2\,\mathrm{d}x}{x^2} = 0$ divide numerator by x^2 $-x^{-2}y^2\,\mathrm{d}x + 2x^{-1}y\,\mathrm{d}y = 0$ convert to M, N form $M(x,y)\,\mathrm{d}x + N(x,y)\,\mathrm{d}y = 0$ for comparison $\dfrac{\partial M}{\partial y} = \dfrac{\partial}{\partial y}\left[-x^{-2}y^2\right] = -2x^{-2}y$ treat $-x^{-2}$ as constant $\dfrac{\partial N}{\partial x} = \dfrac{\partial}{\partial x}\left[2x^{-1}y\right] = -2x^{-2}y$ treat $2y$ as constant $\dfrac{\partial M}{\partial y} = -2x^{-2}y = \dfrac{\partial N}{\partial x}$, hence, exact	13) Answer: $\frac{y^2}{x} = c$.
14) $\dfrac{2xy\,\mathrm{d}x - x^2\,\mathrm{d}y}{y^2} = 0$ divide numerator by y^2 $2xy^{-1}\,\mathrm{d}x - x^2y^{-2}\,\mathrm{d}y = 0$ convert to M, N form $M(x,y)\,\mathrm{d}x + N(x,y)\,\mathrm{d}y = 0$ for comparison $\dfrac{\partial M}{\partial y} = 2x\dfrac{\partial}{\partial y}\left[y^{-1}\right] = -2xy^{-2}$ treat $2x$ as constant $\dfrac{\partial N}{\partial x} = \dfrac{\partial}{\partial x}\left[-x^2y^{-2}\right] = -2xy^{-2}$ treat $-y^{-2}$ as constant $\dfrac{\partial M}{\partial y} = -2xy^{-2} = \dfrac{\partial N}{\partial x}$, hence, exact	14) Answer: $\frac{x^2}{y} = c$.
15) $-\dfrac{3x^2y\,\mathrm{d}x - x^3\,\mathrm{d}y}{x^6} = 0$ divide numerator by x^6 $-3x^{-4}y\,\mathrm{d}x + x^{-3}y^0\,\mathrm{d}y = 0$ convert to M, N form $M(x,y)\,\mathrm{d}x + N(x,y)\,\mathrm{d}y = 0$ for comparison $\dfrac{\partial M}{\partial y} = \dfrac{\partial}{\partial y}\left[-3x^{-4}y\right] = -3x^{-4}$ treat $-3x^{-4}$ as constant $\dfrac{\partial N}{\partial x} = \dfrac{\partial}{\partial x}\left[x^{-3}\right] = -3x^{-4}$ $\dfrac{\partial M}{\partial y} = -3x^{-4} = \dfrac{\partial N}{\partial x}$, hence, exact	15) Answer: $\frac{y}{x^3} = c$.
16) $\dfrac{y^2\,\mathrm{d}x - 2xy\,\mathrm{d}y}{y^4} = 0$ divide numerator by y^4 $x^0y^{-2}\,\mathrm{d}x - 2xy^{-3}\,\mathrm{d}y = 0$ convert to M, N form $M(x,y)\,\mathrm{d}x + N(x,y)\,\mathrm{d}y = 0$ for comparison $\dfrac{\partial M}{\partial y} = \dfrac{\partial}{\partial y}\left[y^{-2}\right] = -2y^{-3}$ $\dfrac{\partial N}{\partial x} = \dfrac{\partial}{\partial x}\left[-2xy^{-3}\right] = -2y^{-3}$ treat $-2y^{-3}$ as constant $\dfrac{\partial M}{\partial y} = -2y^{-3} = \dfrac{\partial N}{\partial x}$, hence, exact	16) Answer: $\frac{x}{y^2} + \frac{x}{2y^4} = c$.

17) $-\frac{y\,\mathrm{d}x + x\,\mathrm{d}y}{x^2y^2} = 0$ divide numerator by x^2y^2 $-x^{-2}y^{-1}\,\mathrm{d}x - x^{-1}y^{-2}\,\mathrm{d}y = 0$ convert to M, N form $M(x,y)\,\mathrm{d}x + N(x,y)\,\mathrm{d}y = 0$ for comparison $\frac{\partial M}{\partial y} = \frac{\partial}{\partial y}\left[-x^{-2}y^{-1}\right] = x^{-2}y^{-2}$ treat $-x^{-2}$ as constant $\frac{\partial N}{\partial x} = \frac{\partial}{\partial x}\left[-x^{-1}y^{-2}\right] = x^{-2}y^{-2}$ treat $-y^{-2}$ as constant $\frac{\partial M}{\partial y} = x^{-2}y^{-2} = \frac{\partial N}{\partial x},$ hence, exact	17) Answer: $\frac{1}{xy} = c.$
18) $-\frac{y^3\,\mathrm{d}x - xy^2\,\mathrm{d}y}{x^4} = 0$ $-x^{-4}y^3\,\mathrm{d}x + x^{-3}y^2\,\mathrm{d}y = 0$ convert to M, N form $M(x,y)\,\mathrm{d}x + N(x,y)\,\mathrm{d}y = 0$ for comparison $\frac{\partial M}{\partial y} = \frac{\partial}{\partial y}\left[-x^{-4}y^3\right] = -3x^{-4}y^2$ treat $-x^{-4}$ as constant $\frac{\partial N}{\partial x} = \frac{\partial}{\partial x}\left[x^{-3}y^2\right] = -3x^{-4}y^2$ treat y^2 as constant $\frac{\partial M}{\partial y} = -3x^{-4}y^2 = \frac{\partial N}{\partial x},$ hence, exact	18) Answer: $\frac{y^3}{3x^3} = c.$
19) $\mathrm{e}^{3x}\,\mathrm{d}y + 3\mathrm{e}^{3x}y\,\mathrm{d}x = 0$ $M(x,y)\,\mathrm{d}x + N(x,y)\,\mathrm{d}y = 0$ for comparison $\frac{\partial M}{\partial y} = \frac{\partial}{\partial y}\left[3\mathrm{e}^{3x}y\right] = 3\mathrm{e}^{3x}$ $\frac{\partial N}{\partial x} = \frac{\partial}{\partial x}\left[\mathrm{e}^{3x}\right] = 3\mathrm{e}^{3x}$ $\frac{\partial M}{\partial y} = 3\mathrm{e}^{3x} = \frac{\partial N}{\partial x},$ hence, exact	19) $3\mathrm{e}^{3x}y\,\mathrm{d}x + \mathrm{e}^{3x}\,\mathrm{d}y = 0.$ Closed form formula from this chapter: $F(x,y) = \int M(x,y)\,\mathrm{d}x + \int\left[N(x,y) - \frac{\partial}{\partial y}\int M(x,y)\,\mathrm{d}x\right]\mathrm{d}y = C.$ $\int M(x,y)\,\mathrm{d}x = \int 3\mathrm{e}^{3x}y\,\mathrm{d}x$ already in $\int \mathrm{e}^u\,\mathrm{d}u$ form $= y\int \mathrm{e}^{3x}3\,\mathrm{d}x = \mathrm{e}^{3x}y$ $\mathrm{e}^{3x}y + \int\left[\mathrm{e}^{3x} - \frac{\partial}{\partial y}\left(\mathrm{e}^{3x}y\right)\right]\mathrm{d}y = C$ $\mathrm{e}^{3x}y + \int\left[\mathrm{e}^{3x} - \mathrm{e}^{3x}\right]\mathrm{d}y = C$ $\mathrm{e}^{3x}y + \int 0\,\mathrm{d}y = C$ $\mathrm{e}^{3x}y + 0 = C$ $\mathrm{e}^{3x}y = C$

The Return of the Jedi Integrating Factor

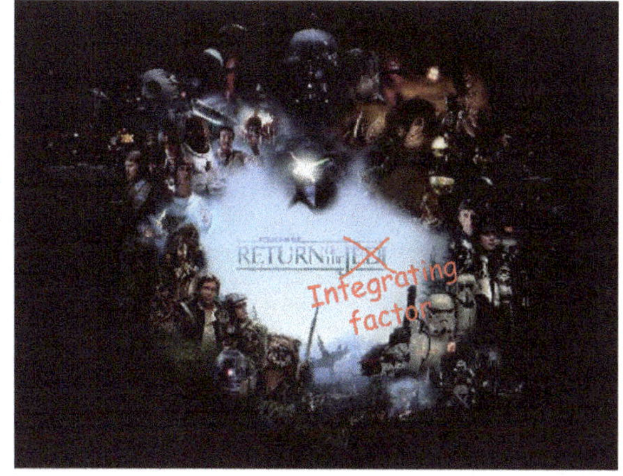

You may recall a major discussion in Chapter 8 about a concept called an integrating factor. There, we learned that any differential equation of the form $\frac{dy}{dx} + P(x)y = Q(x)$ could be transformed by multiplying both sides by $I(x) = e^{P(x)}$ resulting in the left side being a derivative of a product which was easily integrated as $\int (I(x) \times y)' = I(x) \times y + C$.

Review of Chapter 8

$$\frac{dy}{dx} + P(x)y = Q(x)$$

$$I(x)\left[\frac{dy}{dx} + P(x)y\right] = I(x) \times Q(x)$$

$$I(x) \times \frac{dy}{dx} + I(x) \times P(x) \times y = I(x) \times Q(x)$$

$$[I(x) \times y]' = I(x) \times Q(x)$$

$$\int [I(x) \times y]' = \int [I(x) \times Q(x)]$$

Then, after integrating both sides, the path will be clear to solve for y.

$$[I(x) \times y] = \int [I(x) \times Q(x)]$$

$$y = \frac{\int [I(x) \times Q(x)]}{I(x)}$$

Let's get back to Chapter 12 now. If you understood Chapter 8, then you will understand the following. Some differential equations are not in exact form

$$\frac{-y\,dx - x\,dy}{x^2} = 0$$

$$-x^{-2}y\,dx - x^{-1}y^0\,dy = 0$$

compares with $M(x, y)\,dx + N(x, y)\,dy = 0$

but $\frac{\partial M}{\partial y} = \frac{\partial}{\partial y}\left(-x^{-2}y\right) = -x^{-2}$, while $\frac{\partial N}{\partial x} = \frac{\partial}{\partial x}\left(-x^{-1}y^0\right) = x^{-2}$

Since $\frac{\partial M}{\partial y} \neq \frac{\partial N}{\partial x}$, $\frac{-y\,dx - x\,dy}{x^2} = 0$ is not an exact differential equation.

However, by multiplying both sides of the original equation by a magic term called an integrating factor *($-x^2$ for the above equation)*,

$$-x^2 \times \left(\frac{-y\,dx - x\,dy}{x^2}\right) = -x^2 \times 0$$

$$y\,dx + x\,dy = 0$$

compares with $M(x, y)\,dx + N(x, y)\,dy = 0$

and $\frac{\partial M}{\partial y} = \frac{\partial}{\partial y}(y) = 1$, while $\frac{\partial N}{\partial x} = \frac{\partial}{\partial x}(x) = 1$

Since $\frac{\partial M}{\partial y} = \frac{\partial N}{\partial x}$, $y\,dx + x\,dy = 0$ is an exact differential equation.

we transform the original equation to exact form, which can then be solved by the techniques taught in Chapter 12.

Chapter 12 Review

What follows is a primitive chart to help beginning students (the target audience for this book) differentiate and mentally organize their thoughts to this point in the book. This chart is simplistic because many equations can be solved by more than one technique.

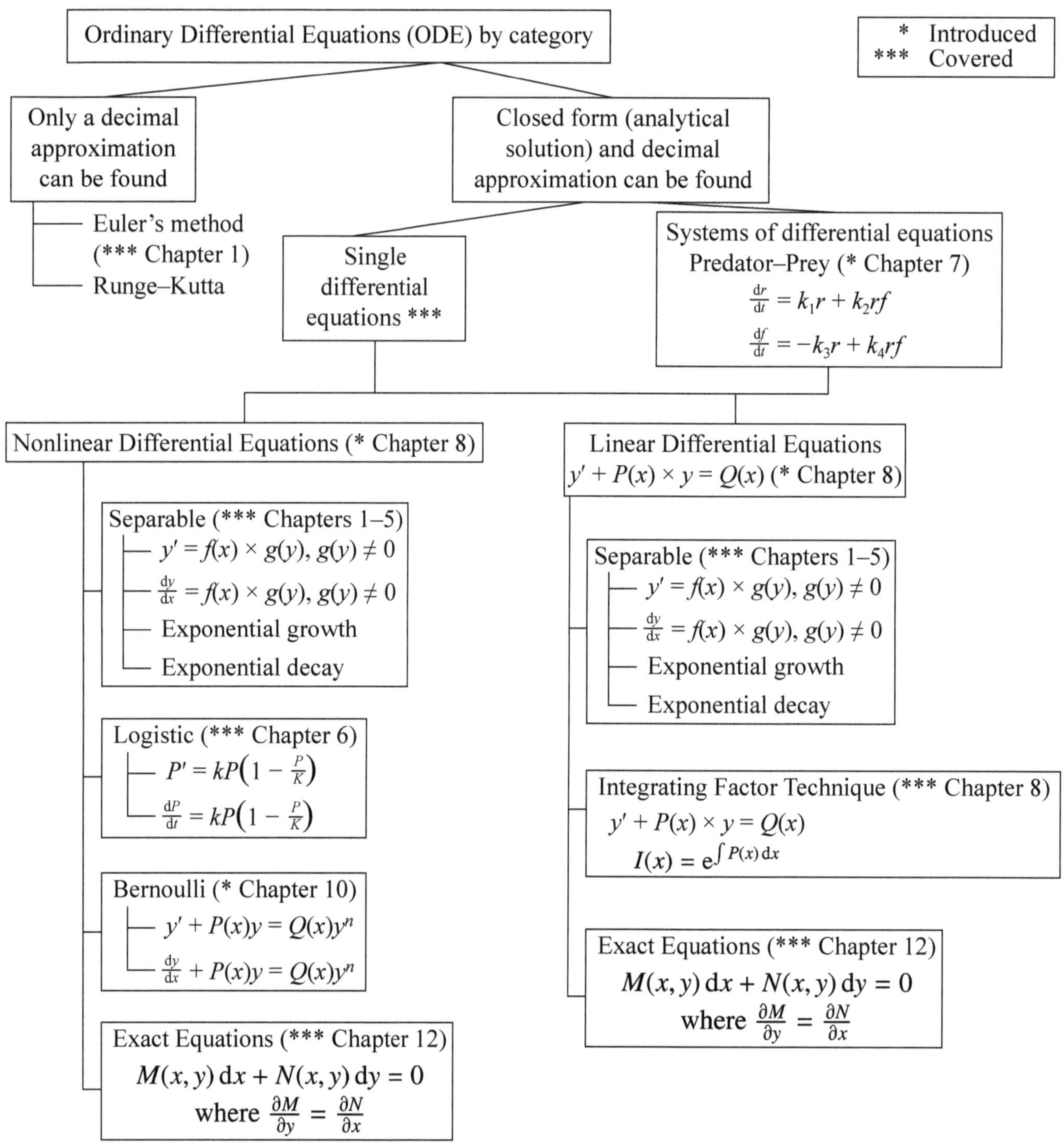

An exact differential equation has the form

$$(\text{function of } x \,\&\, y)\,dx + (\text{function of } x \,\&\, y)\,dy = 0$$

or symbolically

$$M(x, y)\,dx \quad + \quad N(x, y)\,dy \quad = 0$$

and $\frac{\partial M}{\partial y} = \frac{\partial N}{\partial x}$. There are specific techniques for solving exact differential equations that always work.

Chapter	Type of Differential Equation	Generic form, y'	Generic form, $\frac{dy}{dx}$
1–5	Separable (esp. growth & decay)	$y' = f(x) \times g(y)$	$\frac{dy}{dx} = f(x) \times g(y)$
6	Logistic $P(x) = \frac{K}{(A \times e^{-kt}+1)}$, where $A = \frac{K-P_0}{P_0}$	$P' = kP(1 - \frac{P}{K})$	$\frac{dP}{dt} = kP(1 - \frac{P}{K})$
7	Systems of Differential Equations, Predator–Prey		
8	Linear	$y' + P(x) \times y = Q(x)$	$\frac{dy}{dx} + P(x) \times y = Q(x)$
9	Differential Equations in Physics		
10	Bernoulli	$y' + P(x)y = Q(x)y^n$	$\frac{dy}{dx} + P(x) \times y = Q(x)y^n$
12	Exact	$M(x, y) + N(x, y)y'(x) = 0$ where $\frac{\partial M}{\partial y} = \frac{\partial N}{\partial x}$	$M(x, y)\,dx + N(x, y)\,dy = 0$ or $M(x, y) + N(x, y)\frac{dy}{dx} = 0$, where $\frac{\partial M}{\partial y} = \frac{\partial N}{\partial x}$

The Murder Mystery Method for Identifying and Solving Exact Differential Equations

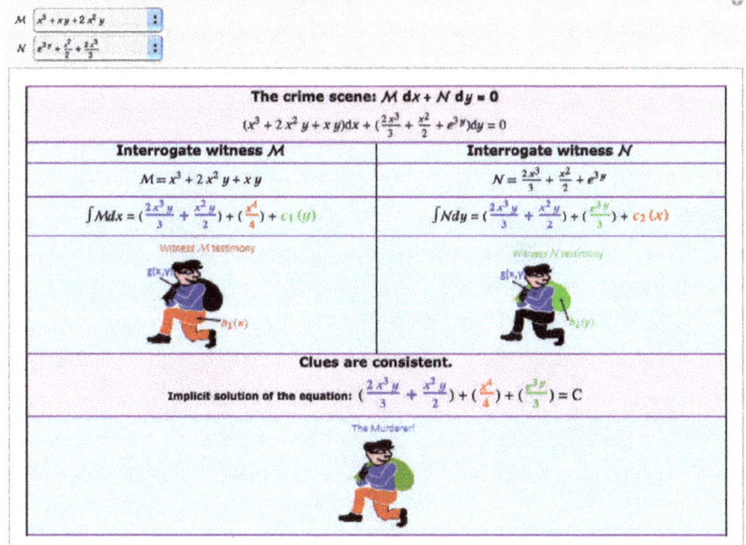

Online Application

Visit demonstrations.wolfram.com/TheMurderMysteryMethodForIdentifyingAndSolvingExactDifferent for a great little application that will demonstrate some of the concepts from this chapter. "The Murder Mystery Method for Identifying and Solving Exact Differential Equations" from the Wolfram Demonstrations Project. Contributed by José Luis Gómez-Muñoz, Roxana Ramirez-Herrera, Jezahel Lara-Sandoval and Edgar Fernández-Vergara.

Chapter 13
Homogeneous First-Order Substitution Technique

In algebra a linear equation was in the form $ax^1 + by^1 = c$ or $y^1 = mx^1 + b$. The dependent variable, y, and the independent variable(s) have a power of one. In Chapter 13, a linear differential equation is a differential equation in which the dependent variable, y, and all its derivatives have a power of zero or one. For example,

$$\frac{\mathrm{d}\mathbf{y}}{\mathrm{d}x}^{1} + P(x) \times \mathbf{y}^{1} = Q(x) \qquad \text{first-order } \textbf{linear differential equation}$$

$$\mathbf{y}'^{1} + P(x)\mathbf{y}^{1} = Q(x)$$

$$\frac{\mathrm{d}^2\mathbf{y}}{\mathrm{d}x^2}^{1} + P(x) \times \frac{\mathrm{d}\mathbf{y}}{\mathrm{d}x}^{1} + Q(x) \times \mathbf{y}^{1} = R(x) \qquad \text{second-order } \textbf{linear differential equation}$$

$$\mathbf{y}''^{1} + P(x)\mathbf{y}'^{1} + Q(x)\mathbf{y}^{1} = R(x)$$

$$\frac{\mathrm{d}^3\mathbf{y}}{\mathrm{d}x^3}^{1} + P(x) \times \frac{\mathrm{d}^2\mathbf{y}}{\mathrm{d}x^2}^{1} + Q(x) \times \frac{\mathrm{d}\mathbf{y}}{\mathrm{d}x}^{1} + R(x) \times \mathbf{y}^{1} = S(x) \qquad \text{third-order } \textbf{linear differential equation}$$

$$\mathbf{y}'''^{1} + P(x)\mathbf{y}''^{1} + Q(x)\mathbf{y}'^{1} + R(x)\mathbf{y}^{1} = S(x)$$

The term *homogeneous* has come through history with two totally different meanings in differential equations. This can be very confusing to a beginning differential equation student.

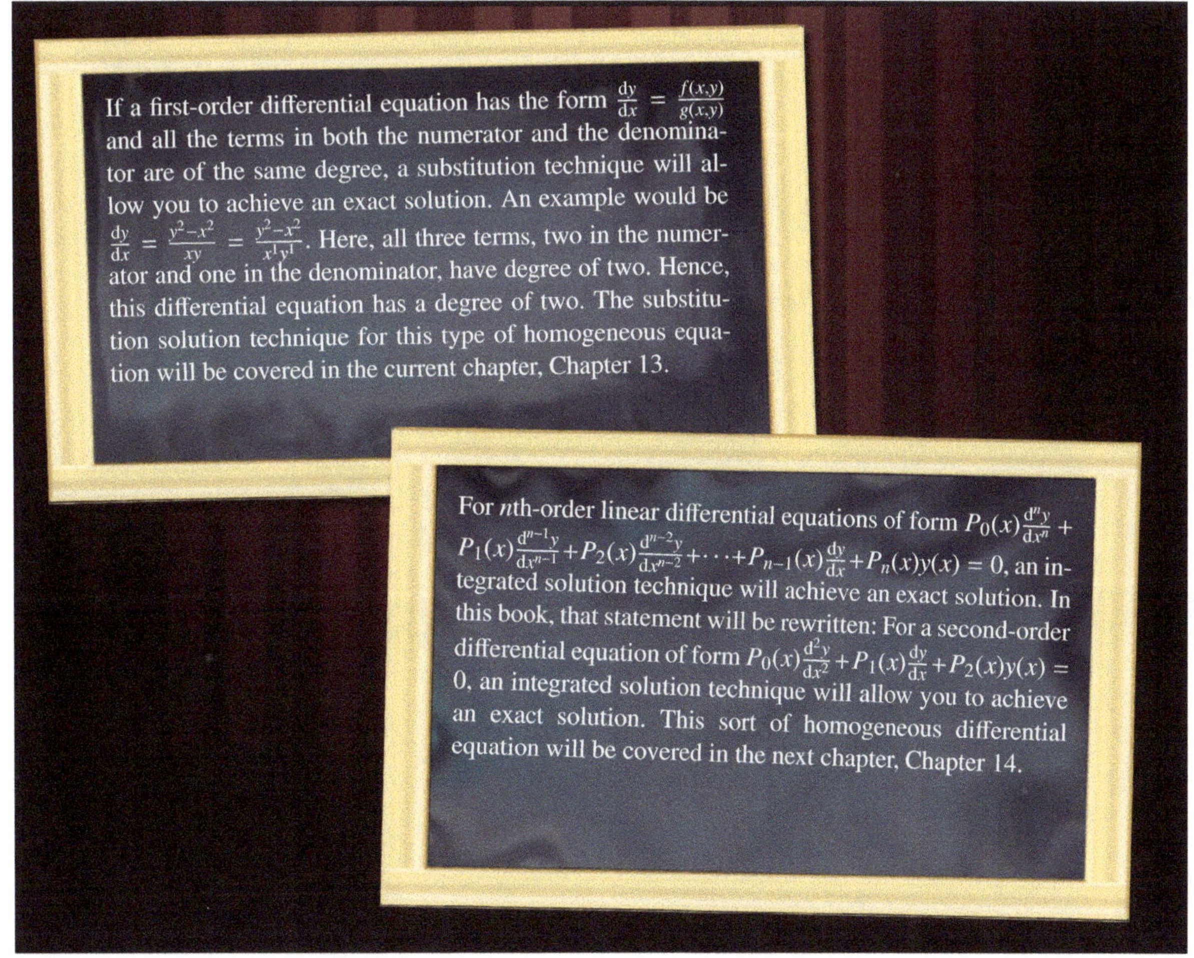
If a first-order differential equation has the form $\frac{\mathrm{d}y}{\mathrm{d}x} = \frac{f(x,y)}{g(x,y)}$ and all the terms in both the numerator and the denominator are of the same degree, a substitution technique will allow you to achieve an exact solution. An example would be $\frac{\mathrm{d}y}{\mathrm{d}x} = \frac{y^2-x^2}{xy} = \frac{y^2-x^2}{x^1y^1}$. Here, all three terms, two in the numerator and one in the denominator, have degree of two. Hence, this differential equation has a degree of two. The substitution solution technique for this type of homogeneous equation will be covered in the current chapter, Chapter 13.
For nth-order linear differential equations of form $P_0(x)\frac{\mathrm{d}^n y}{\mathrm{d}x^n} + P_1(x)\frac{\mathrm{d}^{n-1}y}{\mathrm{d}x^{n-1}} + P_2(x)\frac{\mathrm{d}^{n-2}y}{\mathrm{d}x^{n-2}} + \cdots + P_{n-1}(x)\frac{\mathrm{d}y}{\mathrm{d}x} + P_n(x)y(x) = 0$, an integrated solution technique will achieve an exact solution. In this book, that statement will be rewritten: For a second-order differential equation of form $P_0(x)\frac{\mathrm{d}^2y}{\mathrm{d}x^2} + P_1(x)\frac{\mathrm{d}y}{\mathrm{d}x} + P_2(x)y(x) = 0$, an integrated solution technique will allow you to achieve an exact solution. This sort of homogeneous differential equation will be covered in the next chapter, Chapter 14.

Homogeneous, First Order, Differential Equation Technique

$$\text{Solve } \frac{\mathrm{d}y}{\mathrm{d}x} = \frac{y^2 - x^2}{xy} = \frac{y^2 - x^2}{x^1 y^1} \ldots \text{ i.e., } \frac{f(x,y)}{g(x,y)}.$$

Step 1: Verify that each term in the numerator (y^2 and x^2) and each term in the denominator (x^1y^1) have the same degree. In this case, each term has degree two.

Step 2: Because the requirement in Step 1 has been met, it is possible to transform the equation such that it involves either $\frac{y}{x}$ or $\frac{x}{y}$, which will be replaced with an arbitrary substitution of v to greatly simplify the equation.

$$\frac{\mathrm{d}y}{\mathrm{d}x} = \frac{y^2 - x^2}{xy}$$

$$\frac{\mathrm{d}y}{\mathrm{d}x} = \frac{y^2 - x^2}{xy} \times \frac{\frac{1}{x^2}}{\frac{1}{x^2}}$$

$$\frac{\mathrm{d}y}{\mathrm{d}x} = \frac{\frac{y^2}{x^2} - \frac{x^2}{x^2}}{\frac{xy}{x^2}} = \frac{\frac{y^2}{x^2} - 1}{\frac{y}{x}} = \frac{\left(\frac{y}{x}\right)^2 - 1}{\frac{y}{x}}$$

Notice that the original equation has been transformed to have variables of form $\frac{y}{x}$. This is possible and was anticipated because the original differential equation was formed in such a way that all the terms in numerator and denominator had the same degree: in this case, two.

Step 3: Simplify the appearance of the differential equation above by substituting $v = \frac{y}{x}$. It needs to be repeated that this step follows logically from Steps 1 and 2 above and was anticipated as soon as the equation's form was recognized as *first-order homogeneous.*

$$\frac{\mathrm{d}y}{\mathrm{d}x} = \frac{\left(\frac{y}{x}\right)^2 - 1}{\frac{y}{x}} = \frac{v^2 - 1}{v} \qquad \text{(equation 1)}$$

Step 4: Since $v = \frac{y}{x}$, $y = vx$. Implicitly differentiate $y = vx$ in terms of x:

$$\frac{\mathrm{d}y}{\mathrm{d}x} = v\frac{\mathrm{d}x}{\mathrm{d}x} + x\frac{\mathrm{d}v}{\mathrm{d}x} = v + x\frac{\mathrm{d}v}{\mathrm{d}x}. \qquad \text{(equation 2)}$$

Step 5: Applying the transitive property from equation 1 to equation 2,

$$\frac{\mathrm{d}y}{\mathrm{d}x} = \frac{v^2 - 1}{v} = v + x\frac{\mathrm{d}v}{\mathrm{d}x}$$

$$\frac{v^2 - 1}{v} = v + x\frac{\mathrm{d}v}{\mathrm{d}x}$$

$$v - \frac{1}{v} = v + x\frac{\mathrm{d}v}{\mathrm{d}x}$$

$$-\frac{1}{v} = x\frac{\mathrm{d}v}{\mathrm{d}x} \quad \text{subtract } v \text{ from both sides}$$

$$-\frac{\mathrm{d}x}{v} = x\,\mathrm{d}v \quad \text{This looks very much like a separable differential equation!!}$$

$$\mathrm{d}x = -xv\,\mathrm{d}v$$

$$x^{-1}\,\mathrm{d}x = -v\,\mathrm{d}v$$

Step 6: Variables and differentials are separated. Integrate both sides.

$$\int -v\,\mathrm{d}v = \int x^{-1}\,\mathrm{d}x \quad \text{symmetric property, preparing to solve for } v$$

$$\frac{-v^2}{2} + c_1 = \ln|x| + c_2$$

$$-\frac{v^2}{2} = \ln|x| + c_2 - c_1$$

$$-\frac{v^2}{2} = \ln|x| + c_3 \quad \text{let } c_3 = c_2 - c_1$$

$$v^2 = -2\ln|x| + c_4 \quad \text{let } c_4 = -2c_3$$

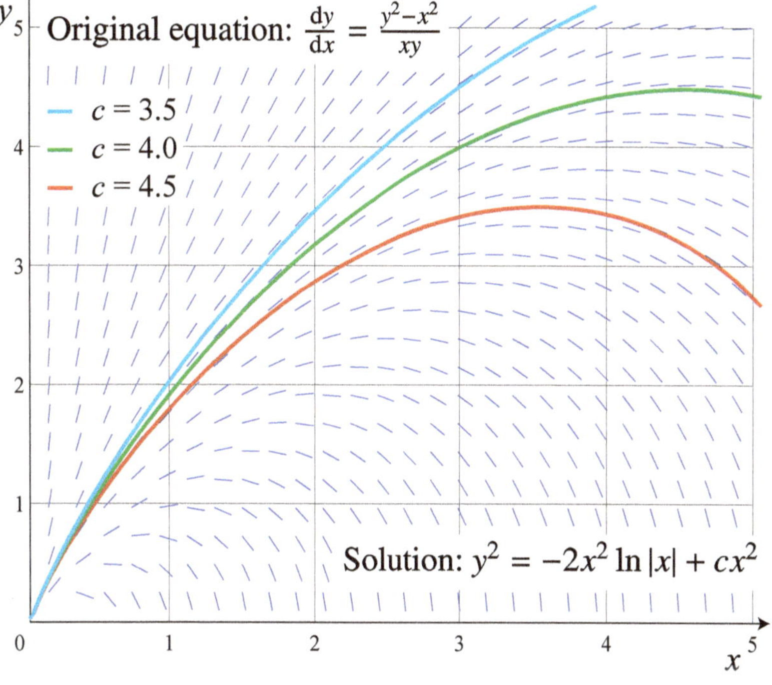

Step 7: From above, $v = \frac{y}{x}$, therefore $v^2 = \frac{y^2}{x^2}$.

$$\frac{y^2}{x^2} = -2\ln|x| + c_4$$

$$y^2 = -2x^2\ln|x| + c_4x^2$$

Repetition and drill in recognizing homogeneous equations of the form $\frac{\mathrm{d}y}{\mathrm{d}x} = \frac{f(x,y)}{g(x,y)}$.

1) $\frac{\mathrm{d}y}{\mathrm{d}x} = \frac{y^3-2x^3}{xy^2}$	y^3 has degree three, $2x^3$ has degree three, xy^2 has degree three. Therefore, $\frac{\mathrm{d}y}{\mathrm{d}x}$ is homogeneous and can be solved by substitution using $v = \frac{y}{x}$ or $v = \frac{x}{y}$ or $y = vx$ or $x = vy$.
2) $\frac{\mathrm{d}y}{\mathrm{d}x} = \frac{5x^4+2y^4}{xy^3}$	$5x^4$ has degree four, $2y^4$ has degree four, xy^3 has degree four. Therefore, $\frac{\mathrm{d}y}{\mathrm{d}x}$ is homogeneous and can be solved by substitution using $v = \frac{y}{x}$ or $v = \frac{x}{y}$ or $y = vx$ or $x = vy$.
3) $\frac{\mathrm{d}y}{\mathrm{d}x} = \frac{2xy}{y^2-x^2}$	$2xy$ has degree two, y^2 has degree two, x^2 has degree two. Therefore, $\frac{\mathrm{d}y}{\mathrm{d}x}$ is homogeneous and can be solved by substitution using $v = \frac{y}{x}$ or $v = \frac{x}{y}$ or $y = vx$ or $x = vy$.
4) $\frac{\mathrm{d}y}{\mathrm{d}x} = \frac{2x^5+y^5}{x^2y^3}$	$2x^5$ has degree five, y^5 has degree five, x^2y^3 has degree five. Therefore, $\frac{\mathrm{d}y}{\mathrm{d}x}$ is homogeneous and can be solved by substitution using $v = \frac{y}{x}$ or $v = \frac{x}{y}$ or $y = vx$ or $x = vy$.
5) $\left(x^3 - 3y^3\right)\mathrm{d}x + 2x^2y\,\mathrm{d}y = 0$ $2x^2y\,\mathrm{d}y = -\left(x^3 - 3y^3\right)\mathrm{d}x$ $\frac{\mathrm{d}y}{\mathrm{d}x} = \frac{-(x^3-3y^3)}{2x^2y}$	x^3 has degree three, $3y^3$ has degree three, $2x^2y$ has degree three. Therefore, $\frac{\mathrm{d}y}{\mathrm{d}x}$ is homogeneous and can be solved by substitution using $v = \frac{y}{x}$ or $v = \frac{x}{y}$ or $y = vx$ or $x = vy$.

Chapter 13 Review

Chapter 13 has shown how to solve a *first-order homogeneous* differential equation by substitution so that the equation of the form $\frac{dy}{dx} = \frac{f(x,y)}{g(x,y)}$ is transformed into a separable differential equation.

Chapter	Type of Differential Equation	Generic form, y'	Generic form, $\frac{dy}{dx}$
1–5	Separable (esp. growth & decay)	$y' = f(x) \times g(y)$	$\frac{dy}{dx} = f(x) \times g(y)$
6	Logistic $P(x) = \frac{K}{(A\times e^{-kt}+1)}$, where $A = \frac{K-P_0}{P_0}$	$P' = kP(1 - \frac{P}{K})$	$\frac{dP}{dt} = kP(1 - \frac{P}{K})$
7	Systems of Differential Equations, Predator–Prey	$r' = \beta r - k_1 rf$ $f' = -\alpha f + k_2 rf$	$\frac{dr}{dt} = \beta r - k_1 rf$ $\frac{df}{dt} = -\alpha f + k_2 rf$
8	Linear	$y' + P(x) \times y = Q(x)$	$\frac{dy}{dx} + P(x) \times y = Q(x)$
9	Differential Equations in Physics		
10	Bernoulli	$y' + P(x)y = Q(x)y^n$	$\frac{dy}{dx} + P(x) \times y = Q(x)y^n$
12	Exact	$M(x, y) + N(x, y)y'(x) = 0$ where $\frac{\partial M}{\partial y} = \frac{\partial N}{\partial x}$	$M(x, y)\,dx + N(x, y)\,dy = 0$ or $M(x, y) + N(x, y)\frac{dy}{dx} = 0$, where $\frac{\partial M}{\partial y} = \frac{\partial N}{\partial x}$
13	First-Order Homogeneous	$y' = \frac{f(x,y)}{g(x,y)}$	$\frac{dy}{dx} = \frac{f(x,y)}{g(x,y)}$, where all terms in f and g are of the same degree
14	Second-Order Homogeneous	$y''(x) + by'(x) + cy(x) = 0$	$\frac{d^2y}{dx^2} + b\frac{dy}{dx} + cy = 0$

The technique of cleverly substituting to change an unfamiliar differential equation form into a familiar one is often useful. We have seen two examples of the technique now.

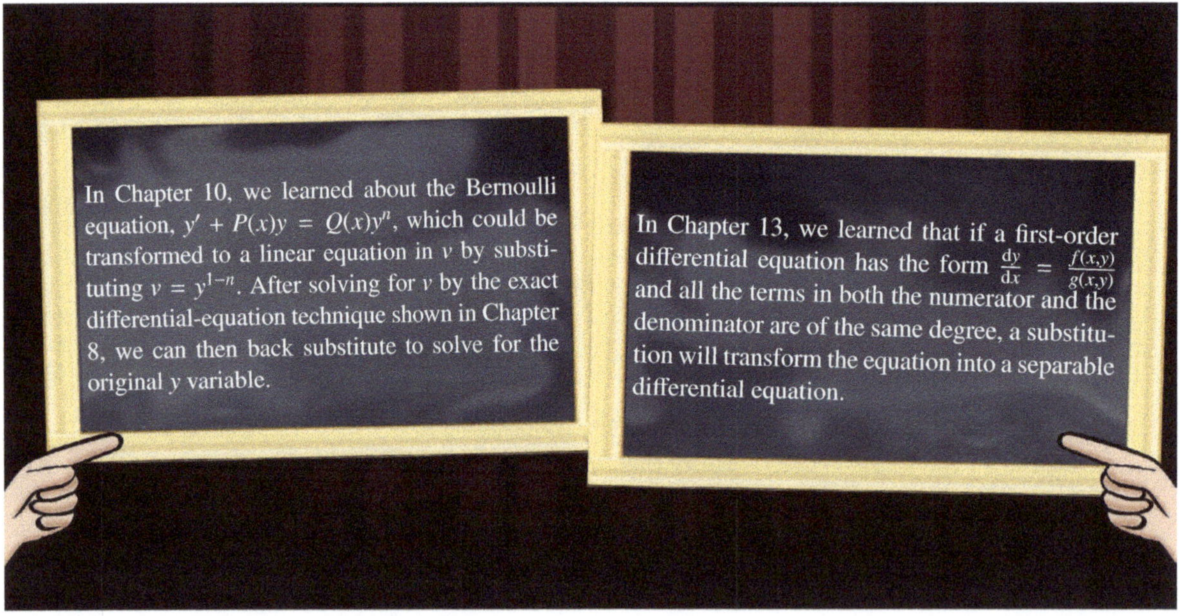

In Chapter 10, we learned about the Bernoulli equation, $y' + P(x)y = Q(x)y^n$, which could be transformed to a linear equation in v by substituting $v = y^{1-n}$. After solving for v by the exact differential-equation technique shown in Chapter 8, we can then back substitute to solve for the original y variable.
In Chapter 13, we learned that if a first-order differential equation has the form $\frac{dy}{dx} = \frac{f(x,y)}{g(x,y)}$ and all the terms in both the numerator and the denominator are of the same degree, a substitution will transform the equation into a separable differential equation.

What follows is a primitive chart to help beginning students (the target audience for this book) differentiate and mentally organize their thoughts to this point in the book. This chart is simplistic because many equations can be solved by more than one technique.

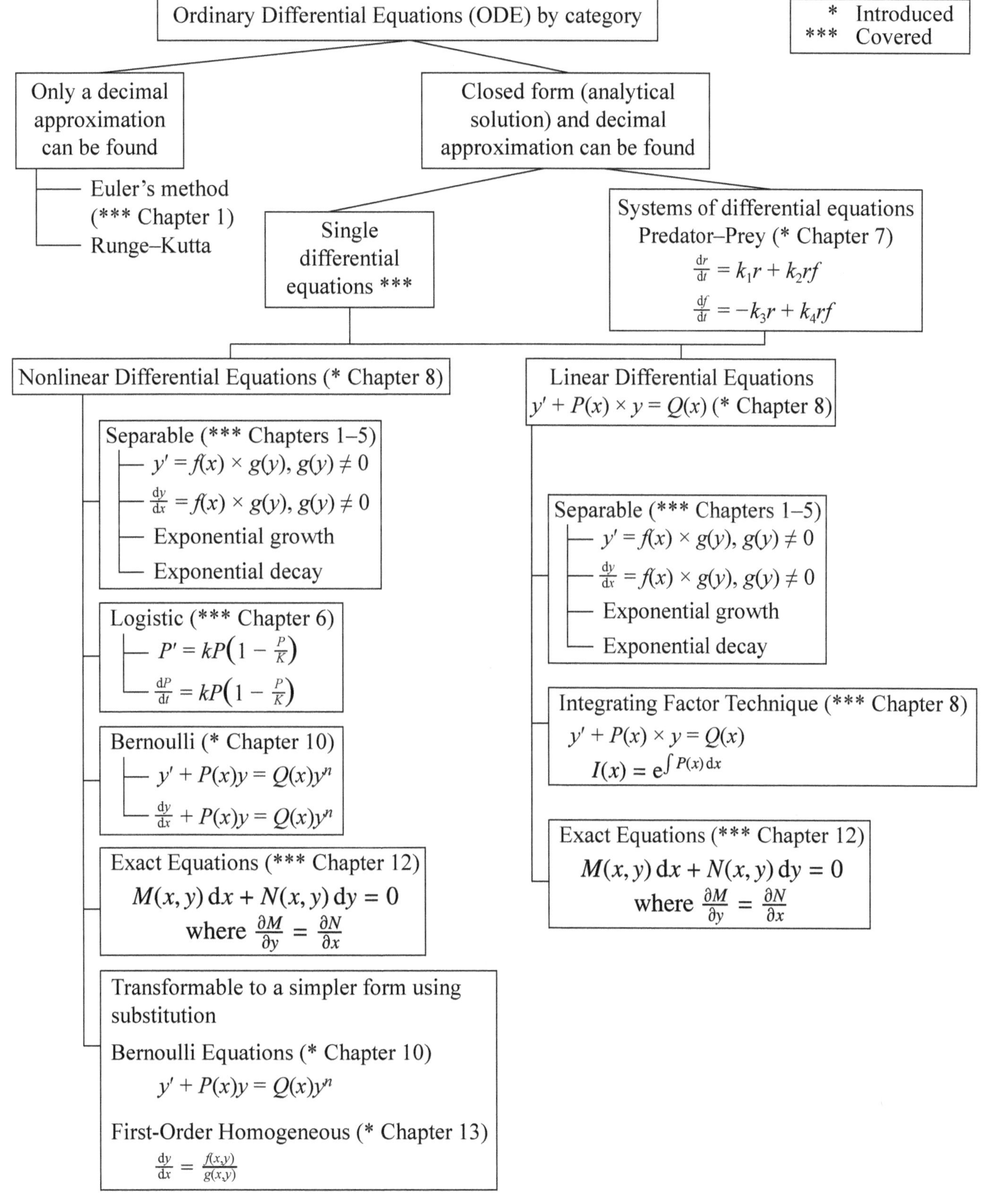

Chapter 14
Homogeneous Second-Order Linear O.D.E. Solution Technique with Constant Coefficients

In algebra a general linear equation was in the form of $y^1 = mx + b$. The dependent variable, y, has a power of one. A linear differential equation is a differential equation in which the dependent variable, y, and all its derivatives, have a power of zero or one. *Linear differential equations are also organized by their order*:

$$\frac{\mathrm{d}\mathbf{y}}{\mathrm{d}x}^1 + P(x) \times \mathbf{y}^1 = Q(x)$$ first-order **linear differential equation**

$$\mathbf{y}'^1 + P(x)\mathbf{y}^1 = Q(x)$$

$$\frac{\mathrm{d}^2\mathbf{y}}{\mathrm{d}x^2}^1 + P(x) \times \frac{\mathrm{d}\mathbf{y}}{\mathrm{d}x}^1 + Q(x) \times \mathbf{y}^1 = R(x)$$ second-order **linear differential equation**

$$\mathbf{y}''^1 + P(x)\mathbf{y}'^1 + Q(x)\mathbf{y}^1 = R(x)$$

$$\frac{\mathrm{d}^3\mathbf{y}}{\mathrm{d}x^3}^1 + P(x) \times \frac{\mathrm{d}^2\mathbf{y}}{\mathrm{d}x^2}^1 + Q(x) \times \frac{\mathrm{d}\mathbf{y}}{\mathrm{d}x}^1 + R(x) \times \mathbf{y}^1 = S(x)$$ third-order **linear differential equation**

$$\mathbf{y}'''^1 + P(x)\mathbf{y}''^1 + Q(x)\mathbf{y}'^1 + R(x)\mathbf{y}^1 = S(x)$$

The term "homogeneous differential equation" has come through history with two totally different meanings in mathematics. This can be very confusing to a beginning differential-equation student.

Homogeneous First-Order Differential Equation Technique, Chapter 13

If a first-order differential equation has the form $\frac{\mathrm{d}y}{\mathrm{d}x} = \frac{f(x,y)}{g(x,y)}$ and all the terms in both the numerator and the denominator are of the same degree, a substitution technique will allow you to achieve an exact solution. An example would be $\frac{\mathrm{d}y}{\mathrm{d}x} = \frac{y^2-x^2}{xy} = \frac{y^2-x^2}{x^1y^1}$. Here, all three terms, two in the numerator and one in the denominator, have degree of two. Hence, this differential equation has a degree of two. The substitution solution technique for this type of "first-order homogeneous" equation was covered in Chapter 13.

Homogeneous Second-Order Differential Equation Technique, Chapter 14

For nth-order linear differential equations of the form

$$F_0(x)\frac{\mathrm{d}^n y}{\mathrm{d}x^n} + F_1(x)\frac{\mathrm{d}^{n-1} y}{\mathrm{d}x^{n-1}} + F_2(x)\frac{\mathrm{d}^{n-2} y}{\mathrm{d}x^{n-2}} + \cdots + F_{n-1}(x)\frac{\mathrm{d}y}{\mathrm{d}x} + F_n(x)y(x) = G(x),$$

a different solution technique will achieve an exact solution. In this book, that nth-order equation becomes a second-order linear differential equation of form

$$F_0(x)\frac{\mathrm{d}^2 y}{\mathrm{d}x^2} + F_1(x)\frac{\mathrm{d}y}{\mathrm{d}x} + F_2(x)y = G(x) \quad \text{or} \quad F_0y''(x) + F_1y'(x) + F_2y(x) = G(x),$$

an integrated-solution technique will allow you to achieve an exact solution. **Here each F_k coefficient represents a function in x. For a beginner's approach to homogeneous differential equations, it is traditional to assume constants for all the F_k coefficients and 0 for G(x). So, for beginners, the form necessary to apply the homogeneous-differential-equation solution technique would be: $Ay''(x) + By'(x) + Cy(x) = 0$.** If you are having a déjà-vu feeling right now, that the differential equation just shown has a form that looks very much like the algebraic equation $Ax^2 + Bx + C = 0$, that is good!!!!! Hold that thought!!

What follows is a well-intended attempt by the author to explain **why** the second-order homogeneous solution technique works. If you are only interested in **how** to solve such equations, jump ahead two pages.

A specific example of a second-order homogeneous linear differential equation $Ay''(x)+By'(x)+Cy(x)=0$ with $A=1$, $B=2$, and $C=-24$ would look like $1y''(x)+2y'(x)-24y(x)=0$. This means that the second derivative of $y(x)+$ twice the first derivative of $y(x)-24$ times the function $y(x)$ is zero. After a short digression, this will be the differential equation we solve.

A great technique for understanding new math ideas is to back away from the problem at hand, posit a simpler version of that problem, analyze and solve the easier problem, then take what you have learned and try to apply it to the original, more difficult problem. An easier version of the problem $1y''(x)+2y'(x)-24y(x)=0$ would be $2y'-24y(x)=0$. Hmm. Remembering that the goal of a differential equation is to solve for its $y(x)$, we get $-24y(x)=-2y'$ so $y(x)=\frac{1}{12}y'(x)$. Hmmm. Do we know any functions that are $\frac{1}{12}$ of their first derivative? Recalling that $\frac{d}{dx}e^u=e^u\,du$, if $y(x)=e^{12x}$, then $y'(x)=12e^{12x}=12\times y(x)$. Since the derivative $y'(x)=12\times y(x)$, $y(x)$ is $\frac{1}{12}$ of its derivative, $y'(x)$.

Let's check by substituting $y(x)=e^{12x}$ and $y'(x)=12e^{12x}$ into the original differential equation. See the work at right. In general, if $y(x)=e^{kx}$ then $y'(x)=ke^{kx}=k\times y(x)$. Let's do that again with a slight twist. The differential equation $2y'-24y(x)=0$ looks like the algebraic equation $2r-24=0$, so $r=12$. Let $y(x)=e^{rx}$ so $y'=re^{rx}$. For $r=12$ we have $y(x)=e^{12x}$ and $y'=12e^{12x}$. These are, of course, the same as the answers we got before when working only with the differential equation.

$$2y'-24y(x)=0 \quad \text{original}$$
$$2\left(12e^{12x}\right)-24\left(e^{12x}\right)\stackrel{?}{=}0$$
$$24e^{12x}-24e^{12x}\stackrel{?}{=}0$$
$$0=0 \quad \text{check!}$$

By solving the differential equation $2y'(x)-24y(x)=0$ as though it were a linear algebraic equation—i.e., $2r-24=0$, so $r=12$—we got information that allowed us to solve that differential equation. Now we have a clue as to how to attack the original specific differential equation, $1y''(x)+2y'(x)-24y(x)=0$.

Substituting into the original differential equation from the chart at right:

$$y(x)=e^{rx}$$
$$y'(x)=re^{rx}$$
$$y''(x)=r^2e^{rx}$$

$$1y''(x)+2y'(x)-24y(x)=0 \quad \text{original differential equation}$$
$$1r^2e^{rx}+2\times re^{rx}-24e^{rx}=0 \quad \text{substitute values from chart}$$
$$e^{rx}\left(1r^2+2r-24\right)=0$$
$$\frac{e^{rx}\left(1r^2+2r-24\right)}{e^{rx}}=\frac{0}{e^{rx}} \quad \text{ok because } e^{rx}\neq 0$$
$$1r^2+2r-24=0$$
$$(r+6)(r-4)=0$$
$$r=-6,\ r=4 \quad \text{two answers this time because the equation was quadratic}$$

Check by substituting these two values of r into the original equation

r	$1y''(x)$	$+$	$2y'(x)$	$-$	$24y(x)$	$=$	0	
	$1r^2e^{rx}$	$+$	$2\times re^{rx}$	$-$	$24e^{rx}$	$=$	0	
-6	$1\times(-6)^2e^{-6x}$	$+$	$(2\times -6)e^{-6x}$	$-$	$24e^{-6x}$	$=$	0	
	$36e^{-6x}$	$-$	$12\times e^{-6x}$	$-$	$24e^{-6x}$	$=$	0	
					0	$=$	0	check!
4	$1\times(4)^2e^{4x}$	$+$	$(2\times 4)e^{4x}$	$-$	$24e^{4x}$	$=$	0	
	$16e^{4x}$	$+$	$8e^{4x}$	$-$	$24e^{4x}$	$=$	0	
					0	$=$	0	check!

$1y''(x)+2y'(x)-24y(x)=0$ when $r=-6$ and also when $r=4$. There are two answers this time because the equation was quadratic. *Somehow these two r values, −6 and 4, should help us in solving the original differential equation.*

From our study back in Chapters 1–3, we should remember that for every differential equation where the right-hand side is equal to zero, there is a family of solutions: infinitely many solutions. We need a general equation for that. We have yet to achieve that goal. Recall the original generic differential equation form we are studying in Chapter 14, $Ay''(x) + By'(x) + Cy(x) = 0$. By definition, the solution to this differential equation would be $y(x)$. Would $c_1 \times y(x)$ also be a solution to the original differential equation? That is, would a multiple of the solution to the differential equation also be a solution? If so we should be able to substitute it into the original equation.

$Ay''(x) + By'(x) + Cy(x) = 0$	homogeneous differential equation
$A \times [c_1 \times y''(x)] + B \times [c_1 \times y'(x)] + C \times [c_1 \times y(x)] \stackrel{?}{=} 0$	arbitrary substitution of a constant
$c_1 \times [Ay''(x) + By'(x) + Cy(x)] \stackrel{?}{=} 0$	factoring out the constant c_1
$c_1(0) = 0$	substituting from the original
$0 = 0$	

Yes, if a function $y(x)$ is a solution to a homogeneous linear differential equation, then $c_1 \times y(x)$ will also be a solution. From that little digression, if $y = e^{rx}$ is a solution, then $y = Ce^{rx}$ is also a solution. If $y = e^{-6x}$ is a solution, then $y = c_1e^{-6x}$ is also a solution. If $y = e^{4x}$ is a solution, then $y = c_2e^{4x}$ is also a solution.

"If $y(x)$ is the solution to a homogeneous second-order differential equation, then $c \times y(x)$ is also a solution to that second-order differential equation. This is saying that if a solution is of the form $y = e^{rx}$ then constant multiples of e^{rx} (that is, ce^{rx}) are also solutions!"

Assume that, in addition to $y(x)$ being a solution to the differential equation, there is a function $f(x)$ that is also a solution to that same equation. It turns out that the sum $y(x) + f(x)$ will also be a solution.

$$Ay''(x) + By'(x) + Cy(x) = 0 \quad \text{and} \quad Af''(x) + Bf'(x) + Cf(x) = 0$$

Add the left sides of these two equations and also the right, then group terms:

$[Ay''(x) + By'(x) + Cy(x)] + [Af''(x) + Bf'(x) + Cf(x)] = 0$	add the left side and right sides
$Ay''(x) + Af''(x) + By'(x) + Bf'(x) + Cy(x) + Cf(x) = 0$	group A, B, and C terms
$A[y''(x) + f''(x)] + B[y'(x) + f'(x)] + C[y(x) + f(x)] = 0$	factor out A, B, and C
$A\phi''(x) + B\phi'(x) + C\phi(x) = 0$	where $\phi(x) = y(x) + f(x)$, etc.

Therefore, if functions $y(x)$ and $f(x)$ are both solutions to a differential equation then $\phi(x) = y(x) + f(x)$ is also a solution. **This is called the "linearity principle."**

The two ideas shown above are grouped together.

Given a second-order linear homogeneous differential equation $Ay''(x) + By'(x) + Cy(x) = 0$:

1. If $f(x)$ and $g(x)$ are both solutions, then $[f(x) + g(x)]$ is also a solution. This is called a *linear combination* of the funtions (The sum of solutions is also a solution.)

2. If $f(x)$ is a solution, then $c \times f(x)$ is also a solution. (A constant multiple of a solution is also a solution.)

Finally, returning to the original differential equation, $y''(x) + 2y'(x) - 24y(x) = 0$, if $y_1 = e^{-6x}$ is a solution, then $y_1 = c_1e^{-6x}$ is also a solution. If $y_2 = e^{4x}$ is a solution, then $y_2 = c_2e^{4x}$ is also. Furthermore, we can state that the solution $y(x) = c_1e^{-6x} + c_2e^{4x}$ is also. Check our work:

$y = c_1e^{-6x}$	$y = c_2e^{4x}$	$y = c_1e^{-6x} + c_2e^{4x}$
$y' = -6c_1e^{-6x}$	$y' = 4c_2e^{4x}$	$y' = -6c_1e^{-6x} + 4c_2e^{4x}$
$y'' = 36c_1e^{-6x}$	$y'' = 16c_2e^{4x}$	$y'' = 36c_1e^{-6x} + 16c_2e^{4x}$

$1y''(x) + 2y'(x) - 24y(x) = 0$ Test $y_1 = c_1e^{-6x}$	$1y''(x) + 2y'(x) - 24y(x) = 0$ Test $y_2 = c_2e^{4x}$	$1y''(x) + 2y'(x) - 24y(x) = 0$ Test $y_3 = c_1e^{-6x} + c_2e^{4x}$
$36c_1e^{-6x} + 2\left(-6c_1e^{-6x}\right) - 24c_1e^{-6x} = 0$ $36c_1e^{-6x} - 12c_1e^{-6x} - 24c_1e^{-6x} = 0$ $0 = 0$	$16c_2e^{4x} + 2\left(4c_2e^{4x}\right) - 24c_2e^{4x} = 0$ $16c_2e^{4x} + 8c_2e^{4x} - 24c_2e^{4x} = 0$ $0 = 0$	$\left(36c_1e^{-6x} + 16c_2e^{4x}\right) + 2\left(-6c_1e^{-6x} + 4c_2e^{4x}\right)$ $- 24\left(c_1e^{-6x} + c_2e^{4x}\right) = 0$ $36c_1e^{-6x} + 16c_2e^{4x} - 12c_1e^{-6x} + 8c_2e^{4x}$ $- 24c_1e^{-6x} - 24c_2e^{4x} = 0$

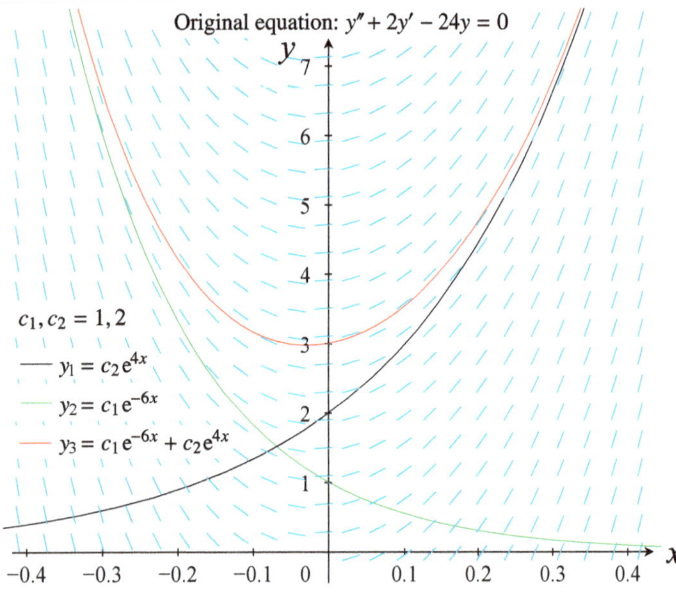

If y_1 and y_2 are both solutions, then $y_3 = y_1 + y_2$ is also a solution. Note that the slope field follows the complete solution and only applies to selected intervals of the partial solutions.

The graph shows that as $x \to \infty$, the c_1e^{-6x} term decays and the c_2e^{4x} term dominates the linear combination causing the graph to go to infinity. Similarly, as $x \to -\infty$, the c_2e^{4x} term decays and the c_1e^{-6x} term dominates the linear combination causing the graph to go to infinity.

Practice Solving a Second-Order Homogeneous Differential Equation #1

Solve the equation $y'' + 2y' - 8y = 0$.

Step 1: Notice that the differential equation is in the form $Ay''(x) + By'(x) + Cy(x) = 0$. The given differential equation is in standard form for a second-order homogeneous equation.

Step 2: Using the differential coefficients 1, 2, and −8 to form a quadratic algebraic equation, we get $r^2 + 2r - 8 = 0$. Evaluating the discriminant, we get $2^2 - (4 \times 1 \times -8) = 4 + 32 = 36 > 0$. That means that the general solution to the equation will be $y_{\text{gen}} = c_1e^{r_1x} + c_2e^{r_2x}$.

Step 3: Solve the quadratic algebraic equation $1r^2 + 2r - 8 = 0$: $(r + 4)(r - 2) = 0$. Hence, $r = -4$ and 2.

Step 4: Substitute the *r* values into $y_{\text{gen}} = c_1e^{r_1x} + c_2e^{r_2x}$:

$$y_{\text{gen}} = c_1e^{-4x} + c_2e^{2x}.$$

$y(x) = c_1e^{-4x} + c_2e^{2x}$
$y'(x) = -4c_1e^{-4x} + 2c_2e^{2x}$
$y''(x) = (-4)^2c_1e^{-4x} + (2)^2c_2e^{2x}$
$= 16c_1e^{-4x} + 4c_2e^{2x}$

That's it! The equation y_{gen} is a family of differential equations with two parameters that satisfies the original differential equation.

$$1y''(x) \quad + \quad 2y'(x) \quad - \quad 8y(x) \quad = 0 \text{ check by substituting for } y, y', \text{ and } y''$$

$$\left(16c_1e^{-4x} + 4c_2e^{2x}\right) + 2\left(-4c_1e^{-4x} + 2c_2e^{2x}\right) - 8\left(c_1e^{-4x} + c_2e^{2x}\right) \stackrel{?}{=} 0$$

$$16c_1e^{-4x} + 4c_2e^{2x} - 8c_1e^{-4x} + 4c_2e^{2x} - 8c_1e^{-4x} - 8c_2e^{2x} \stackrel{?}{=} 0$$

$$\left(16c_1e^{-4x} - 8c_1e^{-4x} - 8c_1e^{-4x}\right) + \left(4c_2e^{2x} + 4c_2e^{2x} - 8c_2e^{2x}\right) \stackrel{?}{=} 0$$

$$0 \quad + \quad 0 \quad = 0 \text{ check}$$

Changing the equation above to an IVP, solve the differential equation $y'' + 2y' - 8y = 0$ given that $y(0) = 2$ and $y'(0) = 6$. (Author's note: Two initial conditions must be given for a second-order differential equation.)

$$y(x) = c_1e^{-4x} + c_2e^{2x}, \text{ so } y(0) = c_1e^{-4\times 0} + c_2e^{2\times 0} = 2.$$

Therefore, $c_1 + c_2 = 2$. (1)

$$y'(x) = -4c_1e^{-4x} + 2c_2e^{2x}, \text{ so } y'(0) = -4c_1e^{-4\times 0} + 2c_2e^{2\times 0} = 6.$$

Therefore, $-4c_1 + 2c_2 = 6$. (2)

$4 \times [c_1 + c_2] = 4 \times 2$ make an executive decision as to which variable to eliminate (1)

$4c_1 + 4c_2 = 8$ eliminate the c_1 terms by "adding" the two equations (1)

$-4c_1 + 2c_2 = 6$ (2)

$$6c_2 = 14$$

$$c_2 = \frac{7}{3}$$

$c_1 + c_2 = 2$ (1)

$$c_1 + \frac{7}{3} = \frac{6}{3}$$

$$c_1 = -\frac{1}{3}$$

Hence, for the general differential equation $y'' + 2y' - 8y = 0$, the specific solution for the given IVP conditions is $y_{gen} = \left(-\frac{1}{3}e^{-4x} + \frac{7}{3}e^{2x}\right)$. Note that $\lim_{x\to\infty}\left(-\frac{1}{3}e^{-4x} + \frac{7}{3}e^{2x}\right) = 0 + \infty = \infty$ and $\lim_{x\to-\infty}\left(-\frac{1}{3}e^{-4x} + \frac{7}{3}e^{2x}\right) = -\infty + 0 = -\infty$.

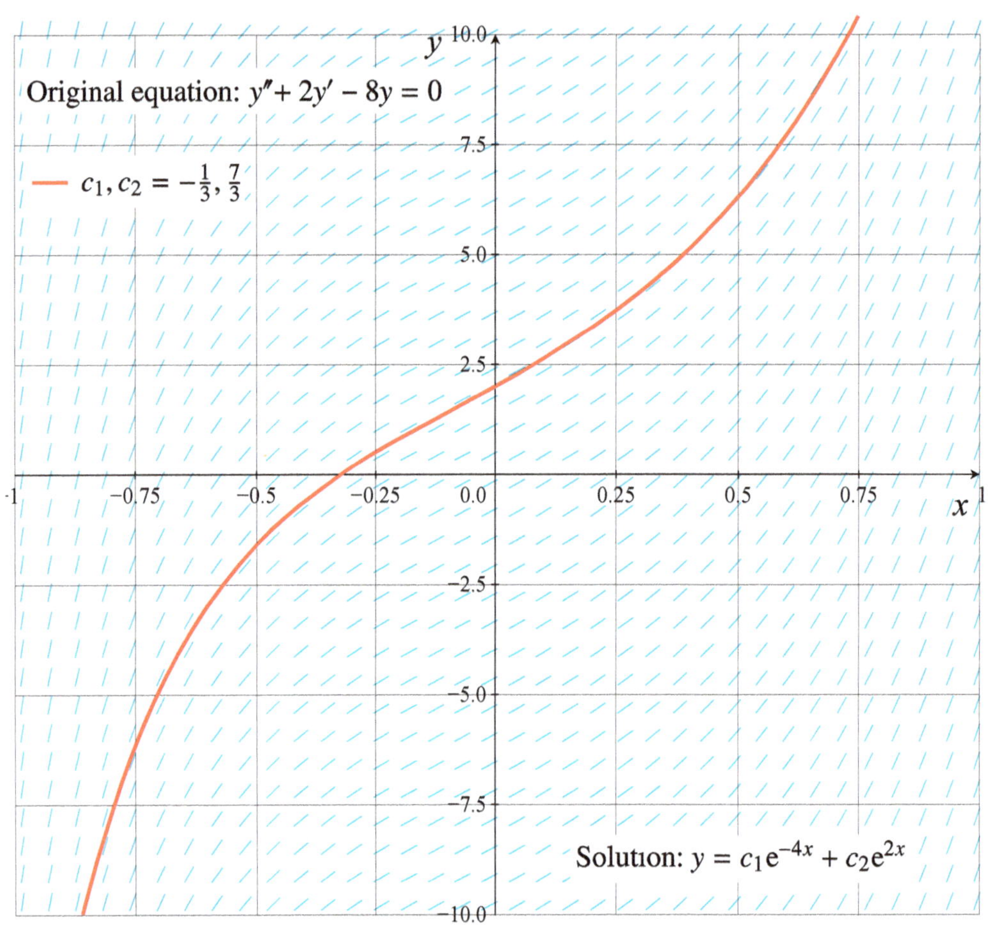

Notice from the graph that the c_1e^{-4x} term decays to zero as $x \to \infty$ while the c_2e^{2x} term dominates the linear combination causing the graph to go to infinity. A similar process occurs as $x \to -\infty$.

Practice Solving a Second-Order Homogeneous Differential Equation #2

Solve the equation $y'' + 5y' + 6y = 0$.

Step 1: Notice that the differential equation is in the form $Ay''(x) + By'(x) + Cy(x) = 0$. The given differential equation is in standard form for a second-order homogeneous equation.

Step 2: Using the differential coefficients 1, 5, and 6 to form a quadratic algebraic equation, we get $r^2 + 5r + 6 = 0$. Evaluating the discriminant, we get $5^2 - (4 \times 1 \times 6) = 25 - 24 = 1 > 0$. That means that the general solution to the equation will be $y_{\text{gen}} = c_1e^{r_1x} + c_2e^{r_2x}$.

Step 3: Solve the quadratic algebraic equation $1r^2 + 5r + 6 = 0$: $(r + 2)(r + 3) = 0$. Hence, $r = -2$ and -3.

Step 4: Substitute the r values into $y_{\text{gen}} = c_1e^{r_1x} + c_2e^{r_2x}$:

$$y_{\text{gen}} = c_1e^{-2x} + c_2e^{-3x}.$$

$$y(x) = c_1e^{-2x} + c_2e^{-3x}$$
$$y'(x) = -2c_1e^{-2x} - 3c_2e^{-3x}$$
$$y''(x) = (-2)^2c_1e^{-2x} + (-3)^2c_2e^{-3x} = 4c_1e^{-2x} + 9c_2e^{-3x}$$

That's it! The equation y_{gen} is a family of solutions with two parameters that satisfies the original differential equation.

$$y''(x) \quad + \quad 5y'(x) \quad + \quad 6y(x) \quad = 0 \text{ check by substituting for } y, y', \text{ and } y''$$
$$\left(4c_1e^{-2x} + 9c_2e^{-3x}\right) + 5\left(-2c_1e^{-2x} - 3c_2e^{-3x}\right) + 6\left(c_1e^{-2x} + c_2e^{-3x}\right) \stackrel{?}{=} 0$$
$$4c_1e^{-2x} + 9c_2e^{-3x} - 10c_1e^{-2x} - 15c_2e^{-3x} + 6c_1e^{-2x} + 6c_2e^{-3x} \stackrel{?}{=} 0$$
$$\left(4c_1e^{-2x} - 10c_1e^{-2x} + 6c_1e^{-2x}\right) + \left(9c_2e^{-3x} - 15c_2e^{-3x} + 6c_2e^{-3x}\right) \stackrel{?}{=} 0$$
$$0 \quad + \quad 0 \quad = 0 \text{ check}$$

Changing the equation above to an IVP, solve the differential equation $y'' + 5y' + 6y = 0$ given that $y(0) = 1$ and $y'(0) = 8$. (Author's note: Two initial conditions must be given for a second-order differential equation.)

$$y(x) = c_1 e^{-2x} + c_2 e^{-3x}, \text{ so } y(0) = c_1 e^{-2\times 0} + c_2 e^{-3\times 0} = 1.$$

Therefore, $c_1 + c_2 = 1$ (1)

$$y'(x) = -2c_1 e^{-2x} - 3c_2 e^{-3x}, \text{ so } y'(0) = -2c_1 e^{-2\times 0} - 3c_2 e^{-3\times 0} = 8.$$

Therefore, $-2c_1 - 3c_2 = 8$ (2)

$2[c_1 + c_2] = 2 \times 1$ (1)

$-2c_1 - 3c_2 = 8$ (2)

$2c_1 + 2c_2 = 2$ (1)

$-c_2 = 10$

$c_2 = -10$

$c_1 + c_2 = 1$

$c_1 - 10 = 1$

$c_1 = 11$

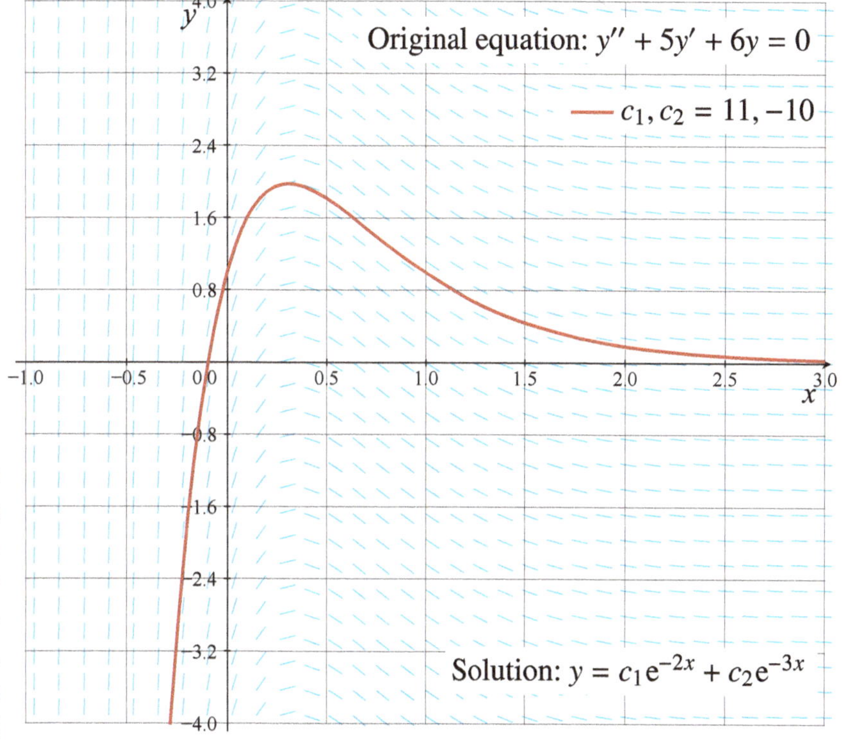

Hence, $y = 11e^{-2x} - 10e^{-3x}$.

Notice from the graph that as $x \to \infty$ that both terms in the linear combination decay to zero causing the asymptotic approach to $y = 0$. This is called *damping to a sink*. One of the beauties of mathematics is the omnipresence of patterns. Before proceeding on with the discussion of solving second-order linear differential equations, let's take a break to reflect on a neat pattern that may not be obvious to you.

The differential equation $n\frac{dy}{dx} + my = 0$, where n and m are scalars, can be solved using the separable technique:

$$n\frac{dy}{dx} + my = 0$$

$$\frac{dy}{dx} = \frac{-m}{n}y$$

$$y^{-1}\,dy = \frac{-m}{n}\,dx$$

$$\int y^{-1}\,dy = \frac{-m}{n}\int dx$$

$$\ln y + c_1 = \frac{-m}{n}x + c_2$$

$$\ln y = \frac{-m}{n}x + c_2 - c_1$$

$$\ln y = \frac{-m}{n}x + c_3$$

$$e^{\ln y} = e^{\frac{-m}{n}x + c_3}$$

$$y = e^{\frac{-m}{n}x} \times e^{c_3}$$

$$y = ce^{\frac{-m}{n}x}$$

Finally, $y_{gen} = ce^{rx}$, where $r = \frac{-m}{n}$.

In Chapter 14, we have just learned that, for $Ay''(x)+By'(x)+Cy(x) = 0$, $y_{gen} = c_1 e^{r_1 x} + c_2 e^{r_2 x}$, where r (i.e., r_1 & r_2) $= \frac{-B\pm\sqrt{B^2-4AC}}{2A}$. Let's review and then look forward a bit.

First order	Second order	Third order
$Ay'(x) + By(x) = 0$	$Ay''(x) + By'(x) + Cy(x) = 0$	$Ay'''(x) + By''(x) + Cy'(x) + Dy(x) = 0$
$y_{gen} = c_1 e^{rx}$, where r $= \frac{-B}{A}$	$y_{gen} = c_1 e^{r_1 x} + c_2 e^{r_2 x}$, where r (i.e. r_1 & r_2) $= \frac{-B\pm\sqrt{B^2-4AC}}{2A}$	$y_{gen} =$?????

It is not unreasonable to anticipate that the solutions to a third-order homogeneous differential equation, $Ay'''(x)+By''(x)+Cy'(x)+Dy(x) = 0$, might be $y_{gen} = c_1 e^{r_1 x}+c_2 e^{r_2 x}+c_3 e^{r_3 x}$, where r_1, r_2, and r_3 are solutions to the polynomial equation $Ar^3 + Br^2 + Cr + D = 0$. It turns out that this inference, with a few caveats, is correct.

A philosophy of teaching held by some teachers and math-book authors is that, when introducing new ideas, one must be complete in the presentation. Everything there is to know about a subject must be taught all at once. The result is much like sticking a fire hose in the mouth of someone who is thirsty. The philosophy of this author is to teach as needed for the immediate situation and then to teach other material when the occasion arises.

Up to this point, the author has carefully avoided linear second-order homogeneous differential equation problems whose associated quadratic equation had a discriminant with a double or negative root. You may recall your algebra teacher saying that quadratic equations can have two distinct real roots, two equal real roots, or two imaginary roots depending on whether the value of the discriminant ($b^2 - 4ac$) was positive, zero, or negative. Well, the same concerns hold when solving second-order linear homogeneous differential equations, although the vocabulary changes slightly; in differential equations, one talks about the *characteristic equation* instead of the quadratic equation, but the concerns are the same as for algebra. One treats second-order linear homogeneous differential equations differently depending on the value of the discriminant of the *characteristic equation.* If it evaluates to a positive number, you use one solution technique, a different technique for zero, and yet a third for negative.

Second-Order Linear Homogeneous Differential Equation with Constant Coefficients

$$Ay''(x) + By'(x) + Cy(x) = 0$$

Form the characteristic equation for the given homogeneous differential equation: $Ar^2 + Br + C = 0$. Depending on the sign of the characteristic equation's discriminant, substitute in the appropriate general solution form as shown below.

$B^2 - 4AC > 0$	$B^2 - 4AC = 0$	$B^2 - 4AC < 0$
$y_{gen} = c_1 e^{r_1 x} + c_2 e^{r_2 x}$	$y_{gen} = c_1 e^{rx} + c_2 x e^{rx}$	$y_{gen} = c_1 e^{\alpha x}\cos\beta x + c_2 e^{\alpha x}\sin\beta x$
	Be careful. This formula is different from the one at left!	See Appendix D for a discussion of this integrated form.

If you are given IVP conditions, you must substitute the given information and solve for c_1 and c_2.

Practice Solving a Second-Order Homogeneous Differential Equation #3

Solve the equation $y'' - 6y' + 9y = 0$ as an IVP with $y(0) = 4$ and $y'(0) = 7$.

Step 1: Notice that the differential equation is in the form $Ay''(x) + By'(x) + Cy(x) = 0$. Hence, the given differential equation is in standard form for a second-order homogeneous equation.

Step 2: Using the differential coefficients 1, −6, and 9 to form a quadratic algebraic equation, we get $r^2 - 6r + 9 = 0$. Evaluating the discriminant, we get $(-6)^2 - (4 \times 1 \times 9) = 0$: **The general solution to the equation will be $y_{gen} = c_1 e^{rx} + c_2 x e^{rx}$**

Step 3: Solve the quadratic algebraic equation $r^2 - 6r + 9 = 0$. $(r - 3)(r - 3) = 0$. Hence, $r = 3$ and 3.

Step 4: Substitute the r values into $y_{gen} = c_1 e^{rx} + c_2 x e^{rx}$:

$$y_{gen} = c_1 e^{3x} + c_2 x e^{3x}.$$

That's it! y_{gen} is a family of differential equations that satisfies the original differential equation.

$$y(x) = c_1 e^{3x} + c_2 x e^{3x}$$
$$y'(x) = 3c_1 e^{3x} + 3c_2 x e^{3x} + c_2 e^{3x}$$
$$y''(x) = (3)^2 c_1 e^{3x} + 3\left(3c_2 x e^{3x} + c_2 e^{3x}\right) + 3c_2 e^{3x}$$
$$= 9c_1 e^{3x} + 9c_2 x e^{3x} + 6c_2 e^{3x}$$

$$y'' - 6y' + 9y = 0$$
$$\left(9c_1 e^{3x} + 9c_2 x e^{3x} + 6c_2 e^{3x}\right) - 6\left(3c_1 e^{3x} + 3c_2 x e^{3x} + c_2 e^{3x}\right) + 9\left(c_1 e^{3x} + c_2 x e^{3x}\right) = 0$$
$$\left(9c_1 e^{3x} - 18c_1 e^{3x} + 9c_1 e^{3x}\right) + \left(9c_2 x e^{3x} - 18c_2 x e^{3x} + 9c_2 x e^{3x}\right) + \left(6c_2 e^{3x} - 6c_2 e^{3x}\right) = 0$$
$$0 + 0 + 0 = 0 \text{ check}$$

Now, solve the differential equation $y'' - 6y' + 9y = 0$ given that $y(0) = 4$ and $y'(0) = 7$. (Author's note: Two initial conditions must be given for a second-order differential equation.)

$$y(x) = c_1 e^{3x} + c_2 x e^{3x}$$
$$y(0) = c_1 e^{3\times 0} + 0c_2 e^{3\times 0}$$
$$4 = c_1 e^0 + 0 \quad \text{IVP } y(0) = 4$$

Therefore, $c_1 = 4$.

Since $y(x) = c_1 e^{3x} + c_2 x e^{3x}$,

$$y'(x) = 3c_1 e^{3x} + c_2(xe^{3x}(3) + [e^{3x} \times 1])$$
$$y'(x) = 3c_1 e^{3x} + c_2(3xe^{3x} + e^{3x})$$
$$y'(0) = 3c_1 e^{3\times 0} + c_2(0 \times e^{3\times 0} + e^{3\times 0})$$
$$7 = 3c_1 e^0 + c_2(0 + e^0) \quad \text{IVP } y'(0) = 7$$
$$7 = 3c_1 + c_2 \times (0 + 1)$$
$$7 = 3c_1 + c_2 \times 1$$
$$7 = 3c_1 + c_2 \quad \text{substitute } c_1 = 4$$
$$7 = 3 \times 4 + c_2$$
$$7 = 12 + c_2$$
$$c_2 = -5$$

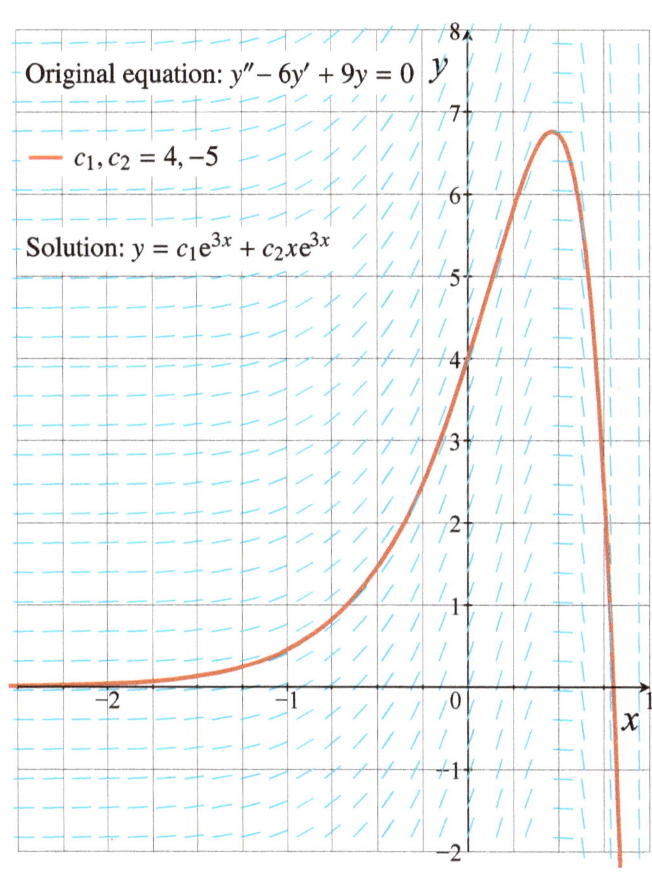

So, using the IVP conditions, $y_{gen} = 4e^{3x} - 5xe^{3x}$.

Note that factoring e^{3x} from the solution results in $y = e^{3x}(4 - 5x)$. As $x \to \infty$, $e^{3x}(4 - 5x)$ decreases and is negative.

Practice Solving a Second-Order Homogeneous Differential Equation #4

Solve the equation $y'' - 4y' + 13y = 0$ as an IVP with $y(0) = 2$ and $y'(0) = 1$.

Step 1: Notice that the differential equation is in the form $Ay''(x) + By'(x) + Cy(x) = 0$. Hence, the given differential equation is in standard form for a second-order homogeneous equation.

Step 2: Using the differential coefficients 1, −4, and 13 to form a quadratic algebraic equation, we get $r^2 - 4r + 13 = 0$. Evaluating the discriminant we get $(-4)^2 - (4 \times 1 \times 13) = 16 - 52 < 0$. That means that **the general solution to the equation will be $y_{gen} = c_1 e^{\alpha x} \cos \beta x + c_2 e^{\alpha x} \sin \beta x$.** (See Appendix D for a discussion of this integrated-solution form.)

Step 3: Solve the quadratic algebraic equation $r^2 - 4r + 13 = 0$: $r = \frac{4 \pm \sqrt{16 - 4\times 1\times 13}}{2} = \frac{4 \pm \sqrt{-36}}{2} = 2 \pm 3i$. Therefore, $\alpha = 2$ and $\beta = 3$.

Step 4: Substitute the $\alpha = 2$ and $\beta = 3$ values from r into $y_{gen} = c_1 e^{\alpha x} \cos \beta x + c_2 e^{\alpha x} \sin \beta x$:

$$y_{gen} = c_1 e^{2x} \cos 3x + c_2 e^{2x} \sin 3x.$$

That's it! y_{gen} is a family of differential equations that satisfies the original differential equation, $y''-4y'+13y=0$. Check if for yourself using the derivatives below if you would like to confirm this result.

$$\begin{aligned}
y(x) &= c_1e^{2x}\cos 3x + c_2e^{2x}\sin 3x \\
y'(x) &= 2c_1e^{2x}\cos 3x - 3c_1e^{2x}\sin 3x + 2c_2e^{2x}\sin 3x + 3c_2e^{2x}\cos 3x \\
y''(x) &= 4c_1e^{2x}\cos 3x - 6c_1e^{2x}\sin 3x - 6c_1e^{2x}\sin 3x - 9c_1e^{2x}\cos 3x \\
&\quad + 4c_2e^{2x}\sin 3x + 6c_2e^{2x}\cos 3x + 6c_2e^{2x}\cos 3x - 9c_2e^{2x}\sin 3x \\
&= -5c_1e^{2x}\cos 3x - 12c_1e^{2x}\sin 3x - 5c_2e^{2x}\sin 3x + 12c_2e^{2x}\cos 3x
\end{aligned}$$

Now, solve the differential equation $y'' - 4y' + 13y = 0$ given that $y(0) = 2$ and $y'(0) = 1$. (Author's note: Two initial conditions must be given for a second-order differential equation.)

$$\begin{aligned}
y(x) &= c_1e^{2x}\cos 3x + c_2e^{2x}\sin 3x \\
2 = y(0) &= c_1e^{2\times 0}\cos(3\times 0) + c_2e^{2\times 0}\sin(3\times 0) \\
2 &= c_1\times 1\times\cos 0 + c_2\times 1\times\sin(0) \\
2 &= c_1\times 1\times 1 + c_2\times 1\times 0 \\
2 &= c_1 + 0
\end{aligned}$$

Therefore, $c_1 = 2$.

$$\begin{aligned}
y'(x) &= 2c_1e^{2x}\cos 3x - 3c_1e^{2x}\sin 3x + 2c_2e^{2x}\sin 3x + 3c_2e^{2x}\cos 3x \\
1 = y'(0) &= 2c_1e^{2\times 0}\cos(3\times 0) - 3c_1e^{2\times 0}\sin(3\times 0) + 2c_2e^{2\times 0}\sin(3\times 0) + 3c_2e^{2\times 0}\cos(3\times 0) \\
1 &= 2\times 2\times e^0\cos(0) - 3\times 2\times e^0\sin(0) + 2c_2e^0\times\sin(0) + 3c_2e^0\cos(0) \\
1 &= (4\times 1\times 1) - (6\times 1\times 0) + (2c_2\times 1\times 0) + (3c_2\times 1\times 1) \\
1 &= 4 - 0 + 3c_2 \\
-3 &= 3c_2 \\
c_2 &= -1
\end{aligned}$$

The equation $y'' - 4y' + 13y = 0$ as an IVP with $y(0) = 2$ and $y'(0) = 1$ has general solution $y = c_1e^{2x}\cos 3x + c_2e^{2x}\sin 3x$ and particular solution $y = 2e^{2x}\cos 3x - e^{2x}\sin 3x$. See the graph on the next page.

Author's note: The graph at right looks as though the graph is going to infinity as $x \to 2$, but this is misleading. By factoring out the e^{2x} term in the equation graphed above, one gets $y = e^{2x}(c_1\cos 3x + c_2\sin 3x)$ which means that the graph should have shown a section where the graph cycled. I changed both the domain and range of my MATLAB code (See Appendix A) in an attempt to show that cycling but was never able to achieve the desired graph. By the time the range was extended sufficiently to show the entire trig function cycle, the graph was terribly distorted. That is the nature of exponential growth—e^{2x}.

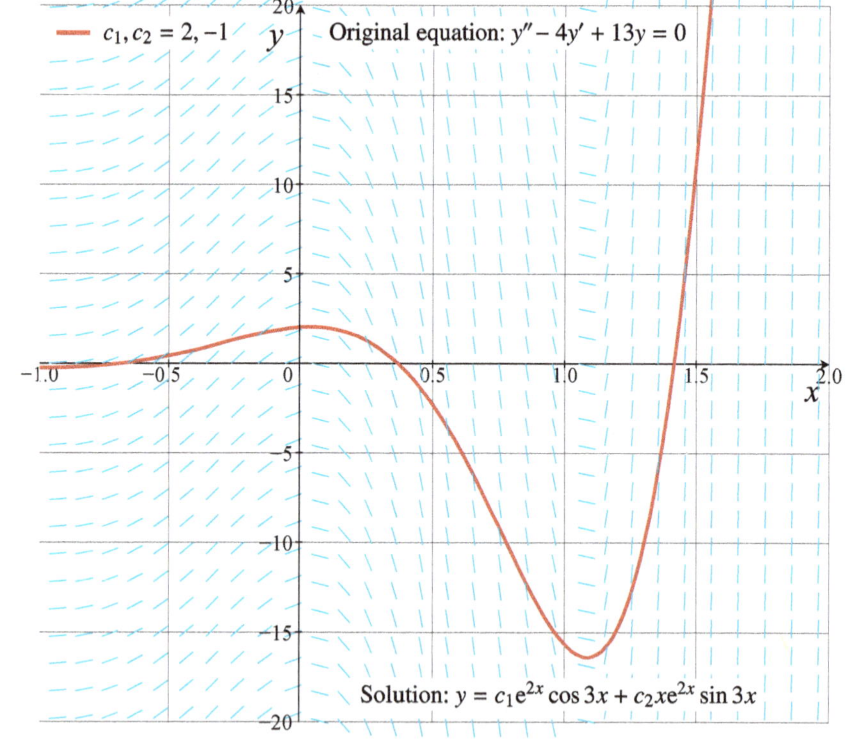

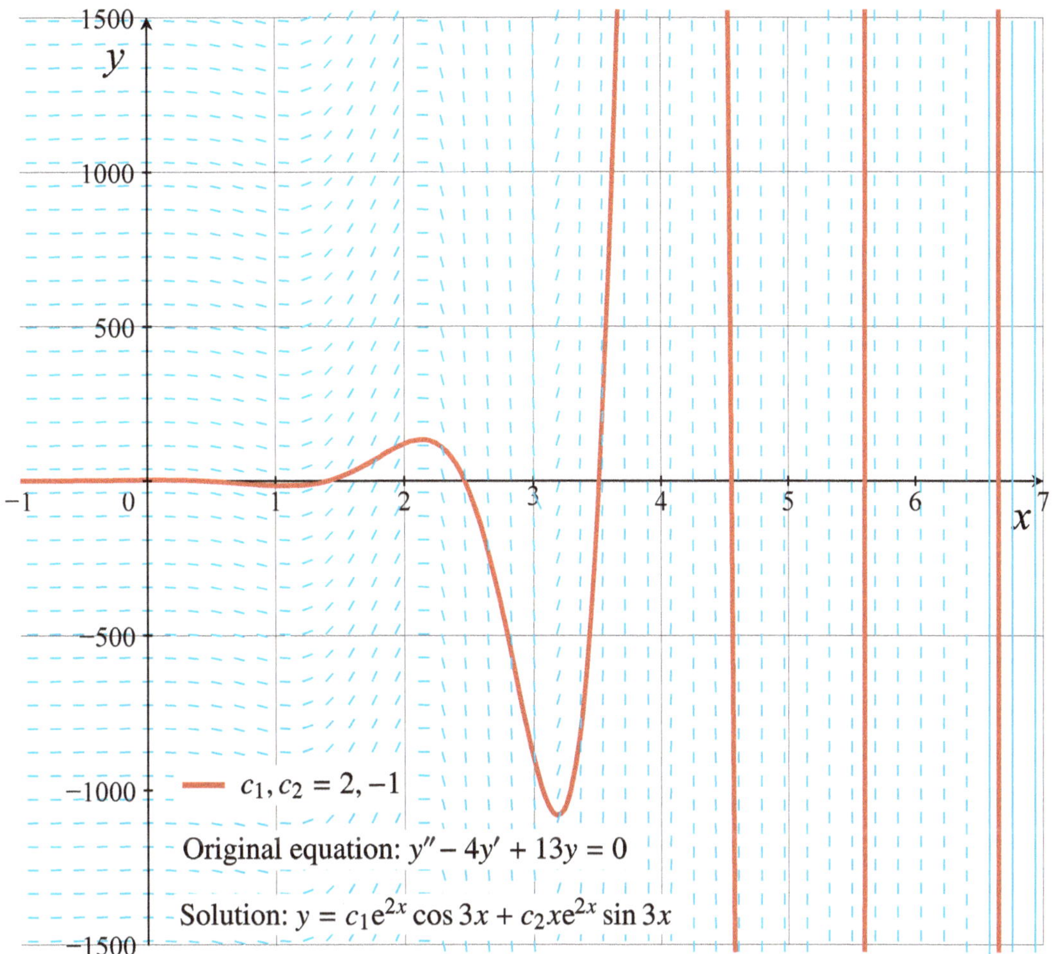

Author's confession: I did not "process" anything in second-order homogeneous problem #4. I merely changed the formula and data as appropriate and worked through the formula. That is the beauty of working with integrated-solution forms.

Chapter 14 Review

The solution to the differential-equation form $Ay''(x)+By'(x)+Cy(x) = 0$ was taught. The three integrated/algebraic solution forms used on such equations were presented and demonstrated. The solution for when $b^2 - 4ac > 0$ was discussed at length, and the solution for when $b^2 - 4ac < 0$ is proved in Appendix D.

Second-Order Linear Homogeneous Differential Equation with Constant Coefficients

$$Ay''(x) + By'(x) + Cy(x) = 0$$

Form the characteristic equation for the given homogeneous differential equation: $Ar^2 + Br + C = 0$. Depending on the sign of the characteristic equation's discriminant, substitute in the appropriate general solution form as shown below.

$B^2 - 4AC > 0$	$B^2 - 4AC = 0$	$B^2 - 4AC < 0$
$y_{gen} = c_1 e^{r_1 x} + c_2 e^{r_2 x}$	$y_{gen} = c_1 e^{rx} + c_2 x e^{rx}$ Be careful. This formula is different from the one at left!	$y_{gen} = c_1 e^{\alpha x} \cos \beta x + c_2 e^{\alpha x} \sin \beta x$ See Appendix D for a discussion of this integrated form.

A Linear Homogeneous Second-Order Differential Equation with Constant Coefficients

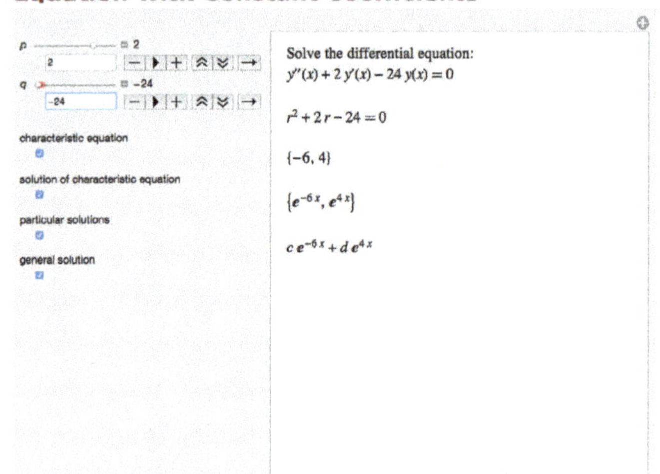

Online Application

Visit demonstrations.wolfram.com/ALinearHomogeneousSecondOrderDifferentialEquationWithConstan for a great little application that will demonstrate some of the concepts from this chapter. "A Linear Homogeneous Second Order Differential Equation With Constant Coefficients" from the Wolfram Demonstrations Project. Contributed by Izidor Hafner.

Chapter 15
Hooke's Law and Homogeneous Differential Equations

"Hooke's law is a principle of physics that states that the force, F, needed to extend or compress a spring by some distance x is proportional to that distance. That is: $F = kx$, where k is a constant factor characteristic of the spring, its stiffness. The law is named after 17th century British physicist Robert Hooke. He first stated the law in 1660 as a Latin anagram. He published the solution of his anagram in 1678 as: *ut tensio, sic vis* ('as the extension, so the force' or 'the extension is proportional to the force')" (Wikipedia).

Hooke's law comes up in many different places in mathematics and especially physics.

1. Algebra

A 10 lb weight stretches a spring 7 in. Assuming direct proportion, how much will a 12 lb weight stretch that same spring?

$$F = kx$$
$$10 = k(7)$$
$$k = \frac{10}{7}\frac{\text{lb}}{\text{in}}$$

$$F = kx$$
$$12 = \frac{10}{7}x$$
$$x = \frac{84}{10}\text{ in}$$

2. Geometry and Calculus

A force of 500 pounds compresses a spring 2 inches from its natural length of 16 inches. Use Hooke's Law, $f = kd$, and the formula for work, $w = fd$, to find the work done in compressing the spring an additional 4 inches.

BigStockPhoto.com

Substituting into Hooke's law, we get $500 = k(2)$, so $k = 250$. Hence $f = 250x$. Using the formula for work, we get $w = (250x)d$. The additional work required to further compress the spring from 2 inches to 6 inches is

By geometry

$$\text{area} = \tfrac{1}{2}b_1h_1 - \tfrac{1}{2}b_2h_2$$
$$= \tfrac{1}{2}(6 \times 1{,}500) - \tfrac{1}{2}(2 \times 500)$$
$$= 4500 - 500 = 4{,}000 \text{ in–lb}$$

Here the geometric approach is much easier than the calculus approach. ***However, had the function been nonlinear, the calculus approach with the FTC would win hands down!!***

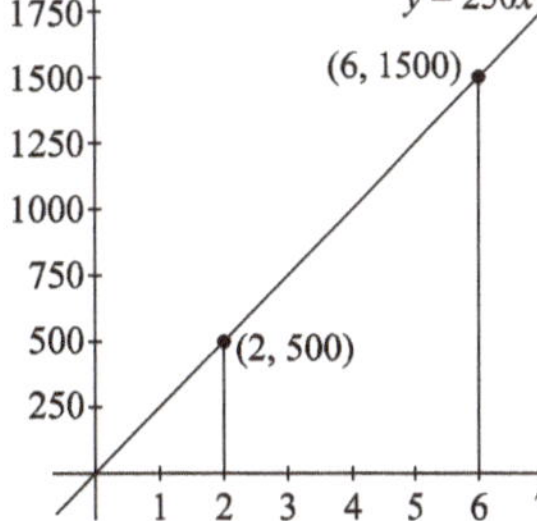

By calculus

$$\text{area} = \int_2^6 250x\,dx$$
$$= 250\int_2^6 x\,dx$$
$$= 250\frac{x^2}{2}\bigg|_2^6 = 125(6^2 - 2^2)$$
$$= 125(36 - 4) = 4{,}000 \text{ in–lb}$$

From *Twenty Key Ideas in Beginning Calculus* by Dan Umbarger (2011).

3. Differential Equations

Often, in a differential equations class, Hooke's law comes up when studying second-order linear homogeneous differential equations of the form

$$Ay'' + By' + Cy = 0.$$

Spring–Mass Systems

The following problem is taken from *A First Course in Differential Equations with Modeling Applications*, Dennis G. Zill, Brooks/Cole Cengage Learning, 2013, pgs. 193–196. Errors, if any, can be attributed to my modifications.

Assume that a spring (shown below) is suspended in a perfect vacuum. Its unstretched length is denoted by l (Fig. 1). If an unspecified weight ($w = m \times g$) is attached to the spring, it is elongated by a length of s (Fig. 2). The position of the attached weight would therefore be at length $(l + s)$, shown below. By Hooke's law, the spring exerts a restoring force, f_w, opposite to the elongation and proportional to the amount of elongation, s. That is, $f_w = -ks$ where k is a constant of proportionality called the spring constant. The restoring force is equal in magnitude, but opposite in direction, to the force exerted by gravity on the object. If a pulling force f_p is applied to the mass, it will be displaced from its equilibrium position by a distance y so that the restoring force of the spring is $f_p + f_w = -k(y+s)$ as shown in the figure below (Fig. 3).

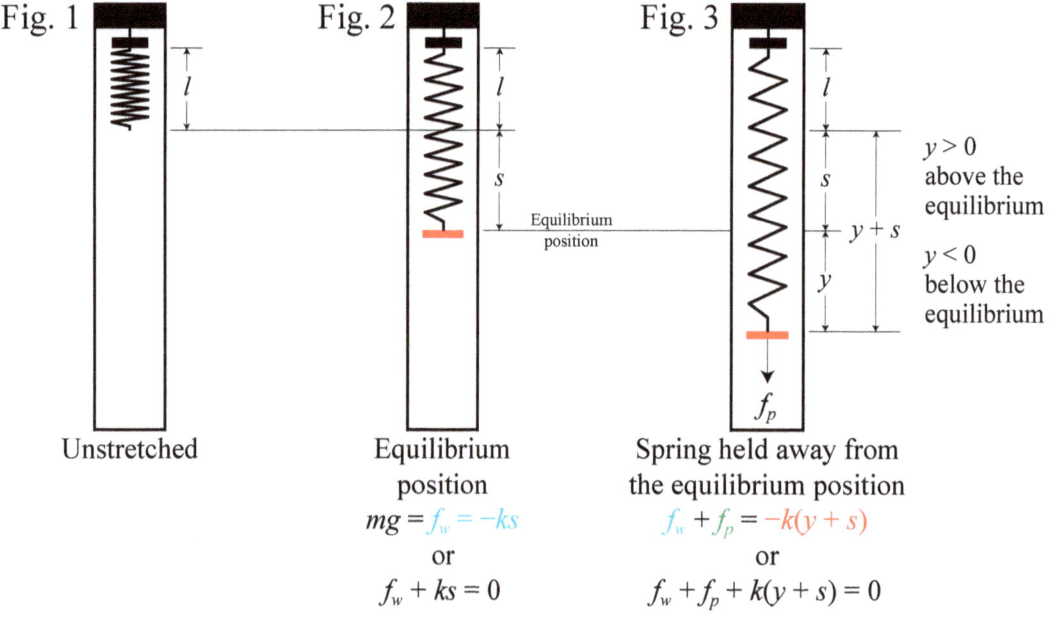

Substituting $f_w = -ks$ into $f_w + f_p = -k(y+s)$

$$\cancel{-ks} + f_p = -ky \cancel{- ks}$$

$$f_p = -ky$$

$$ma = -ky \quad \text{Newton's second law, } F = ma \text{ (Chapter 9)}$$

$$m\frac{d^2y}{dt^2} = -ky \quad a\text{ (acceleration)} = \frac{d^2y}{dt^2} \text{ (Chapter 9)}$$

Therefore, $m\dfrac{d^2y}{dt^2} + ky = 0.$

Because the spring is assumed to be located in a vacuum, the motion will be unaffected by the medium surrounding the spring. The restoring force of the released spring will act to return the object to the equilibrium position: Free motion will result. Note that the final equation of motion is independent of g; gravity only affects the absolute position of the equilibrium (s). According to Newton's second law ($F = ma$), the equation of motion will involve the net restoring force:

$m\dfrac{d^2y}{dt^2} + 0\dfrac{dy}{dt} + ky = 0$	differential equation obtained using Hooke's law and Newton's second law
$Ay'' + By' + Cy = 0$	generic second-order linear homogeneous differential equation

Dividing out the leading coefficient, m,

$$\frac{m\frac{d^2y}{dt^2}}{m} + \frac{0\frac{dy}{dt}}{m} + \frac{ky}{m} = \frac{0}{m}$$
$$\frac{d^2y}{dt^2} + 0\frac{dy}{dt} + \frac{k}{m}y = 0$$

So, $A = 1$, $B = 0$, and $C = \frac{k}{m}$. Anyone who has played with a spring before could probably guess that the mass will bounce up and down after it is released.

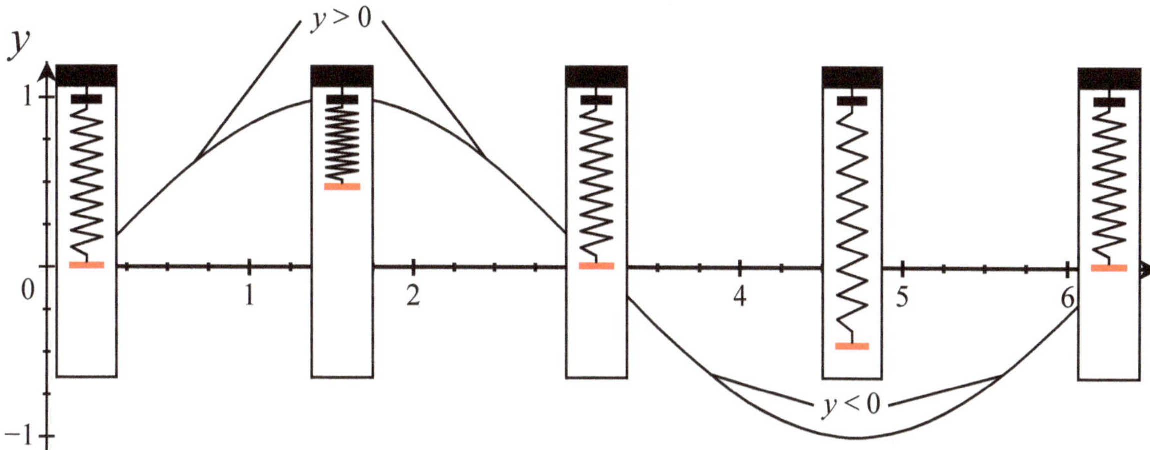

Author's note: Recall from trigonometry that the periods of the sine and cosine functions are 2π. The periods of $\sin(5t)$ and $\cos(5t)$ would, therefore, be $\frac{2\pi}{5}$. See figure below.

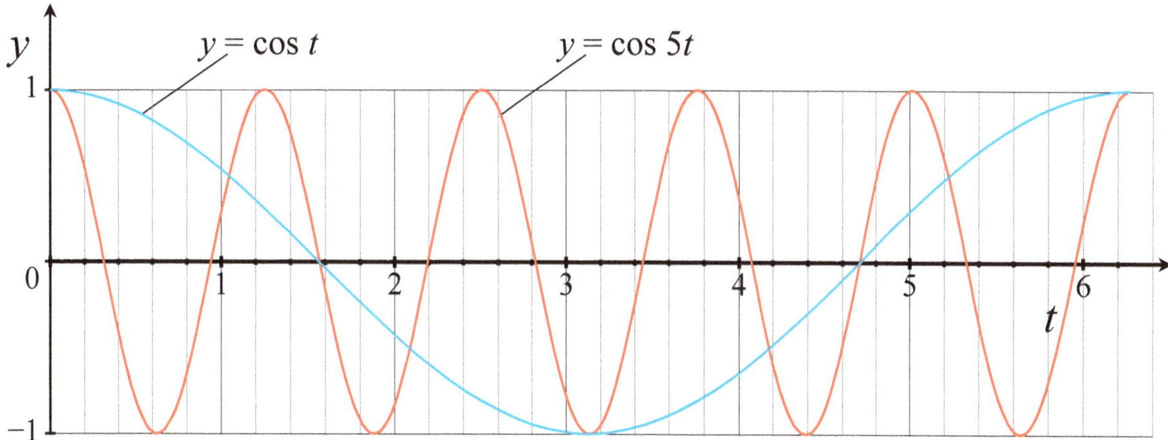

It makes sense that the period of the up and down function resulting from the stretching and releasing of the spring above would be related to 1) the mass of the object attached to the spring and 2) its "stiffness" or constant of proportionality, k. Physics tells us that the period is

$$T = \frac{2\pi}{\sqrt{\frac{k}{m}}}.$$

To simplify this yucky expression, it is common to substitute $\omega = \sqrt{\frac{k}{m}}$. The period, T, would be $\frac{2\pi}{\omega}$ and $\omega^2 = \frac{k}{m}$. Therefore, the equation

$$\frac{d^2y}{dt^2} + 0\frac{dy}{dt} + \frac{k}{m}y = 0 \quad \text{becomes}$$
$$\frac{d^2y}{dt^2} + 0\frac{dy}{dt} + \omega^2 y = 0.$$

Review from Chapter 14:

Second-Order Homogeneous Differential Equation
$Ay''(x) + By'(x) + Cy(x) = 0$
Form the characteristic equation for the given homogenous differential equation: $Ar^2+Br+C = 0$. Solve for r. Depending on the evaluation of the "discriminant" of the characteristic equation substitute in the appropriate general solution form as shown in Chapter 14. For complex r values, $y_{\text{gen}} = c_1 e^{\alpha t}\cos\beta t + c_2 e^{\alpha t}\sin\beta t$

Recopying the differential-equation form from above,

$$1\frac{d^2y}{dt^2} + 0\frac{dy}{dt} + \omega^2 y = 0,$$

the characteristic equation would be $r^2 + 0r + \omega^2 = 0$. Note that $\omega^2 = \frac{k}{m}$ where k is Hooke's spring constant and m is mass.

Applying the quadratic equation,

$$r = \frac{-b \pm \sqrt{b^2-4ac}}{2a} = \frac{-0 \pm \sqrt{0^2 - 4\times 1\times\omega^2}}{2\times 1} = \frac{0 \pm \sqrt{-4\omega^2}}{2} = \frac{0 \pm 2\omega\sqrt{-1}}{2} = 0 \pm \omega i = \alpha + \beta i.$$

Since the discriminant is $B^2 - 4\times A\times C = -4\omega^2 < 0$,
$y_{\text{gen}} = c_1 e^{\alpha t}\cos\beta t + c_2 e^{\alpha t}\sin\beta t$ (Chapter 14),
where $\alpha = 0$ and $\beta = \omega$ from the preceding work.

$$\begin{aligned} y_{\text{gen}} &= c_1 e^{\alpha t}\cos\beta t + c_2 e^{\alpha t}\sin\beta t \\ &= e^{\alpha t}(c_1\cos\beta t + c_2\sin\beta t) \quad \text{factor out the } e^{\alpha t} \\ &= e^{0\times t}(c_1\cos\omega t + c_2\sin\omega t) \quad \text{substitute } \alpha = 0 \text{ and } \beta = \omega \\ &= e^{0}(c_1\cos\omega t + c_2\sin\omega t) \\ &= 1(c_1\cos\omega t + c_2\sin\omega t) \\ &= c_1\cos\omega t + c_2\sin\omega t \quad \text{two-parameter family of solutions to } \frac{d^2y}{dt^2} + 0\frac{dy}{dt} + \omega^2 y = 0 \end{aligned}$$

In the discussion above, assume that a mass weighing 2 lb stretches a spring 6 in—$\frac{1}{2}$ ft. At $t = 0$ the mass is released from a point 8 in below the equilibrium position with an upward velocity of $\frac{4}{3}\frac{\text{ft}}{\text{s}}$. Recall from Chapter 9 that velocity $= v = \frac{dy}{dt}$. Determine the equation of motion.

$$\text{mass} = m = \frac{w}{g} = \frac{2\,\text{lb}}{32\,\frac{\text{ft}}{\text{s}^2}} = \frac{1}{16}\,\text{slug} \quad (\text{a } \textit{slug} \text{ is the measure of mass in customary US units})$$

By Hooke's law, $F = ks$, so $2\,\text{lb} = k\times\frac{1}{2}\,\text{ft}$, so $k = 4\,\frac{\text{lb}}{\text{ft}}$.

$$\omega = \sqrt{\frac{k}{m}} = \sqrt{\frac{4}{\frac{1}{16}}} = \sqrt{64} = 8\,\frac{\text{rad}}{\text{s}}$$

So, substituting into y_{gen} above, we get

$$y_{\text{gen}} = c_1\cos 8t + c_2\sin 8t.$$

The IVP (initial value problem) information above will allow us to solve for c_1 and c_2. The problem condition states that the mass is released at **a position 8 in below the equilibrium**—that is, $y(0) = -\frac{2}{3}$ ft—and that the initial **upward velocity** is $\frac{4}{3}\frac{\text{ft}}{\text{s}}$—that is, $y'(0) = \frac{4}{3}\frac{\text{ft}}{\text{s}}$. This velocity is positive because the spring will be moving upward.

First a quick review of the relevant information. From Chapter 9, you may recall that, if you had a function representing position, you could take its derivative and find a function for velocity.

t (time) $\rightarrow\rightarrow\rightarrow$	p (position) at time t $p = -4.9t^2 + v_0 t + p_o$ $\rightarrow\rightarrow\rightarrow\rightarrow\rightarrow$	v(velocity) $= \frac{dp}{dt}$ $v_f = -9.8t + v_0$ $v_f = \frac{dp}{dt}$

Using the notation from the example of Hooke's law, $y(t)$ = position and $y'(t)$ = velocity.

Initial Value Conditions:
$p = y(t) = -\frac{2}{3}$ ft and $v = y'(t) = \frac{4}{3}\frac{\text{ft}}{\text{s}}$,
when $t = 0$.

$$y(t) = c_1 \cos 8t + c_2 \sin 8t$$
$$y(0) = c_1 \cos(8 \times 0) + c_2 \sin(8 \times 0)$$
$$-\frac{2}{3} = c_1 \cos(0) + c_2 \sin(0)$$
$$-\frac{2}{3} = c_1 \times 1 + c_2 \times 0$$
$$-\frac{2}{3} = c_1$$
$$c_1 = -\frac{2}{3}$$

$$y'(t) = c_1 \times [-\sin(8t) \times 8] + c_2 \times \cos(8t) \times 8$$
$$y'(0) = -[c_1 \times \sin(8 \times 0) \times 8] + [c_2 \times \cos(8 \times 0) \times 8]$$
$$\frac{4}{3} = -\left(-\frac{2}{3}\right) \times \sin(0) \times 8 + c_2 \times \cos(0) \times 8$$
$$\frac{4}{3} = \frac{2}{3} \times (0 \times 8) + c_2 \times (1 \times 8)$$
$$\frac{4}{3} = 0 + c_2 \times 8$$
$$\frac{4}{3} = 8c_2$$
$$\frac{4}{3} \times \frac{1}{8} = 8c_2 \times \frac{1}{8}$$
$$c_2 = \frac{1}{6}$$

Summarizing, for $\frac{d^2y}{dt^2} + 0\frac{dy}{dt} + \omega^2 y = 0$ where $\omega = \sqrt{\frac{k}{m}}$ so $\omega^2 = \frac{k}{m}$,

$$y_{\text{gen}} = c_1 e^{\alpha t} \cos \beta t + c_2 e^{\alpha t} \sin \beta t,$$
$$y(t) = c_1 e^{0 \times t} \cos \omega t + c_2 e^{0 \times t} \sin \omega t \quad \text{substitute } \alpha = 0 \text{ and } \beta = \omega$$
$$y(t) = c_1 e^{0} \cos \omega t + c_2 e^{0} \sin \omega t$$
$$y(t) = c_1 \times 1 \times \cos \omega t + c_2 \times 1 \times \sin \omega t$$
$$y(t) = c_1 \cos \omega t + c_2 \sin \omega t$$
$$y(t) = c_1 \cos 8t + c_2 \sin 8t, \quad \text{solve for } c_1 \text{ and } c_2 \text{ using the given IVP conditions}$$
$$y(t) = -\frac{2}{3} \cos 8t + \frac{1}{6} \sin 8t$$

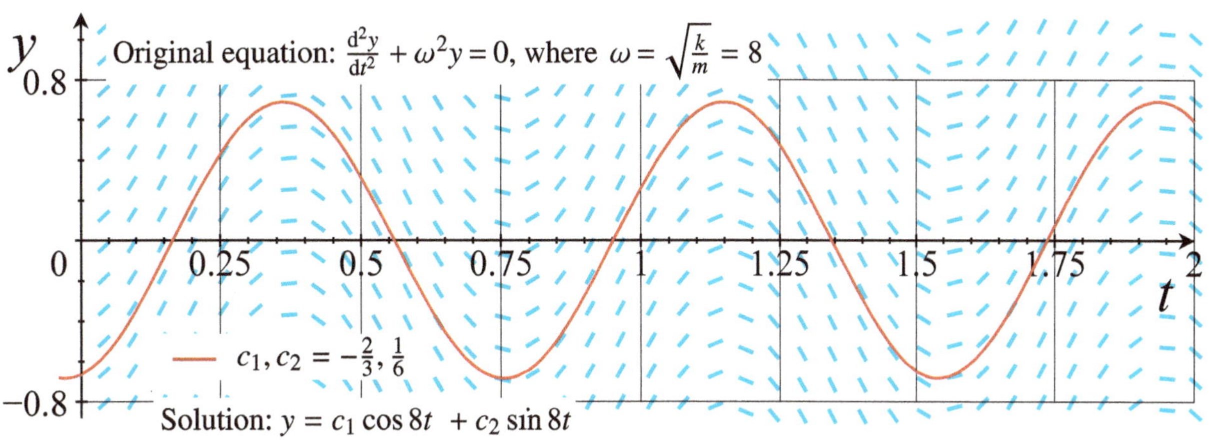

Here, when the spring is stretched beyond the equilibrium, the initial displacement is at its most negative. See the figure below. When the spring is released it will be pulled back toward the equilibrium. Due to the mass's inertia, it proceeds past the equilibrium toward its unstretched position, the most positive displacement.

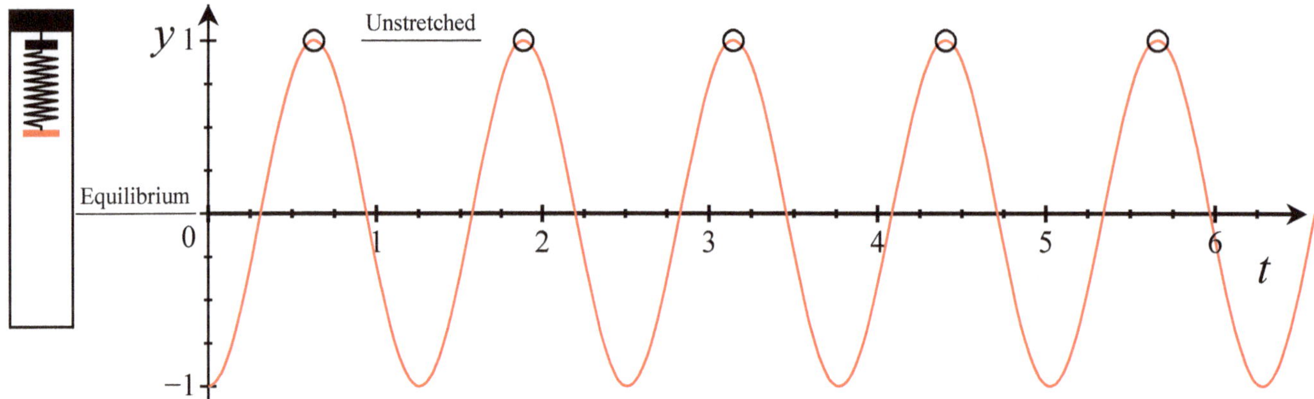

The spring passes equilibrium as it travels between positive and negative amplitudes.

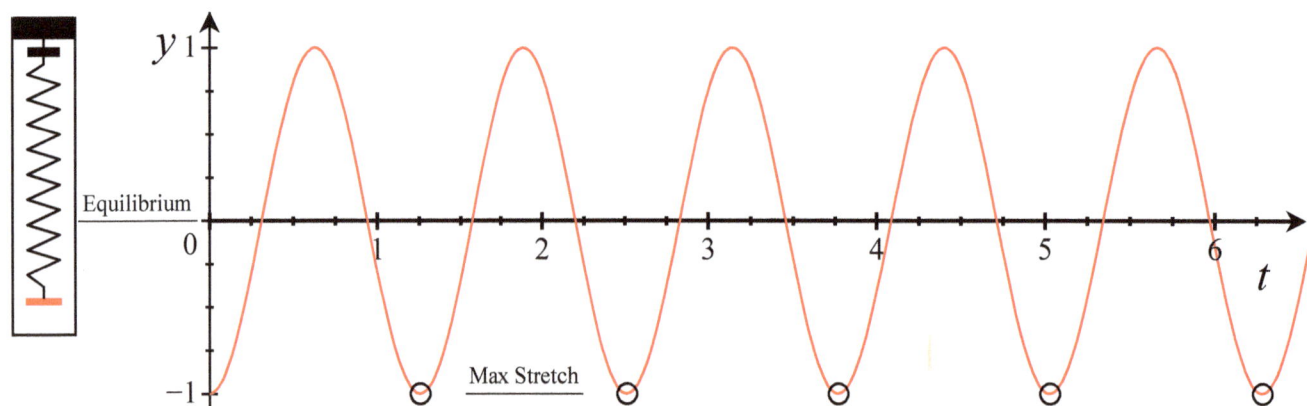

The infinite domain of each of the above graphs can be captured in a "phase plane." Two are shown on the next page. The one on the left shows the infinite solutions that exist for the second-order linear homogeneous differential equation $y'' + \omega^2 y = 0$ whose general solutions are $0 \pm \omega i$. The one on the right shows the specific solution chosen from the family of solutions that satisfies the IVP stated above.

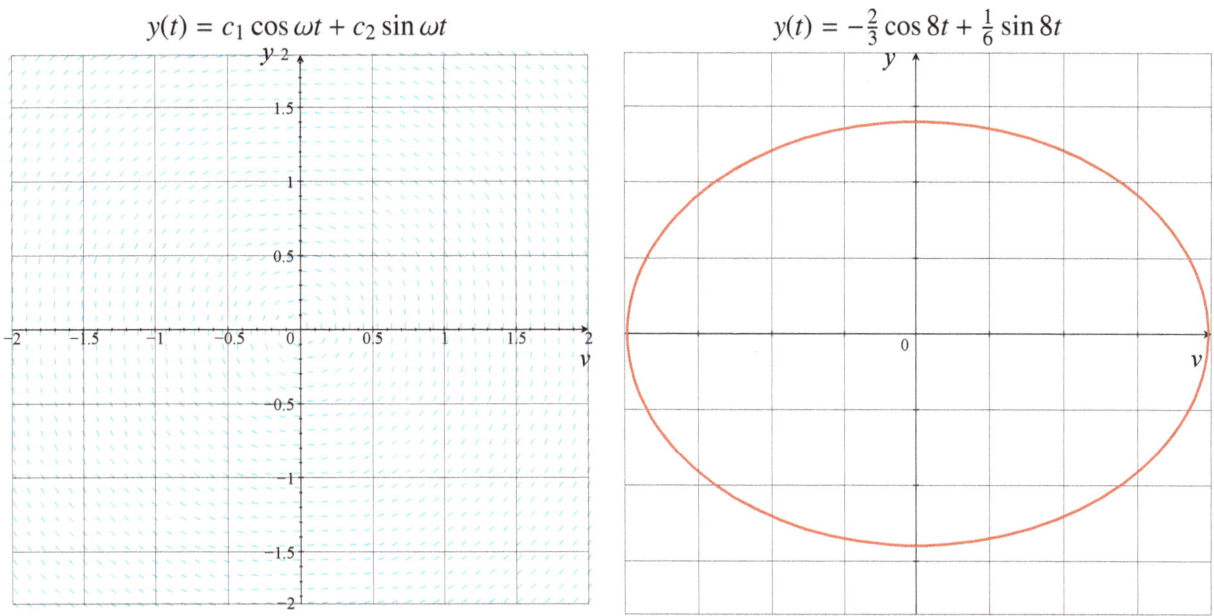

The figure below shows the graph of both position and velocity over time. Recall that $\frac{dy}{dt}$ = velocity, so Points 1 and 2 in the graph show that when the mass reaches the extreme points of its travel its velocity will be zero. The object will speed up as it is pulled back toward the equilibrium position due to the force of the spring. Then, once the object reaches its equilibrium position (i.e. when displacement is zero), the velocity attains its maximum positive and negative values (Points 3 and 4, respectively). Once it passes this point, the force of the spring will now point in the other direction slowing the object as it once again reaches a position of maximum displacement now on the other side of the equilibrium. Absent damping forces, the spring will continue on this sinusoidal motion indefinitely. Notice that the velocity and displacement are always *90° out of phase* from one another. When one reaches its maximum amplitude, the other will be zero. Recall from trigonometry that 90° out of phase refers to the position graph crossing the t-axis one quarter of the way through the cycle after the velocity graph crosses the t-axis.

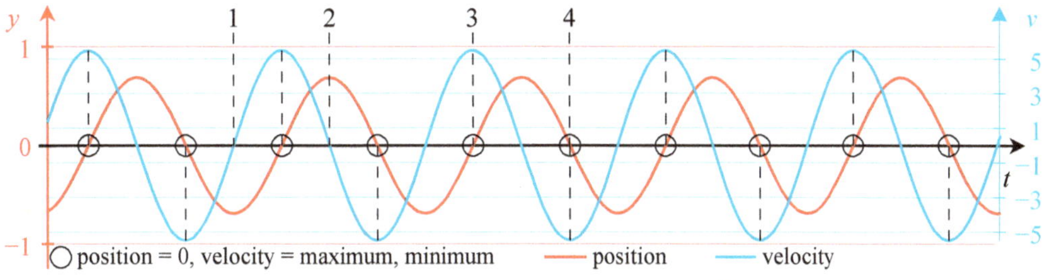

One of the neat things about modern computers is that you can change parameters around and do experiments much like the experiments in a physical science class right on the computer screen. There are many software programs and websites, both free and paid, that make this very easy. Following is an example of making the spring twice as stiff as in the previous case (replace k with $2k$). As we might expect, with a stiffer spring the resulting spring force will be larger causing the object to oscillate faster and reach greater speeds. Thus, the amplitude of the velocity–time graph will be greater than it was before, and both velocity–time and displacement–time will have higher frequencies.

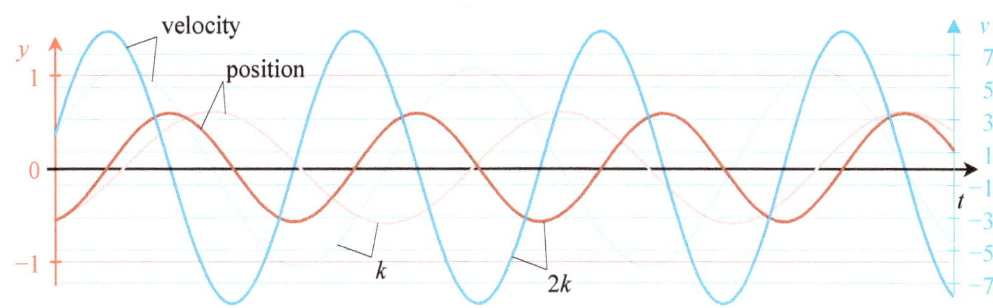

Or, returning the spring constant to its original value, we could double the mass. With increased inertial mass of the object, the same force will result in a lower acceleration, causing the spring to oscillate more slowly than before. We also see in the velocity–time graph that the amplitude of the velocity will be reduced, as we would expect with a more massive object.

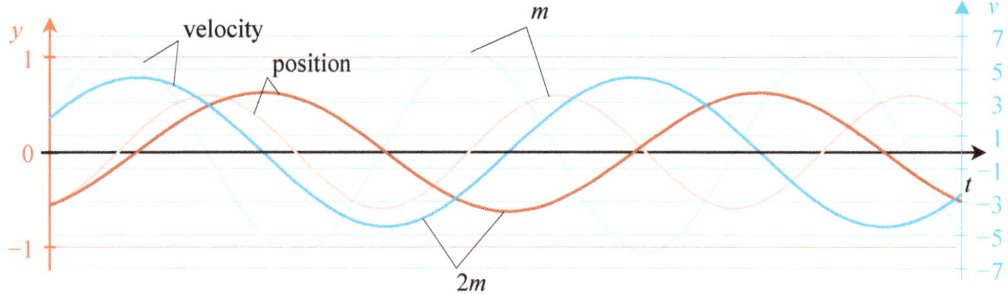

Here is a thought experiment for you. If a black belt in karate strikes a fixed board, what happens? $F = ma$ right? The force of the blow will break the board, right? What would happen if the same experiment were conducted underwater? ($F = ma$ still, but what will happen to the value "a"?) Do you agree that the board would not break?

In the original scenario, the force of the blow will be great because the person's hand will be moving through the air, which offers little resistance. However, in the second scenario the blow will be minimal because the person's hand will move more slowly through the water due to the water's resistance. These two scenarios are different because the different materials (air and water) have different viscosity. A greater viscosity generates more resistance. The result is that the motion will be dampened.

Similarly, when the spring action took place in a vacuum, the homogeneous differential equation used above for the undampened spring motion was $\frac{d^2y}{dt^2} + \frac{k}{m}y = 0$. When there is resistance due to viscosity from the medium in which the spring is acting, the homogeneous differential equation is $\frac{d^2y}{dt^2} + \frac{\beta}{m}\frac{dy}{dt} + \frac{k}{m}y = 0$, where β is the dampening factor and the dampening force is proportional to the instantaneous velocity. Although the math is more difficult, the concepts and technique for solving such an equation would be the same. A dampened spring action might look like the figure at right although there are other possibilities.

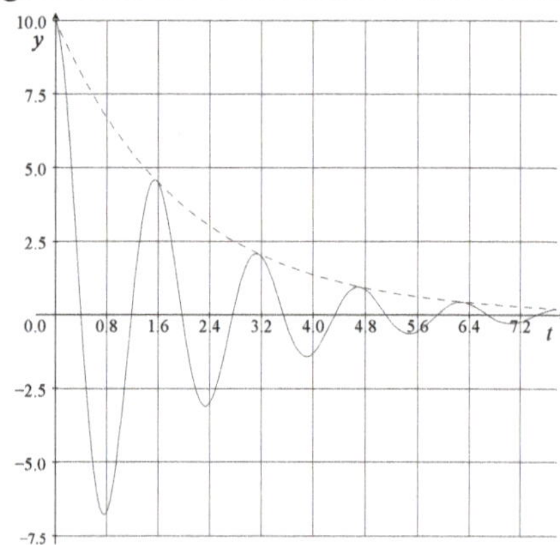

Example: A mass weighing 8 lb stretches a spring 2 ft. Assuming that a dampening force equal to twice the instantaneous velocity acts on the system (with the appropriately adjusted dimensions), determine the equation of motion if the mass is initially released from the equilibrium position with a downward velocity of $5\,\frac{\text{ft}}{\text{s}}$.

From Hooke's law, $F = ky$, we get $8 = k(2)$ hence $k = \frac{8}{2} = 4\,\frac{\text{lb}}{\text{ft}}$. Since $w = mg$, $m = \frac{8\,\text{lb}}{32\,\frac{\text{ft}}{\text{s}^2}} = \frac{1}{4}$ slug. Thus, the differential equation of motion would be

$$\frac{d^2y}{dt^2} + \frac{\beta}{m}\frac{dy}{dt} + \frac{k}{m}y = 0 \quad \text{where } \beta = 2 \text{ is the dampening factor}$$

$$\frac{d^2y}{dt^2} + \frac{2}{\frac{1}{4}}\frac{dy}{dt} + \frac{4}{\frac{1}{4}}y = 0$$

$$\frac{d^2y}{dt^2} + 8\frac{dy}{dt} + 16y = 0$$

The characteristic equation would be $1r^2 + 8r + 16 = 0 \to (r+4)(r+4) = 0$, so $r = -4$. From Chapter 14, the general solution to the differential equation is

$$y_{\text{gen}} = (c_1 + c_2 t)e^{rt}.$$

Hence, $y_{\text{gen}} = c_1 e^{-4t} + c_2 t e^{-4t}$. Assume IVP conditions $y_0 = 0\,\text{ft}$ and $y_0' = -5\,\frac{\text{ft}}{\text{s}}$.

> $B^2 - 4AC = 0$
>
> $y_{\text{gen}} = c_1 e^{rt} + c_2 x e^{rt}$
>
> Be careful to include the x factor in the second addend.

$$y_0 = (c_1 + c_2 \times 0)e^{-4\times 0}$$
$$0 = c_1 \times 1 + 0$$
$$c_1 = 0$$

So

$$y_{\text{gen}} = (0 + c_2 t)e^{-4t}$$
$$y_{\text{gen}} = c_2 t e^{-4t}$$
$$(y_{\text{gen}})' = c_2 t e^{-4t}(-4) + e^{-4t}(c_2)$$
$$= e^{-4t}c_2(-4t + 1)$$
$$(y_0)' = e^{-4\times 0}c_2(-4\times 0 + 1)$$
$$-5 = 1 \times c_2(1)$$
$$c_2 = -5$$

Finally, for $\frac{d^2y}{dt^2} + 8\frac{dy}{dt} + 16y = 0$ and the IVP conditions mentioned above,

$$y = 0 \times e^{-4t} + \left(-5te^{-4t}\right) = -5te^{-4t}.$$

As t gets larger, what happens to e^{-4t}? Do you recall a "sink" from Chapter 4? Why is there a sink at $y = 0$ in the slope field above? There are two answers: 1) intuitive and 2) mathematical. You need to understand both of them. 1) the viscosity of the medium dampens the spring motion and 2) e^{-4t} as $t \to \infty$ will decay to zero. Hence, $-5t \times e^{-4t}$ will decay to zero as $t \to \infty$.

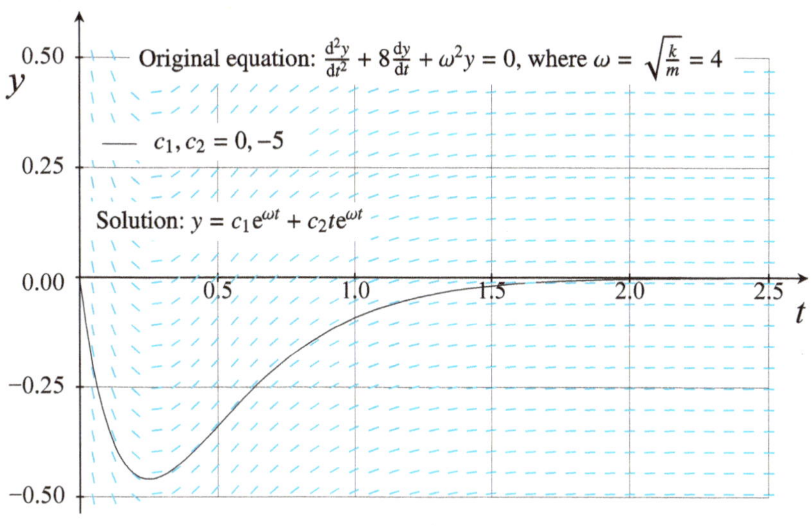

Electrical Circuits

In physics, there is a topic called LRC-Series Circuits. In the figure at right i denotes current in the LRC-series electrical circuit. The figure also shows the voltage drops across the inductor, resistor, and capacitor.

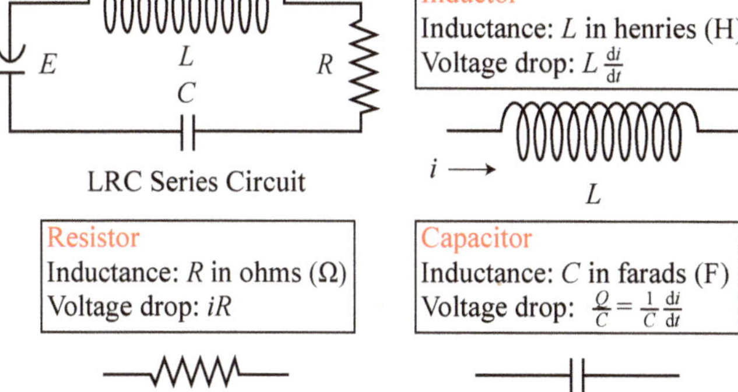

By Kirchhoff's laws (see Chapter 5), the sum of these voltages is equal to the voltage impressed on the circuit, $E(t)$. This is shown as

$$L\frac{di}{dt} + Ri + \frac{1}{c}Q = E(t).$$

But if the charge $Q(t)$ on the capacitor is related to the current $i(t)$ by $i = \frac{dQ}{dt}$ then the equation above becomes

$$L\frac{d^2Q}{dt^2} + R\frac{dQ}{dt} + \frac{1}{c}Q = E(t).$$

If $E(t) = 0$, the electrical vibrations of the circuit are said to be free and the equation above becomes

$$L\frac{d^2Q}{dt^2} + R\frac{dQ}{dt} + \frac{1}{c}Q = 0.$$

Comparing $L\frac{d^2Q}{dt^2}+R\frac{dQ}{dt}+\frac{1}{c}Q = 0$ with $Ay''+By'+Cy = 0$ we notice that the differential equation forms that result from working with Kirchhoff's law are the exact same differential equation forms that often result from working with Ohm's laws: second-order linear differential equations. Except for terminology and physical interpretations of the four terms in the linear differential equation,

$$a\frac{d^2y}{dt^2} + b\frac{dy}{dt} + cy = g(t),$$

the mathematics of an electrical series circuit is identical to that of a vibrating spring–mass system.

Chapter 15 Review

Chapter 14 discussed what a second-order homogeneous differential equation, $Ay'' + By' + Cy = 0$, looked like and how it is solvable using an algebraic or integrated form. Chapter 15 discussed the second-order differential equations that arise in the application of Hooke's Law, and a specific problem was presented and discussed at length. Lastly, it was also noted that equations using Kirchhoff's law were also of the form of a second-order differential equation.

Partial Derivatives
First-Order Homogeneous Differential Equations
Second-Order Homogeneous Differential Equations
Exact Differential Equations
Feel-in' smart now.
Feel-in' smart now.
Feel-in' smart now.

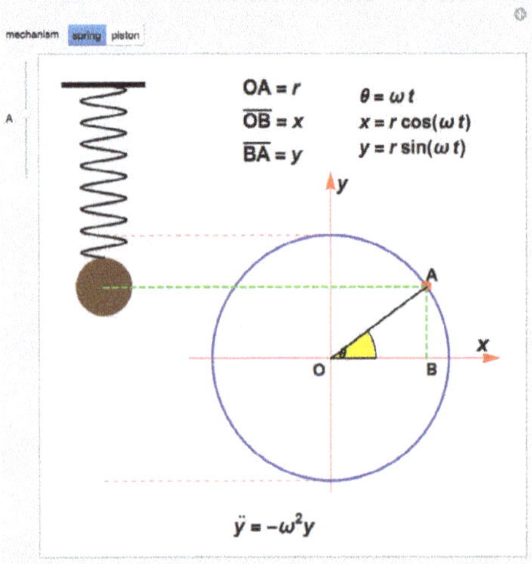

Online Application

Visit demonstrations.wolfram.com/SimpleHarmonicMotion for a great little application that will demonstrate some of the concepts from this chapter. "Simple Harmonic Motion" from the Wolfram Demonstrations Project. Contributed by Paul Rosemand (Cegep de l'Outaouais, Gatineau, Quebec).

Chapter 16
Predator–Pursuit Problems

The Shooter, the Hunter and the Coordinator

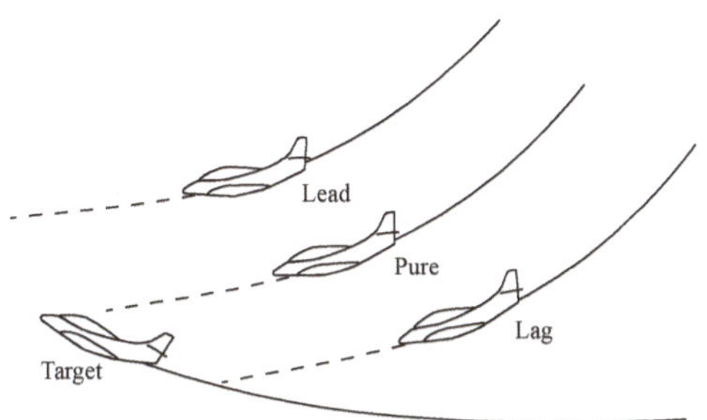

- Pure pursuit: Direction of your velocity is always towards the target.
- Lead pursuit: Direction of your velocity is always ahead of the target.
- Lag pursuit: Direction of your velocity is always behind the target.

"Pursuit curves have different uses. It becomes more pronounced when the target is moving along a curved path. For example, lead pursuit is necessary for all long range sniper kills. This is because it takes time for the bullets to travel through space. Sometimes snipers have to shoot from as far as 2 miles. If the target is that far, you cannot just aim and shoot. All kind of ungodly things come into play here! If you shoot where the target is right now the target will no longer be there when the bullets arrive. Therefore, you must predict the future flight path of the target in order to put the ammunition into the target. This is the essence of aerial combat as well. The ability to predict where the target will be in the future is crucial to all facets of aerial combat."* All this passage translates quickly to all the "shoot-em up" video games that many young people love so much. Predator is to prey as missile (bullet) is to target.

Suppose a dog is located on the origin of a 600×400 Java graphics screen and a rabbit is located at (60, 300).† The rabbit starts running due east at a constant speed. The dog sees the rabbit and gives chase.‡ After every second, the dog adjusts his direction so as to keep pointing at the rabbit. The pursuit would like the one at right.

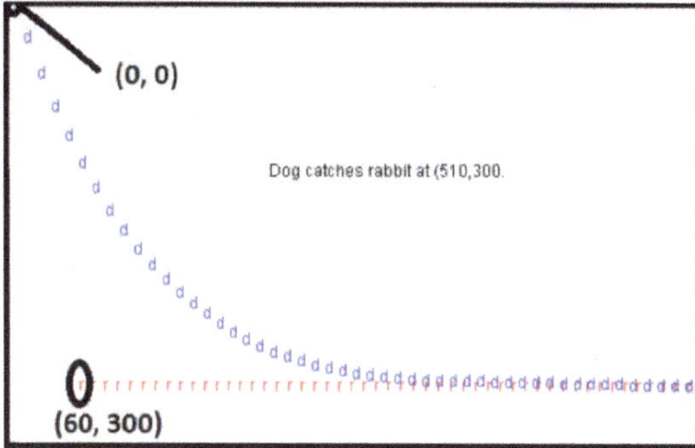

*http://prateekvjoshi.com/2013/01/09/the-clairvoyant-curves-of-pursuit/

†The coordinate system of the output screen for a Java problem locates the origin (0, 0) at the top left corner. Hence when y increases you go down the screen, not up.

‡For the sake of simple programming, the x component of the dog's speed is assumed constant, but the y component actually decreases as the dog turns more to the east.

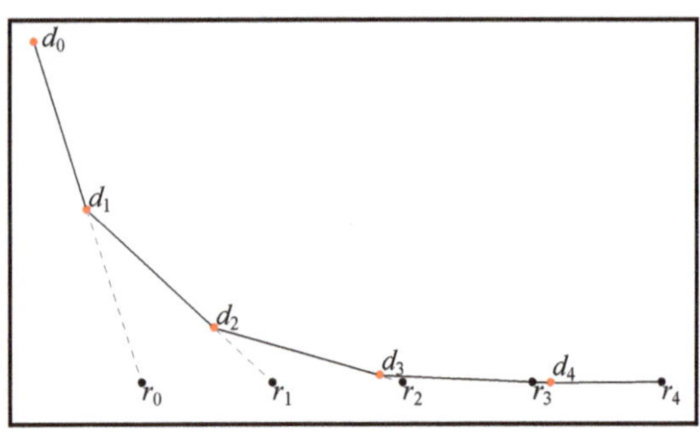

The dog's directional readjustment is shown to the left. The rabbit's y value stays constant at 300 while its x value increases at a constant rate. The x component of the dog's speed increases at a constant rate while its y value is projected at the slope of the angle connecting the dog's current position and the rabbit's current position, $M = \frac{Y_{\text{rabbit}} - Y_{\text{dog}}}{X_{\text{rabbit}} - X_{\text{dog}}}$. This is repeated until the dog catches the rabbit. With increasingly small increments, this discrete approximation of the path will approach the continuous movement of the real dog.

```
while (dog has not caught rabbit)
{
  move rabbit horizontally one position
  calculate dog's next position on the line toward rabbit's current position
  move dog toward rabbit
}
```

The result of this code is discrete output. You may recall our discussion of discrete data versus continuous data in Chapters 2 and 3. The purpose of differential equations is to be able to work with continuous data.

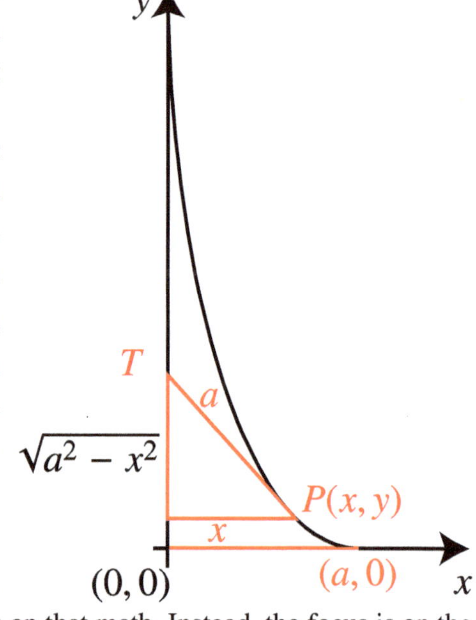

Consider the figure at right. A point P is dragged along the x–y plane by a string PT of constant length a. Initially T is at the origin and P is on the x-axis at point $(a, 0)$. T is then moved up the y-axis. Find the resulting equation for the path of P.

Because T and P are connected by a string, T can only pull P straight toward it. Therefore, the string will always be tangent to the curve followed by P. So, the slope of the string is equal to the derivative of the curve followed by P: $m = \frac{\mathrm{d}y}{\mathrm{d}x} = -\frac{\sqrt{a^2-x^2}}{x}$. This is a differential equation that, when solved, results in $y = a \ln\left(\frac{a+\sqrt{a^2-x^2}}{x}\right) - \sqrt{a^2 - x^2}$,[§] which is the equation describing the dog's path in the previous discussion. Note that this is an equation in y that yields continuous data. In a graphics program, we have pixels with discrete addresses, but one gets the appearance of continuous data if the resolution is high enough.

The steps between $\frac{\mathrm{d}y}{\mathrm{d}x} = -\frac{\sqrt{a^2-x^2}}{x}$ and its solution, $y = a \ln\left(\frac{a+\sqrt{a^2-x^2}}{x}\right) - \sqrt{a^2 - x^2}$, are numerous and involve a great deal of trig and algebra. Today with graphical programs such as MATLAB, there is not as much emphasis on that math. Instead, the focus is on the interpretation of graphical data that results from running prewritten code.

The image below shows a commercial software program demonstrating the pursuit of the dog, red line starting at $(0, 0)$, and the rabbit, blue line starting at $(5, 0)$.

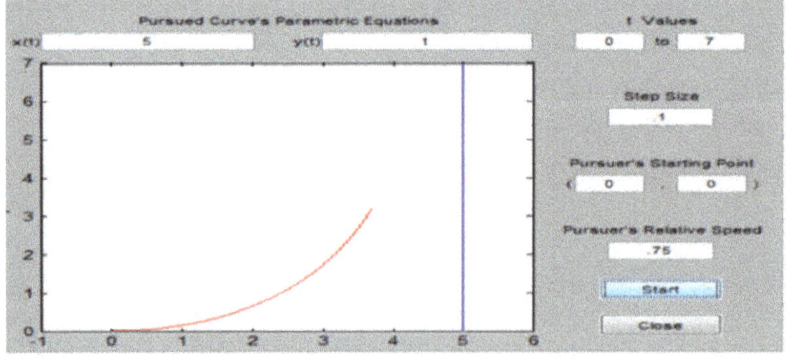

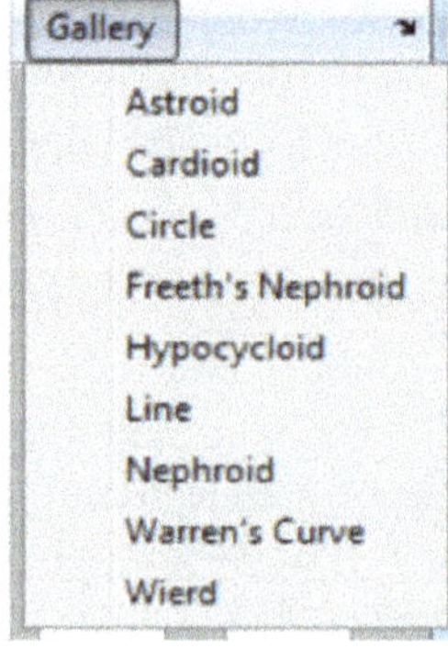

[§]This is a standard integral and can be looked up in a short table of integrals.

Many, many computer games use the math introduced here to allow a bullet or missile to be launched toward an adversary or target of some kind. As long as the coordinates of the predator/bullet/missile and the coordinates of the prey/target are known, the math that is used by the programmer will be effective in seeking, converging on, and contacting the target. This is demonstrated in the following examples using commonly known math curves. Again the red line indicates the pursuing object while the blue line indicates the pursued object.

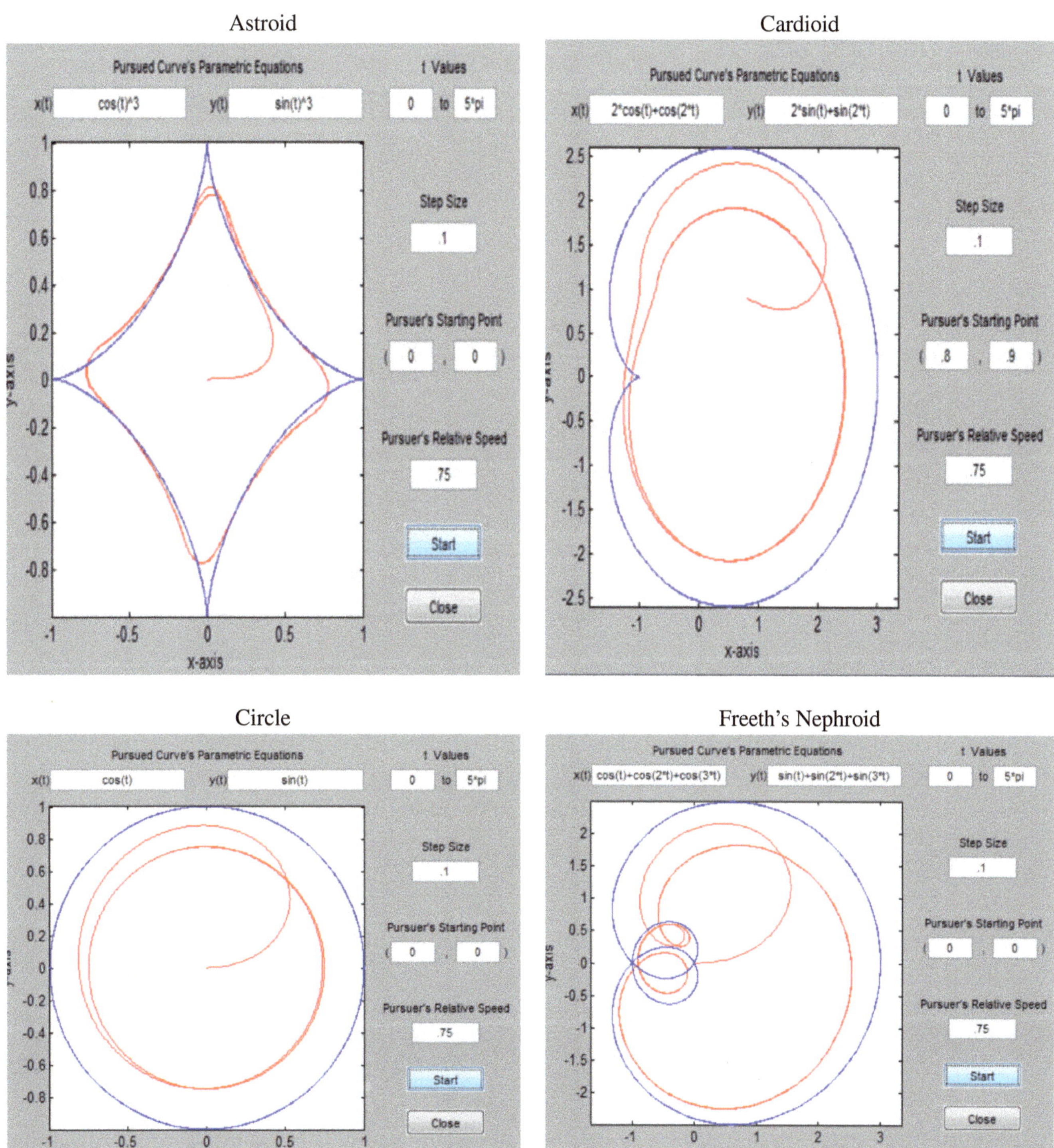

MATLAB also allows a person to create pursuit curves in three space.

A Pursuit Curve on a 3D Surface

Graphing in 3D opens up the possibility of graphing a pursuit curve on a 3D surface. In order to do this, the surface must be defined as a function of x and y. That way the pursuit curve will be directly graphed on the surface. In this example, the lion verses gazelle model will be used. The lion is pursuing the gazelle at 30% of the gazelle's speed and the surface equation is $Z = 20 \sin x \cos y$.

Our equations are as follows:

$$p(t) = \cos t$$
$$q(t) = \sin 3t$$
$$r(t) = 20 \sin x \cos y$$

Solving

$$x = \cos t$$
$$y = \sin 3t$$
$$z = 20 \sin x \cos y$$
$$\frac{dx}{dt} = -\sin t$$
$$\frac{dy}{dt} = 3 \cos 3t$$
$$\frac{dz}{dt} = 20 \cos x \frac{dx}{dt} \cos y + 20 \sin x(-\sin y)\frac{dy}{dt}$$

and coding in MATLAB (see Appendix A) produces the following graph.[¶]

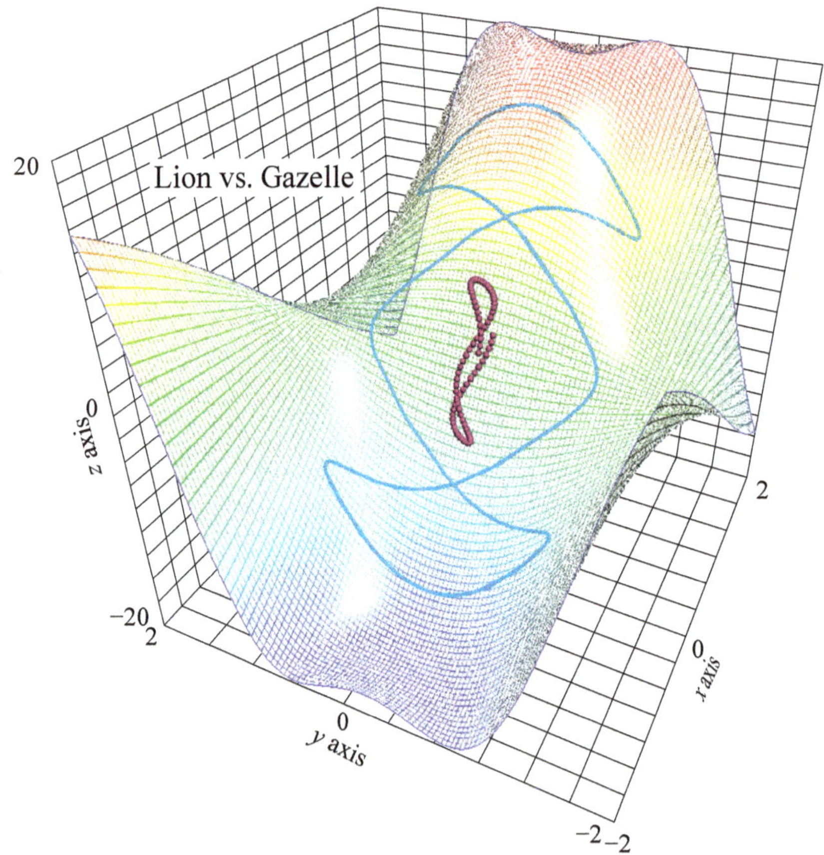

[¶]*Pursuit Curves* Katy Steiner, Jonah Franchi 5/10/2011 http://home2.fvcc.edu/~dhicketh/DiffEqns/Spring11projects/Jonah_Franchi_Katy_Steiner/Diff%20EQ%20Project.pdf

Chapter 16 Review

Chapters 7 and 16 are examples of systems of differential equations. Chapter 16 discussed how the differential equation of a target object can be used to guide a pursuing object. This would be useful if you wanted a supply ship to chase after and connect with the space station, or if you wanted to land a space ship on the moon, or if you were trying to get an antiballistic missile to intercept an incoming missile. Also predator–pursuit is used in programming a great many computer games.

Online Application

Predator-Prey Equations

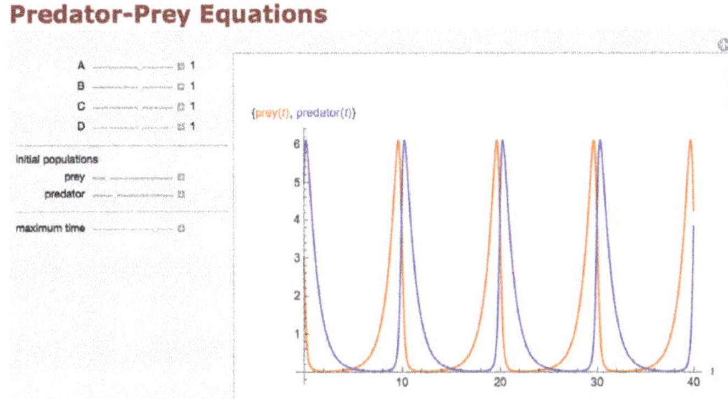

Visit demonstrations.wolfram.com/FlyingToTheMoon, demonstrations.wolfram.com/PredatorPreyEquations, and demonstrations.wolfram.com/PursuitCurves for three great little applications that will demonstrate some of the concepts from this chapter. "Predator Prey Equations" from the Wolfram Demonstrations Project. Contributed by Eric W. Weisstein. "Pursuit Curves" from the Wolfram Demonstrations Project. Contributed by Stan Wagon. "Flying to the Moon" from the Wolfram Demonstrations Project. Contributed by Michael Trott.

Flying to the Moon

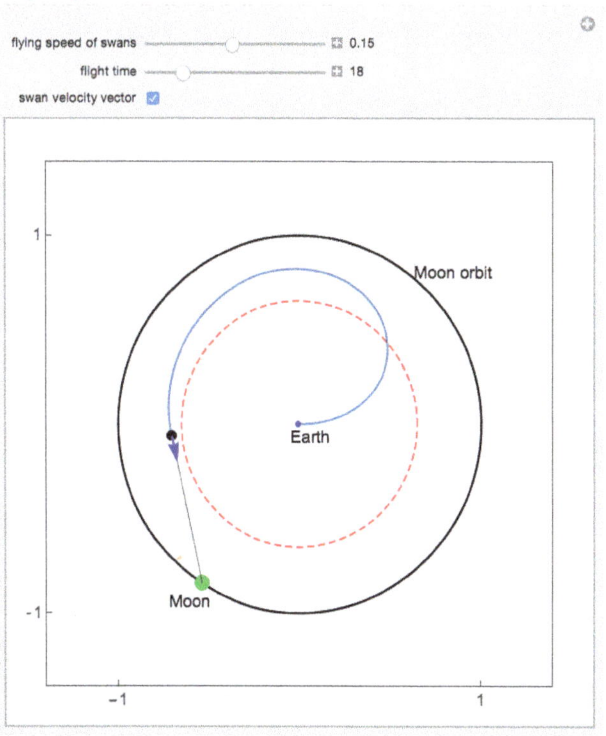

Pursuit Curves

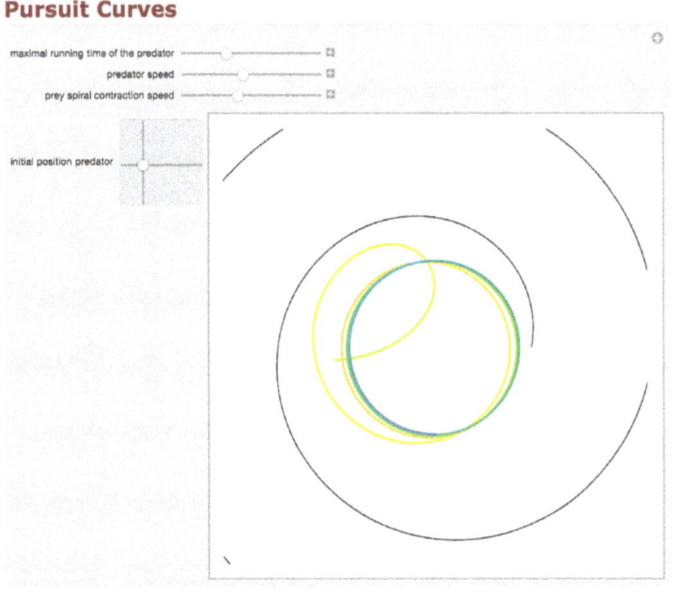

Chapter 17
Review of Algebraic Systems of Equations

The next major topic covered in this book is solving systems of linear, homogeneous, constant-coefficient differential equations. The concept of systems of differential equations has been briefly discussed twice, in Chapters 7 and 16. Following is the general form of such equations.

$$\begin{aligned}
x_1' &= a_{1,1}x_1 + a_{1,2}x_2 + a_{1,3}x_3 + \cdots + a_{1,n}x_n \\
x_2' &= a_{2,1}x_1 + a_{2,2}x_2 + a_{2,3}x_3 + \cdots + a_{2,n}x_n \\
x_3' &= a_{3,1}x_1 + a_{3,2}x_2 + a_{3,3}x_3 + \cdots + a_{3,n}x_n \\
&\vdots \\
x_n' &= a_{n,1}x_1 + a_{n,2}x_2 + a_{n,3}x_3 + \cdots + a_{n,n}x_n
\end{aligned}$$

Together several skills make up the solution technique to solve such a system of equations. Scan through the following figure and see if you already know the requisite skills and vocabulary. If so you should just jump over to Chapter 20. If any of these skills and vocabulary are not familiar, take a look at the indicated chapter.

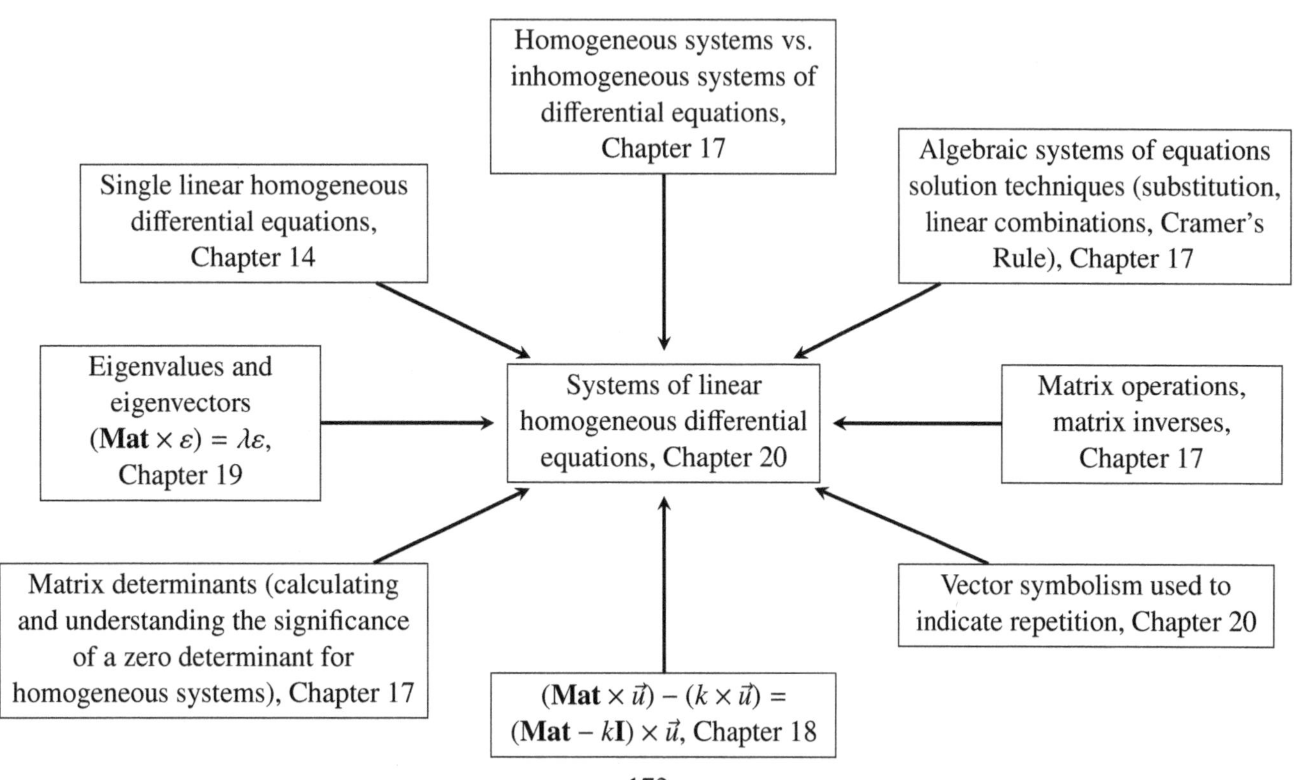

Systems of Algebraic Equations Not Passing through the Origin

In algebra, you were introduced to three types of 2×2 systems of equations.

1. $3x + 2y = 6$

$6x + 4y = 12$

The determinant of the system of coefficients is $(3\times4)-(6\times2) = 0$. Multiplying Row 1 by 2 yields the second row, meaning that the equations are equivalent as can be seen from their intercepts.

$3x + 2y = 6$ or $6x + 4y = 12$	
x	y
0	3
2	0

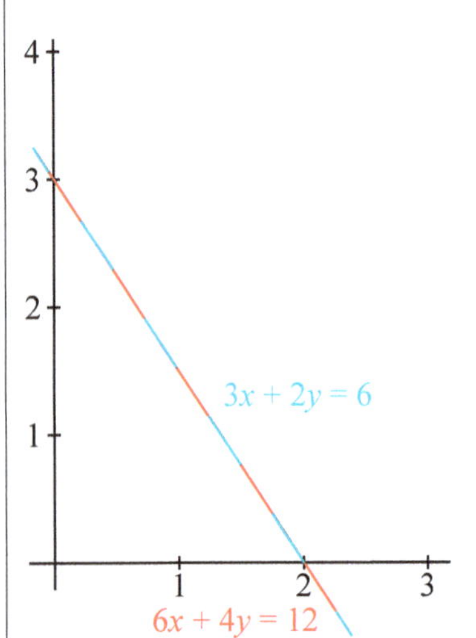

Here the determinant is zero. There are infinite solutions and *each row of the coefficient matrix is a multiple of the other.*

2. $3x + 2y = 6$

$6x + 4y = 8$

The determinant of the system of coefficients is $(3 \times 4) - (6 \times 2) = 12 - 12 = 0$. The Row 2 coefficients are again multiples of Row 1, but the two equations have different x- and y-intercepts. The lines are parallel but distinct.

$3x + 2y = 6$		$6x + 4y = 8$	
x	y	x	y
0	3	0	2
2	0	$\frac{4}{3}$	0

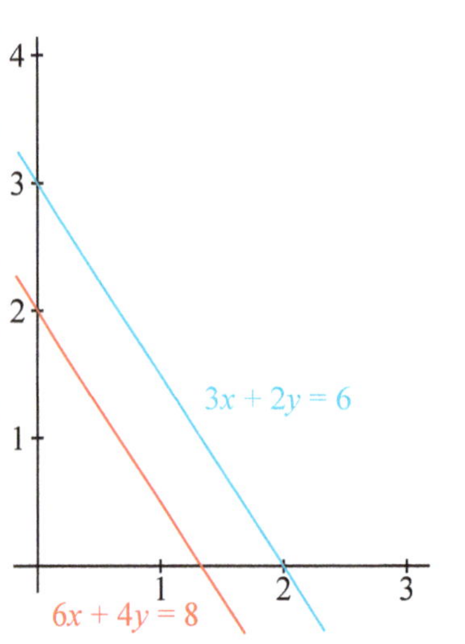

Here the determinant is zero. There are no solutions and *each row of the coefficient matrix is a multiple of the other.*

3. $3x + 2y = 6$

$4x + 1y = 4$

The determinant of the system of coefficients is $(3 \times 1) - (4 \times 2) = 3 - 8 = -5$. The two equations have different slopes and different x- and y-intercepts, *and* they cross.

$3x + 2y = 6$		$4x + 1y = 4$	
x	y	x	y
0	3	0	4
2	0	1	0

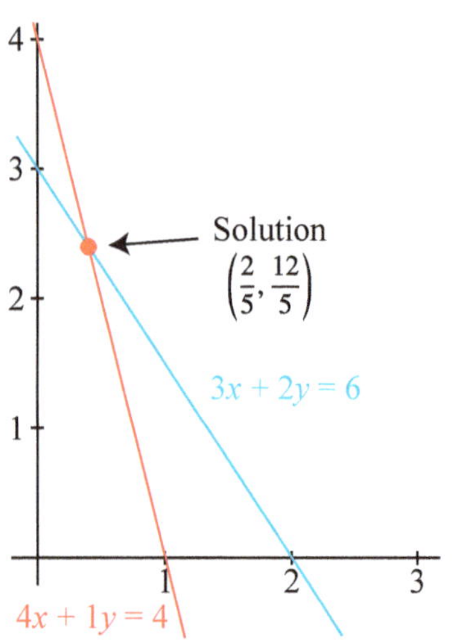

Here the determinant is **not** zero. There is one unique solution and the *rows of the coefficient matrix are not multiples of each other.*

None of the lines in the systems above pass through the origin. That is, there is a nonzero constant term in each one (6 and 12, 6 and 8, and 6 and 4). They are called *inhomogeneous systems* of equations. As shown in the systems above, if a determinant of a 2×2 system is zero,* then the coefficients of one equation are a single multiple of the other coefficients.† However, sometimes, as shown above, a determinant of zero means infinite solutions and sometimes a determinant of zero means no solutions. Here are important facts to know about 2×2 inhomogeneous systems of equations.

1. If the determinant of the coefficient matrix of an inhomogeneous system of equations is zero, the system has infinite or no solutions and the rows of the system matrix are multiples of each other.

2. If the determinant of the coefficient matrix of an inhomogeneous system of equations is not zero, the system has one solution and the rows of the system matrix are not multiples of each other.

*An arithmetic combination of diagonal products, the determinant tells us useful things about a matrix.

†This is somewhat reminiscent of the fact that the discriminant ($b^2 - 4ac$) of a quadratic equation will tell you about the nature of the roots of that equation.

Systems of Algebraic Equations Passing through the Origin

The three systems on the previous page are modified here so that all linear combinations sum to zero. These sorts of systems of equations are called homogeneous equations. Replacing the numbers on the right does not impact the slopes of the lines, nor does it impact the determinant, but it does impact the x- and y-intercepts.

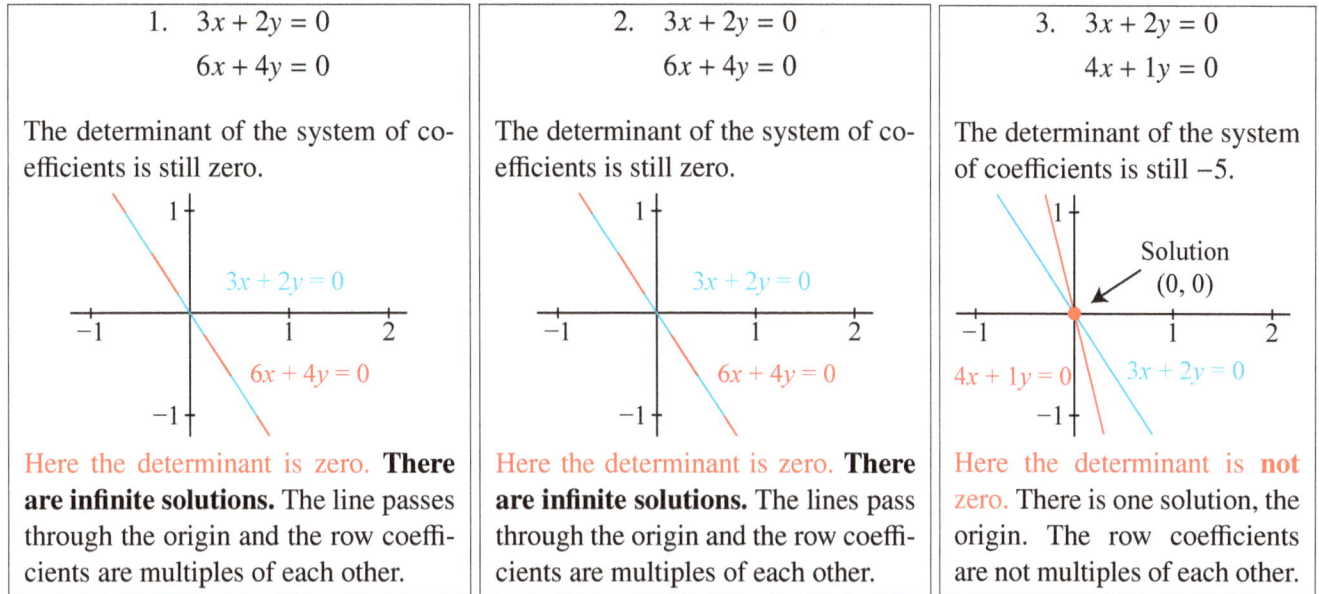

1. $3x + 2y = 0$
   $6x + 4y = 0$

The determinant of the system of coefficients is still zero.

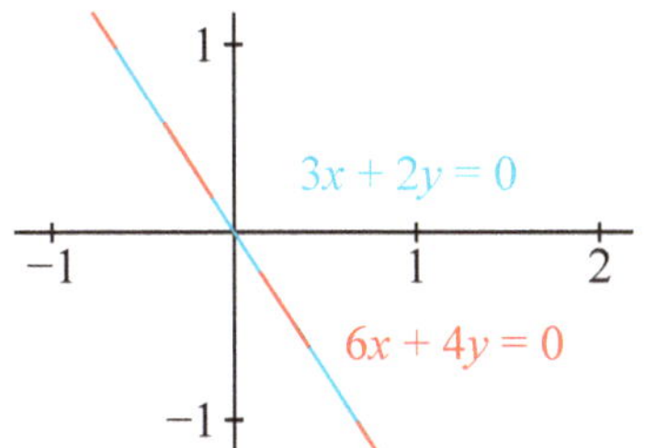

Here the determinant is zero. **There are infinite solutions.** The line passes through the origin and the row coefficients are multiples of each other.

2. $3x + 2y = 0$
   $6x + 4y = 0$

The determinant of the system of coefficients is still zero.

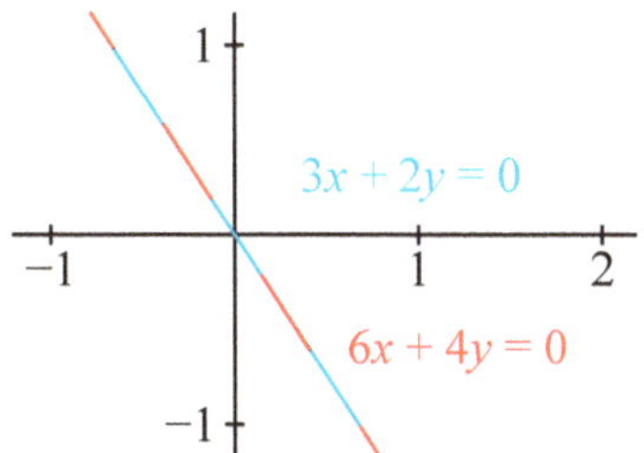

Here the determinant is zero. **There are infinite solutions.** The lines pass through the origin and the row coefficients are multiples of each other.

3. $3x + 2y = 0$
   $4x + 1y = 0$

The determinant of the system of coefficients is still −5.

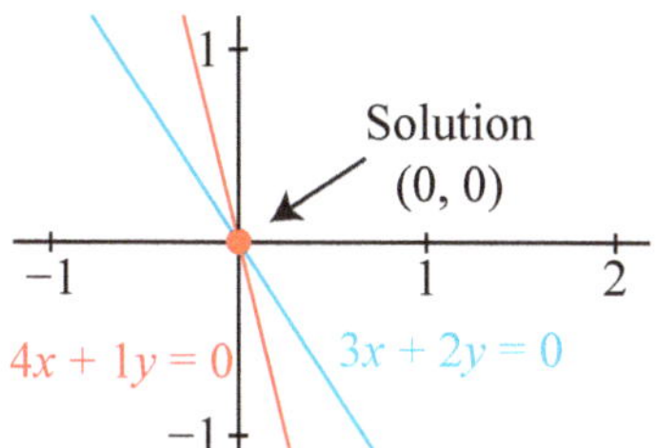

Here the determinant is **not** zero. There is one solution, the origin. The row coefficients are not multiples of each other.

There are two important facts to know about homogeneous systems of equations.

1. If the determinant of a 2×2 constant-coefficient matrix of a homogeneous system is zero, the system has infinite solutions and the rows of the system matrix are multiples of each other. If a system of equations has infinite solutions and the rows of the system matrix are multiples of each other, then its determinant is zero. (This statement could be stated in "if-and-only-if" form.)

2. If the determinant of a 2×2 constant-coefficient matrix of a homogeneous system of equations is not zero, the system has one solution, $(0, 0)$, and the rows of the system matrix are not multiples of each other. If a system of equations has a single solution, then the rows of the system matrix are not multiples of each other and the determinant is not zero. (This statement could be stated in "if-and-only-if" form.)

Each of the above graphs involved two variables and could be graphed in two-space on an x–y-axis. Let's try an experiment. Add another variable to the equation $3x + 2y = 0$, perhaps $3x + 2y - z = 0$. How do you graph such an equation? Well, three variables means we need to have three dimensions to graph our equation, right? Traditionally, when working with three space, the axes are labeled as shown at right. The version at left has the same right-handed orientation, but retains the position of the x- and y-axes found when working in two space.

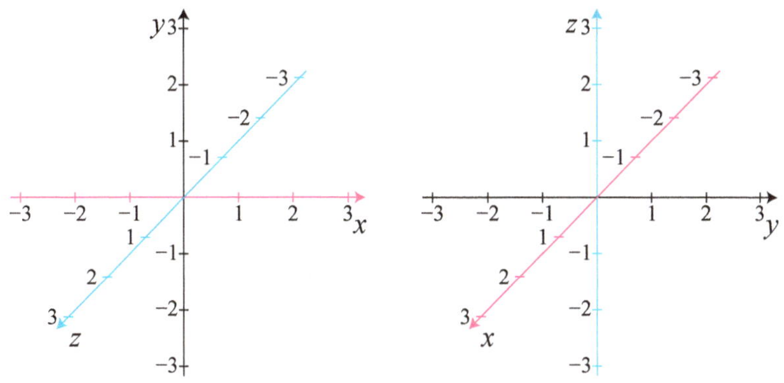

Proof that det(**Mat**) = 0 iff the rows are multiples of each other.

Let Row 2 be a multiple of Row 1.

$$\mathbf{Mat} = \begin{pmatrix} a & b \\ ka & kb \end{pmatrix}$$

$$\begin{aligned} \det(\mathbf{Mat}) &= (a \times kb) - (ka \times b) \\ &= akb - akb \\ &= 0 \end{aligned}$$

Let Row 2 be not a multiple of Row 1.

$$\mathbf{Mat} = \begin{pmatrix} a & b \\ ka & (k+i)b \end{pmatrix}$$

$$\begin{aligned} \det(\mathbf{Mat}) &= (a \times (k+i)b) - (ka \times b) \\ &= akb + aib - akb \\ &= aib \end{aligned}$$

Therefore, if $a, b, i \neq 0$, the determinant cannot be zero.

In graphing $3x + 2y - z = 0$, we would get $3x + 2y = z$ for all z values. If you can visualize in three space, you may see a slanted plane crossing the z axis. Because it is difficult to see what is happening when a three-space object is shown on a two-space surface it is often the case that, when studying a differential equation involving three variables, we use what is known as a phase-plane graph as introduced back in Chapter 4.

```
clc
clear
format
x = [-0: .2: 15];
y = [-5: .2: 5];
[xx, yy] = meshgrid(x,y);
zz = 3 .* xx + 2 .* yy
figure surf(xx, yy, zz);
hold on;
title('z = 3x + 2y');
xlabel('x axis');
ylabel('y axis');
zlabel('z axis');
```

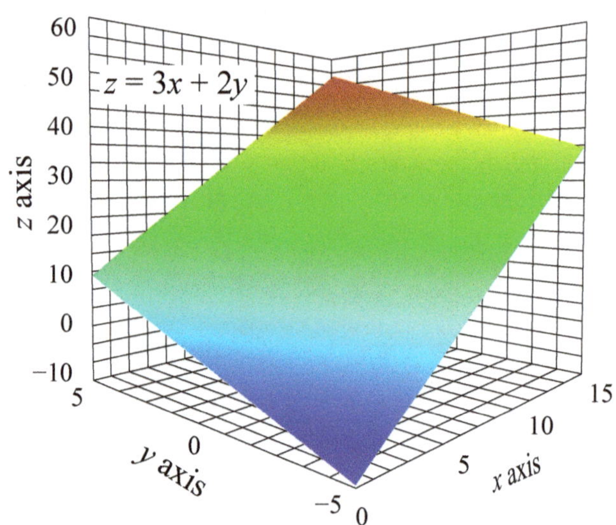

In closing, it should be noted that the equation $3x + 2y = 0$ could have been written as either $y = -\frac{3}{2}x$ or $f(x) = -\frac{3}{2}x$, emphasizing the fact that one of the two variables is dependent on the other. The dependent variable is y. The independent variable is x.

Similarly, the equation, $3x+2y-z = 0$, could have been written as $z = 3x+2y$ or $f(x, y) = 3x+2y$, emphasizing the fact that there are three variables with one being dependent on the other two. The dependent variable is z. The two independent variables are x and y.

Algebra Review, Solving Systems of Linear Equations

In algebra, we are taught how to solve two equations with two unknowns using three different techniques: substitution, linear combinations (aka. elimination), and Cramer's rule. Since there are two variables, the goal is to isolate the two variables and solve for both of them.

Substitution Technique for Solving Two Equations with Two Unknowns	Linear-Combination Technique for Solving Two Equations with Two Unknowns
$3x + 2y = 8 \quad (1)$ $4x + 1y = 6 \quad (2)$	$3x + 2y = 8 \quad (1)$ $4x + 1y = 6 \quad (2)$
From (2), we get $y = 6 - 4x$. Substitute the binomial equivalent of y from (2) into (1).	Multiply both sides of (2) by −2.
$3x + 2(6 - 4x) = 8$ $3x + 12 - 8x = 8$ $-5x = -4$ $x = \frac{4}{5}$	$3x + 2y = 8 \quad (3)$ $-8x - 2y = -12 \quad (4)$ Add (3) to (4). $-5x = -4$ $x = \frac{4}{5}$

Back substitute $x = \frac{4}{5}$ into either (1) or (2).

$$\begin{array}{cc} (1) & (2) \\ 3\left(\frac{4}{5}\right) + 2y = 8 & 4\left(\frac{4}{5}\right) + y = 6 \\ \frac{12}{5} + 2y = \frac{40}{5} & \frac{16}{5} + y = \frac{30}{5} \\ 2y = \frac{28}{5} & y = \frac{14}{5} \\ y = \frac{14}{5} & \end{array}$$

Solution: $x = \frac{4}{5}$, $y = \frac{14}{5}$. This can also be shown as

$$\begin{pmatrix} x \\ y \end{pmatrix} = \begin{pmatrix} \frac{4}{5} \\ \frac{14}{5} \end{pmatrix}.$$

Back substitute $x = \frac{4}{5}$ into either (1) or (2).

$$\begin{array}{cc} (1) & (2) \\ 3\left(\frac{4}{5}\right) + 2y = 8 & 4\left(\frac{4}{5}\right) + y = 6 \\ \frac{12}{5} + 2y = \frac{40}{5} & \frac{16}{5} + y = \frac{30}{5} \\ 2y = \frac{28}{5} & y = \frac{14}{5} \\ y = \frac{14}{5} & \end{array}$$

Solution: $x = \frac{4}{5}$, $y = \frac{14}{5}$. This can also be shown as

$$\begin{pmatrix} x \\ y \end{pmatrix} = \begin{pmatrix} \frac{4}{5} \\ \frac{14}{5} \end{pmatrix}.$$

Both the substitution technique and the linear-combination technique work nicely on systems with two equations and two unknowns. However, when extending these two techniques to three equations with three unknowns, four equations with four unknowns, etc., the tedium of tracking the resulting details and steps is likely to lead to mistakes. It would be nice to have a technique that, although seemingly more difficult initially, can be easily extended to solve n equations with n unknowns using computer code. Cramer's Rule is that technique.

Solving Algebraic Systems Using Cramer's Rule

The basic idea behind Cramer's Rule is that each variable is equal to the ratio of two determinants. The denominator in each ratio is the same, the determinant of the matrix of the coefficients of the variables in the system of equations. The numerator of each ratio is the determinant of a different matrix constructed from that matrix of coefficients. However, the column corresponding to the variable of interest is replaced with the constant values on the right-hand sides of the system of equations.

$$3x + 2y + 1z = 8 \quad (1)$$

$$6x + 5y + 4z = 9 \quad (2)$$

$$0x + 8y + 7z = 0 \quad (3)$$

$$x = \frac{\begin{vmatrix} 8 & 2 & 1 \\ 9 & 5 & 4 \\ 0 & 8 & 7 \end{vmatrix}}{\begin{vmatrix} 3 & 2 & 1 \\ 6 & 5 & 4 \\ 0 & 8 & 7 \end{vmatrix}} = \frac{(8\cdot5\cdot7)+(2\cdot4\cdot0)+(1\cdot9\cdot8)-(0\cdot5\cdot1)-(8\cdot4\cdot8)-(7\cdot9\cdot2)}{(3\cdot5\cdot7)+(2\cdot4\cdot0)+(1\cdot6\cdot8)-(0\cdot5\cdot1)-(8\cdot4\cdot3)-(7\cdot6\cdot2)} = \frac{280+0+72-0-256-126}{105+0+48-0-96-84} = \frac{-30}{-27} = \frac{10}{9}$$

$$y = \frac{\begin{vmatrix} 3 & 8 & 1 \\ 6 & 9 & 4 \\ 0 & 0 & 7 \end{vmatrix}}{\begin{vmatrix} 3 & 2 & 1 \\ 6 & 5 & 4 \\ 0 & 8 & 7 \end{vmatrix}} = \frac{(3\cdot9\cdot7)+(8\cdot4\cdot0)+(1\cdot6\cdot0)-(0\cdot9\cdot1)-(0\cdot4\cdot3)-(7\cdot6\cdot8)}{(3\cdot5\cdot7)+(2\cdot4\cdot0)+(1\cdot6\cdot8)-(0\cdot5\cdot1)-(8\cdot4\cdot3)-(7\cdot6\cdot2)} = \frac{189+0+0-0-0-336}{105+0+48-0-96-84} = \frac{-147}{-27} = \frac{49}{9}$$

$$z = \frac{\begin{vmatrix} 3 & 2 & 8 \\ 6 & 5 & 9 \\ 0 & 8 & 0 \end{vmatrix}}{\begin{vmatrix} 3 & 2 & 1 \\ 6 & 5 & 4 \\ 0 & 8 & 7 \end{vmatrix}} = \frac{(3\cdot5\cdot0)+(2\cdot9\cdot0)+(8\cdot6\cdot8)-(0\cdot5\cdot8)-(8\cdot9\cdot3)-(0\cdot6\cdot2)}{(3\cdot5\cdot7)+(2\cdot4\cdot0)+(1\cdot6\cdot8)-(0\cdot5\cdot1)-(8\cdot4\cdot3)-(7\cdot6\cdot2)} = \frac{0+0+384-0-216-0}{105+0+48-0-96-84} = \frac{168}{-27} = -\frac{56}{9}$$

The solution is $\left(\frac{10}{9}, \frac{49}{9}, -\frac{56}{9}\right)$ or $\begin{pmatrix} \frac{10}{9} \\ \frac{49}{9} \\ -\frac{56}{9} \end{pmatrix}$.

Solving Systems of Equations Using Matrix Algebra

Hopefully, you were taught a concept called *matrix multiplication* in algebra. Here is a quick review: If the number of columns of Matrix **A** equals the number of rows of Matrix **B**, multiplication is defined for the two matrices. Here are three (equivalent) ways to think of matrix multiplication.

Matrix Multiplication #1: "Linear Combination of Columns"

$$\begin{pmatrix} -\frac{1}{5} & \frac{2}{5} \\ \frac{4}{5} & -\frac{3}{5} \end{pmatrix}\begin{pmatrix} 3 & 2 \\ 4 & 1 \end{pmatrix} = \left[\begin{pmatrix} -\frac{1}{5} \\ \frac{4}{5} \end{pmatrix}\times 3 + \begin{pmatrix} \frac{2}{5} \\ -\frac{3}{5} \end{pmatrix}\times 4 \quad \begin{pmatrix} -\frac{1}{5} \\ \frac{4}{5} \end{pmatrix}\times 2 + \begin{pmatrix} \frac{2}{5} \\ -\frac{3}{5} \end{pmatrix}\times 1\right]$$

$$= \left[\begin{pmatrix} -\frac{3}{5} \\ \frac{12}{5} \end{pmatrix} + \begin{pmatrix} \frac{8}{5} \\ -\frac{12}{5} \end{pmatrix} \quad \begin{pmatrix} -\frac{2}{5} \\ \frac{8}{5} \end{pmatrix} + \begin{pmatrix} \frac{2}{5} \\ -\frac{3}{5} \end{pmatrix}\right]$$

$$= \begin{bmatrix} \frac{5}{5} & \frac{0}{5} \\ \frac{0}{5} & \frac{5}{5} \end{bmatrix}$$

$$= \begin{pmatrix} 1 & 0 \\ 0 & 1 \end{pmatrix}$$

Matrix Multiplication #2: "Dot Product of All Column–Row Combinations" (often taught in HS)

$$\begin{pmatrix} -\frac{1}{5} & \frac{2}{5} \\ \frac{4}{5} & -\frac{3}{5} \end{pmatrix}\begin{pmatrix} 3 & 2 \\ 4 & 1 \end{pmatrix} = \begin{pmatrix} -\frac{1}{5}\times 3 + \frac{2}{5}\times 4 & -\frac{1}{5}\times 2 + \frac{2}{5}\times 1 \\ \frac{4}{5}\times 3 + -\frac{3}{5}\times 4 & \frac{4}{5}\times 2 + -\frac{3}{5}\times 1 \end{pmatrix} = \begin{pmatrix} -\frac{3}{5} + \frac{8}{5} & -\frac{2}{5} + \frac{2}{5} \\ \frac{12}{5} - \frac{12}{5} & \frac{8}{5} - \frac{3}{5} \end{pmatrix} = \begin{pmatrix} \frac{5}{5} & 0 \\ 0 & \frac{5}{5} \end{pmatrix} = \begin{pmatrix} 1 & 0 \\ 0 & 1 \end{pmatrix}$$

Matrix Multiplication #3: "Linear Combination of Rows."

$$\begin{pmatrix} -\frac{1}{5} & \frac{2}{5} \\ \frac{4}{5} & -\frac{3}{5} \end{pmatrix}\begin{pmatrix} 3 & 2 \\ 4 & 1 \end{pmatrix} = \begin{bmatrix} -\frac{1}{5}\times\begin{pmatrix} 3 & 2 \end{pmatrix} + \frac{2}{5}\times\begin{pmatrix} 4 & 1 \end{pmatrix} \\ \frac{4}{5}\times\begin{pmatrix} 3 & 2 \end{pmatrix} + -\frac{3}{5}\times(4,1) \end{bmatrix} = \begin{bmatrix} \begin{pmatrix} -\frac{3}{5} & -\frac{2}{5} \end{pmatrix} + \begin{pmatrix} \frac{8}{5} & \frac{2}{5} \end{pmatrix} \\ \begin{pmatrix} \frac{12}{5} & \frac{8}{5} \end{pmatrix} + \begin{pmatrix} -\frac{12}{5} & -\frac{3}{5} \end{pmatrix} \end{bmatrix} = \begin{pmatrix} -\frac{3}{5} + \frac{8}{5} & -\frac{2}{5} + \frac{2}{5} \\ \frac{12}{5} + -\frac{12}{5} & \frac{8}{5} + -\frac{3}{5} \end{pmatrix} = \begin{pmatrix} 1 & 0 \\ 0 & 1 \end{pmatrix}$$

Just as 1 is the multiplicative identity for real numbers (multiplying any number by 1 gives that number), a matrix such as $\begin{pmatrix} 1 & 0 \\ 0 & 1 \end{pmatrix}$, whose elements are all zero except the ones on the major diagonal (upper left to lower right), is known as the *identity matrix* because multiplying any matrix by a properly sized identity matrix results in the original matrix. The identity matrix of any size is commonly given the symbol **I**.

$$5\times 1 = 5 \qquad \frac{8}{11}\times 1 = \frac{8}{11} \qquad -3\times 1 = -3 \quad a\times 1 = a$$

$$\begin{pmatrix} 3 & 5 \\ -4 & 2 \end{pmatrix}\times\begin{pmatrix} 1 & 0 \\ 0 & 1 \end{pmatrix} = \begin{pmatrix} 3 & 5 \\ -4 & 2 \end{pmatrix} \quad \begin{pmatrix} a & b \\ c & d \end{pmatrix}\times\begin{pmatrix} 1 & 0 \\ 0 & 1 \end{pmatrix} = \begin{pmatrix} a & b \\ c & d \end{pmatrix}$$

If the result of multiplying two matrices together is the identity matrix, we say that each matrix is the *multiplicative inverse* of the other.

$$\frac{8}{11}\times\frac{11}{8} = \frac{88}{88} = 1 \quad \text{scalar multiplicative inverses}$$

$$\frac{8}{11} = \left(\frac{11}{8}\right)^{-1}, \left(\frac{8}{11}\right)^{-1} = \frac{11}{8}$$

$$\begin{pmatrix} -\frac{1}{2} & 1 \\ -\frac{3}{2} & 2 \end{pmatrix}\begin{pmatrix} 4 & -2 \\ 3 & -1 \end{pmatrix} = \begin{pmatrix} 1 & 0 \\ 0 & 1 \end{pmatrix} \quad \text{matrix multiplicative inverses}$$

$$\begin{pmatrix} -\frac{1}{2} & 1 \\ -\frac{3}{2} & 2 \end{pmatrix} = \begin{pmatrix} 4 & -2 \\ 3 & -1 \end{pmatrix}^{-1}, \begin{pmatrix} -\frac{1}{2} & 1 \\ -\frac{3}{2} & 2 \end{pmatrix}^{-1} = \begin{pmatrix} 4 & -2 \\ 3 & -1 \end{pmatrix}$$

The inverse of a matrix (if it exists) can be found as follows. Put the matrix whose inverse is desired side by side with an identity matrix of the same dimensions:

$$\left(\begin{array}{cc:cc} 3 & 2 & 1 & 0 \\ 4 & 1 & 0 & 1 \end{array}\right). \quad \text{Here we wish to find the multiplicative inverse of } \begin{pmatrix} 3 & 2 \\ 4 & 1 \end{pmatrix}.$$

Perform row operations that will transform the matrix on the left into an identity matrix. Whatever row operation you perform on the left matrix, perform it also on the right matrix at the same time. If certain conditions are met (the determinant of a square coefficient matrix is not equal to zero), then, when you are finished, the matrix on the right, originally an identity matrix, will be transformed into the matrix inverse of the original left-hand matrix.

So, let's try it! Anticipating the need to transform the 4 in cell (2, 1) to a zero, we multiply Row 1 by 4 and Row 2 by −3.

$$\left(\begin{array}{cc:cc} 3 & 2 & 1 & 0 \\ 4 & 1 & 0 & 1 \end{array}\right) \text{ becomes } \left(\begin{array}{cc:cc} 12 & 8 & 4 & 0 \\ -12 & -3 & 0 & -3 \end{array}\right)$$

Keeping Row 1 the same add Row 1 to Row 2.

$$\left(\begin{array}{cc:cc} 12 & 8 & 4 & 0 \\ 0 & 5 & 4 & -3 \end{array}\right)$$

Anticipating the need to reduce the 8 in cell (1, 2) (above) to zero, we multiply Row 1 by 5 and Row 2 by −8.

$$\left(\begin{array}{cc:cc} 60 & 40 & 20 & 0 \\ 0 & -40 & -32 & 24 \end{array}\right)$$

Keeping Row 2 above the same add Row 2 to Row 1.

$$\left(\begin{array}{cc:cc} 60 & 0 & -12 & 24 \\ 0 & -40 & -32 & 24 \end{array}\right)$$

Keeping mind the goal to transform the matrix on the left into an identity matrix, $\begin{pmatrix} 1 & 0 \\ 0 & 1 \end{pmatrix}$, divide Row 1 by 60 and divide Row 2 by −40.

$$\left(\begin{array}{cc:cc} 1 & 0 & -\frac{12}{60} & \frac{24}{60} \\ 0 & 1 & \frac{-32}{-40} & \frac{24}{-40} \end{array}\right) \text{ or } \left(\begin{array}{cc:cc} 1 & 0 & -\frac{1}{5} & \frac{2}{5} \\ 0 & 1 & \frac{4}{5} & -\frac{3}{5} \end{array}\right)$$

Thus,

$$\begin{pmatrix} 3 & 2 \\ 4 & 1 \end{pmatrix}^{-1} = \begin{pmatrix} -\frac{1}{5} & \frac{2}{5} \\ \frac{4}{5} & -\frac{3}{5} \end{pmatrix}.$$

It is important to note here that sometimes multiplication of matrices is commutative,

$$\begin{pmatrix} -\frac{1}{5} & \frac{2}{5} \\ \frac{4}{5} & -\frac{3}{5} \end{pmatrix}\begin{pmatrix} 3 & 2 \\ 4 & 1 \end{pmatrix} = \begin{pmatrix} 1 & 0 \\ 0 & 1 \end{pmatrix} = \begin{pmatrix} 3 & 2 \\ 4 & 1 \end{pmatrix}\begin{pmatrix} -\frac{1}{5} & \frac{2}{5} \\ \frac{4}{5} & -\frac{3}{5} \end{pmatrix},$$

but not always

$$\begin{pmatrix} 3 & 0 \\ -1 & 4 \end{pmatrix}\begin{pmatrix} 1 & 2 \\ 0 & 3 \end{pmatrix} = \begin{pmatrix} 3 & 6 \\ -1 & 10 \end{pmatrix} \neq \begin{pmatrix} 1 & 8 \\ -3 & 12 \end{pmatrix} = \begin{pmatrix} 1 & 2 \\ 0 & 3 \end{pmatrix}\begin{pmatrix} 3 & 0 \\ -1 & 4 \end{pmatrix}.$$

The system of equations solved previously was

$$3x + 2y = 8 \tag{1}$$

$$4x + 1y = 6. \tag{2}$$

We solved this system using substitution, linear combination, and Cramer's Rule. It can also be solved using matrix algebra.

Solving an Algebraic Equation	Solving a System of Equations Using Matrix Algebra
	Transform into matrix algebra notation. $3x + 2y = 8$ $4x + 1y = 6$
Scalar coefficient × scalar variable = scalar product $\frac{2}{3}x = 5$	Coefficient matrix × variable vector = product vector $\begin{pmatrix}3 & 2\\4 & 1\end{pmatrix}\begin{pmatrix}x\\y\end{pmatrix} = \begin{pmatrix}8\\6\end{pmatrix}$
The goal here is to isolate the scalar variable, x. Multiply both sides by the inverse of the coefficient.	The goal here is to isolate the variable vector, $\begin{pmatrix}x\\y\end{pmatrix}$. Multiply both sides by the inverse of the coefficient matrix.
$\frac{3}{2}\times\frac{2}{3}x = \frac{3}{2}\times 5$	$\begin{pmatrix}-\frac{1}{5} & \frac{2}{5}\\ \frac{4}{5} & -\frac{3}{5}\end{pmatrix}\begin{pmatrix}3 & 2\\4 & 1\end{pmatrix}\begin{pmatrix}x\\y\end{pmatrix} = \begin{pmatrix}-\frac{1}{5} & \frac{2}{5}\\ \frac{4}{5} & -\frac{3}{5}\end{pmatrix}\begin{pmatrix}8\\6\end{pmatrix}$
$1x = \frac{15}{2}$	$\begin{pmatrix}1 & 0\\0 & 1\end{pmatrix}\begin{pmatrix}x\\y\end{pmatrix} = \begin{pmatrix}-\frac{1}{5}\times 8 + \frac{2}{5}\times 6\\ \frac{4}{5}\times 8 \times -\frac{3}{5}\times 6\end{pmatrix}$
$x = \frac{15}{2}$	$\begin{pmatrix}x\\y\end{pmatrix} = \begin{pmatrix}-\frac{8}{5} + \frac{12}{5}\\ \frac{32}{5} + -\frac{18}{5}\end{pmatrix} = \begin{pmatrix}\frac{4}{5}\\ \frac{14}{5}\end{pmatrix}$
One variable, one solution. Check our work:	Two variables, one solution. Check our work:
$\frac{2}{3}x = 5$ $\frac{2}{3}\times\frac{15}{2} = 5$ $\frac{15}{3} = 5$ ck!	$3x + 2y = 8$ $4x + 1y = 6$ $3\times\frac{4}{5} + 2\times\frac{14}{5} = \frac{12}{5} + \frac{28}{5} = \frac{40}{5} = 8$ ck! $4\times\frac{4}{5} + 1\times\frac{14}{5} = \frac{16}{5} + \frac{14}{5} = \frac{30}{5} = 6$ ck!

While this matrix solution technique seems much more complex than the other two, it turns out, with the aid of computers, to be more easily extended into larger systems of equations than the other solution techniques. In solving simple algebraic equations *and* systems of two equations with two unknowns, there are three possibilities.

1. $0x = 0$, ***infinite solutions*** $\frac{0x}{0} = \frac{0}{0}$, indeterminate solution, ($x = 5$, -1, etc.)	1. One equation is the same as or a multiple of the other: det(**Coef Mat**) = 0. $3x + 2y = 5 \qquad 6x + 4y = 10$ This means that the lines are the same. (They both have the same slope and same y-intercept.) There will be ***infinite solutions***. Note that the determinant of such a system is zero: $(3\times 4) - (6\times 2) = 0$. There will be no inverse of the coefficient matrix.
2. $0x = k$ with $k \neq 0$, ***no solution*** (can't divide k by 0) $\frac{0x}{0} = \frac{k}{0}$, undefined	2. The equations have the same slope but different y-intercepts: det(**Coef Mat**) = 0. $3x + 2y = 8 \quad \text{slope} = m = -\frac{3}{2} \qquad 6x + 4y = 5 \quad \text{slope} = m = -\frac{3}{2}$ This means that the lines are parallel, they have the same slope. Because the lines are parallel and have different intercepts, there will be ***no solution***. Note that the determinant of such a system is zero: $(3\times 4) - (6\times 2) = 0$. There will be no inverse of the coefficient matrix.
3. $ax = k$ with $a \neq 0$, ***one solution***: $x = \frac{k}{a}$	3. Neither condition above holds: det(**Coef Mat**) ≠ 0. $3x + 2y = 8 \qquad 4x + 1y = 6$ This means that the lines are distinct and not parallel, so they do intersect. There will be ***one solution***.

What we saw as

$3x + 2y = 8$ $4x + 1y = 6$	would instead be presented in a university class as	$a_{1,1}x_1 + a_{1,2}x_2 + a_{1,3}x_3 + \cdots + a_{1,n}x_n = p_1$ $a_{2,1}x_1 + a_{2,2}x_2 + a_{2,3}x_3 + \cdots + a_{2,n}x_n = p_2$ $a_{3,1}x_1 + a_{3,2}x_2 + a_{3,3}x_3 + \cdots + a_{3,n}x_n = p_3$ $\vdots$ $a_{n,1}x_1 + a_{n,2}x_2 + a_{n,3}x_3 + \cdots + a_{n,n}x_n = p_n$

demonstrating that the technique is not restricted to a two-by-two system of equations. The matrix form of that system of equations

$\begin{pmatrix}3 & 2\\4 & 1\end{pmatrix}\begin{pmatrix}x\\y\end{pmatrix} = \begin{pmatrix}8\\6\end{pmatrix}$	would instead be presented as	$\begin{pmatrix}a_{11} & a_{12} & \cdots & a_{1n}\\a_{21} & a_{22} & \cdots & a_{2n}\\\vdots & \vdots & \ddots & \vdots\\a_{n1} & a_{n2} & \cdots & a_{nn}\end{pmatrix}\begin{pmatrix}x_1\\x_2\\\vdots\\x_n\end{pmatrix} = \begin{pmatrix}p_1\\p_2\\\vdots\\p_n\end{pmatrix}$

Coef Mat × (variable vector) = product vector

Again, this illustrates that the technique is not restricted to a two-by-two system of equations. So, when you see something like the following statement,

$$\begin{aligned}
(\textbf{Coef Mat})(\text{variable vector}) &= \text{product vector}\\
(\textbf{Coef Mat})^{-1} \times (\textbf{Coef Mat})(\text{variable vector}) &= (\textbf{Coef Mat})^{-1} \times \text{product vector}\\
\left[(\textbf{Coef Mat})^{-1} \times (\textbf{Coef Mat})\right](\text{variable vector}) &= (\textbf{Coef Mat})^{-1} \times \text{product vector}\\
(\text{identity matrix})(\text{variable vector}) &= \text{solution vector}\\
\text{vector variable} &= \text{vector values}
\end{aligned}$$

you will understand that this algorithm is not restricted to a two-by-two system of equations: It could be applied to n-by-n systems of equations for $n \in N$!!!

Finally, up to this point, we've only looked at matrices with numbers as entries, but the entries in a matrix can be functions as well.

$$f(t) = \begin{pmatrix}f_{11}(t) & f_{12}(t) & \cdots & f_{1n}(t)\\f_{21}(t) & f_{22}(t) & \cdots & f_{2n}(t)\\\vdots & \vdots & \ddots & \vdots\\f_{n1}(t) & f_{n2}(t) & \cdots & f_{nn}(t)\end{pmatrix} \qquad f'(t) = \begin{pmatrix}f'_{11}(t) & f'_{12}(t) & \cdots & f'_{1n}(t)\\f'_{21}(t) & f'_{22}(t) & \cdots & f'_{2n}(t)\\\vdots & \vdots & \ddots & \vdots\\f'_{n1}(t) & f'_{n2}(t) & \cdots & f'_{nn}(t)\end{pmatrix}$$

Looking at only the first column of the two matrices, we get a vector of functions.

$$f(t) = \begin{pmatrix}f_1(t)\\f_2(t)\\\vdots\\f_n(t)\end{pmatrix} \qquad f'(t) = \begin{pmatrix}f'_1(t)\\f'_2(t)\\\vdots\\f'_n(t)\end{pmatrix}$$

By seeing this vector form (a vector of functions), we prepare ourselves for the later appearance of vectors where each vector element itself is a function indicating a differential equation, or where each vector element is itself a solution to a differential equation (a row) in a system of equations.

Chapter 17 Review

This chapter discussed inhomogeneous and homogeneous systems of equations. *The presentation emphasized the fact that the row coefficients of homogeneous equations are always multiples of each other when the determinant of the coefficient matrix is zero.*

Four techniques for solving systems of algebraic equations were reviewed.

1. Substitution
2. Linear Combination
3. Cramer's Rule
4. Matrix Algebra
5. Matrix Notation
   (a) Multiplying matrices, row and column combinations
   (b) Matrix inverses

Chapter 18
Epistemology: Axioms (Postulates) vs. Theorems

Wikipedia defines *epistemology* as the "branch of philosophy concerned with the nature and scope of knowledge"; it is the "theory of knowledge." Put concisely, it is the study of knowledge and justified belief. It questions what knowledge is, how it can be acquired, and the extent to which knowledge pertinent to any given subject or entity can be verified.

One philosopher, Rene Descartes, was so skeptical about knowledge in general that he doubted his own existence. For centuries, humans assumed that the sun revolved around the earth. They justified this belief with the fact that they saw the sun arching across the sky. The Catholic Church justified this same belief with information from their religious books. The Italian scientist Galileo, however, based upon empirical data and observations, declared otherwise, ***"Eppur si muove"*** ("and yet it (the earth) moves").

Two thousand plus years ago, a famous mathematician named Euclid combined undefined terms (point, line, space, etc.) with "common notions" (which are today called *postulates*) and created theorems. The theorems were "proved" and then those theorems could be combined with the undefined terms and "common notions" to prove even more theorems. The result of Euclid's work was an enormous body of logically consistent knowledge called Euclidean Geometry. The point here is that the "common notions" were just accepted (you have to start somewhere), but the theorems were proved.

Some sixteen hundred years after Euclid, two more mathematicians, Sir Isaac Newton and Gottfried Leibniz, combined common notions (which we call properties: the associative property, the commutative property, etc.) together with definitions to create theorems. Again the proved theorems were used to create even more theorems. The result of their work was an enormous body of knowledge called calculus. The point here is that the properties were just accepted (you have to start somewhere), but the theorems were proved.

The *commutative property* of the multiplication of numbers is taught in algebra as: $a \times b = b \times a$. By lifting the restriction that the operands must be numbers, allowing operations on matrices and vectors, it is possible that entirely new applications of the commutative property shown above can be derived as a theorem. The following proof is important to justifying concepts introduced in Chapters 19 and 20 and Appendices F, G and H.

Commutative Property in Algebra	Matrix Algebra Theorem Proved Using Operation Definitions from Matrix Algebra
$a \times b = b \times a$	*Proof that vector* $\times$ *scalar* $\stackrel{?}{=}$ *scalar* $\times$ *vector.* $\vec{v} \times k = \begin{pmatrix} v_1 \\ v_2 \end{pmatrix} k$ multiplication of a 2×1 matrix and a 1×1 matrix is compatible and defined $= \begin{pmatrix} v_1 \times k \\ v_2 \times k \end{pmatrix} = \begin{pmatrix} k \times v_1 \\ k \times v_2 \end{pmatrix}$ commutative property of reals $= k \times \begin{pmatrix} v_1 \\ v_2 \end{pmatrix}$ defined matrix operation $\vec{v} \times k = k \times \vec{v}$ □

The *distributive property* of multiplication over the addition of real numbers is most frequently seen in the form $a(b + c) = ab + ac$. However, it is often seen in reversed form $ab + ac = a(b + c)$.

$a(b + c) = ab + ac$	distributive property
$ab + ac = a(b + c)$	reverse distributive property (factoring out a common term)

These distributive property forms assume that the operands are all scalars and that multiplication is being distributed over addition. By lifting the restriction to multiplication and addition of scalars, allowing operations on matrices and vectors, it is possible that entirely new applications of the distributive form shown above can be derived. The following proof is important to justify concepts introduced in Chapter 19 and Appendices F, G and I.

Distributive Property in Algebra	Matrix Algebra Theorem Proved Using Operation Definitions from Matrix Algebra
$c(a - b) = (c \times a) - (c \times b)$ $(a \times b) - (a \times c) = a(b - c)$ Analogous to matrix proof at right.	*Proof that* (**Mat** × vector) – (scalar × vector) $\stackrel{?}{=}$ (**Mat** – scalar × **I**) × vector. $\begin{pmatrix} a & b \\ c & d \end{pmatrix}\begin{pmatrix} v_1 \\ v_2 \end{pmatrix} - k\begin{pmatrix} v_1 \\ v_2 \end{pmatrix} = \begin{pmatrix} av_1 + bv_2 \\ cv_1 + dv_2 \end{pmatrix} - \begin{pmatrix} kv_1 \\ kv_2 \end{pmatrix}$ matrix operations $= \begin{pmatrix} [av_1 + bv_2] - kv_1 \\ [cv_1 + dv_2] - kv_2 \end{pmatrix}$ subtract vectors $= \begin{pmatrix} [av_1 - kv_1] + bv_2 \\ cv_1 + [dv_2 - kv_2] \end{pmatrix}$ associative, commutative properties $= \begin{pmatrix} (a-k)v_1 + bv_2 \\ cv_1 + (d-k)v_2 \end{pmatrix}$ factor out v_1 and v_2 $= \begin{pmatrix} a-k & b \\ c & d-k \end{pmatrix} \times \begin{pmatrix} v_1 \\ v_2 \end{pmatrix}$ multiply matrices, 2×2 by 2×1 $= \left[\begin{pmatrix} a & b \\ c & d \end{pmatrix} - \begin{pmatrix} k & 0 \\ 0 & k \end{pmatrix}\right] \times \begin{pmatrix} v_1 \\ v_2 \end{pmatrix}$ subtract matrices $= \left[\begin{pmatrix} a & b \\ c & d \end{pmatrix} - k\begin{pmatrix} 1 & 0 \\ 0 & 1 \end{pmatrix}\right] \times \begin{pmatrix} v_1 \\ v_2 \end{pmatrix}$ factor out k $= \left[\begin{pmatrix} a & b \\ c & d \end{pmatrix} - k\mathbf{I}\right] \times \begin{pmatrix} v_1 \\ v_2 \end{pmatrix}$ definition of **I** $\begin{pmatrix} a & b \\ c & d \end{pmatrix}\begin{pmatrix} v_1 \\ v_2 \end{pmatrix} - k\begin{pmatrix} v_1 \\ v_2 \end{pmatrix} = \left[\begin{pmatrix} a & b \\ c & d \end{pmatrix} - k\mathbf{I}\right] \times \begin{pmatrix} v_1 \\ v_2 \end{pmatrix}$ Yes! (**Mat** × vector) – (scalar × vector) = (**Mat** – scalar × **I**) × vector. □

Chapter 18 Review

Chapter 18 reviewed the difference between an axiom (or postulate) and a theorem, and demonstrated the following two matrix theorems for use in Chapter 19 and Appendices F, G and H.

vector × scalar = scalar × vector

$$\vec{v} \times k = k \times \vec{v}$$

(**Mat** × vector) – (scalar × vector) = (**Mat** – scalar × **I**) × vector

$$\begin{pmatrix} a & b \\ c & d \end{pmatrix}\begin{pmatrix} v_1 \\ v_2 \end{pmatrix} - k\begin{pmatrix} v_1 \\ v_2 \end{pmatrix} = \left[\begin{pmatrix} a & b \\ c & d \end{pmatrix} - k\mathbf{I}\right] \times \begin{pmatrix} v_1 \\ v_2 \end{pmatrix}$$

Chapter 19

Eigenvalues, Eigenvectors, and $\mathbf{Mat} \times \vec{\varepsilon} = \lambda \times \vec{\varepsilon}$

Next we demonstrate what eigenvalues and eigenvectors of a matrix are. Then we show how to obtain the eigenvalues and eigenvectors. Observe, in the left and right text boxes, two expressions involving the multiplication of a vector, $\begin{pmatrix} v_1 \\ v_2 \end{pmatrix}$, by a given matrix, $\begin{pmatrix} 3 & 2 \\ 4 & 1 \end{pmatrix}$, and multiplication of that same vector, $\begin{pmatrix} v_1 \\ v_2 \end{pmatrix}$, by a scalar.

$$\begin{pmatrix} 3 & 2 \\ 4 & 1 \end{pmatrix} \times \begin{pmatrix} 1 \\ -2 \end{pmatrix} \stackrel{?}{=} -1 \times \begin{pmatrix} 1 \\ -2 \end{pmatrix}$$

$$\begin{pmatrix} 3 \times 1 + 2 \times (-2) \\ 4 \times 1 + 1 \times (-2) \end{pmatrix} \stackrel{?}{=} \begin{pmatrix} -1 \\ 2 \end{pmatrix}$$

$$\begin{pmatrix} 3 - 4 \\ 4 - 2 \end{pmatrix} \stackrel{?}{=} \begin{pmatrix} -1 \\ 2 \end{pmatrix}$$

$$\begin{pmatrix} -1 \\ 2 \end{pmatrix} = \begin{pmatrix} -1 \\ 2 \end{pmatrix} \quad \text{ck!}$$

Notice that the matrix $\begin{pmatrix} 3 & 2 \\ 4 & 1 \end{pmatrix}$ times the vector $\begin{pmatrix} 1 \\ -2 \end{pmatrix}$ is the same as the product of -1 and that same vector. For a square matrix, such as $\begin{pmatrix} 3 & 2 \\ 4 & 1 \end{pmatrix}$ above, a slick technique allows you to determine that vector—$\begin{pmatrix} 1 \\ -2 \end{pmatrix}$, called the eigenvector, $\vec{\varepsilon}$—and that value—-1, called the eigenvalue, λ. If the eigenvector and eigenvalues are correctly chosen, the result will be

$$\mathbf{Mat} \times \vec{\varepsilon} = \lambda \times \vec{\varepsilon}.$$

$$\begin{pmatrix} 3 & 2 \\ 4 & 1 \end{pmatrix} \times \begin{pmatrix} 1 \\ 1 \end{pmatrix} \stackrel{?}{=} 5 \times \begin{pmatrix} 1 \\ 1 \end{pmatrix}$$

$$\begin{pmatrix} 3 \times 1 + 2 \times (1) \\ 4 \times 1 + 1 \times (1) \end{pmatrix} \stackrel{?}{=} \begin{pmatrix} 5 \\ 5 \end{pmatrix}$$

$$\begin{pmatrix} 3 + 2 \\ 4 + 1 \end{pmatrix} \stackrel{?}{=} \begin{pmatrix} 5 \\ 5 \end{pmatrix}$$

$$\begin{pmatrix} 5 \\ 5 \end{pmatrix} = \begin{pmatrix} 5 \\ 5 \end{pmatrix} \quad \text{ck!}$$

Notice that the matrix $\begin{pmatrix} 3 & 2 \\ 4 & 1 \end{pmatrix}$ times the vector $\begin{pmatrix} 1 \\ 1 \end{pmatrix}$ is the same as the product of 5 and that same vector. For a square matrix, such as $\begin{pmatrix} 3 & 2 \\ 4 & 1 \end{pmatrix}$ above, a slick technique allows you to determine that vector—$\begin{pmatrix} 1 \\ 1 \end{pmatrix}$, called the eigenvector, $\vec{\varepsilon}$—and that value—5, called the eigenvalue, λ. If the eigenvector and eigenvalues are correctly chosen, the result will be

$$\mathbf{Mat} \times \vec{\varepsilon} = \lambda \times \vec{\varepsilon}.$$

The values $\vec{\varepsilon}$ and λ will eventually (Chapter 20) be used to substitute into an integrated equation solution form—$\vec{x} = c_1\vec{\varepsilon}_1 e^{\lambda_1 t} + c_2\vec{\varepsilon}_2 e^{\lambda_2 t}$—for a system of differential equations whose constant coefficients are used to form the matrix for which we will calculate the eigenvalues and eigenvectors.

Algorithm to Find the Eigenvalues of $\begin{pmatrix} 3 & 2 \\ 4 & 1 \end{pmatrix}$

Assuming an eigenvector, $\vec{\varepsilon}$, and eigenvalue, λ, exist for a given coefficient matrix (this will be the case for square matrices) how do you find them?

$$\mathbf{Mat} \times \vec{\varepsilon} = \lambda \times \vec{\varepsilon} \quad \text{from previous page}$$

$$(\mathbf{Mat} \times \vec{\varepsilon}) - (\lambda \times \vec{\varepsilon}) = 0 \quad \text{dimension error is introduced}$$

$$(\mathbf{Mat} \times \vec{\varepsilon}) - (\lambda \times \vec{\varepsilon}) = \begin{pmatrix} 0 \\ 0 \end{pmatrix} \quad \text{dimension error corrected}$$

$$(\mathbf{Mat} - \lambda\mathbf{I}) \times \vec{\varepsilon} = \begin{pmatrix} 0 \\ 0 \end{pmatrix}$$

This is not "factoring out a common term" (the reverse distributive property, $ab + ac = a(b + c)$). That property applies to scalars. This step is justified by a theorem proved in Chapter 18. Note that we have to include the identity matrix (**I**) to avoid introducing another dimension error.

$$\left[\begin{pmatrix} 3 & 2 \\ 4 & 1 \end{pmatrix} - \lambda\begin{pmatrix} 1 & 0 \\ 0 & 1 \end{pmatrix}\right] \times \vec{\varepsilon} = \begin{pmatrix} 0 \\ 0 \end{pmatrix} \quad (\mathbf{Mat} - \lambda\mathbf{I}) \times \vec{\varepsilon} = \begin{pmatrix} 0 \\ 0 \end{pmatrix}$$

$$\left[\begin{pmatrix} 3 & 2 \\ 4 & 1 \end{pmatrix} - \begin{pmatrix} \lambda & 0 \\ 0 & \lambda \end{pmatrix}\right] \times \vec{\varepsilon} = \begin{pmatrix} 0 \\ 0 \end{pmatrix} \quad (\mathbf{Mat} - \lambda\mathbf{I}) \times \vec{\varepsilon} = \begin{pmatrix} 0 \\ 0 \end{pmatrix}$$

$$\begin{pmatrix} 3-\lambda & 2-0 \\ 4-0 & 1-\lambda \end{pmatrix} \times \vec{\varepsilon} = \begin{pmatrix} 0 \\ 0 \end{pmatrix} \quad (\mathbf{Mat} - \lambda\mathbf{I}) \times \vec{\varepsilon} = \begin{pmatrix} 0 \\ 0 \end{pmatrix}$$

$$\begin{pmatrix} 3-\lambda & 2 \\ 4 & 1-\lambda \end{pmatrix} \times \vec{\varepsilon} = \begin{pmatrix} 0 \\ 0 \end{pmatrix} \quad (\mathbf{Mat} - \lambda\mathbf{I}) \times \vec{\varepsilon} = \begin{pmatrix} 0 \\ 0 \end{pmatrix}$$

$$\begin{pmatrix} 3-\lambda & 2 \\ 4 & 1-\lambda \end{pmatrix} \times \begin{pmatrix} x_1 \\ x_2 \end{pmatrix} = \begin{pmatrix} 0 \\ 0 \end{pmatrix} \quad (\mathbf{Mat} - \lambda\mathbf{I}) \times \vec{\varepsilon} = \begin{pmatrix} 0 \\ 0 \end{pmatrix}$$

In Chapter 17, we discussed that the equation $0x = 0$ meant there were infinitely many solutions. Likewise, the equation $ax = 0$ for $a = 0$ meant there were infinitely many solutions.

Hopefully, you can see that $\begin{pmatrix} 3-\lambda & 2 \\ 4 & 1-\lambda \end{pmatrix} \times \begin{pmatrix} x_1 \\ x_2 \end{pmatrix} = \begin{pmatrix} 0 \\ 0 \end{pmatrix}$ is a vector form of the equation $ax = 0$. Since Chapter 17 just discussed that $ax = 0$ with $a = 0$ guarantees infinite solutions, det(**Mat** − λ**I**) = 0 ***will guarantee that there are infinitely many x-vector solutions,*** $\begin{pmatrix} x_1 \\ x_2 \end{pmatrix}$, ***that will solve the equation*** $(\mathbf{Mat} - \lambda\mathbf{I})\begin{pmatrix} x_1 \\ x_2 \end{pmatrix} = \begin{pmatrix} 0 \\ 0 \end{pmatrix}$, ***and that the rows of the coefficient matrix will be multiples of each other.*** By setting det(**Mat** − λ**I**) = 0 (for our specific matrix, $\det\begin{vmatrix} 3-\lambda & 2 \\ 4 & 1-\lambda \end{vmatrix} = 0$) and solving for λ, we find the eigenvalues for our particular matrix, $\begin{pmatrix} 3 & 2 \\ 4 & 1 \end{pmatrix}$. *Evaluate the* *determinant of the matrix and set it equal to 0.* Solve. The result will be the eigenvalues for the system matrix.

$$\det\begin{bmatrix}3-\lambda & 2\\ 4 & 1-\lambda\end{bmatrix}=0$$
$$(3-\lambda)(1-\lambda)-4(2)=0$$
$$(3-3\lambda-\lambda+\lambda^2)-8=0$$
$$\lambda^2-4\lambda-5=0$$
$$(\lambda-5)(\lambda+1)=0$$
$$\lambda=5,-1$$

The eigenvalues, λ_1 and λ_2, for the matrix $\begin{pmatrix}3 & 2\\ 4 & 1\end{pmatrix}$ are 5 and −1.

Algorithm to Find the Eigenvectors of $\begin{pmatrix}3 & 2\\ 4 & 1\end{pmatrix}$ for Eigenvalues $\lambda_1 = 5$ and $\lambda_2 = -1$

By substituting, respectively, the two λ values obtained above, together with our given matrix and the Identity matrix for the order of the given matrix, into the equation $(\mathbf{Mat} - \lambda\mathbf{I}) \times \vec{\varepsilon} = \begin{pmatrix}0\\0\end{pmatrix}$ we can solve for $\vec{\varepsilon}$ where $\vec{\varepsilon}$ is the vector describing a line through the origin. Substitute, respectively, the eigenvalues 5 and −1 into the equation shown above: $\begin{pmatrix}3-r & 2\\ 4 & 1-r\end{pmatrix}\begin{pmatrix}x_1\\x_2\end{pmatrix}=\begin{pmatrix}0\\0\end{pmatrix}$. Solve for $\begin{pmatrix}x_1\\x_2\end{pmatrix}$ using row operations.

Eigenvalue $\lambda = 5$	Eigenvalue $\lambda = -1$
$(\mathbf{Mat} - \lambda\mathbf{I}) \times \vec{\varepsilon} = \begin{pmatrix}0\\0\end{pmatrix}$	$(\mathbf{Mat} - \lambda\mathbf{I}) \times \vec{\varepsilon} = \begin{pmatrix}0\\0\end{pmatrix}$
$\left[\begin{pmatrix}3 & 2\\ 4 & 1\end{pmatrix}-5\begin{pmatrix}1 & 0\\ 0 & 1\end{pmatrix}\right]\begin{pmatrix}x_1\\x_2\end{pmatrix}=\begin{pmatrix}0\\0\end{pmatrix}$	$\left[\begin{pmatrix}3 & 2\\ 4 & 1\end{pmatrix}-(-1)\begin{pmatrix}1 & 0\\ 0 & 1\end{pmatrix}\right]\begin{pmatrix}x_1\\x_2\end{pmatrix}=\begin{pmatrix}0\\0\end{pmatrix}$
$\left[\begin{pmatrix}3 & 2\\ 4 & 1\end{pmatrix}-\begin{pmatrix}5 & 0\\ 0 & 5\end{pmatrix}\right]\begin{pmatrix}x_1\\x_2\end{pmatrix}=\begin{pmatrix}0\\0\end{pmatrix}$	$\left[\begin{pmatrix}3 & 2\\ 4 & 1\end{pmatrix}+\begin{pmatrix}1 & 0\\ 0 & 1\end{pmatrix}\right]\begin{pmatrix}x_1\\x_2\end{pmatrix}=\begin{pmatrix}0\\0\end{pmatrix}$
$\begin{pmatrix}3-5 & 2\\ 4 & 1-5\end{pmatrix}\begin{pmatrix}x_1\\x_2\end{pmatrix}=\begin{pmatrix}0\\0\end{pmatrix}$	$\begin{pmatrix}3+1 & 2\\ 4 & 1+1\end{pmatrix}\begin{pmatrix}x_1\\x_2\end{pmatrix}=\begin{pmatrix}0\\0\end{pmatrix}$
$\begin{pmatrix}-2 & 2\\ 4 & -4\end{pmatrix}\begin{pmatrix}x_1\\x_2\end{pmatrix}=\begin{pmatrix}0\\0\end{pmatrix}$ divide top row by −2, bottom by 4	$\begin{pmatrix}4 & 2\\ 4 & 2\end{pmatrix}\begin{pmatrix}x_1\\x_2\end{pmatrix}=\begin{pmatrix}0\\0\end{pmatrix}$ divide top row by 2, bottom by 2
$\begin{pmatrix}1 & -1\\ 1 & -1\end{pmatrix}\begin{pmatrix}x_1\\x_2\end{pmatrix}=\begin{pmatrix}0\\0\end{pmatrix}$ identical rows	$\begin{pmatrix}2 & 1\\ 2 & 1\end{pmatrix}\begin{pmatrix}x_1\\x_2\end{pmatrix}=\begin{pmatrix}0\\0\end{pmatrix}$ identical rows
There should be no surprise that the row coefficients that result are multiples of each other ensuring infinite solutions to the system. (This suggests that the last step was unnecessary.) Why is this the case?	There should be no surprise that the row coefficients that result are multiples of each other ensuring infinite solutions to the system. (This suggests that the last step was unnecessary.) Why is this the case?
$x_1 - x_2 = 0$	$2x_1 + x_2 = 0$
$x_1 = x_2$	$x_2 = -2x_1$
Choose arbitrarily for x_1 or x_2. Say, $x_2 = 1$, so $x_1 = 1$.	Choose arbitrarily for x_1 or x_2. Say, $x_1 = 1$, so $x_2 = -2$.
$\vec{x}=\begin{pmatrix}x_1\\x_2\end{pmatrix}=\begin{pmatrix}1\\1\end{pmatrix}$ or actually any vector of the form $k\begin{pmatrix}1\\1\end{pmatrix}$	$\vec{x}=\begin{pmatrix}x_1\\x_2\end{pmatrix}=\begin{pmatrix}1\\-2\end{pmatrix}$ or actually any vector of the form $k\begin{pmatrix}1\\-2\end{pmatrix}$

For eigenvalue $r = 5$ the eigenvector is $k\begin{pmatrix}1\\1\end{pmatrix}$ $\begin{pmatrix}3&2\\4&1\end{pmatrix}\begin{pmatrix}k\\k\end{pmatrix} \overset{?}{=} 5\begin{pmatrix}k\\k\end{pmatrix}$ $\begin{pmatrix}5k\\5k\end{pmatrix} = \begin{pmatrix}5k\\5k\end{pmatrix}$ ck!	For eigenvalue $r = -1$ the eigenvector is $k\begin{pmatrix}1\\-2\end{pmatrix}$ $\begin{pmatrix}3&2\\4&1\end{pmatrix}\begin{pmatrix}k\\-2k\end{pmatrix} \overset{?}{=} -1\begin{pmatrix}k\\-2k\end{pmatrix}$ $\begin{pmatrix}-k\\2k\end{pmatrix} = \begin{pmatrix}-k\\2k\end{pmatrix}$ ck!

Chapter 19 Review

In other words, for any square matrix $\begin{pmatrix}a&b\\c&d\end{pmatrix}$ it can be shown that a vector $\begin{pmatrix}e\\f\end{pmatrix}$ and a scalar g, exist such that $\begin{pmatrix}a&b\\c&d\end{pmatrix}\begin{pmatrix}e\\f\end{pmatrix} = g\begin{pmatrix}e\\f\end{pmatrix}$ where $\begin{pmatrix}e\\f\end{pmatrix}$ and g are known respectively as the eigenvector and eigenvalue of the matrix $\begin{pmatrix}a&b\\c&d\end{pmatrix}$: **Mat** × eigenvector = (scalar or complex number) × eigenvector. **Mat** $\times \vec{\varepsilon} = \lambda \times \vec{\varepsilon}$ transforms to $(\mathbf{Mat} - \lambda\mathbf{I})\vec{\varepsilon} = \begin{pmatrix}0\\0\end{pmatrix}$, where $\vec{\varepsilon}$ denotes an eigenvector and λ denotes an eigenvalue.

www.merriam-webster.com/dictionary/eigenvalue

eigenvalue

noun | ei·gen·val·ue | \ˈī-gən-ˌval-(ˌ)yü\

Definition of EIGENVALUE Popularity: Bottom 30% of words

: a scalar associated with a given linear transformation of a vector space and having the property that there is some nonzero vector which when multiplied by the scalar is equal to the vector obtained by letting the transformation operate on the vector; *especially* : a root of the characteristic equation of a matrix

www.merriam-webster.com/dictionary/eigenvector

eigenvector

noun | ei·gen·vec·tor | \-ˌvek-tər\

Definition of EIGENVECTOR Popularity: Bottom 30% of words

: a nonzero vector that is mapped by a given linear transformation of a vector space onto a vector that is the product of a scalar multiplied by the original vector — called also *characteristic vector*

See the following for a nice visual discussion of eigenvectors, eigenvalues and their applications

setosa.io/ev/eigenvectors-and-eigenvalues

Chapter 20

Solving Systems of Differential Equations

Chapter 7 introduced the concept of systems of differential equations that looked like the text box at right and used graphing software to graph these equations for k_1, k_2, k_3, and k_4. In algebra, we studied systems of equations primarily to find a point of intersection. In differential equations, we also find points of intersection, but as shown in Chapters 7 and 16, we can use the system to study how changes in one equation can impact another.

$$\frac{dr}{dt} = k_1 r - k_2 r f \qquad \frac{df}{dt} = -k_3 f + k_4 r f$$

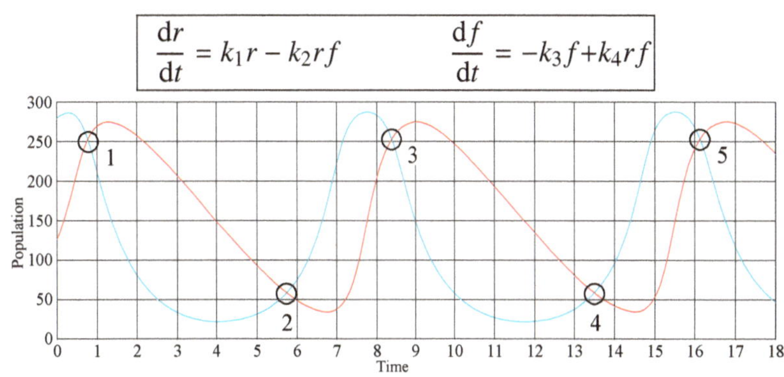

A specific system of differential equations might be like the following.

$$\frac{dx}{dt} = 3x + 2y \qquad (1)$$
$$\frac{dy}{dt} = 4x + 1y \qquad (2)$$

Transform 2, solving for x to substitute it back into 1 below, mimicking the algebraic method of substitution of Chapter 17 in the review of solving systems of algebraic equations.

$$\frac{dy}{dt} - y = 4x \quad \text{solve (2) for } x$$

$$x = \frac{1}{4}\frac{dy}{dt} - \frac{1}{4}y \quad \text{(2) has been solved for } x$$

Rewriting 1, above, $\frac{dx}{dt} = 3x + 2y$, as $\frac{d}{dt}x = 3x + 2y$, and substituting $x = \frac{1}{4}\frac{dy}{dt} - \frac{1}{4}y$ into both x values results in $\frac{d}{dt}\left(\frac{1}{4}\frac{dy}{dt} - \frac{1}{4}y\right) = 3\left(\frac{1}{4}\frac{dy}{dt} - \frac{1}{4}y\right) + 2y$.

$$\frac{d}{dt}\left(\frac{1}{4}\frac{dy}{dt}\right) - \frac{d}{dt}\left(\frac{1}{4}y\right) = \frac{3}{4}\frac{dy}{dt} - \frac{3}{4}y + 2y$$
$$\frac{1}{4}\frac{d^2y}{dt} - \frac{1}{4}\frac{dy}{dt} = \frac{3}{4}\frac{dy}{dt} - \frac{3}{4}y + 2y$$
$$\frac{d^2y}{dt} - \frac{dy}{dt} = 3\frac{dy}{dt} - 3y + 8y \quad \text{multiply by 4 to eliminate all denominators}$$
$$\frac{d^2y}{dt} - 4\frac{dy}{dt} = 5y$$
$$\frac{d^2y}{dt} - 4\frac{dy}{dt} - 5y = 0$$

Through a substitution similar to what is taught when solving algebraic systems of equations, we have arrived at a single, second-order, differential equation. Well, bust my britches! That looks like a second-order linear homogeneous differential equation (Chapter 14): $Ay'' + By' + Cy = 0!$ $\frac{d^2y}{dt} - 4\frac{dy}{dt} - 5y = 0$.

The discriminant of the characteristic equation for r using $A = 1$, $B = -4$, and $C = -5$ ($1r^2 - 4r - 5 = 0$) is $B^2 - 4AC = (-4)^2 - 4\times1\times-5 = 16+20 = 36 > 0$. Hence, from Chapter 14, the general solution for $y(t)$ above would be $y(t) = c_1e^{r_1t} + c_2e^{r_2t}$ with $r = \frac{-(-4)\pm\sqrt{(-4)^2-4\times1\times(-5)}}{2(1)} = \frac{4\pm\sqrt{16+20}}{2} = \frac{4\pm\sqrt{36}}{2} = \frac{4\pm6}{2} = 5, -1$. So, $y(t) = c_1e^{5t} + c_2e^{-1t}$.

It's always good to check your work!

$$y(t) = c_1e^{5t} + c_2e^{-1t}$$
$$y'(t) = 5c_1e^{5t} - c_2e^{-1t}$$
$$y''(t) = 25c_1e^{5t} + c_2e^{-1t}$$

$$\frac{d^2y}{dt} - 4\frac{dy}{dt} - 5y = 0$$
$$\left(25c_1e^{5t} + c_2e^{-1t}\right) - 4\left(5c_1e^{5t} - c_2e^{-1t}\right) - 5\left(c_1e^{5t} + c_2e^{-1t}\right) \stackrel{?}{=} 0$$
$$25c_1e^{5t} + c_2e^{-1t} - 20c_1e^{5t} + 4c_2e^{-1t} - 5c_1e^{5t} - 5c_2e^{-1t} \stackrel{?}{=} 0$$
$$0 = 0$$

All this work is just to solve for $y(t)$. Hopefully, it will not take much to convince you that solving systems of three, four, and more differential equations using the substitution technique would not be a lot of fun. Do you recall from Chapter 17 where we reviewed how to solve systems of algebraic equations using matrix algebra? It actually was not a great improvement over solving by substitution or elimination. However, that chapter pointed out that, using computers or calculators to do the nasty part (finding inverse matrices and multiplying matrices), the matrix-algebra technique was not so bad and the potential for extending that technique to 3×3, 4×4, ..., $n \times n$ systems was tremendous. Keep that thought as you progress through this chapter.

Special Math Symbolism Abstracting Repetition

Since grade school, we have seen many examples where the idea of repetition has been masked using special symbolism.

1. $5 + 5 + 5 + 5 + 5 + 5 = 6 \times 5$. Here, the repeated addition is abstracted as multiplication.

$$\overbrace{5 + 5 + 5 + \cdots + 5}^{n \text{ terms}} = n \times 5$$

2. $5 \times 5 \times 5 \times 5 \times 5 \times 5 = 5^6$. Here, the repeated multiplication is abstracted as exponentiation.

$$\overbrace{5 \times 5 \times 5 \times \cdots \times 5}^{n \text{ terms}} = 5^n$$

3. $x_1 + x_2 + x_3 + \cdots + x_n = \sum_{i=1}^{n} x_i$. Here, the Greek capital sigma ($\sum$) indicates repeated addition.
4. $x_1 \times x_2 \times x_3 \times \cdots \times x_n = \prod_{i=1}^{n} x_i$. Here, the Greek capital pi ($\prod$) indicates repeated multiplication.
5. $\log b^n = \log(\overbrace{b \times b \times b \times \cdots \times b}^{n \text{ terms}}) = \overbrace{\log b + \log b + \log b + \cdots + \log b}^{n \text{ terms}} = n \log b$, product of n terms.
6. Previously, we learned that the solution to a single differential equation, $\frac{dx}{dt} = \ldots$, looked like $x =$ "solution to differential equation."

For the system of n differential equations

$$\frac{d}{dt}x_1 = a_{1,1}t_1 + a_{1,2}t_2 + \cdots + a_{1,n}t_n$$
$$\frac{d}{dt}x_2 = a_{2,1}t_1 + a_{2,2}t_2 + \cdots + a_{2,n}t_n$$
$$\vdots$$
$$\frac{d}{dt}x_n = a_{n,1}t_1 + a_{n,2}t_2 + \cdots + a_{n,n}t_n,$$

there will be n solutions, each of the form

$$x_1 = \text{solution family 1}$$
$$x_2 = \text{solution family 2}$$
$$\vdots$$
$$x_n = \text{solution family } n,$$

which, can be abstracted as

$$\vec{x} = \begin{pmatrix} \text{solution family 1} \\ \text{solution family 2} \\ \vdots \\ \text{solution family } n \end{pmatrix}$$

representing n families of solutions.

Anticipating this abstraction of the set of family solutions to a system of differential equations will benefit you greatly as you move forward.

Preteaching the Formula to Solve a System of Differential Equations by Connecting to Previous Work (Solving Separable Differential Equations)

It may be helpful to start the discussion of how to solve a system of differential equations by looking at the trivial case, a 1×1 system of equations. Let $\frac{dx}{dt} = Ax$ where A represents a 1×1 matrix; a 1×1 matrix is, by definition, a scalar, so $\frac{dx}{dt} = (a)x$ where (a) represents a scalar:

$$\frac{dx}{dt} = ax \quad \text{this is a separable differential equation}$$

$$x^{-1}\,dx = a\,dt$$

$$\int x^{-1}\,dx = a\int dt$$

$$\ln x + c_1 = at + c_2$$

$$\ln x = at + c_3$$

$$e^{\ln x} = e^{at+c_3}$$

$$x = e^{at} \times e^{c_3}$$

$$x = ce^{at} \quad (\text{letting } c = e^{c_3})$$

The solution to a 1×1 system of differential equations whose solution was shown above to be $x = ce^{at}$ looks suspiciously similar to each of the addends of the solution to a second-order differential equation (with real and distinct characteristic values, r_1 and r_2), as was taught in Chapter 14: $x = c_1e^{r_1t} + c_2e^{r_2t}$. That is a sort of "that's curious" observation.

Something very clever is about to happen as we transition from Chapter 14 to Chapter 20. In Chapter 14, we solved a single differential equation for its family of solutions: one differential equation, one family of solutions. If $Ax'' + Bx' + Cx = 0$ then $x = c_1e^{r_1t} + c_2e^{r_2t}$. The variables for the characteristic equation in Chapter 14, r_1 and r_2, are obtained by solving the quadratic equation $Ar^2 + Br + C = 0$. A 2×2 system of differential equations results in a family of solutions of the form $x = c_1e^{\lambda_1t} + c_2e^{\lambda_2t}$ **for each row** in the system. The variables for the characteristic equation in Chapters 19 and 20, λ_1 and λ_2, are obtained by processing the determinant of the coefficient matrix of the differential equation. To be able to store each of the families of solutions (one for each row of the system) we will have to understand that the solution to a system of equations will be, not a **single** family of solutions, but **many** of them: a "vector of families of solutions." This is shown as follows for a 2×2 system of equations.

$$\frac{d}{dt}x_1 = ax_1 + bx_2 \qquad x_1 = c_1e^{\lambda_1t} + c_2e^{\lambda_2t} \qquad (1)$$

$$\frac{d}{dt}x_2 = cx_1 + dx_2 \qquad x_2 = c_1e^{\lambda_1t} + c_2e^{\lambda_2t} \qquad (2)$$

$\vec{x} = \begin{pmatrix} x_1 \\ x_2 \end{pmatrix}$ with two families of solutions, each of the form $c_1e^{\lambda_1t} + c_2e^{\lambda_2t}$

The formula to solve a system of differential equations such as that shown at right is $\vec{x} = c_1\vec{\varepsilon}_1e^{\lambda_1t} + c_2\vec{\varepsilon}_2e^{\lambda_2t}$ (assuming λ_1 and λ_2 are reals and $\lambda_1 \neq \lambda_2$, as proved in Appendices F and G), where c_1 and c_2 are constants of integration, $\vec{\varepsilon}_1$ and $\vec{\varepsilon}_2$ are eigenvectors of the system, and λ_1 and λ_2 are eigenvalues of the system.

$$\frac{dx_1}{dt} = ax_1 + bx_2 \qquad (1)$$

$$\frac{dx_2}{dt} = cx_1 + dx_2 \qquad (2)$$

The eigenvector symbols, $\vec{\varepsilon}_1$ and $\vec{\varepsilon}_2$, in the equation above $\vec{x} = c_1\vec{\varepsilon}_1e^{\lambda_1t} + c_2\vec{\varepsilon}_2e^{\lambda_2t}$, are a mathematician's terse, but very succinct, indication that more than one family of solutions exists. Compare the following two formulas:

Chapter 14, Linear Second-Order Differential Equations

$$Ax'' + Bx' + Cx = 0$$

Assuming $B^2 - 4AC > 0$, the solution is $x = c_1 e^{r_1 t} + c_2 e^{r_2 t}$, where r_1 and r_2 were obtained by solving the quadratic equation $Ar^2 + Br + C = 0$.

Chapter 20, 2 × 2 System of Linear Differential Equations with Constant Coefficients

$$x' = \begin{pmatrix} x_1' \\ x_2' \end{pmatrix} = \begin{pmatrix} ax_1 + bx_2 \\ cx_1 + dx_2 \end{pmatrix} = \begin{pmatrix} a & b \\ c & d \end{pmatrix} \begin{pmatrix} x_1 \\ x_2 \end{pmatrix}$$

For real and distinct eigenvalues, the solution is $\vec{x} = c_1 \vec{\varepsilon}_1 e^{\lambda_1 t} + c_2 \vec{\varepsilon}_2 e^{\lambda_2 t}$, where λ_1 and λ_2 are obtained by processing the determinant of the system's coefficient matrix (see Chapter 19), $\vec{\varepsilon}_1$ and $\vec{\varepsilon}_2$ are obtained by solving from **Mat** $\times \vec{\varepsilon}_1 = \lambda_1 \times \vec{\varepsilon}_1$ and **Mat** $\times \vec{\varepsilon}_2 = \lambda_2 \times \vec{\varepsilon}_2$.

The following three text boxes summarize the preceding discussion.

For the system of two differential equations with real, distinct eigenvalues,

$$\frac{d}{dt}x_1 = ax_1 + bx_2$$
$$\frac{d}{dt}x_2 = cx_1 + dx_2$$

there will be two solutions each of the form

$$c_1 e^{\lambda_1 t} + c_2 e^{\lambda_2 t}$$

which, using eigenvector notation, can be represented in terse and compact solution form as

$$\vec{x} = c_1 \vec{\varepsilon}_1 e^{\lambda_1 t} + c_2 \vec{\varepsilon}_2 e^{\lambda_2 t}$$

Here we see one of the many things that make math difficult for many people to learn. The term *vector* is taking on multiple meanings based on the context. Your teacher and text seem to just take it for granted that everyone knows this and are unlikely to point it out.

Vector: 1) A quantity that indicates direction as well as magnitude, $\vec{x} = (4,5)$ (Chapters 20 and 21). 2) A quantity indicating multiple data, $\vec{x} = \begin{pmatrix} 3 \\ 5 \\ 2 \end{pmatrix}$ (Chapter 17). 3) Multiple functions or multiple differentials, $\vec{x} = \begin{pmatrix} f(x_1) \\ f(x_2) \end{pmatrix}$ or $\vec{x}' = \begin{pmatrix} f'(x_1) \\ f'(x_2) \end{pmatrix}$ (Chapter 17).

Solving a System of Differential Equations Using Matrices

The system of differential equations shown previously,

$$\frac{dx}{dt} = 3x + 2y \qquad (1)$$
$$\frac{dy}{dt} = 4x + 1y, \qquad (2)$$

can be rewritten as

$$\frac{d}{dt}x_1 = 3x_1 + 2x_2 \qquad (1)$$
$$\frac{d}{dt}x_2 = 4x_1 + 1x_2, \qquad (2)$$

which can be rewritten in matrix and vector notation as

$$\begin{pmatrix} \frac{d}{dt}x_1 \\ \frac{d}{dt}x_2 \end{pmatrix} = \begin{pmatrix} 3 & 2 \\ 4 & 1 \end{pmatrix} \begin{pmatrix} x_1 \\ x_2 \end{pmatrix}$$

$$3x_1 + 2x_2 = 8 \quad \text{recopied and modified}$$
$$4x_1 + 1x_2 = 6 \quad \text{from above}$$

Matrix notation allows for extension into $n \times n$ systems. Transform into matrix algebra notation: Coefficient matrix × variable vector = product vector.

$$\begin{pmatrix} 3 & 2 \\ 4 & 1 \end{pmatrix} \begin{pmatrix} x_1 \\ x_2 \end{pmatrix} = \begin{pmatrix} 8 \\ 6 \end{pmatrix} \qquad (1)$$

multiply both sides by the inverse coefficient matrix

$$\begin{pmatrix} -\frac{1}{5} & \frac{2}{5} \\ \frac{4}{5} & -\frac{3}{5} \end{pmatrix} \begin{pmatrix} 3 & 2 \\ 4 & 1 \end{pmatrix} \begin{pmatrix} x_1 \\ x_2 \end{pmatrix} = \begin{pmatrix} -\frac{1}{5} & \frac{2}{5} \\ \frac{4}{5} & -\frac{3}{5} \end{pmatrix} \begin{pmatrix} 8 \\ 6 \end{pmatrix} \qquad (2)$$

$$\begin{pmatrix} 1 & 0 \\ 0 & 1 \end{pmatrix} \begin{pmatrix} x_1 \\ x_2 \end{pmatrix} = \begin{pmatrix} \frac{4}{5} \\ \frac{14}{5} \end{pmatrix} \qquad (3)$$

$$\begin{pmatrix} x_1 \\ x_2 \end{pmatrix} = \begin{pmatrix} \frac{4}{5} \\ \frac{14}{5} \end{pmatrix}$$

The goal here is to solve for $\vec{x}$ or $\begin{pmatrix} x_1 \\ x_2 \end{pmatrix}$, giving two families of solutions. When we encountered a system of linear equations (shown at right), we 1) converted to matrix form, 2) found the inverse of the coefficient matrix, and 3) multiplied both sides by the inverse of the coefficient matrix. That required that we stop what we were doing, obtain the inverse of the coefficient matrix, and then come back and use it to obtain the identity matrix on the left of the equation and the solution vector on the right.

Solving systems of differential equations is loosely analogous to the process of solving systems of algebraic equations. However, **instead of finding and using the inverse of the coefficient matrix in your solution process (see text box above), you will, using the coefficient matrix, find the eigenvalues and eigenvectors of the coefficient matrix (see Chapter 19!!!!) and use them in your solution process.** It is really helpful in math to see that new ideas are often just variations of old ideas.

Solving a System of Algebraic Equations	Solving a System of Differential Equations
$3x + 2y = 8$ recopied from above $4x + 1y = 6$	$\frac{d}{dt}x_1 = 3x_1 + 2x_2$ (1) $\frac{d}{dt}x_2 = 4x_1 + 1x_2$ (2)
Transform into matrix algebra notation: Coefficient matrix × variable vector = product vector. $\begin{pmatrix} 3 & 2 \\ 4 & 1 \end{pmatrix} \begin{pmatrix} x \\ y \end{pmatrix} = \begin{pmatrix} 8 \\ 6 \end{pmatrix}$	Transform into matrix algebra notation: Derivative vector = coefficient matrix × variable vector $\begin{pmatrix} \frac{d}{dt}x_1 \\ \frac{d}{dt}x_2 \end{pmatrix} = \begin{pmatrix} 3 & 2 \\ 4 & 1 \end{pmatrix} \begin{pmatrix} x_1(t) \\ x_2(t) \end{pmatrix}$
Next we need to isolate the vector variable: $\begin{pmatrix} x \\ y \end{pmatrix} = \begin{pmatrix} ? \\ ? \end{pmatrix}$	Next, for each derivative, we need to solve for a family of solutions. You may recall from Chapter 17 that vector positions need not always refer to numbers. They can be functions as well. $\begin{pmatrix} x_1(t) \\ x_2(t) \end{pmatrix} = \begin{pmatrix} ? \\ ? \end{pmatrix}$
Find the multiplicative inverse of the coefficient matrix.	Find the eigenvalues and eigenvectors for the given system of linear differential equations.

Refer to Chapter 17 to justify the following. $\begin{pmatrix}-\frac{1}{5} & \frac{2}{5}\\ \frac{4}{5} & -\frac{3}{5}\end{pmatrix}\begin{pmatrix}3 & 2\\ 4 & 1\end{pmatrix}=\begin{pmatrix}1 & 0\\ 0 & 1\end{pmatrix}$	**Refer to Chapter 19 for the following eigenvector and eigenvalue data.** An eigenvector of $\begin{pmatrix}3 & 2\\ 4 & 1\end{pmatrix}$ is $\begin{pmatrix}1\\ 1\end{pmatrix}$ when the eigenvalue is 5. An eigenvector of $\begin{pmatrix}3 & 2\\ 4 & 1\end{pmatrix}$ is $\begin{pmatrix}1\\ -2\end{pmatrix}$ when the eigenvalue is -1.
Multiply both sides by the inverse of the coefficient matrix $\begin{pmatrix}-\frac{1}{5} & \frac{2}{5}\\ \frac{4}{5} & -\frac{3}{5}\end{pmatrix}\begin{pmatrix}3 & 2\\ 4 & 1\end{pmatrix}\begin{pmatrix}x\\ y\end{pmatrix}=\begin{pmatrix}-\frac{1}{5} & \frac{2}{5}\\ \frac{4}{5} & -\frac{3}{5}\end{pmatrix}\begin{pmatrix}8\\ 6\end{pmatrix}$ $\begin{pmatrix}1 & 0\\ 0 & 1\end{pmatrix}\begin{pmatrix}x\\ y\end{pmatrix}=\begin{pmatrix}\frac{4}{5}\\ \frac{14}{5}\end{pmatrix}$	By setting the determinant of (**Coef Mat** $-\lambda\times I)=0$ and solving for λ, we get information that will allow us to determine which of the three solution forms shown in this chapter will be approriate for our data.
Here, on the right, are all scalars. Hence, the product will involve scalars. Variables are all gathered to the left. The path to the solution is all clear. $\begin{pmatrix}x\\ y\end{pmatrix}=\begin{pmatrix}\frac{4}{5}\\ \frac{14}{5}\end{pmatrix}$	Hence, the solution using $\vec{x}=c_1\vec{\varepsilon}_1 e^{\lambda_1 t}+c_2\vec{\varepsilon}_2 e^{\lambda_2 t}$ $\vec{x}=c_1\begin{pmatrix}1\\ 1\end{pmatrix}e^{5t}+c_2\begin{pmatrix}1\\ -2\end{pmatrix}e^{-1t}$ aka $\vec{x}=\begin{pmatrix}x_1(t)\\ x_{2(t)}\end{pmatrix}=\begin{pmatrix}c_1e^{5t}+c_2e^{-t}\\ c_1e^{5t}-2c_2e^{-t}\end{pmatrix}$

Checking your work is always a good idea and in this case helps to reinforce the ideas being demonstrated.

Original System of Differential Equations

$$\frac{d}{dt}x_1(t)=\overbrace{3x_1+2x_2} \qquad (1)$$

$$\frac{d}{dt}x_2(t)=\underbrace{4x_1+1x_2} \qquad (2)$$

Substitute the solutions into the original system.

$$\frac{d}{dt}x_1(t)=\overbrace{3x_1(t)+2x_2(t)}\stackrel{?}{=}5c_1e^{5t}-c_2e^{-t} \quad \text{see box at right} \qquad (1)$$

$$3(c_1e^{5t}+c_2e^{-t})+2(c_1e^{5t}-2c_2e^{-t})\stackrel{?}{=}5c_1e^{5t}-c_2e^{-t} \quad \text{substituting for } x_1 \text{ and } x_2$$

$$3c_1e^{5t}+3c_2e^{-t}+2c_1e^{5t}-4c_2e^{-t}\stackrel{?}{=}5c_1e^{5t}-c_2e^{-t}$$

$$5c_1e^{5t}-c_2e^{-t}=5c_1e^{5t}-c_2e^{-t} \quad \text{ck!}$$

$$\frac{d}{dt}x_2=\underbrace{4x_1(t)+1x_2(t)}\stackrel{?}{=}5c_1e^{5t}+2c_2e^{-t} \quad \text{see box at right} \qquad (2)$$

$$4(c_1e^{5t}+c_2e^{-t})+1(c_1e^{5t}-2c_2e^{-t})\stackrel{?}{=}5c_1e^{5t}+2c_2e^{-t} \quad \text{substituting for } x_1 \text{ and } x_2$$

$$4c_1e^{5t}+4c_2e^{-t}+c_1e^{5t}-2c_2e^{-t})\stackrel{?}{=}5c_1e^{5t}+2c_2e^{-t}$$

$$5c_1e^{5t}+2c_2e^{-t}=5c_1e^{5t}+2c_2e^{-t} \quad \text{ck!}$$

System of Differential Equations
$\begin{pmatrix}\frac{d}{dt}x_1\\ \frac{d}{dt}x_2\end{pmatrix}=\begin{pmatrix}3 & 2\\ 4 & 1\end{pmatrix}\begin{pmatrix}x_1(t)\\ x_2(t)\end{pmatrix}$
After solving for eigenvalue and eigenvector
$\begin{pmatrix}x_1(t)\\ x_{2(t)}\end{pmatrix}=\begin{pmatrix}c_1e^{5t}+c_2e^{-t}\\ c_1e^{5t}-2c_2e^{-t}\end{pmatrix}$
and
$\begin{pmatrix}\frac{d}{dt}x_1\\ \frac{d}{dt}x_2\end{pmatrix}=\begin{pmatrix}5c_1e^{5t}-c_2e^{-t}\\ 5c_1e^{5t}+2c_2e^{-t}\end{pmatrix}$

Recapping, the solution to a 2x2 system of differential equations with constant coefficients and real and distinct determinants, such as

$$\frac{d}{dt}x_1=ax_1+bx_2$$
$$\frac{d}{dt}x_2=cx_1+dx_2,$$

will be

> $\vec{x}=c_1\vec{\varepsilon}_1e^{\lambda_1 t}+c_2\vec{\varepsilon}_2e^{\lambda_2 t}$ (Appendix F and G), where c_1 and c_2 are constants of integration, $\vec{\varepsilon}_1$ and $\vec{\varepsilon}_2$ are eigenvectors of the system, and λ_1 and λ_2 are real and distinct eigenvalues of the system.

This $\vec{\varepsilon}$ symbolism is used to indicate that the system of differential equations has more than one solution family.

Below, the three Chapter 14 solution techniques for solving a second-order homogeneous differential equation are compared with the three Chapter 20 solution techniques for solving an 2*x*2 system of linear homogenous differential equations with constant coefficients.

Second-Order Homogeneous Differential Equation with Constant Coefficients, **Chapter 14**
$$Ax''(t) + Bx'(t) + Cx(t) = 0$$ Form the characteristic equation for the given homogenous differential equation: $Ar^2 + Br + C = 0$. Depending on the evaluation of the "discriminant" of the characteristic equation, substitute in the appropriate general solution form as shown below.

$B^2 - 4 \times A \times C > 0$,	$B^2 - 4 \times A \times C = 0$	$B^2 - 4 \times A \times C < 0$,
where r_1 and $r_2 \in \mathbb{R}$, and $r_1 \neq r_2$. $x = c_1 e^{r_1 t} + c_2 e^{r_2 t}$, where the solution to the characteristic equation is traditionally represented as r.	, where r_1 and $r_2 \in \mathbb{R}$, and $r_1 = r_2$. $x = c_1 e^{rt} + c_2 t e^{rt}$, where the solution to the characteristic equation is traditionally represented as r. Be careful, this formula is different from the one at left.	where r_1 and r_2 are complex. $x = c_1 e^{\alpha t} \cos \beta t + c_2 e^{\alpha t} \sin \beta t$, where the real and imaginary parts of the solution to the characteristic equation are traditionally represented using α and β. See Appendix D for a discussion of this integrated form.
$y_{gen} = c_1 e^{r_1 t} + c_2 e^{r_2 t}$ see Chapter 14	$y_{gen} = c_1 e^{rt} + c_2 t e^{rt}$ see Chapter 14	$y_{gen} = c_1 e^{\alpha t} \cos \beta t + c_2 e^{\alpha t} \sin \beta t$ see Chapter 14

Two-by-Two System of Linear Homogeneous First-Order Differential Equations, **Chapter 20**.
$$\frac{dx_1}{dt} = ax_1(t) + bx_2(t) \quad (1)$$ $$\frac{dx_2}{dt} = cx_1(t) + dx_2(t) \quad (2)$$ Again, there are three possibilities.

Real and Distinct Eigenvalues	Repeated eigenvalues	Complex eigenvalues
Set $\det(A - \lambda\mathbf{I}) = 0$, solve for λ. If λ_1 and $\lambda_2 \in \mathbb{R}$ and $\lambda_1 \neq \lambda_2$, then use λ_1 and λ_2 and matrix **A** to solve for eigenvectors $\vec{\varepsilon}_1$ and $\vec{\varepsilon}_2$. $$\vec{x} = c_1 \vec{\varepsilon}_1 e^{\lambda_1 t} + c_2 \vec{\varepsilon}_2 e^{\lambda_2 t},$$ where the solution to the characteristic equation is traditionally represented as λ. (See Appendices F and G.)	Unfortunately, the solution to a system of differential equations with repeated eigenvalues, $$\vec{x} = c_1 \vec{\varepsilon} e^{\lambda t} + c_2 \left(\vec{\varepsilon} t e^{\lambda t} + \vec{\rho} e^{\lambda t} \right),$$ does not "parallel" the Chapter 14 solution to a second-order differential equation whose characteristic equation has repeated roots. It is rather involved and the reader is referred to *Elementary Differential Equations* by Boyce and DiPrima or http://tutorial.math.lamar.edu/Classes/DE/RepeatedEigenvalues.aspx.	Set $\det(A - \lambda\mathbf{I}) = 0$. Solve for λ. λ_1 and λ_2 will be complex eigenvalues. Let $\lambda_1 = r + ci$, and $\lambda_2 = r - ci$. Use these complex eigenvalues and matrix **A** to solve for complex eigenvectors $\vec{\varepsilon}_1$ and $\vec{\varepsilon}_2$. Vectors $\vec{\varepsilon}_1$ and $\vec{\varepsilon}_2$ may contain complex components. $$\vec{x} = c_1 \vec{\varepsilon}_1 e^{\lambda_1 t} + c_2 \vec{\varepsilon}_2 e^{\lambda_2 t}.$$ Real-valued solutions can be constructed using the identity $$e^{(r+ci)t} = e^{rt}[\cos(ct) + i \sin(ct)].$$

Solving a System of Homogeneous ODEs with Distinct Eigenvalues	***Solving a System of Homogeneous ODEs with Repeated Eigenvalues***	***Solving a System of Homogeneous ODEs with Complex Eigenvalues***
$\frac{\mathrm{d}}{\mathrm{d}t}x_1 = -5x_1 + 1x_2$ $\frac{\mathrm{d}}{\mathrm{d}t}x_2 = 4x_1 - 2x_2$ Transform to matrix notation. $\begin{pmatrix} \frac{dx_1}{dt} \\ \frac{dx_2}{dt} \end{pmatrix} = \begin{pmatrix} -5 & 1 \\ 4 & -2 \end{pmatrix} \begin{pmatrix} x_1(t) \\ x_2(t) \end{pmatrix}$ Substitute into $(\mathbf{Mat} - \lambda\mathbf{I}) \times \vec{\varepsilon} = \begin{pmatrix} 0 \\ 0 \end{pmatrix}$. $\left[\begin{pmatrix} -5 & 1 \\ 4 & -2 \end{pmatrix} - \begin{pmatrix} \lambda & 0 \\ 0 & \lambda \end{pmatrix}\right] \begin{pmatrix} x_1 \\ x_2 \end{pmatrix} = \begin{pmatrix} 0 \\ 0 \end{pmatrix}$ $\begin{pmatrix} -5-\lambda & 1 \\ 4 & -2-\lambda \end{pmatrix} \begin{pmatrix} x_1 \\ x_2 \end{pmatrix} = \begin{pmatrix} 0 \\ 0 \end{pmatrix}$	$\frac{\mathrm{d}}{\mathrm{d}t}x_1 = 7x_1 + 1x_2$ $\frac{\mathrm{d}}{\mathrm{d}t}x_2 = -4x_1 + 3x_2$ Transform to matrix notation. $\begin{pmatrix} \frac{dx_1}{dt} \\ \frac{dx_2}{dt} \end{pmatrix} = \begin{pmatrix} 7 & 1 \\ -4 & 3 \end{pmatrix} \begin{pmatrix} x_1(t) \\ x_2(t) \end{pmatrix}$ Substitute into $(\mathbf{Mat} - \lambda\mathbf{I}) \times \vec{\varepsilon} = \begin{pmatrix} 0 \\ 0 \end{pmatrix}$. $\left[\begin{pmatrix} 7 & 1 \\ -4 & 3 \end{pmatrix} - \begin{pmatrix} \lambda & 0 \\ 0 & \lambda \end{pmatrix}\right] \begin{pmatrix} x_1 \\ x_2 \end{pmatrix} = \begin{pmatrix} 0 \\ 0 \end{pmatrix}$ $\begin{pmatrix} 7-\lambda & 1 \\ -4 & 3-\lambda \end{pmatrix} \begin{pmatrix} x_1 \\ x_2 \end{pmatrix} = \begin{pmatrix} 0 \\ 0 \end{pmatrix}$	$\frac{\mathrm{d}}{\mathrm{d}t}x_1 = 3x_1 - 9x_2$ $\frac{\mathrm{d}}{\mathrm{d}t}x_2 = 4x_1 - 3x_2$ Transform to matrix notation. $\begin{pmatrix} \frac{dx_1}{dt} \\ \frac{dx_2}{dt} \end{pmatrix} = \begin{pmatrix} 3 & -9 \\ 4 & -3 \end{pmatrix} \begin{pmatrix} x_1(t) \\ x_2(t) \end{pmatrix}$ Substitute into $(\mathbf{Mat} - \lambda\mathbf{I}) \times \vec{\varepsilon} = \begin{pmatrix} 0 \\ 0 \end{pmatrix}$. $\left[\begin{pmatrix} 3 & -9 \\ 4 & -3 \end{pmatrix} - \begin{pmatrix} \lambda & 0 \\ 0 & \lambda \end{pmatrix}\right] \begin{pmatrix} x_1 \\ x_2 \end{pmatrix} = \begin{pmatrix} 0 \\ 0 \end{pmatrix}$ $\begin{pmatrix} 3-\lambda & -9 \\ 4 & -3-\lambda \end{pmatrix} \begin{pmatrix} x_1 \\ x_2 \end{pmatrix} = \begin{pmatrix} 0 \\ 0 \end{pmatrix}$
Setting $\det\begin{pmatrix} -5-\lambda & 1 \\ 4 & -2-\lambda \end{pmatrix} = 0$ guarantees that the two rows of the equation will be multiples of each other, yielding infinite solutions. Solving for λ will determine which of the three solution equations will be appropriate for the data in the system of equations. $(-5-\lambda)(-2-\lambda) - (4 \times 1) = 0$ $10 + 5\lambda + 2\lambda + \lambda^2 - 4 = 0$ $\lambda^2 + 7\lambda + 6 = 0$ $(\lambda + 1)(\lambda + 6) = 0$ $\lambda = -1, -6$ The solution form will be the one for distinct eigenvalues. $\vec{x} = c_1\vec{\varepsilon}_1 e^{\lambda_1 t} + c_2\vec{\varepsilon}_2 e^{\lambda_2 t}$	Setting $\det\begin{pmatrix} 7-\lambda & 1 \\ -4 & 3-\lambda \end{pmatrix} = 0$ guarantees that the two rows of the equation will be multiples of each other, yielding infinite solutions. Solving for λ will determine which of the three solution equations will be appropriate for the data in the system of equations. $(7-\lambda)(3-\lambda) - (-4 \times 1) = 0$ $21 - 7\lambda - 3\lambda + \lambda^2 + 4 = 0$ $\lambda^2 - 10\lambda + 25 = 0$ $(\lambda - 5)(\lambda - 5) = 0$ $\lambda = 5, 5$ The solution form for repeated eigenvalues (not taught in this book) will be $\vec{x} = c_1\vec{\varepsilon}e^{\lambda t} + c_2\left(\vec{\varepsilon}te^{\lambda t} + \vec{\rho}e^{\lambda t}\right)$. See *Elementary Differential Equations* by Boyce and DiPrima or http://tutorial.math.lamar.edu/Classes/DE/RepeatedEigenvalues.aspx for more information.	Setting $\det\begin{pmatrix} 3-\lambda & -9 \\ 4 & -3-\lambda \end{pmatrix} = 0$ guarantees that the two rows of the equation will be multiples of each other, yielding infinite solutions. Solving for λ will determine which of the three solution equations will be appropriate for the data in the system of equations. $(3-\lambda)(-3-\lambda) - (4 \times -9) = 0$ $-9 - 3\lambda + 3\lambda + \lambda^2 + 36 = 0$ $\lambda^2 + 27 = 0$ $\lambda^2 = -27$ $\lambda = \sqrt{-27} = \sqrt{3}\sqrt{9}\sqrt{-1} = \pm 3i\sqrt{3}$ The solution form will be the one for complex eigenvalues. $\vec{x} = c_1\vec{\varepsilon}_1 e^{\lambda_1 t} + c_2\vec{\varepsilon}_2 e^{\lambda_2 t}$ for $\lambda_1 = r + ci$, $\lambda_2 = r - ci$, $\vec{\varepsilon}_1 = \alpha + \beta i$, and $\vec{\varepsilon}_2 = \alpha - \beta i$.
Substitute the matrix value and each value of λ into the equation $(\mathbf{Mat} - \lambda\mathbf{I}) \times \vec{\varepsilon} = \begin{pmatrix} 0 \\ 0 \end{pmatrix}$ and solve for $\vec{\varepsilon}$. For $\lambda_1 = -1$, $\begin{pmatrix} -5-(-1) & 1 \\ 4 & -2-(-1) \end{pmatrix} \begin{pmatrix} x_1 \\ x_2 \end{pmatrix} = \begin{pmatrix} 0 \\ 0 \end{pmatrix}$ $\begin{pmatrix} -4 & 1 \\ 4 & -1 \end{pmatrix} \begin{pmatrix} x_1 \\ x_2 \end{pmatrix} = \begin{pmatrix} 0 \\ 0 \end{pmatrix}$	Substitute the matrix value and λ into the equation $(\mathbf{Mat} - \lambda\mathbf{I}) \times \vec{\varepsilon} = \begin{pmatrix} 0 \\ 0 \end{pmatrix}$ and solve for $\vec{\varepsilon}$. For $\lambda = 5$, $\begin{pmatrix} 7-5 & 1 \\ -4 & 3-5 \end{pmatrix} \begin{pmatrix} x_1 \\ x_2 \end{pmatrix} = \begin{pmatrix} 0 \\ 0 \end{pmatrix}$ $\begin{pmatrix} 2 & 1 \\ -4 & -2 \end{pmatrix} \begin{pmatrix} x_1 \\ x_2 \end{pmatrix} = \begin{pmatrix} 0 \\ 0 \end{pmatrix}$	Substitute the matrix value and each value of λ into the equation $(\mathbf{Mat} - \lambda\mathbf{I}) \times \vec{\varepsilon} = \begin{pmatrix} 0 \\ 0 \end{pmatrix}$ and solve for $\vec{\varepsilon}$. For $\lambda_1 = 3i\sqrt{3}$ $\begin{pmatrix} 3 - 3i\sqrt{3} & -9 \\ 4 & -3 - 3i\sqrt{3} \end{pmatrix} \begin{pmatrix} x_1 \\ x_2 \end{pmatrix} = \begin{pmatrix} 0 \\ 0 \end{pmatrix}$

Solving a System of Homogeneous ODEs with Distinct Eigenvalues continued	***Solving a System of Homogeneous ODEs with Repeated Eigenvalues continued***	***Solving a System of Homogeneous ODEs with Complex Eigenvalues continued***
Given how λ was determined, it is not surprising that the two rows of the system above are multiples of each other. Ignoring Row 2 yields $-4x_1 + x_2 = 0$ $x_2 = 4x_1$ So $\begin{pmatrix}x_1\\x_2\end{pmatrix} = k\begin{pmatrix}1\\4\end{pmatrix}$. For $\lambda_1 = -1$, $\vec{\varepsilon}_1 = \begin{pmatrix}1\\4\end{pmatrix}$. Similarly, for $\lambda_2 = -6$, $\vec{\varepsilon}_2 = \begin{pmatrix}-1\\1\end{pmatrix}$.	Given how λ was determined, it is not surprising that the two rows of the system above are multiples of each other. Ignoring Row 2 yields $2x_1 + x_2 = 0$ $x_2 = -2x_1$ So $\begin{pmatrix}x_1\\x_2\end{pmatrix} = k\begin{pmatrix}1\\-2\end{pmatrix}$. For $\lambda = 5$, $\vec{\varepsilon} = \begin{pmatrix}1\\-2\end{pmatrix}$.	Given how λ was determined, it is not surprising that the two rows of the system above are multiples of each other. Ignoring Row 2 yields $\left(3 - 3i\sqrt{3}\right)x_1 - 9x_2 = 0$ $(3 - 3i\sqrt{3})x_1 = 9x_2$ $9x_2 = (3 - 3i\sqrt{3})x_1$ $x_2 = \dfrac{1 - i\sqrt{3}}{3}x_1$ So $\begin{pmatrix}x_1\\x_2\end{pmatrix} = k\begin{pmatrix}1\\\frac{1-i\sqrt{3}}{3}\end{pmatrix}$ $\begin{pmatrix}x_1\\x_2\end{pmatrix} = k\begin{pmatrix}3\\1 - i\sqrt{3}\end{pmatrix}$. For $\lambda_1 = 3i\sqrt{3}$, $\vec{\varepsilon}_1 = \begin{pmatrix}3\\1 - i\sqrt{3}\end{pmatrix}$. Similarly, for $\lambda_2 = -3i\sqrt{3}$, $\vec{\varepsilon}_2 = \begin{pmatrix}3\\1 + i\sqrt{3}\end{pmatrix}$.
The general solution, $\vec{x} = c_1\vec{\varepsilon}_1 e^{\lambda_1 t} + c_2\vec{\varepsilon}_2 e^{\lambda_2 t}$, would be $\vec{x} = c_1\begin{pmatrix}1\\4\end{pmatrix}e^{-1t} + c_2\begin{pmatrix}-1\\1\end{pmatrix}e^{-6t}$. It would be instructive for you to take $\begin{pmatrix}x_1\\x_2\end{pmatrix}$ shown here, calculate $\begin{pmatrix}x_1'\\x_2'\end{pmatrix}$, and back substitute into the original system.	This eigenvalue and its corresponding eigenvector is then substituted into the general solution for repeated eigenvalues from *Elementary Differential Equations* by Boyce and DiPrima or http://tutorial.math.lamar.edu/Classes/DE/RepeatedEigenvalues.aspx: $\vec{x} = c_1\vec{\varepsilon}e^{\lambda t} + c_2\left(\vec{\varepsilon}te^{\lambda t} + \vec{\rho}e^{\lambda t}\right)$. It would be instructive for you to take $\begin{pmatrix}x_1\\x_2\end{pmatrix}$ shown here, calculate $\begin{pmatrix}x_1'\\x_2'\end{pmatrix}$, and back substitute into the original system.	The general solution, $\vec{x} = c_1\vec{\varepsilon}_1 e^{\lambda_1 t} + c_2\vec{\varepsilon}_2 e^{\lambda_2 t}$, would be $\vec{x} = c_1\begin{pmatrix}3\\1 - i\sqrt{3}\end{pmatrix}e^{3it\sqrt{3}}$ $+ c_2\begin{pmatrix}3\\1 + i\sqrt{3}\end{pmatrix}e^{-3it\sqrt{3}}$. It would be instructive for you to take $\begin{pmatrix}x_1\\x_2\end{pmatrix}$ shown here, calculate $\begin{pmatrix}x_1'\\x_2'\end{pmatrix}$, and back substitute into the original system.

Author's Note: The complex numbers in both the vector and exponential of the rightmost solution can be eliminated using Euler's formulas (see Chapter 14 and Appendices D and E). The final answer with real numbers is*

$$\vec{x} = c_1\begin{pmatrix}3\cos\left(3t\sqrt{3}\right)\\ \cos\left(3t\sqrt{3}\right) + \sqrt{3}\sin\left(3t\sqrt{3}\right)\end{pmatrix} + c_2\begin{pmatrix}3\sin\left(3t\sqrt{3}\right)\\ \sin\left(3t\sqrt{3}\right) - \sqrt{3}\cos\left(3t\sqrt{3}\right)\end{pmatrix}.$$

*Paul Dawkins, tutorial.math.lamar.edu/Classes/DE/ComplexRoots.aspx

It will be helpful to connect what we are doing now with what we discussed back in Chapter 17. There we connected concepts in beginning algebra to concepts in linear algebra.

Important Concepts in Beginning Algebra

1.	2.	3.
$0x = 0$	$0x = k$	$ax = k$
Infinite/indeterminate solutions	no solution (cannot divide k by 0)	one solution
$\frac{0x}{0} = \frac{0}{0}$, ($x = 5, -1$, etc.)	$\frac{0x}{0} = \frac{k}{0}$, undefined	$x = \frac{k}{a}$

Recall that we are taught that we should not divide by zero. Each of the first two equations above involves a division by the scalar zero. The equation at the right (above) does not.

News flash! In each of the three equations shown above the coefficient, variable, and product are all assumed to be scalars. The equation $ax = k$, with a, x, and k all scalars, can be thought of as analogous to $\mathbf{A}\vec{x} = \vec{k}$ where $\mathbf{A}$ is a 1×1 matrix, $\vec{x}$ is a 1×1 vector, and $\vec{k}$ is also a 1×1 vector. In the following equations, we will also be studying $\mathbf{A}\vec{x} = \vec{k}$, but $\mathbf{A}$, $\vec{x}$, and $\vec{k}$ will not be restricted to being 1×1 in dimension. By lifting the restriction on the dimensions of the coefficient matrix, by analogy, the determinant of the matrix must not be zero.

Important Concepts in Linear Algebra

$3x + 2y = 0$ not 6*

$6x + 4y = 0$ not 12*

$$\begin{pmatrix} 3 & 2 \\ 6 & 4 \end{pmatrix}\begin{pmatrix} x \\ y \end{pmatrix} = \begin{pmatrix} 0 \\ 0 \end{pmatrix} \quad \text{or any} \begin{pmatrix} p \\ 2p \end{pmatrix}$$

$$\det\begin{pmatrix} 3 & 2 \\ 6 & 4 \end{pmatrix} = (3)(4) - (2)(6) = 0$$

infinite/indeterminant solutions

*Chapter 17's vector, $\begin{pmatrix} 6 \\ 12 \end{pmatrix}$, is now $\begin{pmatrix} 0 \\ 0 \end{pmatrix}$ for consistency with $0x = 0$ above.

3x + 2y = 0

6x + 4y = 0

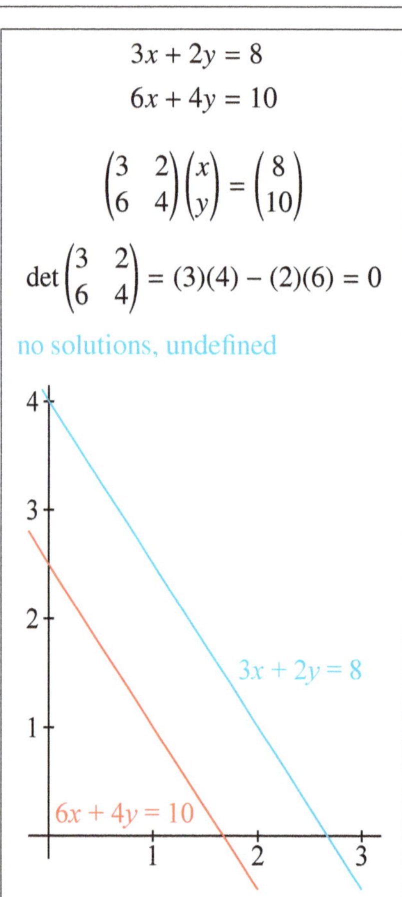

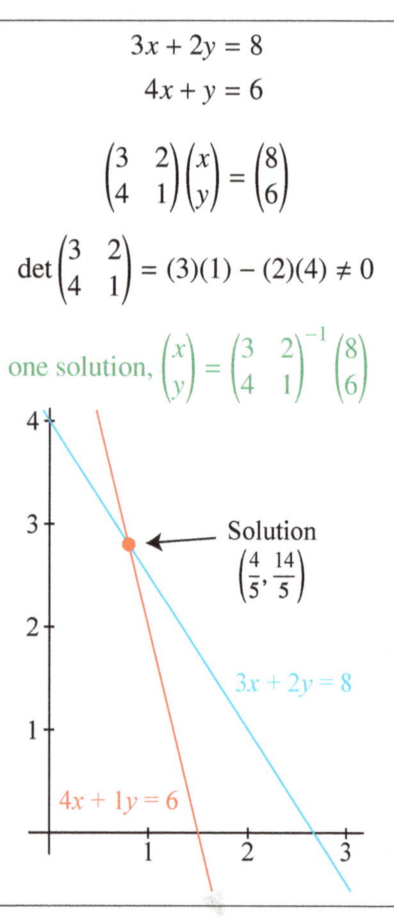

Your math teacher will not tell you the following because they are so smart that it will not occur to them that this comparison will help you in organizing and thinking about all this new knowledge, skills, and vocabulary. The three graphs above show three things that can happen when you have two linear algebraic equations with two unknowns: 1) The two lines can be the same yielding infinite solutions, 2) The two lines can have the same slope but different intercepts in which case there are no solutions, and 3) the two lines can intersect in which case there is one solution. The first two cases usually do not provide any sort of information that would be useful so mostly when we work with systems of 2×2 algebraic equations we are interested in only case three, when the two lines intersect. There is a test called the determinant test which allows us to determine, without a graph, if our system is a useful case-three type of system. If the system coefficients are $\begin{pmatrix} a & b \\ c & d \end{pmatrix}$, then the determinant is defined to be $ad-bc$. If $\det\begin{pmatrix} a & b \\ c & d \end{pmatrix} = ad-bc \neq 0$ (page 198, right), then the system has a unique solution. Otherwise, it has either infinite or no solutions.

These scenarios are (roughly) repeated similarly when working with systems of linear differential equations. Again, the first two cases usually provide little information that would be useful so, when we work with systems of 2×2 differential equations, we mostly are interested in only case three shown in the graph below right. There is a test called the Wronskian determinant test which allows us to see if our system is a useful case-three type of system without seeing a graph. If the two solutions to a 2×2 system of equations are f_1 and f_2, the Wronskian determinant is calculated as $W(f_1, f_2) = \det\begin{pmatrix} f_1 & f_2 \\ f_1' & f_2' \end{pmatrix} = f_1 f_2' - f_2 f_1'$ and if $W(f_1, f_2) \neq 0$, then the system of differential equations will yield usable data as shown in the graph below right. Another way to think about this is to say that the two lines in the left and middle graphs below are constant multiples of each other (you will hear the term *linearly dependent*) at each x, whereas the two lines in the graph below right are not (you will hear the term *linearly independent*) at each x.

$f_1(x) = \sin(x) \quad f_2 = \sin(x)$ $f_1'(x) = \cos(x) \quad f_2' = \cos(x)$ $W\begin{vmatrix} \sin(x) & \sin(x) \\ \cos(x) & \cos(x) \end{vmatrix} = \sin(x)\cos(x)$ $- \sin(x)\cos(x) = 0$ Linearly dependent.	$f_1(x) = 1.5\sin(x) \quad f_2 = \sin(x)$ $f_1'(x) = 1.5\cos(x) \quad f_2' = \cos(x)$ $W\begin{vmatrix} 1.5\sin(x) & \sin(x) \\ 1.5\cos(x) & \cos(x) \end{vmatrix}$ $= 1.5\sin(x)\cos(x) - 1.5\sin(x)\cos(x)$ $= 0$ Linearly dependent.	$f_1(x) = \sin(x+1.5)+0.5 \quad f_2 = \sin(x)$ $f_1'(x) = \cos(x+1.5) \quad f_2' = \cos(x)$ $W\begin{vmatrix} \sin(x+1.5)+0.5 & \sin(x) \\ \cos(x+1.5) & \cos(x) \end{vmatrix}$ $= [\sin(x+1.5)+0.5]\cos(x)$ $- \sin(x)\cos(x+1.5) \neq 0$ Linearly independent.

Each of the graphs below reflects arbitrarily chosen IVP conditions to select a specific differential equation from a family of differential equations.

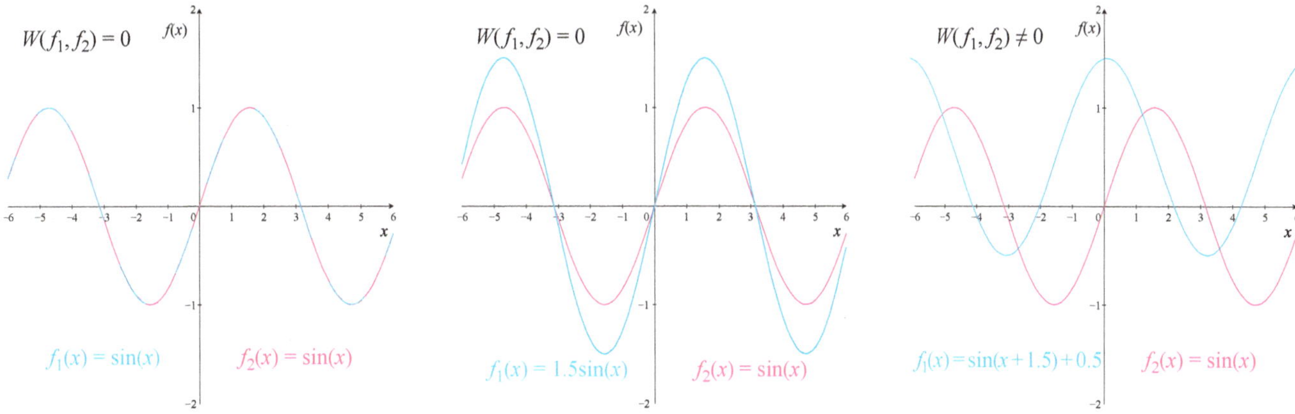

Glove is to hand as shoe is to foot. The determinant of the coefficient matrix of an algebraic system of equations is analogous to the Wronskian of a pair of functions. There is a distinction however. In basic linear algebra, in 2×2 square systems, the relation is an "if and only if " one. The equations are linearly independent if and only if det(**Coef Mat**) $\neq 0$. This isn't the case with the Wronskian. Still, whenever it is possible it is helpful for beginners to make connections to previous mathematics.

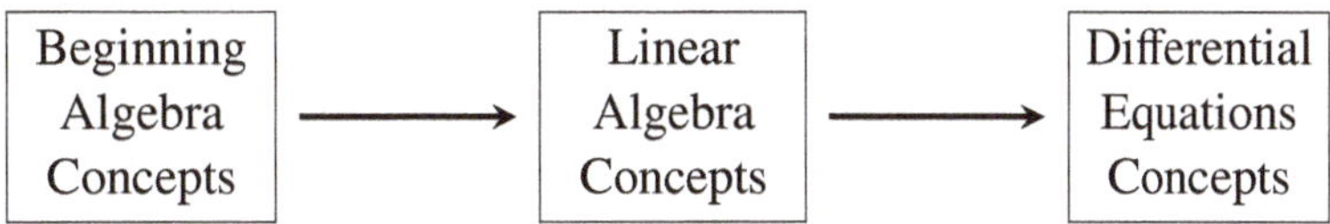

Or, as Monsieur Jourdain stated back in Chapter 1, "Par ma foi! Il y a plus de quarante ans que je dis de la prose sans que j'en susse rien, et je vous suis le plus obligé du monde de m'avoir appris cela."

Since we have seen the idea of the *Wronskian determinant test* it might be good to show it being applied. To save time and work, let's reuse a system of differential equations that we solved earlier in Chapter 20.

$$\frac{\mathrm{d}}{\mathrm{d}t}x_1 = -5x_1 + 1x_2$$
$$\frac{\mathrm{d}}{\mathrm{d}t}x_2 = 4x_1 - 2x_2$$

and using the following solution also from Chapter 20

$$\vec{x} = c_1\begin{pmatrix}1\\4\end{pmatrix}\mathrm{e}^{-1t} + c_2\begin{pmatrix}-1\\1\end{pmatrix}\mathrm{e}^{-6t}.$$

We are calculating a Wronskian for a 2×2 system of ODEs,

$$\vec{x}\,'(t) = \begin{pmatrix}\frac{\mathrm{d}}{\mathrm{d}t}x_1\\ \frac{\mathrm{d}}{\mathrm{d}t}x_1\end{pmatrix} = \begin{pmatrix}ax_1 + bx_2\\ cx_1 + dx_2\end{pmatrix},$$

whose solutions are $x_1(t)$ and $x_2(t)$. The Wronskian is defined as

$$W\begin{vmatrix}x_1(t) & x_2(t)\\ x_1'(t) & x_2'(t)\end{vmatrix} = x_1(t) \times x_2'(t) - x_2(t) \times x_1'(t).$$

Repeating the solution from earlier, we have

$$\vec{x} = c_1\begin{pmatrix}1\\4\end{pmatrix}\mathrm{e}^{-1t} + c_2\begin{pmatrix}-1\\1\end{pmatrix}\mathrm{e}^{-6t},$$

from which we get

$$x_1 = c_1(1)\mathrm{e}^{-t} + c_2(-1)\mathrm{e}^{-6t} \qquad x_2 = c_1(4)\mathrm{e}^{-t} + c_2(1)\mathrm{e}^{-6t}$$
$$x_1' = -c_1\mathrm{e}^{-t} + 6c_2\mathrm{e}^{-6t} \qquad x_2' = -4c_1\mathrm{e}^{-t} - 6c_2\mathrm{e}^{-6t}.$$

So,

$$W(x_1(t), x_2(t)) = \begin{vmatrix}c_1\mathrm{e}^{-t} - c_2\mathrm{e}^{-6t} & 4c_1\mathrm{e}^{-t} + c_2\mathrm{e}^{-6t}\\ -c_1\mathrm{e}^{-t} + 6c_2\mathrm{e}^{-6t} & -4c_1\mathrm{e}^{-t} - 6c_2\mathrm{e}^{-6t}\end{vmatrix}$$
$$= \left[\left(c_1\mathrm{e}^{-t} - c_2\mathrm{e}^{-6t}\right)\left(-4c_1\mathrm{e}^{-t} - 6c_2\mathrm{e}^{-6t}\right)\right] - \left[\left(4c_1\mathrm{e}^{-t} + c_2\mathrm{e}^{-6t}\right)\left(-c_1\mathrm{e}^{-t} + 6c_2\mathrm{e}^{-6t}\right)\right]$$
$$= \left[-4c_1^2\mathrm{e}^{-2t} - 6c_1c_2\mathrm{e}^{-7t} + 4c_1c_2\mathrm{e}^{-7t} + 6c_2^2\mathrm{e}^{-12t}\right] - \left[-4c_1^2\mathrm{e}^{-2t} + 24c_1c_2\mathrm{e}^{-7t} - c_1c_2\mathrm{e}^{-7t} + 6c_2^2\mathrm{e}^{-12t}\right]$$
$$= \left[-4c_1^2\mathrm{e}^{-2t} - 2c_1c_2\mathrm{e}^{-7t} + 6c_2^2\mathrm{e}^{-12t}\right] - \left[-4c_1^2\mathrm{e}^{-2t} + 23c_1c_2\mathrm{e}^{-7t} + 6c_2^2\mathrm{e}^{-12t}\right]$$
$$= \cancel{-4c_1^2\mathrm{e}^{-2t}} - 2c_1c_2\mathrm{e}^{-7t} + \cancel{6c_2^2\mathrm{e}^{-12t}} + \cancel{4c_1^2\mathrm{e}^{-2t}} - 23c_1c_2\mathrm{e}^{-7t} - \cancel{6c_2^2\mathrm{e}^{-12t}}$$
$$= -25c_1c_2\mathrm{e}^{-7t}$$

So, $W(x_1(t), x_2(t)) = -25c_1c_2\mathrm{e}^{-7t}$ and, if both c_1 and c_2 are not zero, then the Wronskian is not zero and the functions are linearly independent.

Lastly, for work in a university differential equations class, it is important to connect the notation introduced in this chapter on a 2×2 system of differential equations with the notation used for $n \times n$ systems of differential equations. **The symbolism $v_i^{\lambda_j}$ below indicates the ith component of the eigenvector associated with λ_j; that is, $\vec{\mathcal{E}}_j$. The component parts, $v_1^{\lambda_j}$, $v_2^{\lambda_j}$, etc., form the entire eigenvector $\vec{\mathcal{E}}_j$.**

For the 2 × 2 system of differential equations with real, distinct eigenvalues,

$$\frac{d}{dt}x_1 = ax_1 + bx_2$$
$$\frac{d}{dt}x_2 = cx_1 + dx_2$$

substitute the matrix $\begin{pmatrix} a & b \\ c & d \end{pmatrix}$ into the equation $(\text{Mat} - \lambda \text{I})\vec{\mathcal{E}} = \begin{pmatrix} 0 \\ 0 \end{pmatrix}$ and solve for the two solutions of the eigenvalue λ: λ_1 and λ_2.

There will be two solutions for the system, each of the form

$$\kappa_1 e^{\lambda_1 t} + \kappa_2 e^{\lambda_2 t}$$

$$x_1 = c_1 v_1^{\lambda_1} e^{\lambda_1 t} + c_2 v_1^{\lambda_2} e^{\lambda_2 t}$$
$$x_2 = c_1 v_2^{\lambda_1} e^{\lambda_1 t} + c_2 v_2^{\lambda_2} e^{\lambda_2 t},$$

which, after solving for the eigenvector for each λ, can be represented as

$$\vec{x} = c_1 \begin{pmatrix} v_1^{\lambda_1} \\ v_2^{\lambda_1} \end{pmatrix} e^{\lambda_1 t} + c_2 \begin{pmatrix} v_1^{\lambda_2} \\ v_2^{\lambda_2} \end{pmatrix} e^{\lambda_2 t},$$

or

$$\vec{x} = c_1 \vec{\mathcal{E}}_1 e^{\lambda_1 t} + c_2 \vec{\mathcal{E}}_2 e^{\lambda_2 t}.$$

For the $n \times n$ system of differential equations with real, distinct eigenvalues,

$$\frac{d}{dt}x_1 = a_{1,1}x_1 + a_{1,2}x_2 + \cdots + a_{1,n}x_n$$
$$\frac{d}{dt}x_2 = a_{2,1}x_1 + a_{2,2}x_2 + \cdots + a_{2,n}x_n$$
$$\vdots$$
$$\frac{d}{dt}x_n = a_{n,1}x_1 + a_{n,2}x_2 + \cdots + a_{n,n}x_n,$$

substitute the matrix $\begin{pmatrix} a_{1,1} & a_{1,2} & \cdots & a_{1,n} \\ a_{2,1} & a_{2,2} & \cdots & a_{2,n} \\ \vdots & \vdots & \ddots & \vdots \\ a_{n,1} & a_{n,2} & \cdots & a_{n,n} \end{pmatrix}$ into the equation $(\text{Mat} - \lambda \text{I})\vec{\mathcal{E}} = \begin{pmatrix} 0 \\ \vdots \\ 0 \end{pmatrix}$ and solve for the n solutions of the eigenvalue λ: $\lambda_1, \lambda_2, \ldots, \lambda_n$.

There will be n solutions for the system, each of the form

$$\kappa_1 e^{\lambda_1 t} + \kappa_2 e^{\lambda_2 t} + \cdots + \kappa_n e^{\lambda_n t}$$

$$x_1 = c_1 v_1^{\lambda_1} e^{\lambda_1 t} + c_2 v_1^{\lambda_2} e^{\lambda_2 t} + \cdots + c_n v_1^{\lambda_n} e^{\lambda_n t}$$
$$x_2 = c_1 v_2^{\lambda_1} e^{\lambda_1 t} + c_2 v_2^{\lambda_2} e^{\lambda_2 t} + \cdots + c_n v_2^{\lambda_n} e^{\lambda_n t}$$
$$\vdots$$
$$x_n = c_1 v_n^{\lambda_1} e^{\lambda_1 t} + c_2 v_n^{\lambda_2} e^{\lambda_2 t} + \cdots + c_n v_n^{\lambda_n} e^{\lambda_n t},$$

which, after solving for the eigenvector for each λ, can be represented as

$$\vec{x} = c_1 \begin{pmatrix} v_1^{\lambda_1} \\ v_2^{\lambda_1} \\ \vdots \\ v_n^{\lambda_1} \end{pmatrix} e^{\lambda_1 t} + c_2 \begin{pmatrix} v_1^{\lambda_2} \\ v_2^{\lambda_2} \\ \vdots \\ v_n^{\lambda_2} \end{pmatrix} e^{\lambda_2 t} + \cdots + c_n \begin{pmatrix} v_1^{\lambda_n} \\ v_2^{\lambda_n} \\ \vdots \\ v_n^{\lambda_n} \end{pmatrix} e^{\lambda_n t},$$

or

$$\vec{x} = c_1 \vec{\mathcal{E}}_1 e^{\lambda_1 t} + c_2 \vec{\mathcal{E}}_2 e^{\lambda_2 t} + \cdots + c_n \vec{\mathcal{E}}_n e^{\lambda_2 t}.$$

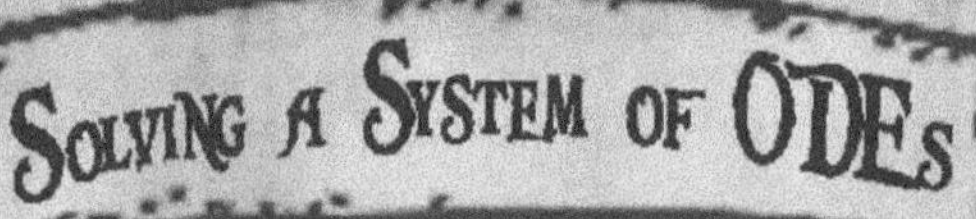

Original System of Differential Equations

1.

$$\frac{dx_1}{dt} = ax_1 + bx_2 \quad (1)$$

$$\frac{dx_2}{dt} = cx_1 + dx_2 \quad (2)$$

2. System of Differential Equations Converted to Matrix Form

$$\begin{pmatrix} \frac{d}{dt}x_1 \\ \frac{d}{dt}x_2 \end{pmatrix} = \begin{pmatrix} a & b \\ c & d \end{pmatrix}\begin{pmatrix} x_1 \\ x_2 \end{pmatrix}$$

3. Find the eigenvalues λ_1 and λ_2 of the coefficient matrix $\begin{pmatrix} a & b \\ c & d \end{pmatrix}$ using the identity equation $\mathbf{Mat} \times \vec{\varepsilon} = \lambda \times \vec{\varepsilon}$.

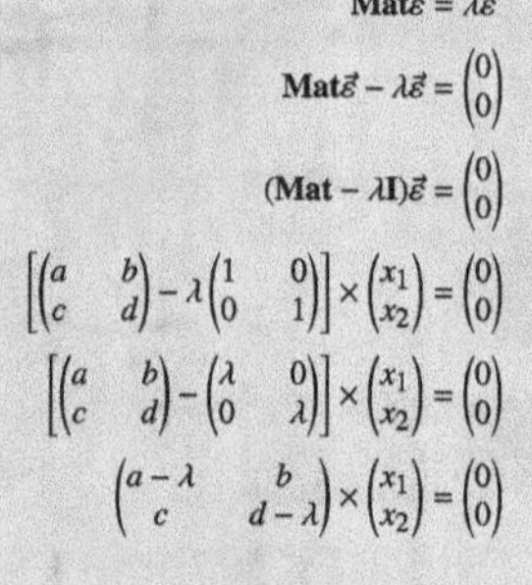

$$\mathbf{Mat}\vec{\varepsilon} = \lambda\vec{\varepsilon}$$

$$\mathbf{Mat}\vec{\varepsilon} - \lambda\vec{\varepsilon} = \begin{pmatrix} 0 \\ 0 \end{pmatrix}$$

$$(\mathbf{Mat} - \lambda\mathbf{I})\vec{\varepsilon} = \begin{pmatrix} 0 \\ 0 \end{pmatrix}$$

$$\left[\begin{pmatrix} a & b \\ c & d \end{pmatrix} - \lambda\begin{pmatrix} 1 & 0 \\ 0 & 1 \end{pmatrix}\right] \times \begin{pmatrix} x_1 \\ x_2 \end{pmatrix} = \begin{pmatrix} 0 \\ 0 \end{pmatrix}$$

$$\left[\begin{pmatrix} a & b \\ c & d \end{pmatrix} - \begin{pmatrix} \lambda & 0 \\ 0 & \lambda \end{pmatrix}\right] \times \begin{pmatrix} x_1 \\ x_2 \end{pmatrix} = \begin{pmatrix} 0 \\ 0 \end{pmatrix}$$

$$\begin{pmatrix} a-\lambda & b \\ c & d-\lambda \end{pmatrix} \times \begin{pmatrix} x_1 \\ x_2 \end{pmatrix} = \begin{pmatrix} 0 \\ 0 \end{pmatrix}$$

Find the λ values such that $\det\begin{pmatrix} a-\lambda & b \\ c & d-\lambda \end{pmatrix} = 0$. That will guarantee that there are nonzero x-vector solutions $\begin{pmatrix} x_1 \\ x_2 \end{pmatrix}$ that will solve the original equation.

4. Substitute the given coefficient matrix and the eigenvalues (λ_1 and λ_2) obtained in Step 3 into the equation $\mathbf{Mat}\vec{\varepsilon} - \lambda\vec{\varepsilon} = \begin{pmatrix} 0 \\ 0 \end{pmatrix}$.

$$\begin{pmatrix} a-\lambda & b \\ c & d-\lambda \end{pmatrix} \times \begin{pmatrix} x_1 \\ x_2 \end{pmatrix} = \begin{pmatrix} 0 \\ 0 \end{pmatrix}$$

Set up an algebraic system of equations and solve for x_1 and x_2 using techniques reviewed in Chapter 17. There will be infinite solutions for x_1 and x_2 or $\vec{\varepsilon}$.

5. Depending on the values for λ, substitue into one of the equations below.

1. Real and distinct eigenvalues (Appendices F and G):
   $$\vec{x} = c_1\vec{\varepsilon}_1 e^{\lambda_1 t} + c_2\vec{\varepsilon}_2 e^{\lambda_2 t}.$$
2. Repeated eigenvalues:
   $$\vec{x} = c_1\vec{\varepsilon}e^{\lambda t} + c_2\left(\vec{\varepsilon}te^{\lambda t} + \vec{\rho}e^{\lambda t}\right).$$
3. Imaginary eigenvalues: λ_1 and λ_2 will be complex eigenvalues. Let $\lambda_1 = r + ci$ and $\lambda_2 = r - ci$. Let $\vec{\varepsilon}_1 = \alpha + \beta i$ and $\vec{\varepsilon}_1 = \alpha - \beta i$.
   $$\vec{x} = c_1\vec{\varepsilon}_1 e^{\lambda_1 t} + c_2\vec{\varepsilon}_2 e^{\lambda_2 t}.$$

6. If you have IVP information, use it to solve for the constants of integration.

Chapter 21

Phase-Plane Portraits for Two-by-Two Systems of Linear Homogeneous Differential Equations

The system of algebraic equations

$$x + 2y = 5 \quad (1)$$
$$x + 3y = 6 \quad (2)$$

has nice integer solutions. Hence, it can be solved using a graphical technique by graphing both equations on an x–y-axis:

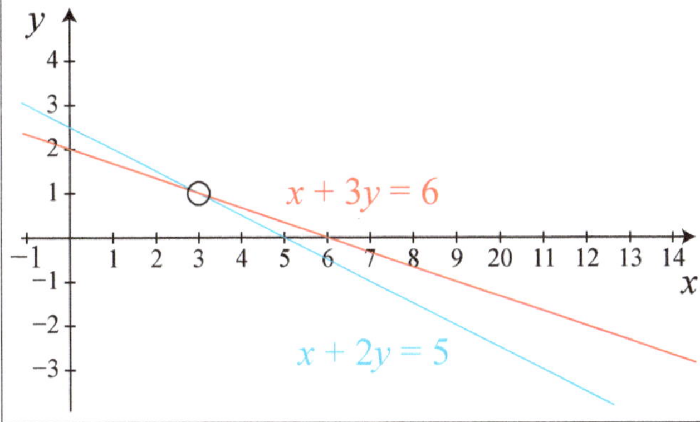

However, the following system of differential equations,

$$\frac{\mathrm{d}}{\mathrm{d}t}x_1 = x_1' = 3x_1 + 2x_2 \quad (1)$$
$$\frac{\mathrm{d}}{\mathrm{d}t}x_2 = x_2' = 4x_1 + x_2, \quad (2)$$

cannot be graphed as the relation of a dependent variable and an independent variable. That is because it is formed from two differential equations with two dependent variables (x_1 and x_2) and one independent variable (t). There are three variables involved here, x_1, x_2, and t, which would require a three-dimensional graph. What is often done is to "ignore" the independent variable, t, and think of the solutions as trajectories—paths obtained by parametric equations—of the two dependent variables (x_1 and x_2). Then the equilibrium (or critical point) solution would be the origin, and the x_1, x_2-plane would be the phase plane.

Before showing how to represent the system in a phase plane, we'll do a quick review of the concept of *scaling a vector*. Basically, scaling a vector stretches or contracts the vector according to a value, k ($k \neq 0$), by which the vector is multiplied: $k\begin{pmatrix}x_1\\x_2\end{pmatrix} = \begin{pmatrix}kx_1\\kx_2\end{pmatrix}$. If $k > 1$, then the magnitude of the vector will be increased. If $k < 1$, the magnitude of the vector will decrease.

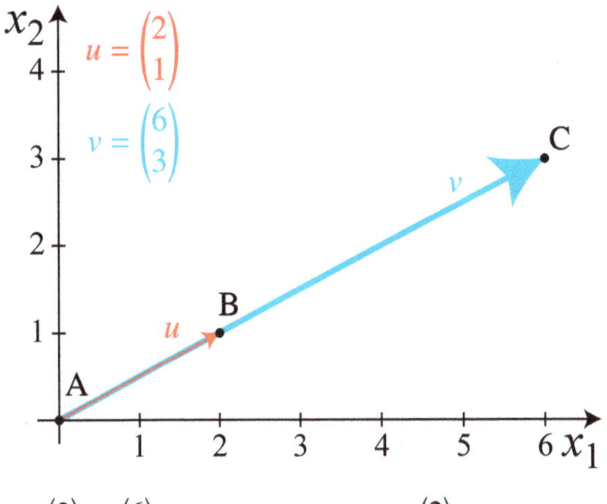

$3\begin{pmatrix}2\\1\end{pmatrix} = \begin{pmatrix}6\\3\end{pmatrix}$, $k = 3 > 1$, vector $\vec{u}$, $\begin{pmatrix}2\\1\end{pmatrix}$, stretches.

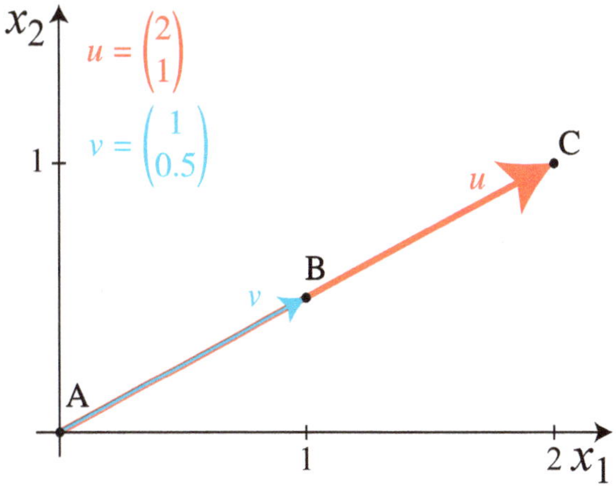

$\frac{1}{2}\begin{pmatrix}2\\1\end{pmatrix} = \begin{pmatrix}1\\\frac{1}{2}\end{pmatrix}$, $k = \frac{1}{2} < 1$, vector $\vec{u}$, $\begin{pmatrix}2\\1\end{pmatrix}$, contracts.

$$\frac{d}{dt}x_1 = x_1' = 3x_1 + 2x_2$$
$$\frac{d}{dt}x_2 = x_2' = 4x_1 + x_2$$

So, let's get back to that phase plane.

$$\vec{x}' = \frac{d}{dt}(\vec{x}) = \begin{pmatrix} 3x_1 + 2x_2 \\ 4x_1 + x_2 \end{pmatrix}$$

$$\vec{x} = \begin{pmatrix} x_1 \\ x_2 \end{pmatrix} = c_1\vec{\varepsilon}_1 e^{\lambda_1 t} + c_1\vec{\varepsilon}_2 e^{\lambda_2 t}$$

where λ_1 and λ_2 are eigenvalues, and $\vec{\varepsilon}_1$ and $\vec{\varepsilon}_2$ and eigenvectors respectively of the system $\begin{pmatrix} 3 & 2 \\ 4 & 1 \end{pmatrix}$ (see Chapter 20 and Appendices F and G)

From Chapters 19 and Chapter 20, an eigenvector of $\begin{pmatrix} 3 & 2 \\ 4 & 1 \end{pmatrix}$ is $\begin{pmatrix} 1 \\ -2 \end{pmatrix}$ when the eigenvalue is -1. An eigenvector of $\begin{pmatrix} 3 & 2 \\ 4 & 1 \end{pmatrix}$ is $\begin{pmatrix} 1 \\ 1 \end{pmatrix}$ when the eigenvalue is 5.

$$\vec{x} = c_1\vec{\varepsilon}_1 e^{\lambda_1 t} + c_2\vec{\varepsilon}_2 e^{\lambda_2 t} \quad \text{(See Appendices F and G.)}$$

$$= c_1\begin{pmatrix} 1 \\ -2 \end{pmatrix}e^{-1t} + c_2\begin{pmatrix} 1 \\ 1 \end{pmatrix}e^{5t} \qquad (3)$$

There are many solutions to the system of differential equations, but five of them are special. Those special solutions arise when c_1 or c_2 (or both) equal zero. The result is that one or both terms is eliminated from the original binomial.

One of the five is called a *steady-state* solution. It is the point $(0, 0)$. The other four special solutions are special because their *contour* will result in an open interval on a *straight line.*

Steady-state special-case solution #1: c_1 and c_2 are both zero.

$$\vec{x} = 0\begin{pmatrix} 1 \\ -2 \end{pmatrix}e^{-1t} + 0\begin{pmatrix} 1 \\ 1 \end{pmatrix}e^{5t} = \begin{pmatrix} 0 \\ 0 \end{pmatrix} + \begin{pmatrix} 0 \\ 0 \end{pmatrix} = \begin{pmatrix} 0 \\ 0 \end{pmatrix}$$

The point $\begin{pmatrix} 0 \\ 0 \end{pmatrix}$ is known as a *critical point* or *equilibrium solution.* When the right side of the equation is $\begin{pmatrix} 0 \\ 0 \end{pmatrix}$, the critical point is always $\begin{pmatrix} 0 \\ 0 \end{pmatrix}$. Imagine standing with your eyes looking down the t-axis, below middle. You would see the x_1–x_2 graph at right.

t	x_1	x_2
−2	0	0
−1	0	0
0	0	0
1	0	0
2	0	0

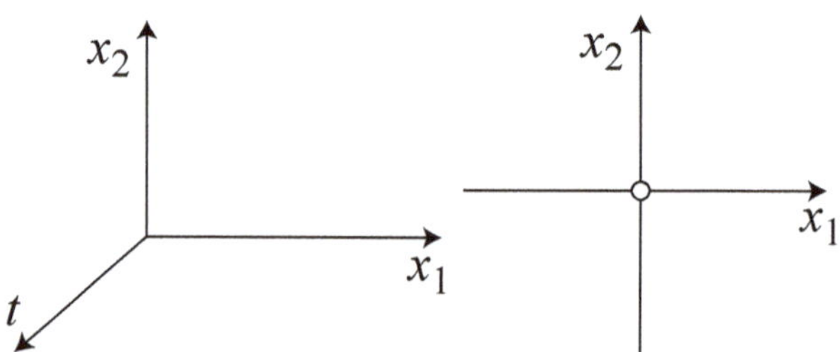

The second and third special-case solutions of this system (cases where one, but not both, of the system components is constant) can be found by letting $c_1 = \pm 1$, $c_2 = 0$.

Substituting $c_1 = +1$ and $c_2 = 0$, into (3) above. #2

$$\vec{x} = c_1\begin{pmatrix} 1 \\ -2 \end{pmatrix}e^{-1t} + c_2\begin{pmatrix} 1 \\ 1 \end{pmatrix}e^{5t}$$

$$= \begin{pmatrix} 1 \\ -2 \end{pmatrix}e^{-t} + 0, \quad c_2 \text{ term eliminated}$$

$$= e^{-t}\begin{pmatrix} 1 \\ -2 \end{pmatrix} \quad \text{emphasizing the scaling factor, } e^{-t}$$

$$\begin{pmatrix} x_1 \\ x_2 \end{pmatrix} = \begin{pmatrix} 1 \\ -2 \end{pmatrix} \quad \text{special solution #2 shown for } t = 0$$

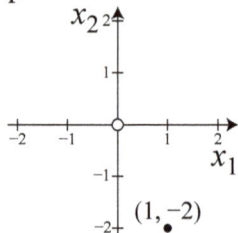

Substituting $c_1 = -1$ and $c_2 = 0$ into (3) above. #3

$$\vec{x} = c_1\begin{pmatrix} 1 \\ -2 \end{pmatrix}e^{-1t} + c_2\begin{pmatrix} 1 \\ 1 \end{pmatrix}e^{5t}$$

$$= -1\begin{pmatrix} 1 \\ -2 \end{pmatrix}e^{-t} + 0, \quad c_2 \text{ term eliminated}$$

$$= e^{-t}\begin{pmatrix} -1 \\ 2 \end{pmatrix} \quad \text{emphasizing the scaling factor, } e^{-t}$$

$$\begin{pmatrix} x_1 \\ x_2 \end{pmatrix} = \begin{pmatrix} -1 \\ 2 \end{pmatrix} \quad \text{special solution #3 shown for } t = 0$$

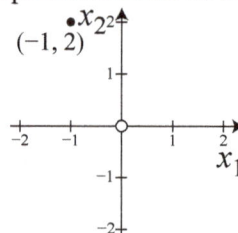

As reviewed on page 203 of this chapter, multiplying a vector by a scaling factor will either expand (if $k > 1$) or shrink (if $0 < k < 1$) the vector.

Analyzing the special-case behavior as $t \to \infty$ for $c_1 = 1$ and $c_2 = 0$.

$$\vec{x} = 1\begin{pmatrix}1\\-2\end{pmatrix}e^{-1t}$$

Rewriting as $\vec{x} = \frac{1}{e^t}\begin{pmatrix}1\\-2\end{pmatrix}$ to emphasize that $\frac{1}{e^t}$ is a scaling factor, we see that, as $t \to \infty$, the vector $\left[\frac{1}{e^t}\begin{pmatrix}1\\-2\end{pmatrix}\right]$ contracts. The amplitude (distance from the origin) decreases; the vector is approaching, but will never reach, (0, 0).

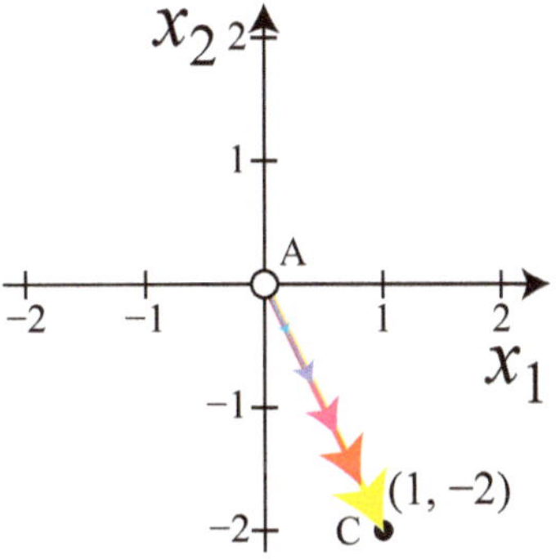

Here, the vector would be shrinking as t gets larger.

Analyzing the special-case behavior as $t \to \infty$ for $c_1 = -1$ and $c_2 = 0$.

$$\vec{x} = -1\begin{pmatrix}1\\-2\end{pmatrix}e^{-1t}$$

Rewriting as $\vec{x} = \frac{1}{e^t}\begin{pmatrix}-1\\2\end{pmatrix}$ to make the scaling factor positive and emphasize that $\frac{1}{e^t}$ is a scaling factor, we see that, as $t \to \infty$, the vector $\left[\frac{1}{e^t}\begin{pmatrix}-1\\2\end{pmatrix}\right]$ contracts. The amplitude (distance from the origin) decreases; the vector is approaching, but will never reach, (0, 0).

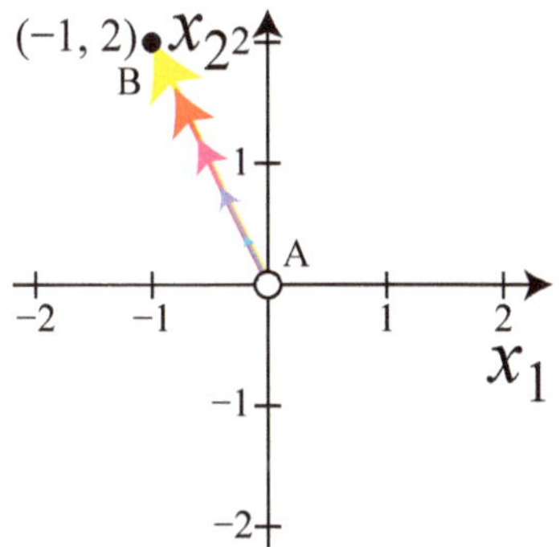

Again, the vector would be shrinking as t gets larger.

Let's use a table to break this down.

t	$e^{-1t}\begin{pmatrix}1\\-2\end{pmatrix}$	$-e^{-1t}\begin{pmatrix}1\\-2\end{pmatrix}$
−2	$e^{-(-2)}\begin{pmatrix}1\\-2\end{pmatrix} = e^2\begin{pmatrix}1\\-2\end{pmatrix} = 7.389\begin{pmatrix}1\\-2\end{pmatrix} = \begin{pmatrix}7.389\\-14.778\end{pmatrix}$	$-e^{-(-2)}\begin{pmatrix}1\\-2\end{pmatrix} = -e^2\begin{pmatrix}1\\-2\end{pmatrix} = -7.389\begin{pmatrix}1\\-2\end{pmatrix} = \begin{pmatrix}-7.389\\14.778\end{pmatrix}$
−1	$e^{-(-1)}\begin{pmatrix}1\\-2\end{pmatrix} = e^1\begin{pmatrix}1\\-2\end{pmatrix} = 2.718\begin{pmatrix}1\\-2\end{pmatrix} = \begin{pmatrix}2.718\\-5.436\end{pmatrix}$ pt. 1	$-e^{-(-1)}\begin{pmatrix}1\\-2\end{pmatrix} = -e^1\begin{pmatrix}1\\-2\end{pmatrix} = -2.718\begin{pmatrix}1\\-2\end{pmatrix} = \begin{pmatrix}-2.718\\5.436\end{pmatrix}$ pt. 1
0	$e^{-0}\begin{pmatrix}1\\-2\end{pmatrix} = 1\begin{pmatrix}1\\-2\end{pmatrix} = \begin{pmatrix}1\\-2\end{pmatrix}$ pt. 2	$-e^{-0}\begin{pmatrix}1\\-2\end{pmatrix} = 1\begin{pmatrix}-1\\2\end{pmatrix} = \begin{pmatrix}-1\\2\end{pmatrix}$ pt. 2
1	$e^{-1}\begin{pmatrix}1\\-2\end{pmatrix} = 0.368\begin{pmatrix}1\\-2\end{pmatrix} = \begin{pmatrix}0.368\\-0.736\end{pmatrix}$ pt. 3	$-e^{-1}\begin{pmatrix}1\\-2\end{pmatrix} = -0.368\begin{pmatrix}1\\-2\end{pmatrix} = \begin{pmatrix}-0.368\\0.736\end{pmatrix}$ pt. 3
2	$e^{-2}\begin{pmatrix}1\\-2\end{pmatrix} = 0.135\begin{pmatrix}1\\-2\end{pmatrix} = \begin{pmatrix}0.135\\-0.270\end{pmatrix}$	$-e^{-2}\begin{pmatrix}1\\-2\end{pmatrix} = -0.135335\begin{pmatrix}1\\-2\end{pmatrix} = \begin{pmatrix}-0.135\\0.270\end{pmatrix}$
10	$e^{-10}\begin{pmatrix}1\\-2\end{pmatrix} = 0.0000454\begin{pmatrix}1\\-2\end{pmatrix} = \begin{pmatrix}0.0000454\\-0.0000908\end{pmatrix} \to (0, 0)$	$-e^{-10}\begin{pmatrix}1\\-2\end{pmatrix} = 0.0000454\begin{pmatrix}1\\-2\end{pmatrix} = \begin{pmatrix}-0.0000454\\0.0000908\end{pmatrix} \to (0, 0)$

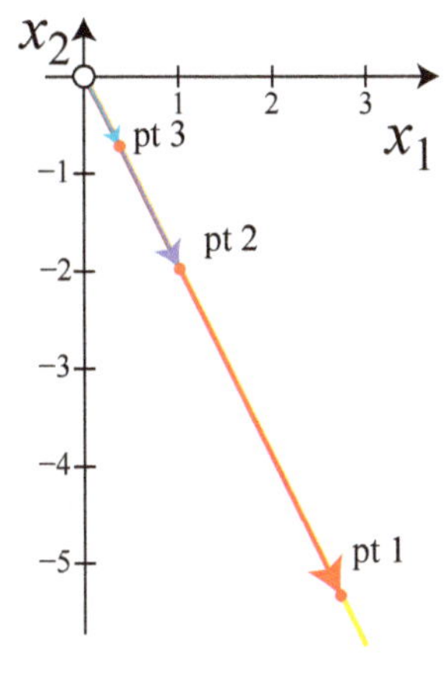

In both cases, the vector is shrinking as $t \to \infty$.

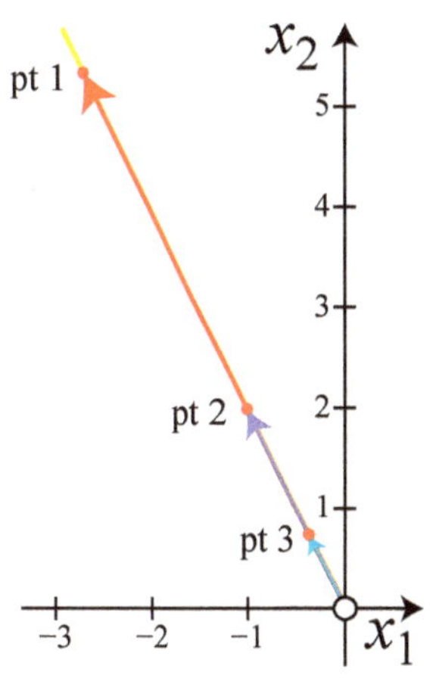

The "lines" being formed here are not the traditional lines as taught in geometry. In geometry a line grows infinitely and unboundedly in two opposite directions. The *line* being discussed here does not extend infinitely in both directions; it is bounded by a point on one side. The line expands and grows forever in both directions but never reaches the point $\begin{pmatrix}0\\0\end{pmatrix}$. To distinguish these infinitely extending-but-bounded-on-one-side lines from traditional Euclidean lines and since they are identified by eigenvectors, they are called *eigenlines* in this book.

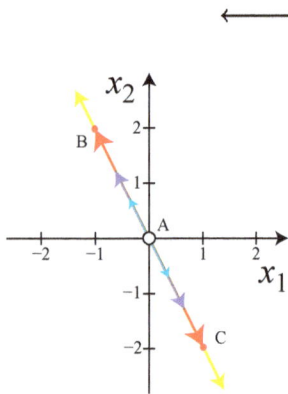

Combining the two graphs above, we get the graph at left. It is worth noting that the two eigenlines shown here have the same slope, do not intersect, and are coplanar. Therefore, *by definition* they are parallel. The point $(0, 0)$ here is called an equilibrium point which is defined in some books as the *intersection of eigenlines*. As discussed above, this is misleading as the eigenlines do not actually intersect.

The two final special-case solutions of this system, #4 and #5 (cases where one, but not both, of the system components is constant), can be found by letting $c_1 = 0$, $c_2 = \pm 1$.

#4 Substituting $c_1 = 0$ and $c_2 = +1$ into

$$\vec{x} = c_1 \begin{pmatrix}1\\-2\end{pmatrix} e^{-1t} + c_2 \begin{pmatrix}1\\1\end{pmatrix} e^{5t}$$

$$\vec{x} = 0 + 1 \begin{pmatrix}1\\1\end{pmatrix} e^{5t} \quad c_1 \text{ term is eliminated}$$

$$\vec{x} = 1 \begin{pmatrix}1\\1\end{pmatrix} e^{5t}$$

$$\vec{x} = e^{5t} \begin{pmatrix}1\\1\end{pmatrix} \quad \text{emphasizing the scaling factor, } e^{5t}$$

$$\begin{pmatrix}x_1\\x_2\end{pmatrix} = \begin{pmatrix}1\\1\end{pmatrix} \quad \text{special-case solution \#4 when } t = 0$$

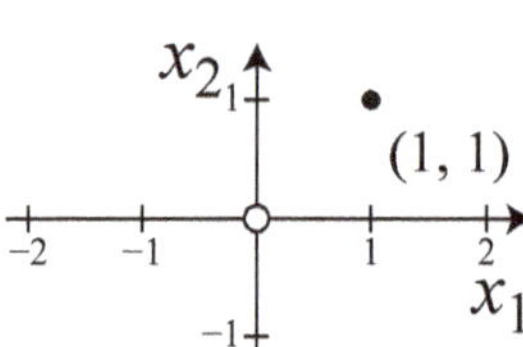

#5 Substituting $c_1 = 0$ and $c_2 = -1$ into

$$\vec{x} = c_1 \begin{pmatrix}1\\-2\end{pmatrix} e^{-1t} + c_2 \begin{pmatrix}1\\1\end{pmatrix} e^{5t}$$

$$\vec{x} = 0 - 1 \begin{pmatrix}1\\1\end{pmatrix} e^{5t} \quad c_1 \text{ term is eliminated}$$

$$\vec{x} = 1 \begin{pmatrix}-1\\-1\end{pmatrix} e^{5t} \quad \text{scaling factor must be positive}$$

$$\vec{x} = e^{5t} \begin{pmatrix}-1\\-1\end{pmatrix} \quad \text{emphasizing the scaling factor, } e^{5t}$$

$$\begin{pmatrix}x_1\\x_2\end{pmatrix} = \begin{pmatrix}-1\\-1\end{pmatrix} \quad \text{special-case solution \#5 when } t = 0$$

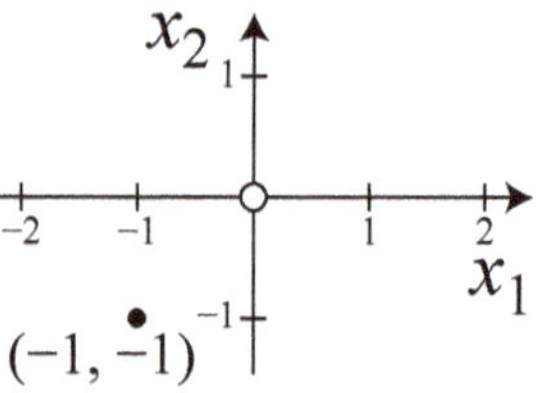

Analyzing the special-case behavior as $t \to \infty$ for $c_1 = 0$ and $c_2 = 1$.

$$\vec{x} = 1 \begin{pmatrix}1\\1\end{pmatrix} e^{5t}$$

Rewriting as $\vec{x} = e^{5t} \begin{pmatrix}1\\1\end{pmatrix}$ to emphasize that e^{5t} is a scaling factor, we see that, as $t \to \infty$, the vector $\left[e^{5t} \begin{pmatrix}1\\1\end{pmatrix}\right]$ expands, getting farther from $(0, 0)$.

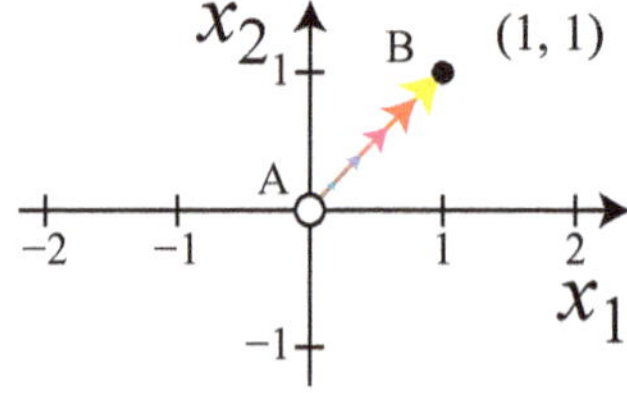

Here, the vector would be expanding as t gets larger.

Analyzing the special-case behavior as $t \to \infty$ for $c_1 = 0$ and $c_2 = -1$.

$$\vec{x} = -1 \begin{pmatrix}1\\1\end{pmatrix} e^{5t}$$

Rewriting as $\vec{x} = e^{5t} \begin{pmatrix}-1\\-1\end{pmatrix}$ to make the scaling factor (e^{5t}) positive, we see that, as $t \to \infty$, the vector $\left[e^{5t} \begin{pmatrix}-1\\-1\end{pmatrix}\right]$ expands, getting farther from $(0, 0)$.

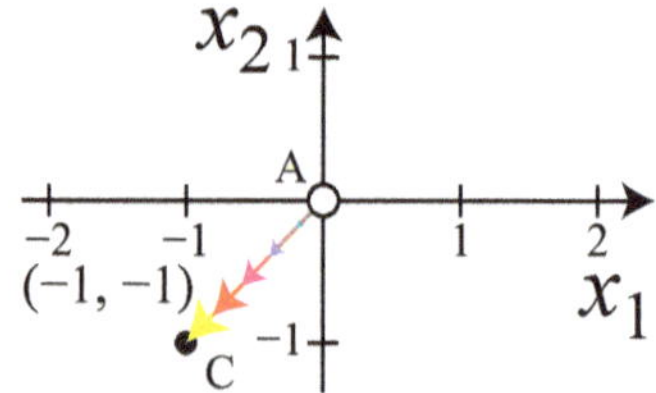

Again, the vector would be expanding as t gets larger.

Again, let's use a table to break this down.

t	$e^{5t}\begin{pmatrix}1\\1\end{pmatrix}$	$-e^{5t}\begin{pmatrix}-1\\-1\end{pmatrix}$
−0.2	$e^{5(-0.2)}\begin{pmatrix}1\\1\end{pmatrix}=e^{-1}\begin{pmatrix}1\\1\end{pmatrix}=0.368\begin{pmatrix}1\\1\end{pmatrix}=\begin{pmatrix}0.368\\0.368\end{pmatrix}$	$e^{5(-0.2)}\begin{pmatrix}-1\\-1\end{pmatrix}=-e^{-1}\begin{pmatrix}-1\\-1\end{pmatrix}=0.368\begin{pmatrix}-1\\-1\end{pmatrix}=\begin{pmatrix}-0.368\\-0.368\end{pmatrix}$
−0.1	$e^{5(-0.1)}\begin{pmatrix}1\\1\end{pmatrix}=e^{-0.5}\begin{pmatrix}1\\1\end{pmatrix}=0.607\begin{pmatrix}1\\1\end{pmatrix}=\begin{pmatrix}0.607\\0.607\end{pmatrix}$ pt. 1	$e^{5(-0.1)}\begin{pmatrix}-1\\-1\end{pmatrix}=e^{-0.5}\begin{pmatrix}-1\\-1\end{pmatrix}=0.607\begin{pmatrix}-1\\-1\end{pmatrix}=\begin{pmatrix}-0.607\\-0.607\end{pmatrix}$ pt. 1
0	$e^{0}\begin{pmatrix}1\\1\end{pmatrix}=1\begin{pmatrix}1\\1\end{pmatrix}=\begin{pmatrix}1\\1\end{pmatrix}$ pt. 2	$e^{0}\begin{pmatrix}-1\\-1\end{pmatrix}=1\begin{pmatrix}-1\\-1\end{pmatrix}=\begin{pmatrix}-1\\-1\end{pmatrix}$ pt. 2
0.1	$e^{5\times0.1}\begin{pmatrix}1\\1\end{pmatrix}=1.649\begin{pmatrix}1\\1\end{pmatrix}=\begin{pmatrix}1.649\\1.649\end{pmatrix}$ pt. 3	$e^{5\times0.1}\begin{pmatrix}-1\\-1\end{pmatrix}=1.649\begin{pmatrix}-1\\-1\end{pmatrix}=\begin{pmatrix}-1.649\\-1.649\end{pmatrix}$ pt. 3
0.2	$e^{5\times0.2}\begin{pmatrix}1\\1\end{pmatrix}=2.718\begin{pmatrix}1\\1\end{pmatrix}=\begin{pmatrix}2.718\\2.718\end{pmatrix}$ pt. 4	$e^{5\times0.2}\begin{pmatrix}-1\\-1\end{pmatrix}=2.718\begin{pmatrix}-1\\-1\end{pmatrix}=\begin{pmatrix}-2.718\\-2.718\end{pmatrix}$ pt. 4
1	$e^{5\times1}\begin{pmatrix}1\\1\end{pmatrix}=148\begin{pmatrix}1\\1\end{pmatrix}=\begin{pmatrix}148\\148\end{pmatrix}\to(\infty,\infty)$	$e^{5\times1}\begin{pmatrix}-1\\-1\end{pmatrix}=148\begin{pmatrix}-1\\-1\end{pmatrix}=\begin{pmatrix}-148\\-148\end{pmatrix}\to(-\infty,-\infty)$

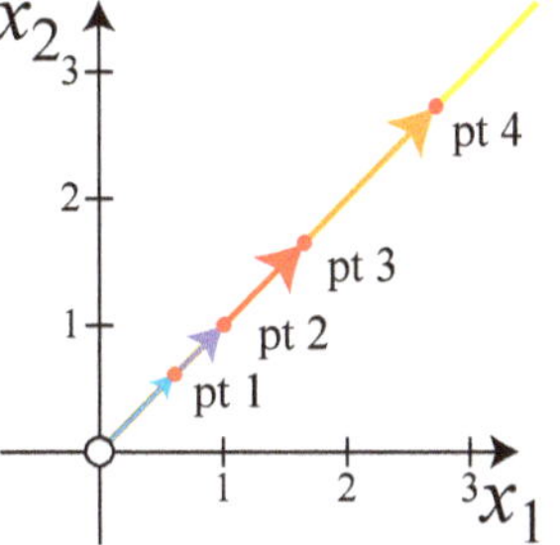

In both cases, the vector is growing as $t \to \infty$.

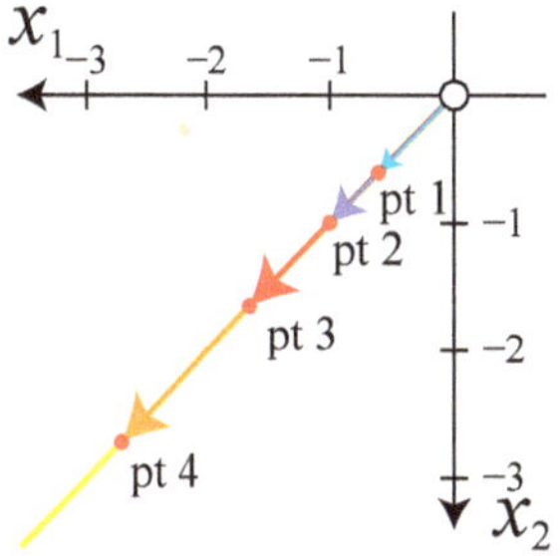

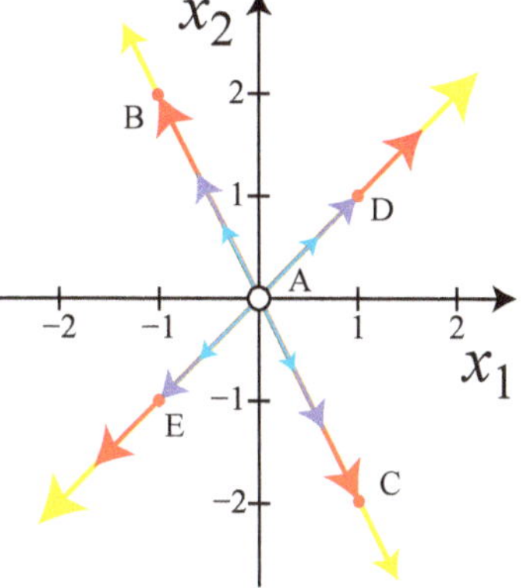

Collecting these graphs, note that the two eigenlines in each pair has the same slope as its mate, they do not intersect, and they are coplanar. Therefore, by definition, we have two pairs of parallel lines. The point (0, 0) here is called an equilibrium point which is defined in some books as the intersection of special case lines. However, as discussed above, this is misleading as those lines do not actually intersect.

A vector function, $\vec{x}(t)$, describes a motion in the x–y plane. It is helpful to think of a vector function as a contour line on a map. The tangent/velocity vector at each point (on the contour line) can be calculated directly from the given system of equations used to compute the tangent/velocity vector using the position vector $\vec{x}(t) = \begin{pmatrix}x(t)\\y(t)\end{pmatrix}$ to compute $\vec{x}\,'(t) = A\vec{x}(t)$. The parametric equations in the vector function tell how the point (x, y) moves in the x–y plane as the time varies. The moving point traces out a curve called the trajectory of the solution. The trajectory is determined by the solution to the given system of equations used to compute the tangent/velocity vector. The tangent vector at each point can be calculated directly from the given vector equation, $\vec{x}(t)$. The x–y plane itself is called the phase plane for the system of differential equations, $\begin{pmatrix}x' = ax + by\\y' = cx + dy\end{pmatrix}$. Now, if both constants in the solution are nonzero, $c_1 \neq 0$ and also $c_2 \neq 0$, we will have a linear combination of the two function behaviors: $c_1\begin{pmatrix}1\\-2\end{pmatrix}e^{-1t} + c_2\begin{pmatrix}1\\1\end{pmatrix}e^{5t}$. As $t \to -\infty$, the solution will be dominated by the portion that has the negative eigenvalue, e^{-t}, because the exponent will be large while the portion with the positive eigenvalue, e^{5t} (as $t \to -\infty$), will decay away. Trajectories as $t \to \infty$ will be "parallel" to the vector $\begin{pmatrix}1\\-2\end{pmatrix}$ and moving in the same direction as that special-case solution vector. Conversely, solutions with large positive t, e^{5t} as $t \to \infty$, will dominate the linear combination as the portion with the negative eigenvalue, e^{-t}, will decay away. Trajectories in this case will be parallel to vector (1, 1) as $t \to \infty$ and moving in the direction of that special-case solution vector. Put more simply, the trajectory (contour line) will start "near" the $c_1\begin{pmatrix}1\\-2\end{pmatrix}$ eigenline, move in toward the origin, and then get closer and closer to the $c_2\begin{pmatrix}1\\1\end{pmatrix}$ eigenline.

Linear Combination Solution

$$c_1\begin{pmatrix}1\\-2\end{pmatrix}e^{-t}+c_2\begin{pmatrix}1\\1\end{pmatrix}e^{5t}\quad\text{for } c_1=\frac{1}{5}\text{ and } c_2=\frac{1}{3}$$

(Author's note: $c_1=\frac{1}{5}$ and $c_2=\frac{1}{3}$ are chosen arbitrarily as IVP examples.)

t	$\frac{1}{5}\begin{pmatrix}1\\-2\end{pmatrix}e^{-t}$	$\frac{1}{3}\begin{pmatrix}1\\1\end{pmatrix}e^{5t}$	$\frac{1}{5}\begin{pmatrix}1\\-2\end{pmatrix}e^{-t}+\frac{1}{3}\begin{pmatrix}1\\1\end{pmatrix}e^{5t}$
−2	$\frac{1}{5}\begin{pmatrix}1\\-2\end{pmatrix}e^{-(-2)}$ $\frac{1}{5}\begin{pmatrix}1\\-2\end{pmatrix}\times 7.389$ $1.478\begin{pmatrix}1\\-2\end{pmatrix}$ $\begin{pmatrix}1.478\\-2.956\end{pmatrix}$	$\frac{1}{3}\begin{pmatrix}1\\1\end{pmatrix}e^{5(-2)}$ $\frac{1}{3}\begin{pmatrix}1\\1\end{pmatrix}\times 0.000$ $0.000\begin{pmatrix}1\\1\end{pmatrix}$ $\begin{pmatrix}0.000\\0.000\end{pmatrix}$	$\begin{pmatrix}1.478\\-2.956\end{pmatrix}$ #1
−1.6	$\frac{1}{5}\begin{pmatrix}1\\-2\end{pmatrix}e^{-(-1.6)}$ $\frac{1}{5}\begin{pmatrix}1\\-2\end{pmatrix}\times 4.953$ $0.991\begin{pmatrix}1\\-2\end{pmatrix}$ $\begin{pmatrix}0.991\\-1.981\end{pmatrix}$	$\frac{1}{3}\begin{pmatrix}1\\1\end{pmatrix}e^{5(-1.6)}$ $\frac{1}{3}\begin{pmatrix}1\\1\end{pmatrix}\times 0.000$ $0.000\begin{pmatrix}1\\1\end{pmatrix}$ $\begin{pmatrix}0.000\\0.000\end{pmatrix}$	$\begin{pmatrix}0.991\\-1.981\end{pmatrix}$ #2
−1	$\frac{1}{5}\begin{pmatrix}1\\-2\end{pmatrix}e^{-(-1)}$ $\frac{1}{5}\begin{pmatrix}1\\-2\end{pmatrix}\times 2.718$ $0.544\begin{pmatrix}1\\-2\end{pmatrix}$ $\begin{pmatrix}0.544\\-1.088\end{pmatrix}$	$\frac{1}{3}\begin{pmatrix}1\\1\end{pmatrix}e^{5(-1)}$ $\frac{1}{3}\begin{pmatrix}1\\1\end{pmatrix}\times 0.007$ $0.002\begin{pmatrix}1\\1\end{pmatrix}$ $\begin{pmatrix}0.002\\0.002\end{pmatrix}$	$\begin{pmatrix}0.546\\-1.086\end{pmatrix}$ #3
0	$\frac{1}{5}\begin{pmatrix}1\\-2\end{pmatrix}e^{-(0)}$ $\frac{1}{5}\begin{pmatrix}1\\-2\end{pmatrix}\times 1.000$ $0.200\begin{pmatrix}1\\-2\end{pmatrix}$ $\begin{pmatrix}0.200\\-0.400\end{pmatrix}$	$\frac{1}{3}\begin{pmatrix}1\\1\end{pmatrix}e^{5(0)}$ $\frac{1}{3}\begin{pmatrix}1\\1\end{pmatrix}\times 1.000$ $0.333\begin{pmatrix}1\\1\end{pmatrix}$ $\begin{pmatrix}0.333\\0.333\end{pmatrix}$	$\begin{pmatrix}0.533\\-0.067\end{pmatrix}$ #4
0.2	$\frac{1}{5}\begin{pmatrix}1\\-2\end{pmatrix}e^{-0.2}$ $\frac{1}{5}\begin{pmatrix}1\\-2\end{pmatrix}\times 0.819$ $0.164\begin{pmatrix}1\\-2\end{pmatrix}$ $\begin{pmatrix}0.164\\-0.328\end{pmatrix}$	$\frac{1}{3}\begin{pmatrix}1\\1\end{pmatrix}e^{5(0.2)}$ $\frac{1}{3}\begin{pmatrix}1\\1\end{pmatrix}\times 2.718$ $0.906\begin{pmatrix}1\\1\end{pmatrix}$ $\begin{pmatrix}0.906\\0.906\end{pmatrix}$	$\begin{pmatrix}1.070\\0.578\end{pmatrix}$ #5
0.3	$\frac{1}{5}\begin{pmatrix}1\\-2\end{pmatrix}e^{-0.3}$ $\frac{1}{5}\begin{pmatrix}1\\-2\end{pmatrix}\times 0.148$ $0.148\begin{pmatrix}1\\-2\end{pmatrix}$ $\begin{pmatrix}0.148\\-0.296\end{pmatrix}$	$\frac{1}{3}\begin{pmatrix}1\\1\end{pmatrix}e^{5(0.3)}$ $\frac{1}{3}\begin{pmatrix}1\\1\end{pmatrix}\times 4.481$ $1.494\begin{pmatrix}1\\1\end{pmatrix}$ $\begin{pmatrix}1.494\\1.494\end{pmatrix}$	$\begin{pmatrix}1.642\\1.198\end{pmatrix}$ #6
0.4	$\frac{1}{5}\begin{pmatrix}1\\-2\end{pmatrix}e^{-0.4}$ $\frac{1}{5}\begin{pmatrix}1\\-2\end{pmatrix}\times 0.670$ $0.134\begin{pmatrix}1\\-2\end{pmatrix}$ $\begin{pmatrix}0.134\\-0.268\end{pmatrix}$	$\frac{1}{3}\begin{pmatrix}1\\1\end{pmatrix}e^{5(0.4)}$ $\frac{1}{3}\begin{pmatrix}1\\1\end{pmatrix}\times 7.389$ $2.463\begin{pmatrix}1\\1\end{pmatrix}$ $\begin{pmatrix}2.463\\2.463\end{pmatrix}$	$\begin{pmatrix}2.597\\2.195\end{pmatrix}$ #7

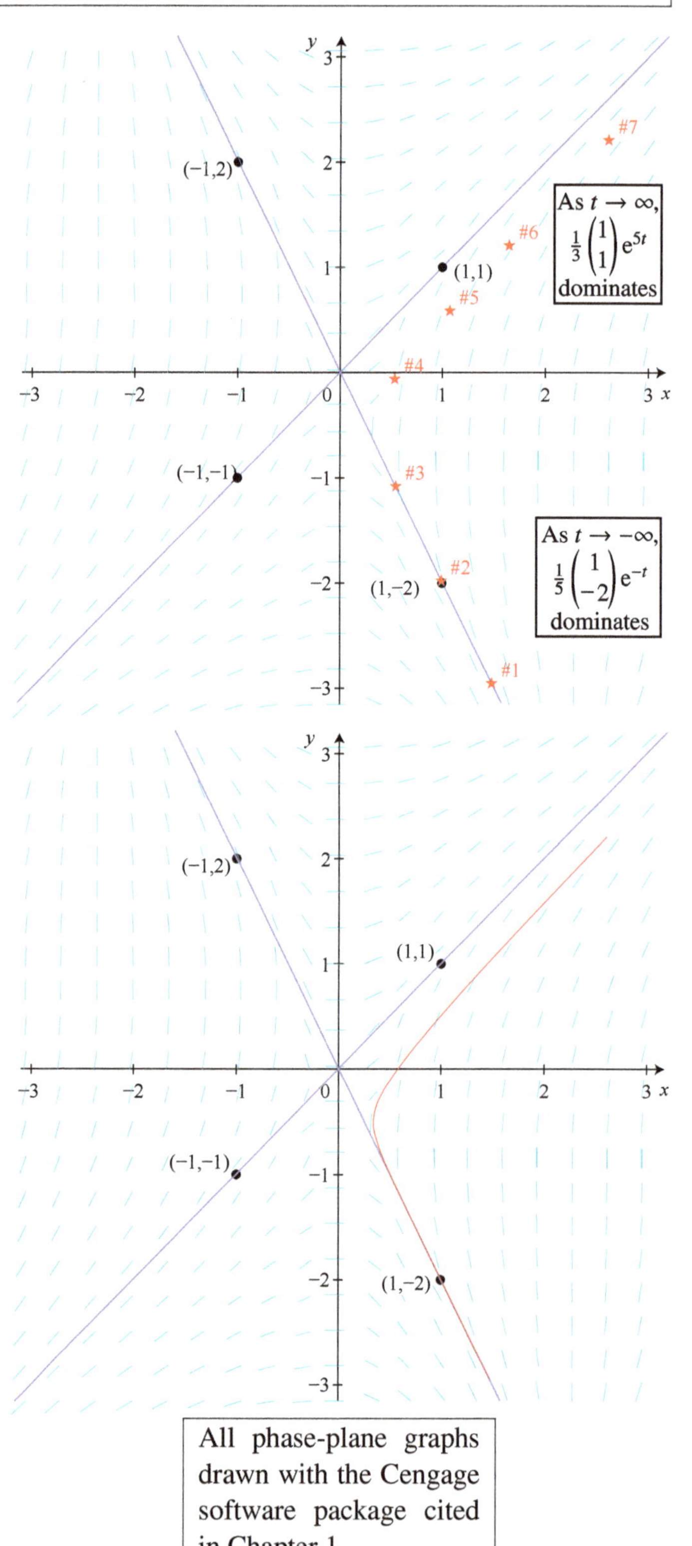

All phase-plane graphs drawn with the Cengage software package cited in Chapter 1.

Here is a Java program to calculate the vector values for $t = -2.0 \ldots 0.4$ with a step of 0.1.

```
import java.text.DecimalFormat;
public class PhasePlanePortrait
{
  public static void main(String args[])
  {
    System.out.println("Hello World");
    double c1 = 1.0/5.0;
    double c2 = 1.0/3.0;
    DecimalFormat df = new DecimalFormat("0.00");
    System.out.print("t eVector1X eVector1Y eVectorCombineVectorXParts");
    System.out.println("eVector2X eVector2Y eVectorCombineVectorYParts");
    for (double t = -2; t < 0.5; t += 0.1)
    {
      double eVector1X = c1 * (1) * Math.exp(-t);
      double eVector1Y = c1 * (-2) * Math.exp(-t);
      double eVector2X = c2 * (1) * Math.exp(5 * t);
      double eVector2Y = c2 * (1) * Math.exp(5 * t);
      double combineVectorXParts = eVector1X + eVector2X;
      double combineVectorYParts = eVector1Y + eVector2Y;
      System.out.println(df.format(t) + " " + df.format(eVector1X) + " " +
      df.format(eVector1Y) + " " +
      df.format(combineVectorXParts) + " " +
      df.format(eVector2X)+ " " +
      df.format(eVector2Y) + " " +
      df.format(combineVectorYParts));
    } // end loop
  } // end main
} // end class
```

$$\text{eVector1} = c_1 \times \begin{pmatrix} 1 \\ -2 \end{pmatrix} e^{-t}$$

$$\text{eVector2} = c_2 \times \begin{pmatrix} 1 \\ 1 \end{pmatrix} e^{5t}$$

Hello World

t	eVector1X	eVector1Y	eVectorCombineVectorXParts	eVector2X	eVector2Y	eVectorCombineVectorYParts
−2.00	1.48	−2.96	1.48	0.00	0.00	−2.96
−1.90	1.34	−2.67	1.34	0.00	0.00	−2.67
−1.80	1.21	−2.42	1.21	0.00	0.00	−2.42
−1.70	1.09	−2.19	1.09	0.00	0.00	−2.19
−1.60	0.99	−1.98	0.99	0.00	0.00	−1.98
−1.50	0.90	−1.79	0.90	0.00	0.00	−1.79
−1.40	0.81	−1.62	0.81	0.00	0.00	−1.62
−1.30	0.73	−1.47	0.73	0.00	0.00	−1.47
−1.20	0.66	−1.33	0.66	0.00	0.00	−1.33
−1.10	0.60	−1.20	0.60	0.00	0.00	−1.20
−1.00	0.54	−1.09	0.55	0.00	0.00	−1.09
−0.90	0.49	−0.98	0.50	0.00	0.00	−0.98
−0.80	0.45	−0.89	0.45	0.01	0.01	−0.88
−0.70	0.40	−0.81	0.41	0.01	0.01	−0.80
−0.60	0.36	−0.73	0.38	0.02	0.02	−0.71
−0.50	0.33	−0.66	0.36	0.03	0.03	−0.63
−0.40	0.30	−0.60	0.34	0.05	0.05	−0.55
−0.30	0.27	−0.54	0.34	0.07	0.07	−0.47
−0.20	0.24	−0.49	0.37	0.12	0.12	−0.37
−0.10	0.22	−0.44	0.42	0.20	0.20	−0.24
0.00	0.20	−0.40	0.53	0.33	0.33	−0.07
0.10	0.18	−0.36	0.73	0.55	0.55	0.19
0.20	0.16	−0.33	1.07	0.91	0.91	0.58
0.30	0.15	−0.30	1.64	1.49	1.49	1.20
0.40	0.13	−0.27	2.60	2.46	2.46	2.19

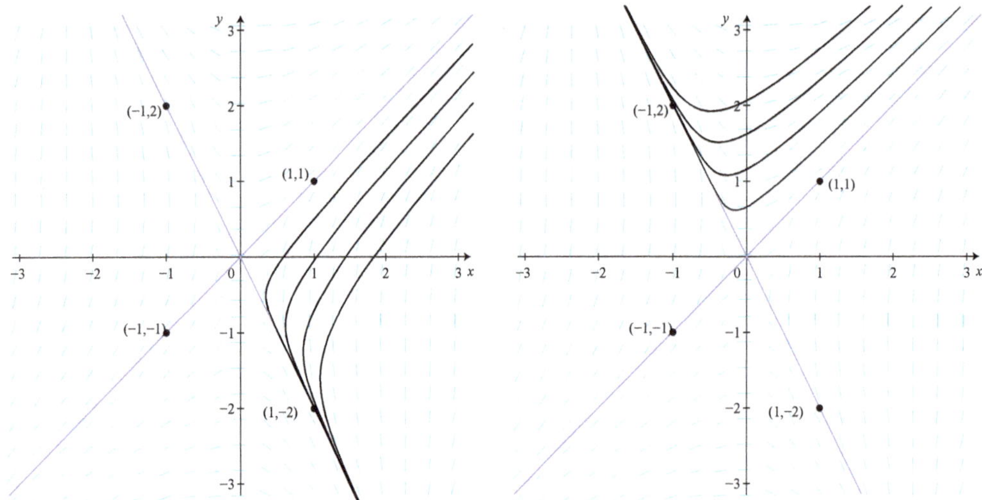

In the graphs of the solutions to the system of ODEs, all solutions move toward the *steady state* solution $(0, 0)$ as t increases from $-\infty$ and then move away from it after t passes zero and moves toward infinity. Such a solution is called a *saddle point* and is said to be *unstable.*

$$\frac{d}{dt}x_1 = x_1' = 3x_1 + 2x_2$$
$$\frac{d}{dt}x_2 = x_2' = 4x_1 + x_2$$

$$\begin{pmatrix} x_1 \\ x_2 \end{pmatrix} = c_1 \begin{pmatrix} 1 \\ -2 \end{pmatrix} e^{-t} + c_2 \begin{pmatrix} 1 \\ 1 \end{pmatrix} e^{5t}$$

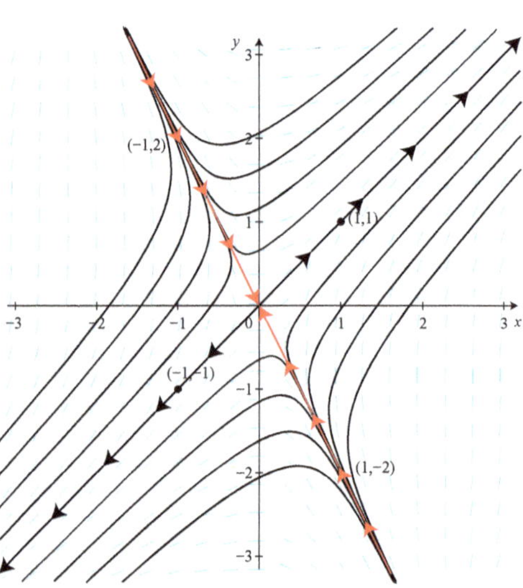

Let's review what we have just done.

1. We used the eigenvector of a system of two-by-two linear differential equations to identify a vector.
2. We used the eigenvalue of the system as a scaling factor to convert the eigenvector to an eigenline, resulting in special-case solutions to the system.
3. We used the product of the eigenvalue and t and our knowledge of exponential growth and exponential decay to sketch in the infinite-solution family for the system.

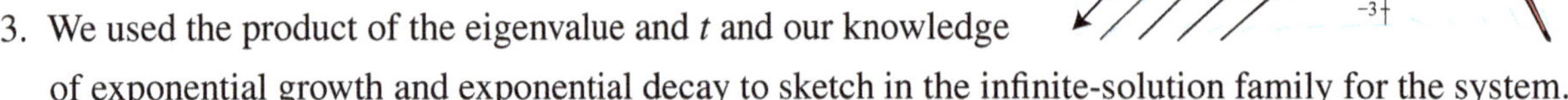

For a two-by-two family of linear homogeneous ODEs, only a finite number of phase plane portraits can result. It is helpful for understanding more advanced systems of equations to look at each of them in turn.

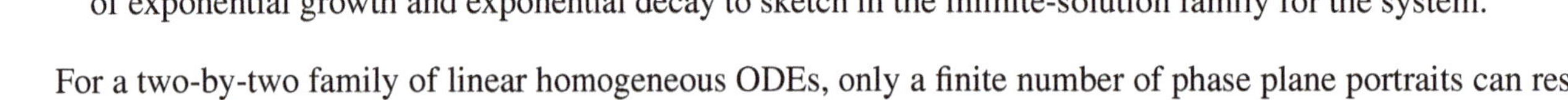

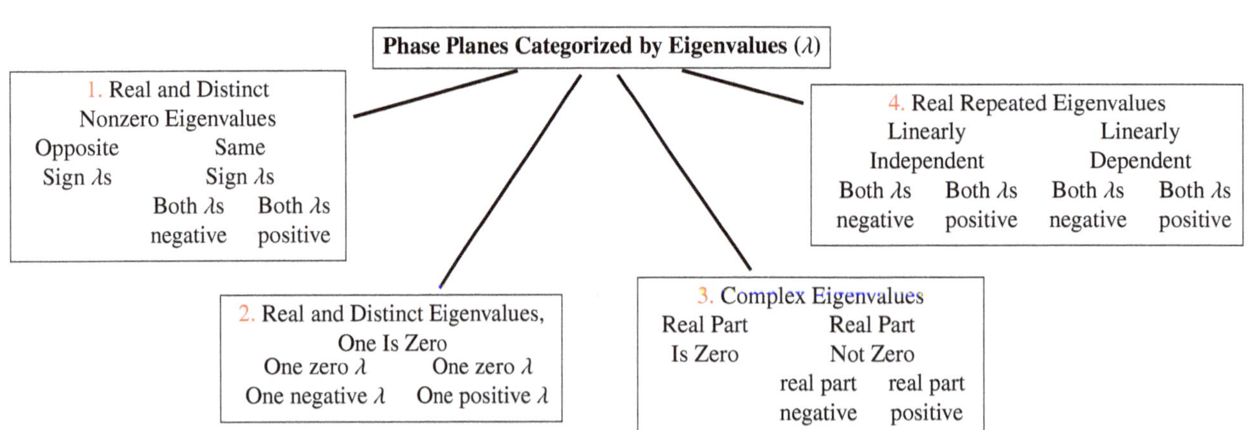

1. Two-by-Two System of Linear Homogeneous Differential Equations with Real and Distinct Nonzero Eigenvalues

	Same Sign λs	
Opposite Sign λs	Both λs Negative	Both λs Positive
$\frac{d}{dt}x_1 = x_1' = 3x_1 + 2x_2$ $\frac{d}{dt}x_2 = x_2' = 4x_1 + x_2$	$\frac{d}{dt}x_1 = x_1' = -5x_1 + x_2$ $\frac{d}{dt}x_2 = x_2' = 4x_1 - 2x_2$	$\frac{d}{dt}x_1 = x_1' = 5x_1 + x_2$ $\frac{d}{dt}x_2 = x_2' = 4x_1 + 2x_2$
$\begin{pmatrix}\frac{d}{dt}x_1\\ \frac{d}{dt}x_2\end{pmatrix} = \begin{pmatrix}3 & 2\\ 4 & 1\end{pmatrix}\begin{pmatrix}x_1\\ x_2\end{pmatrix}$	$\begin{pmatrix}\frac{d}{dt}x_1\\ \frac{d}{dt}x_2\end{pmatrix} = \begin{pmatrix}-5 & 1\\ 4 & -2\end{pmatrix}\begin{pmatrix}x_1\\ x_2\end{pmatrix}$	$\begin{pmatrix}\frac{d}{dt}x_1\\ \frac{d}{dt}x_2\end{pmatrix} = \begin{pmatrix}5 & 1\\ 4 & 2\end{pmatrix}\begin{pmatrix}x_1\\ x_2\end{pmatrix}$
Calculate λ $\det\begin{pmatrix}3-\lambda & 2\\ 4 & 1-\lambda\end{pmatrix} = 0$ $(3-\lambda)(1-\lambda) - (4 \times 2) = 0$ $\lambda^2 - 4\lambda - 5 = 0$ $(\lambda - 5)(\lambda + 1) = 0$ $\lambda = 5, -1$ opposite sign λs	Calculate λ $\det\begin{pmatrix}-5-\lambda & 1\\ 4 & -2-\lambda\end{pmatrix} = 0$ $(-5-\lambda)(-2-\lambda) - (4 \times 1) = 0$ $\lambda^2 + 7\lambda + 6 = 0$ $(\lambda + 6)(\lambda + 1) = 0$ $\lambda = -6, -1$ both λs negative	Calculate λ $\det\begin{pmatrix}5-\lambda & 1\\ 4 & 2-\lambda\end{pmatrix} = 0$ $(5-\lambda)(2-\lambda) - (4 \times 1) = 0$ $\lambda^2 - 7\lambda + 6 = 0$ $(\lambda - 6)(\lambda - 1) = 0$ $\lambda = 6, 1$ both λs positive
Saddle 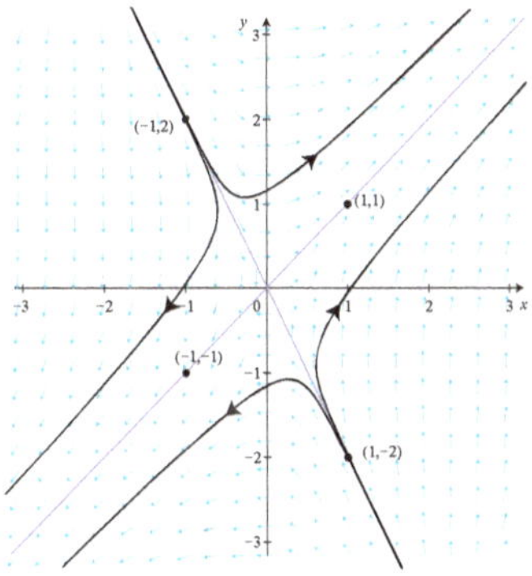	Sink (moving toward equilibrium) 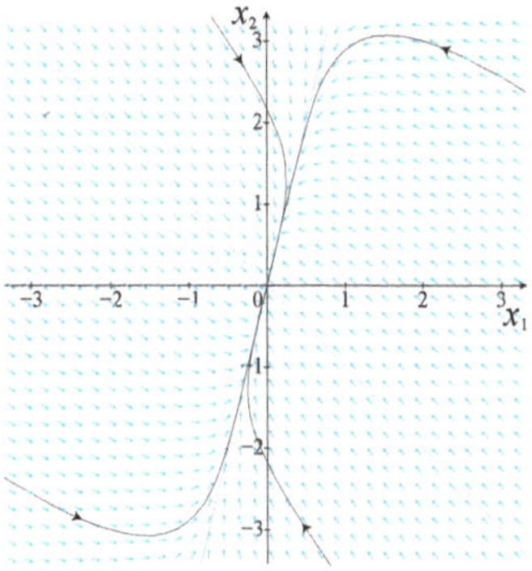	Source (moving away from equilibrium) 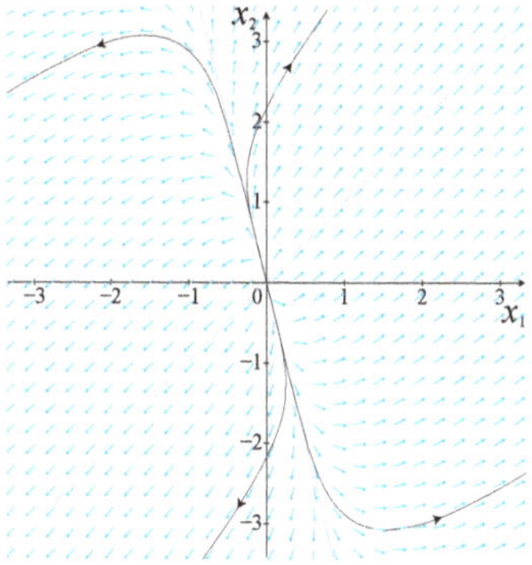

2. Two-by-Two System of Linear Homogeneous Differential Equations with Real and Distinct Eigenvalues, One Zero

One λ is zero, the other is negative	One λ is zero, the other is positive
$\frac{d}{dt}x_1 = x_1' = -2x_1 + 3x_2$ $\frac{d}{dt}x_2 = x_2' = 2x_1 - 3x_2$	$\frac{d}{dt}x_1 = x_1' = 2x_1 + 3x_2$ $\frac{d}{dt}x_2 = x_2' = 2x_1 + 3x_2$
$\begin{pmatrix}\frac{d}{dt}x_1\\ \frac{d}{dt}x_2\end{pmatrix} = \begin{pmatrix}-2 & 3\\ 2 & -3\end{pmatrix}\begin{pmatrix}x_1\\ x_2\end{pmatrix}$	$\begin{pmatrix}\frac{d}{dt}x_1\\ \frac{d}{dt}x_2\end{pmatrix} = \begin{pmatrix}2 & 3\\ 2 & 3\end{pmatrix}\begin{pmatrix}x_1\\ x_2\end{pmatrix}$
Calculate λ $\det\begin{pmatrix}-2-\lambda & 3\\ 2 & -3-\lambda\end{pmatrix} = 0$ $(-2-\lambda)(-3-\lambda) - (2 \times 3) = 0$ $\lambda^2 + 5\lambda = 0$ $\lambda(\lambda + 5) = 0$ $\lambda = 0, -5$	Calculate λ $\det\begin{pmatrix}2-\lambda & 3\\ 2 & 3-\lambda\end{pmatrix} = 0$ $(2-\lambda)(3-\lambda) - (2 \times 3) = 0$ $\lambda^2 - 5\lambda = 0$ $\lambda(\lambda - 5) = 0$ $\lambda = 0, 5$
Negative slope parallel 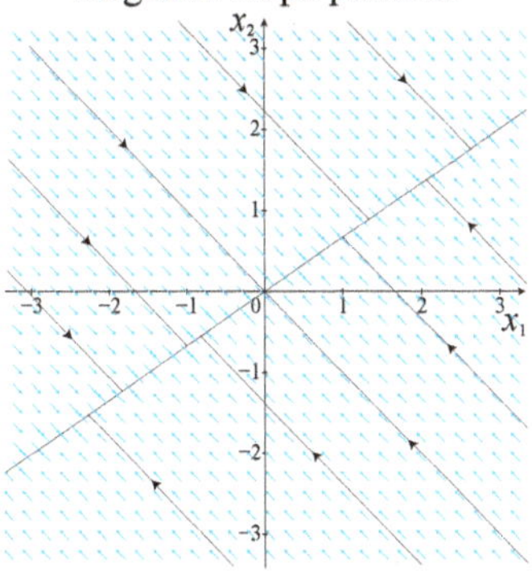	Positive slope parallel 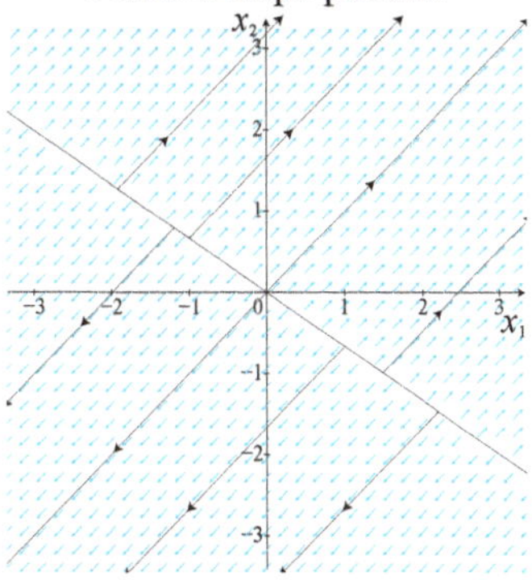

3. Two-by-Two System of Linear Homogeneous Differential Equations with Complex Eigenvalues

	Same Sign λs	
Real Part of λ is zero	Real Part of Both λs Negative	Real Part of Both λs Positive
$\frac{d}{dt}x_1 = x_1' = 3x_1 - 9x_2$ $\frac{d}{dt}x_2 = x_2' = 4x_1 - 3x_2$	$\frac{d}{dt}x_1 = x_1' = -3x_1 - 13x_2$ $\frac{d}{dt}x_2 = x_2' = 5x_1 + x_2$	$\frac{d}{dt}x_1 = x_1' = 3x_1 - 13x_2$ $\frac{d}{dt}x_2 = x_2' = 5x_1 + x_2$
$\begin{pmatrix} \frac{d}{dt}x_1 \\ \frac{d}{dt}x_2 \end{pmatrix} = \begin{pmatrix} 3 & -9 \\ 4 & -3 \end{pmatrix}\begin{pmatrix} x_1 \\ x_2 \end{pmatrix}$	$\begin{pmatrix} \frac{d}{dt}x_1 \\ \frac{d}{dt}x_2 \end{pmatrix} = \begin{pmatrix} -3 & -13 \\ 5 & 1 \end{pmatrix}\begin{pmatrix} x_1 \\ x_2 \end{pmatrix}$	$\begin{pmatrix} \frac{d}{dt}x_1 \\ \frac{d}{dt}x_2 \end{pmatrix} = \begin{pmatrix} 3 & -13 \\ 5 & 1 \end{pmatrix}\begin{pmatrix} x_1 \\ x_2 \end{pmatrix}$
Calculate λ $\det\begin{pmatrix} 3-\lambda & -9 \\ 4 & -3-\lambda \end{pmatrix} = 0$ $(3-\lambda)(-3-\lambda) - (4 \times -9) = 0$ $\lambda^2 + 27 = 0$ $\lambda^2 = -27$ $\lambda = 0 + 3i\sqrt{3}, 0 - 3i\sqrt{3}$	Calculate λ $\det\begin{pmatrix} -3-\lambda & -13 \\ 5 & 1-\lambda \end{pmatrix} = 0$ $(-3-\lambda)(1-\lambda) - (5 \times -13) = 0$ $\lambda^2 + 2\lambda + 62 = 0$ $\lambda = -1 + i\sqrt{61}, -1 - i\sqrt{61}$ real part of both λs negative	Calculate λ $\det\begin{pmatrix} 3-\lambda & -13 \\ 5 & 1-\lambda \end{pmatrix} = 0$ $(3-\lambda)(1-\lambda) - (5 \times -13) = 0$ $\lambda^2 - 4\lambda + 68 = 0$ $\lambda = 2 + 8i, 2 - 8i$ real part of both λs positive
Concentric Ellipses	Spiral Sink (moving toward equilibrium)	Spiral Source (moving away from equilibrium)

4. Two-by-Two System of Linear Homogeneous Differential Equations with Real Repeated Eigenvalues

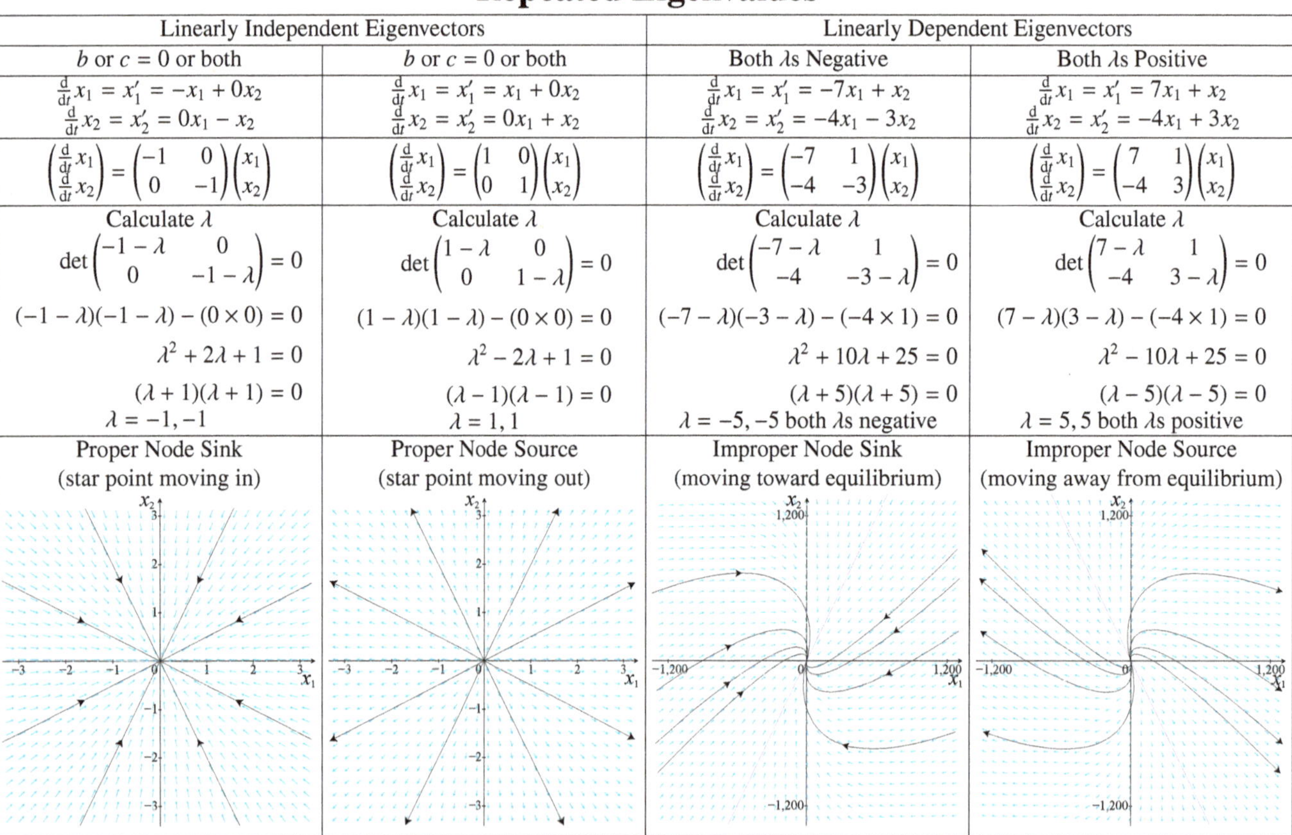

Linearly Independent Eigenvectors		Linearly Dependent Eigenvectors	
b or c = 0 or both	b or c = 0 or both	Both λs Negative	Both λs Positive
$\frac{d}{dt}x_1 = x_1' = -x_1 + 0x_2$ $\frac{d}{dt}x_2 = x_2' = 0x_1 - x_2$	$\frac{d}{dt}x_1 = x_1' = x_1 + 0x_2$ $\frac{d}{dt}x_2 = x_2' = 0x_1 + x_2$	$\frac{d}{dt}x_1 = x_1' = -7x_1 + x_2$ $\frac{d}{dt}x_2 = x_2' = -4x_1 - 3x_2$	$\frac{d}{dt}x_1 = x_1' = 7x_1 + x_2$ $\frac{d}{dt}x_2 = x_2' = -4x_1 + 3x_2$
$\begin{pmatrix} \frac{d}{dt}x_1 \\ \frac{d}{dt}x_2 \end{pmatrix} = \begin{pmatrix} -1 & 0 \\ 0 & -1 \end{pmatrix}\begin{pmatrix} x_1 \\ x_2 \end{pmatrix}$	$\begin{pmatrix} \frac{d}{dt}x_1 \\ \frac{d}{dt}x_2 \end{pmatrix} = \begin{pmatrix} 1 & 0 \\ 0 & 1 \end{pmatrix}\begin{pmatrix} x_1 \\ x_2 \end{pmatrix}$	$\begin{pmatrix} \frac{d}{dt}x_1 \\ \frac{d}{dt}x_2 \end{pmatrix} = \begin{pmatrix} -7 & 1 \\ -4 & -3 \end{pmatrix}\begin{pmatrix} x_1 \\ x_2 \end{pmatrix}$	$\begin{pmatrix} \frac{d}{dt}x_1 \\ \frac{d}{dt}x_2 \end{pmatrix} = \begin{pmatrix} 7 & 1 \\ -4 & 3 \end{pmatrix}\begin{pmatrix} x_1 \\ x_2 \end{pmatrix}$
Calculate λ $\det\begin{pmatrix} -1-\lambda & 0 \\ 0 & -1-\lambda \end{pmatrix} = 0$ $(-1-\lambda)(-1-\lambda) - (0 \times 0) = 0$ $\lambda^2 + 2\lambda + 1 = 0$ $(\lambda+1)(\lambda+1) = 0$ $\lambda = -1, -1$	Calculate λ $\det\begin{pmatrix} 1-\lambda & 0 \\ 0 & 1-\lambda \end{pmatrix} = 0$ $(1-\lambda)(1-\lambda) - (0 \times 0) = 0$ $\lambda^2 - 2\lambda + 1 = 0$ $(\lambda-1)(\lambda-1) = 0$ $\lambda = 1, 1$	Calculate λ $\det\begin{pmatrix} -7-\lambda & 1 \\ -4 & -3-\lambda \end{pmatrix} = 0$ $(-7-\lambda)(-3-\lambda) - (-4 \times 1) = 0$ $\lambda^2 + 10\lambda + 25 = 0$ $(\lambda+5)(\lambda+5) = 0$ $\lambda = -5, -5$ both λs negative	Calculate λ $\det\begin{pmatrix} 7-\lambda & 1 \\ -4 & 3-\lambda \end{pmatrix} = 0$ $(7-\lambda)(3-\lambda) - (-4 \times 1) = 0$ $\lambda^2 - 10\lambda + 25 = 0$ $(\lambda-5)(\lambda-5) = 0$ $\lambda = 5, 5$ both λs positive
Proper Node Sink (star point moving in)	Proper Node Source (star point moving out)	Improper Node Sink (moving toward equilibrium)	Improper Node Source (moving away from equilibrium)

Chapter 21 Review

It is possible to graph systems of equations in algebra in two space on an x–y-axis, but systems of differential equations, such as

$$\frac{\mathrm{d}}{\mathrm{d}t}x_1 = x_1' = 3x_1 + 2x_2 \tag{1}$$

$$\frac{\mathrm{d}}{\mathrm{d}t}x_2 = x_2' = 4x_1 + 1x_2, \tag{2}$$

cannot be graphed in two space as they are not the relation of a dependent variable and an independent variable. That is because such a system is formed of two differential equations with two dependent variables (x_1 and x_2) and one independent variable (t). There are three variables involved here, x_1, x_2,and t which would require a three-dimensional graph. What is often done is to "ignore" the independent variable, t, and think of the solutions as trajectories—(x_1, x_2) paths obtained by parametric equations—of the two dependent variables. Then the equilibrium (or critical point) solution would be the origin and the x_1–x_2 plane in this case would be called the phase plane.
Creating a phase plane graph is accomplished as follows.

1. Use the eigenvector of the two-by-two system linear differential equations to identify a vector.
2. Use the eigenvalue of the system as a scaling factor to convert the eigenvector to an eigenline resulting in special-case solutions to the system.
3. Use the product of the eigenvalue and t and our knowledge of exponential growth and exponential decay to sketch in the infinite solution family for the system.

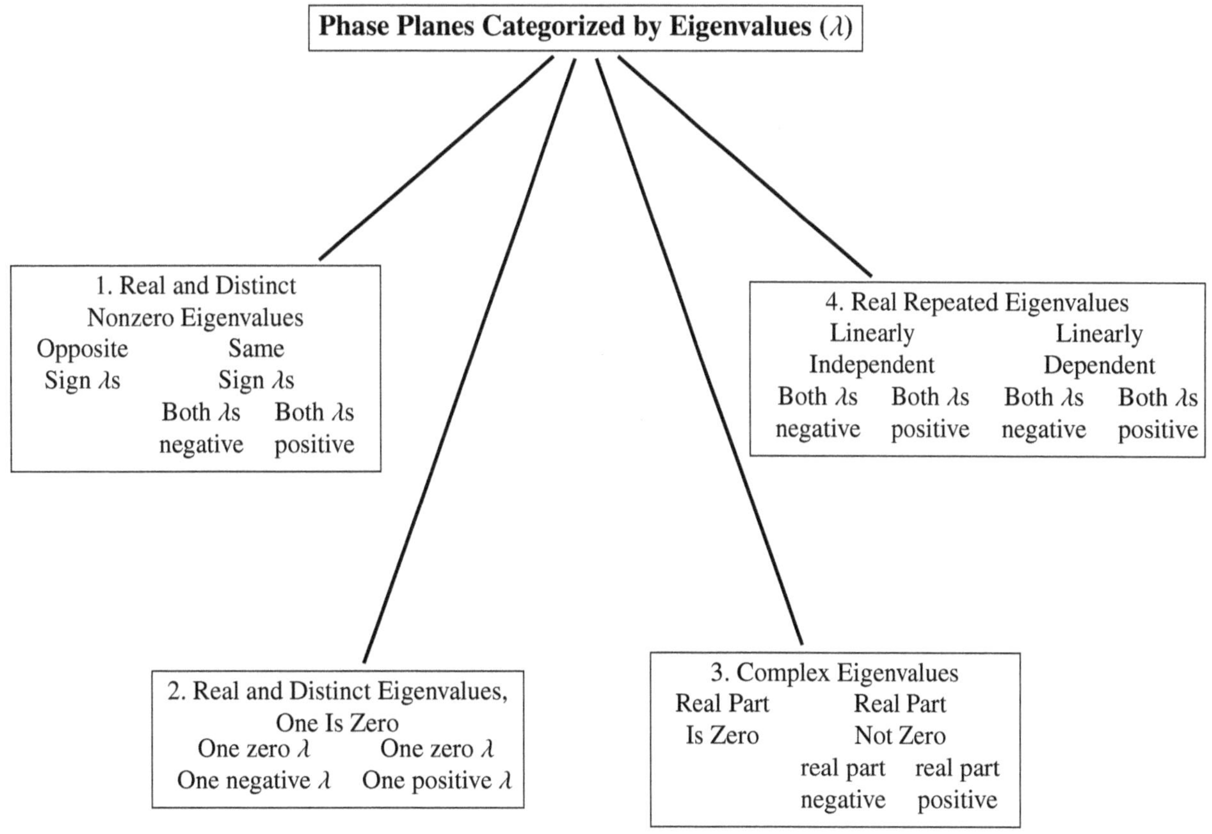

Online Application

Visit demonstrations.wolfram.com/BehaviorOfEquilibriumPointsInTwoDimensionalSystemsOfDifferen, demonstrations.wolfram.com/PhasePortraitsEigenvectorsAndEigenvalues, demonstrations.wolfram.com/VisualizingTheSolutionOfTwoLinearDifferentialEquations, and demonstrations.wolfram.com/LinearODEForms for four great little applications that will demonstrate some of the concepts from this chapter. "Behavior of Equilibrium Points in Two Dimensional Systems of Differential Equations" from the Wolfram Demonstrations Project. Contributed by Akshay Jaggi and Raymond Yuan. "Phase Portraits, Eigenvectors, and Eigenvalues" from the Wolfram Demonstrations Project. Contributed by Stephen Wilkerson and Stanley Florkowski. "Visualizing the Solution of Two Linear Differential Equations" from the Wolfram Demonstrations Project. Contributed by Mikhail Dimitrov Mikhailov. "Linear ODE Forms" from the Wolfram Demonstrations Project. Contributed by Alex S. Dunn.

Behavior of Equilibrium Points in Two-Dimensional Systems of Differential Equations

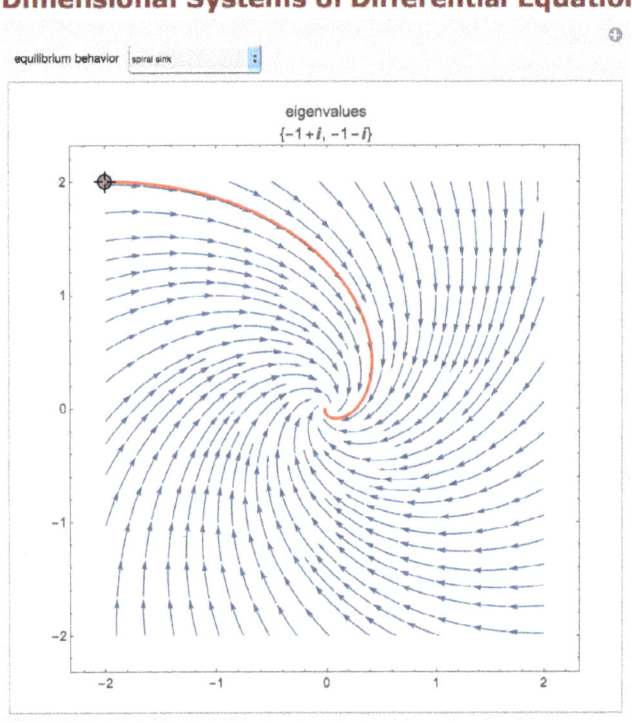

Phase Portraits, Eigenvectors, and Eigenvalues

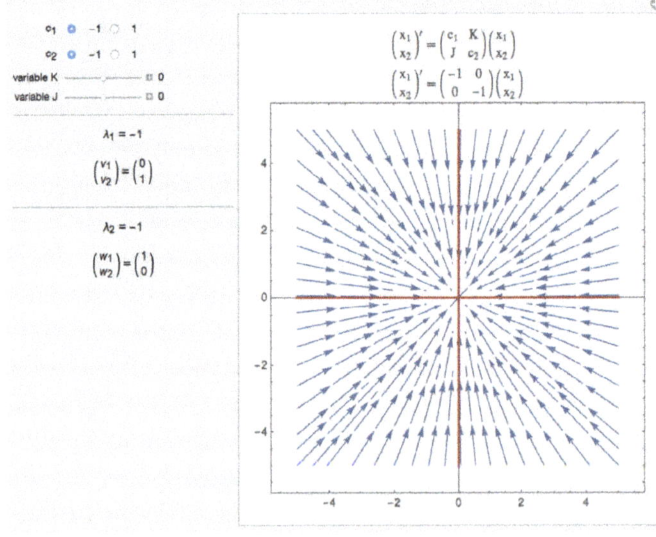

Linear ODE Forms

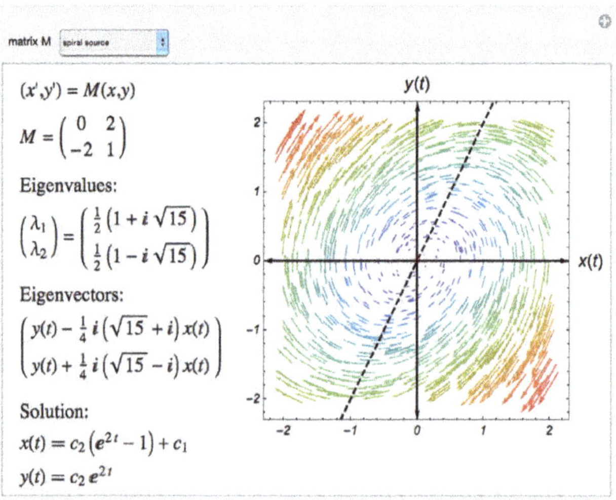

Visualizing the Solution of Two Linear Differential Equations

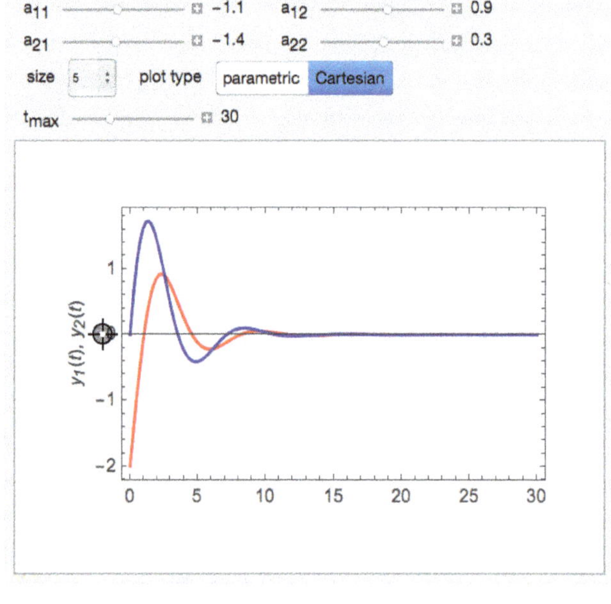

Chapter 22
Background Necessary to Understand the Laplace Transform

A useful technique for solving some forms of differential equations is the Laplace transform. Understanding how that solution technique works requires knowledge of 1) integration by parts, 2) partial fractions and 3) power series. If you feel comfortable with all of those skills, you should feel free to skip ahead to the next chapter.

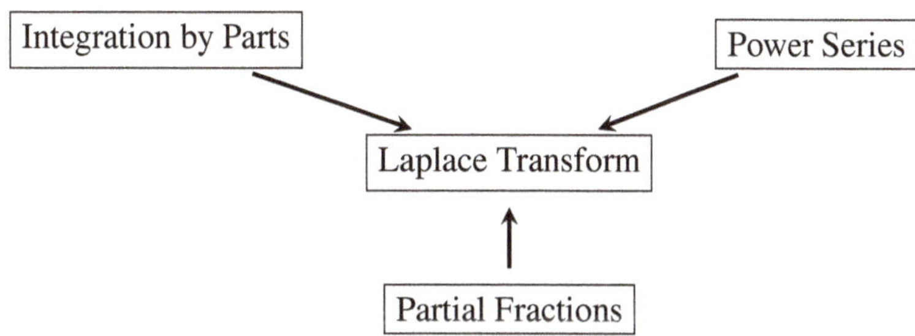

Integration by Parts (Review from Calculus) (Skip if not needed.)

This integration technique is particularly useful in working with integrands that involve a product of algebraic and transcendental functions, for example $\int x^2 \ln x\,\mathrm{d}x$, $\int xe^x\,\mathrm{d}x$, $\int x\sin x\,\mathrm{d}x$, $\int e^x\cos x\,\mathrm{d}x$, etc. We start with the derivative of a product:

$$\frac{\mathrm{d}}{\mathrm{d}x}(uv) = u\frac{\mathrm{d}v}{\mathrm{d}x} + v\frac{\mathrm{d}u}{\mathrm{d}x} = uv' + vu' \quad \text{where both } u \text{ and } v \text{ are differentiable functions}$$

$$\int \frac{\mathrm{d}}{\mathrm{d}x}(uv)\,\mathrm{d}x = \int\left(u\frac{\mathrm{d}v}{\mathrm{d}x} + v\frac{\mathrm{d}u}{\mathrm{d}x}\right)\mathrm{d}x \quad \text{integrate both sides with respect to } x$$

$$uv = \int u\frac{\mathrm{d}v}{\mathrm{d}x}\,\mathrm{d}x + \int v\frac{\mathrm{d}u}{\mathrm{d}x}\,\mathrm{d}x$$

$$uv = \int u\,\mathrm{d}v + \int v\,\mathrm{d}u$$

So

$$\int u\,\mathrm{d}v = uv - \int v\,\mathrm{d}u$$

Here we have an integral expressed in terms of another integral. This integration technique is useful because it is often easier to evaluate the second integral than the original. It is often the case that the most complicated portion of the integral fits a basic integration rule. This is the origin of this technique's utility.

Let's try an example. Find $\int e^x\cos x\,\mathrm{d}x$. How does one integrate such an expression? Treat with $\int u\,\mathrm{d}v = uv - \int v\,\mathrm{d}u$.

Let	u	$= e^x$	and	$\mathrm{d}v$	$= \cos x\,\mathrm{d}x.$
Therefore,	$\mathrm{d}u$	$= e^x\,\mathrm{d}x$	and	v	$= \sin x.$

Hence,

$$\int u\,\mathrm{d}v = uv - \int v\,\mathrm{d}u$$

$$\int [e^x\cos x\,\mathrm{d}x] = e^x\sin x - \int e^x\sin x\,\mathrm{d}x \quad (1)$$

This does not seem to be an improvement, but a second look suggests that, if only we could integrate $e^x \sin x$, we might be getting somewhere. Let's repeat the process above for $\int e^x \sin x$.

Let	u	$= e^x$	and	dv	$= \sin x\, dx.$
Therefore,	du	$= e^x\, dx$	and	v	$= -\cos x.$

Hence,

$$\int u\, dv = uv - \int v\, du$$

$$\int [e^x \sin x\, dx] = -e^x \cos x - \int e^x(-\cos x)\, dx$$

$$= -e^x \cos x + \int e^x \cos x\, dx \quad (2)$$

Substituting (2) into (1), we get

$$\int [e^x \cos x\, dx] = e^x \sin x - \int e^x \sin x\, dx \quad (1)$$

$$= e^x \sin x - \left(-e^x \cos x + \int e^x \cos x\, dx\right)$$

$$= e^x \sin x + e^x \cos x - \int e^x \cos x\, dx.$$

Therefore,

$$2\int e^x \cos x\, dx = e^x \sin x + e^x \cos x$$

Now we have an easily solved expression for $\int e^x \cos x\, dx$:

$$\int e^x \cos x\, dx = \frac{e^x}{2}(\sin x + \cos x) + C.$$

Traditionally, integration by parts is discussed as shown in the preceding text. A different but fun way to look at the topic is shown on page 42 of *Proofs Without Words* by Roger B. Nelsen (Mathematical Association of America,).

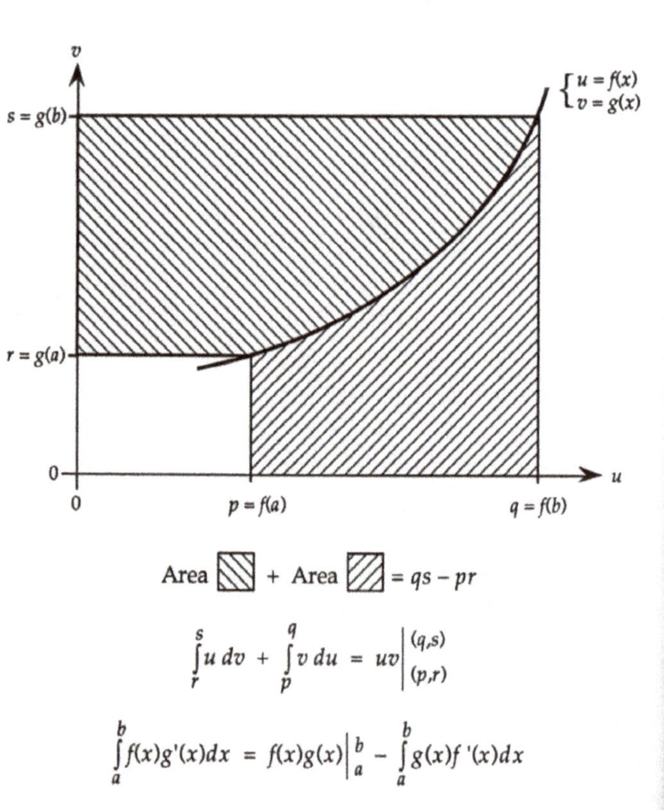

42 Proofs without Words

Integration by Parts

Area ▧ + Area ▨ $= qs - pr$

$$\int_r^s u\, dv + \int_p^q v\, du = uv\Big|_{(p,r)}^{(q,s)}$$

$$\int_a^b f(x)g'(x)dx = f(x)g(x)\Big|_a^b - \int_a^b g(x)f'(x)dx$$

—Richard Courant

Partial Sums (Review from Arithmetic and Calculus) (Skip if not needed.)

1. Solve for X in terms of A, B, C, and D in the following equation.

$$\frac{X}{12} = \frac{A}{1} + \frac{B}{2} + \frac{C}{3} + \frac{D}{4}$$
$$\frac{X}{12} = \frac{A}{1}\left(\frac{12}{12}\right) + \frac{B}{2}\left(\frac{6}{6}\right) + \frac{C}{3}\left(\frac{4}{4}\right) + \frac{D}{4}\left(\frac{3}{3}\right)$$
$$\frac{X}{12} = \frac{12A}{12} + \frac{6B}{12} + \frac{4C}{12} + \frac{3D}{12}$$
$$\frac{X}{12} = \frac{12A + 6B + 4C + 3D}{12}$$
$$X = 12A + 6B + 4C + 3D$$

2. Express $\dfrac{3x - 8}{x^2 - 4x - 5}$ as the sum of two fractions.

$$\frac{3x-8}{x^2-4x-5} = \frac{3x-8}{(x-5)(x+1)} = \frac{A}{x-5} + \frac{B}{x+1}$$
$$[(x-5)(x+1)]\frac{3x-8}{(x-5)(x+1)} = \left[\frac{A}{x-5} + \frac{B}{x+1}\right][(x-5)(x+1)]$$
$$3x - 8 = A(x+1) + B(x-5)$$

$3x - 8 = A(x+1) + B(x-5)$

Let $x = -1$ to eliminate the A term.

$$3(-1) - 8 = A(-1+1) + B(-1-5)$$
$$-11 = 0 + B(-6)$$
$$-11 = -6B$$
$$B = \frac{11}{6}$$

$3x - 8 = A(x+1) + B(x-5)$

Let $x = 5$ to eliminate the B term.

$$3(5) - 8 = A(5+1) + B(5-5)$$
$$7 = 6A + B(0)$$
$$7 = 6A$$
$$A = \frac{7}{6}$$

Check

$$\frac{3x-8}{(x-5)(x+1)} = \frac{A}{x-5} + \frac{B}{x+1}$$
$$\frac{3x-8}{(x-5)(x+1)} \stackrel{?}{=} \frac{\frac{7}{6}}{x-5} + \frac{\frac{11}{6}}{x+1}$$
$$[(x-5)(x+1)]\frac{3x-8}{(x-5)(x+1)} \stackrel{?}{=} \left[\frac{\frac{7}{6}}{x-5} + \frac{\frac{11}{6}}{x+1}\right][(x-5)(x+1)]$$
$$3x - 8 \stackrel{?}{=} \frac{7}{6}(x+1) + \frac{11}{6}(x-5)$$
$$3x - 8 \stackrel{?}{=} \frac{7}{6}x + \frac{7}{6} + \frac{11}{6}x - \frac{55}{6}$$
$$3x - 8 \stackrel{?}{=} \frac{18}{6}x - \frac{48}{6}$$
$$3x - 8 = 3x - 8 \quad \text{Ck}$$

3. Express $\dfrac{-x^2+2x-5}{(x+1)(x+3)^2}$ as the sum of three fractions.

$$\frac{-x^2+2x-5}{(x+1)(x+3)^2} = \frac{A}{x+1} + \frac{B}{x+3} + \frac{C}{(x+3)^2}$$

$$\left[(x+1)(x+3)^2\right]\frac{-x^2+2x-5}{(x+1)(x+3)^2} = \left(\frac{A}{x+1} + \frac{B}{x+3} + \frac{C}{(x+3)^2}\right)\left[(x+1)(x+3)^2\right]$$

$$-x^2+2x-5 = A(x+3)^2 + B(x+1)(x+3) + C(x+1)$$

$$-x^2+2x-5 = A\left(x^2+6x+9\right) + B\left(x^2+4x+3\right) + C(x+1)$$

$$-x^2+2x-5 = Ax^2 + 6Ax + 9A + Bx^2 + 4Bx + 3B + Cx + C$$

$$-x^2+2x-5 = \left(Ax^2+Bx^2\right) + (6Ax+4Bx+Cx) + (9A+3B+C)$$

$$-x^2+2x-5 = (A+B)x^2 + (6A+4B+C)x + (9A+3B+C)$$

Equating terms of the same order, we get three equations in three unknowns:

$$A+B=-1 \qquad (1)$$

$$6A+4B+C=2 \qquad (2)$$

$$9A+3B+C=-5 \qquad (3)$$

Combine (2) and (3) to eliminate C. Pair the resulting equation with (1).

$$3A-B=-7 \qquad ((3)-(2))$$

$$A+B=-1 \qquad (1)$$

$$4A=-8$$

$$A=-2$$

$B=1$ back substitute A into (1) to find B

$C=10$ substitute both A and B into the original (2) or (3) to find C

Power Series (Review from Calculus) (Skip if not needed.)

A series is an addition of a sequence of numbers which are formed predictably so that a pattern is evident. For example, $S_n = (3+7+11+15+\cdots)$ or, in general, $a_1+a_2+a_3+\cdots+(4n-1)$.

A geometric series (geometric progression) is the sum of a sequence in which the ratio of two adjacent terms is the same throughout the sequence. In the progression 2, 6, 18, 54, each successive term is 3 times the one before it. The constant factor is sometimes called the common ratio of the progression. This is indicated generally as: $a+ar+ar^2+ar^3+\cdots+ar^{n-1}+ar^n+\cdots$ where a is the first term of the progression and r is the common ratio. *It is helpful to notice that the ratio of two adjacent terms of the geometric series is r:* $\frac{ar}{a}=r$, $\frac{ar^2}{ar}=r$, $\frac{ar^3}{ar^2}=r, \ldots$.

A power series is a series defined as

1. $\sum_{n=0}^{\infty} a_n x^n = a_0 + a_1x^1 + a_2x^2 + a_3x^3 + \cdots$, $n \in$ nonnegative integers.

   (Obviously this series will converge absolutely to a_0 if $x=0$.)

   Or the shifted version:

2. $\sum_{n=0}^{\infty} a_n(x-c)^n = a_0 + a_1(x-c)^1 + a_2(x-c)^2 + a_3(x-c)^3 + \cdots$, $n \in$ nonnegative integers.

   (Obviously this series will converge absolutely to a_0 if $x=c$.)

Many functions can be represented as infinite power series.

$$\sum_{n=0}^{\infty} a_n x^n = a_0 + a_1 x^1 + a_2 x^2 + a_3 x^3 + \cdots$$

$$\text{Eg. } \sin x = x + \frac{x^3}{3!} + \frac{x^5}{5!} + \frac{x^7}{7!} + \cdots + \frac{x^{2n+1}}{(2n+1)!} + \cdots$$

$$\cos x = 1 - \frac{x^2}{2!} + \frac{x^4}{4!} - \frac{x^6}{6!} + \cdots + (-1)^n \frac{x^{2n}}{(2n)!} + \cdots$$

$$\mathrm{e} = 1 + \frac{1}{1!} + \frac{1}{2!} + \frac{1}{3!} + \frac{1}{4!} + \cdots + \frac{1}{n!} + \cdots$$

$$\mathrm{e}^x = 1 + x + \frac{x^2}{2!} + \frac{x^3}{3!} + \frac{x^4}{4!} + \cdots + \frac{x^n}{n!} + \cdots$$

$$a^x = 1 + x(\ln a) + \frac{(x \ln a)^2}{2!} + \frac{(x \ln a)^3}{3!} + \cdots + \frac{(x \ln a)^n}{n!} + \cdots$$

The definition of a power series stated above,

$$\sum_{n=0}^{\infty} a_n x^n = a_0 + a_1 x^1 + a_2 x^2 + a_3 x^3 + \cdots, \qquad n \in \text{nonnegative integers},$$

involves an infinite summation of distinct terms. If we change the domain from nonnegative integers to $t \in \mathcal{R}$, with $0 < t < \infty$, we would at the same time need to change the sigma notation, $\sum_{n=0}^{\infty} a_n x^n$, to integral notation $\int_{t=0}^{\infty} a_n x^t$. x^t could be changed to a power of e (see box right), resulting in $\int_{t=0}^{\infty} a_n \mathrm{e}^{t \ln x}$. After some more substitutions, we would end up with $\int_0^{\infty} f(t)\mathrm{e}^{-st}\, \mathrm{d}t$, the definition of the Laplace transform. (This is just the kind of integral that can be integrated by the *integration by parts* technique.) Using the definition of the Laplace transform, we will be able to solve many families of differential equations. **That will be shown in the next chapter.** Stay tuned!

> Let $x = \mathrm{e}^{\ln x}$.
> So, $x^t = \left(\mathrm{e}^{\ln x}\right)^t$,
> $x^t = \mathrm{e}^{t \ln x}$.

Chapter 22 Review

Chapter 23
The Laplace Transform

In general, the term transform is associated with change. In the image below left, a toy changes both its appearance and function. In the image at right, a transformer is used to increase or decrease the voltages of alternating current.

In math, the term "transform" means an algorithm, procedure, rule or set of rules which change one set (or many sets) of data into (an)other set(s) of data. The Fourier transform changes data associated with time into data associated with space (e.g., a set of vectors). If you are an engineering major, I totally recommend you to check out the YouTube video at https://www.youtube.com/watch?v=FjmwwDHT98c

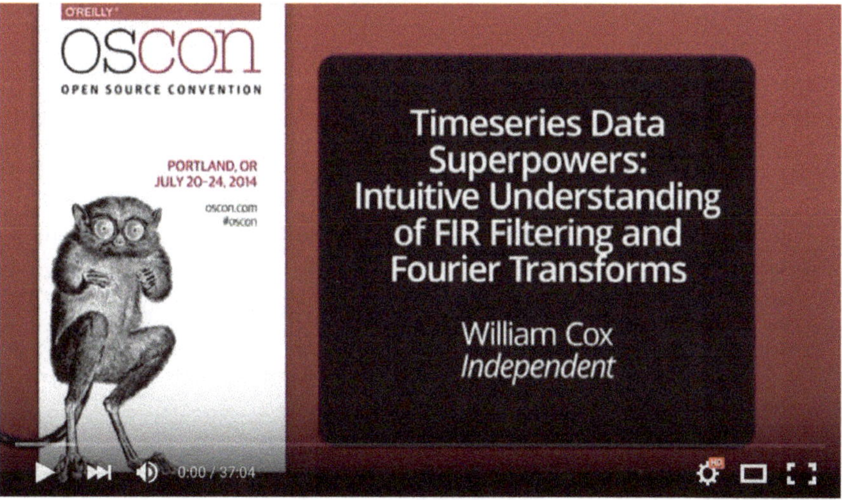

It is helpful to contrast a transform with an operator. With an operator, the data input into an algorithm, procedure, equation, or set of rules is the same in type as the output data.

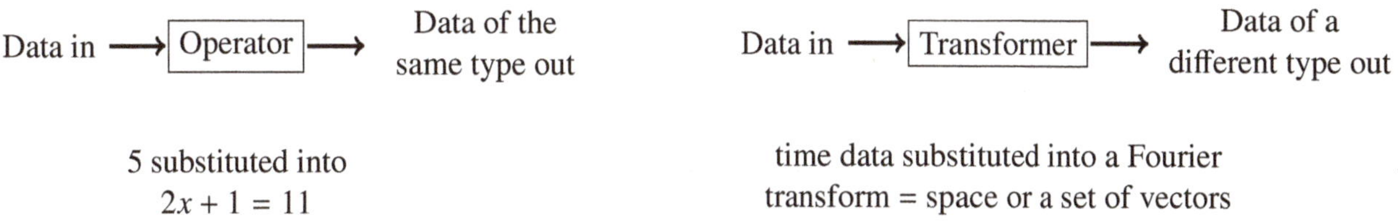

The Laplace transform converts integral and differential equations into algebraic equations. As is often the case, there may be more than one solution technique that will solve a differential equation. Laplace transforms are particularly useful when the differential equation involves transcendental functions, or when the forcing function in the differential equation starts getting more complicated, or when the function is not continuous (but is piecewise continuous as shown below). A piecewise continuous function has a finite number of breaks in it and doesn't blow up to infinity anywhere.

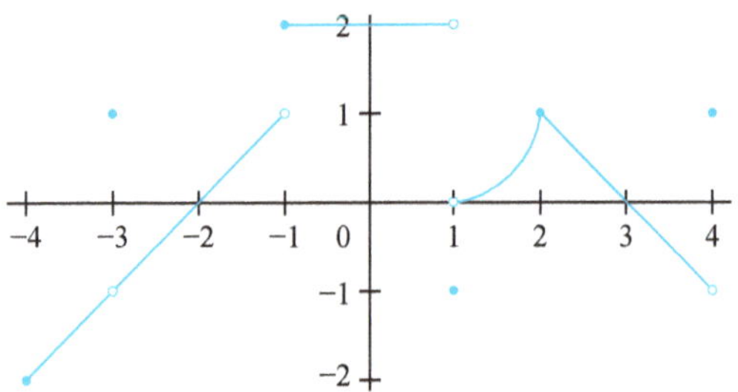

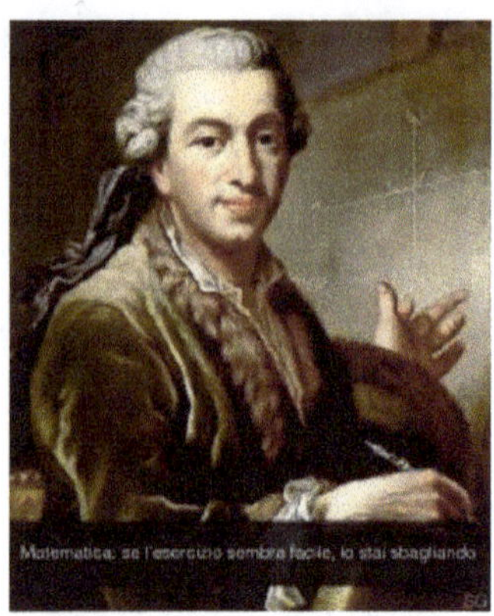

Pierre-Simon Laplace
1749–1827

Laplace transforms find applications in many areas of physics, electrical engineering, control engineering, optics, mathematics, and signal processing. Laplace transforms are a definition. They are an extension of the definition of a power series:

$$\sum_{n=0}^{\infty} a_n x^n = a_0 + a_1 x^1 + a_2 x^2 + a_3 x^3 + \cdots .$$

If the set of coefficients $a_0, a_1, a_2, a_3 \ldots$ is predictable, it can be generated by a function, $a(n)$,

$$\sum_{n=0}^{\infty} a(n) x^n = a_0 + a_1 x^1 + a_2 x^2 + a_3 x^3 + \cdots .$$

Since the name of a function is arbitrary, we can change $a(n)$ to $f(n)$ resulting in

$$\sum_{n=0}^{\infty} f(n) x^n = f_0 + f_1 x^1 + f_2 x^2 + f_3 x^3 + \cdots .$$

This (discrete) summation formula can be converted to a continuous version by changing the variable n ($n \in$ nonnegative integers) to t, ($t \in \Re$ and $0 < t < \infty$). By changing the $\sum$ notation to $\int$ notation: $\sum_{n=0}^{\infty} f(n)x^n$ becomes $\int_0^{\infty} f(t)x^t\,\mathrm{d}t$. We know that, for $\int_0^{\infty} f(t)x^t\,\mathrm{d}t$ to be possible, x^t must converge. From our study of the power series, that implies $|x| < 1$. However, if we allow x values to be negative, then we will be integrating a series with imaginary numbers, so let's require that $0 < x < 1$. It is conventional in math and science to use e wherever possible, so we can write $\int_0^{\infty} f(t)x^t\,\mathrm{d}t$ as $\int_0^{\infty} f(t)\mathrm{e}^{t\ln x}\,\mathrm{d}t$ (see box at left). Since $0 < x < 1$, we deduce that $\ln x < 0$.

> Let $x = \mathrm{e}^{\ln x}$.
>
> So, $x^t = \left(\mathrm{e}^{\ln x}\right)^t$,
>
> $x^t = \mathrm{e}^{t\ln x}$.

Let $s = -\ln x$ (hence $s > 0$ because $\ln x < 0$), so $-s = \ln x$ or $\ln x = -s$. Making all these substitutions,

$$\int_0^{\infty} f(t)x^t = \int_0^{\infty} f(t)\mathrm{e}^{t\ln x} = \int_0^{\infty} f(t)\mathrm{e}^{t(-s)}\,\mathrm{d}t = \int_0^{\infty} f(t)\mathrm{e}^{-st}\,\mathrm{d}t = F(s).$$

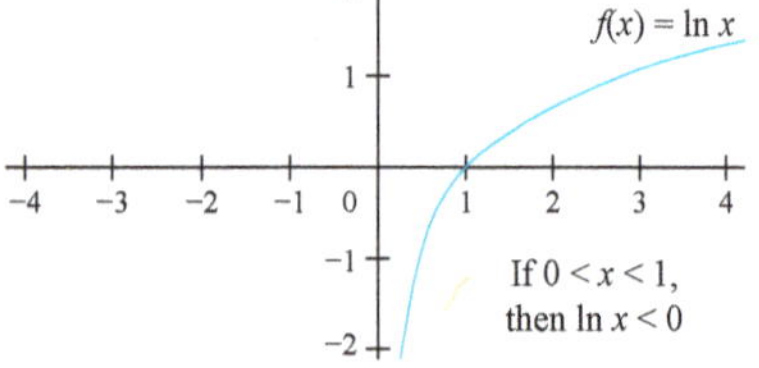

Here, we have a function of t being transformed into a function of s. The function is in s because the t is integrated out. This particular transformation is known as a Laplace transform.* When studying math, it is helpful to most people to understand that new ideas are often extensions or modifications of old ideas. In this case, the Laplace transform is a continuous version of the discrete power series that you have studied previously. Or as Monsieur Jourdain would have said in Chapter 1, "Par ma foi! Il y a plus de quarante ans que je dis de la prose sans que j'en susse rien, et je vous suis le plus obligé du monde de m'avoir appris cela."

---

*For an excellent video about the Laplace transform, see Professor Arthur Mattuck, MIT Math Dept., MIT Open Courseware https://www.youtube.com/watch?v=zvbdoSeGAgI.

Using the integral $\int_0^\infty f(t)e^{-st}\,dt$ shown above, it is traditional to develop a table of Laplace-transform factors for various $f(t)$'s. It would not be harmful to think of such a table as sort of table of "integrals," one for each type of function, $f(t)$. After you fully understand a few of these factors and feel comfortable with the idea, you can just refer to the table and use the fact that some smart person somewhere has done all this work for you; you can just use their work on your particular function for your specific application. For example, in the table at right we will not prove the Laplace transform of Rows 5, 7, or 10. We will just access any published table of such transforms and use that information. This completes the tranformation of the definition of a power series, $\sum_0^\infty a_n x^n$, into the definition of the Laplace transform!

Row	$f(t)$	$F(s)$
1	0	0
2	1	$\frac{1}{s}$
3	c	$\frac{c}{s}$
4	e^{at}	$\frac{1}{s-a}$
5	te^{at}	$\frac{1}{(s-a)^2}$
6	$\sin(at)$	$\frac{a}{s^2+a^2},\ s>0$
7	$\cos(at)$	$\frac{s}{s^2+a^2},\ s>0$
8	$f(t)$	$F(s)$ by definition
9	$f'(t)$	$sF(s)-f(0)$
10	$f''(t)$	$s^2F(s)-sf(0)-f'(0)$

Definition: Suppose $f(t)$ is a piecewise continuous function. The Laplace transform of $f(t)$ is denoted $\mathfrak{L}\{f(t)\}$ and defined as

$$\mathfrak{L}\{f(t)\} = \int_0^\infty e^{-st}f(t)\,dt.$$

There is an alternate notation for Laplace transforms. For the sake of convenience, we will often denote Laplace transforms as

$$\mathfrak{L}\{f(t)\} = F(s).$$

As examples, we establish six of the entries in the Laplace-transform table shown above.

Example 1:

$$\mathfrak{L}\{f(t)\} = F(s) = \int_0^\infty e^{-st}f(t)\,dt$$

Let $f(t) = 0$.

$$\mathfrak{L}\{0\} = \int_0^\infty e^{-st}(0)\,dt$$
$$= 0\int_0^\infty e^{-st}\,dt$$
$$= 0$$
$$\mathfrak{L}\{0\} = 0 \quad \text{Row 1}$$

Example 2:

$$\mathfrak{L}\{f(t)\} = F(s) = \int_0^\infty e^{-st}f(t)\,dt$$

Let $f(t) = 1$.

$$\mathfrak{L}\{1\} = \int_0^\infty e^{-st}(1)\,dt = \int_0^\infty e^{-st}\,dt$$
$$= -\frac{1}{s}\int_0^\infty e^{-st}(-s)\,dt \quad \text{now in } e^u\,du \text{ form}$$
$$= -\frac{1}{s}e^{-st}\Big]_0^\infty = -\frac{1}{s}\frac{1}{e^{st}}\Big]_0^\infty = -\frac{1}{s}(0-1) = \frac{1}{s}$$
$$\mathfrak{L}\{1\} = \frac{1}{s} \quad \text{Row 2}$$

Example 3:

$$\mathfrak{L}\{f(t)\} = F(s) = \int_0^\infty e^{-st}f(t)\,dt$$

Let $f(t) = c$.

$$\text{So, } \mathfrak{L}\{c\} = \int_0^\infty e^{-st}(c)\,dt$$
$$= c\int_0^\infty e^{-st}dt$$
$$= c \times \frac{1}{s} \quad \text{from Example 2}$$
$$\mathfrak{L}\{c\} = \frac{c}{s} = c\mathfrak{L}\{1\} \quad \text{Row 3}$$

Example 4:

$$\mathfrak{L}\{f(t)\} = F(s) = \int_0^\infty e^{-st}f(t)\,dt$$

Let $f(t) = e^{at}$.

$$\mathfrak{L}\{e^{at}\} = \int_0^\infty e^{-st}\left(e^{at}\right)dt$$
$$= \int_0^\infty e^{-st+at}\,dt = \int_0^\infty e^{at-st}\,dt = \int_0^\infty e^{(a-s)t}\,dt$$
$$= \frac{1}{a-s}\int_0^\infty e^{(a-s)t}(a-s)\,dt \quad \text{now in } \int u\,du \text{ form}$$
$$= \frac{1}{a-s}e^{(a-s)t}\Big]_0^\infty$$

If $(a-s) > 0$, there is no limit. If $(a-s) < 0$, $e^{(a-s)t} \to 0$.

$$= \frac{1}{a-s}[0-1] = \frac{1}{s-a}$$
$$\mathfrak{L}\left\{e^{at}\right\} = \frac{1}{s-a} \quad \text{when } s > a \quad \text{Row 4}$$

In Example 4, note that if $a = 0$, $\mathfrak{L}\{e^{at}\} = \mathfrak{L}\left\{e^{0t}\right\} = \mathfrak{L}\{1\} = \frac{1}{s-0} = \frac{1}{s}$, which is consistent with Example 2.

Example 5:

$$\mathfrak{L}\{f(t)\} = F(s) = \int_0^\infty e^{-st} f(t)\,dt$$

Let $f(t) = \sin(at)$.

Very involved work here relegated to Appendix I.

$$\mathfrak{L}\{\sin(at)\} = \frac{a}{s^2 + a^2} \quad \text{Row 6}$$

Example 6:

$$\mathfrak{L}\{f(t)\} = F(s) = \int_0^\infty e^{-st} f(t)\,dt$$

Let $f(t) = f'(t)$.

So, $\mathfrak{L}\{f'(t)\} = \int_0^\infty e^{-st} f'(t)\,dt$.

Let $u = e^{-st}$, so $u' = -se^{-st}\,dt$.

Let $v' = f'(t)\,dt$, so $v = f(t)$.

$$\int uv' = uv - \int vu' \quad \text{see Chapter 22}$$

$$\mathfrak{L}\{f'(t)\} = e^{-st} f(t)\Big]_0^\infty - \int_0^\infty (-se^{-st}) f(t)\,dt$$

$$= e^{-st} f(t)\Big]_0^\infty + s\int_0^\infty e^{-st} f(t)\,dt \quad \text{treating } s \text{ as constant}$$

$$= \frac{f(t)}{e^{st}}\Bigg]_0^\infty + s\int_0^\infty e^{-st} f(t)\,dt$$

$$= \left[0 - \frac{f(0)}{e^0}\right] + s\mathfrak{L}\{f(t)\} = -f(0) + sF(s)$$

$$\mathfrak{L}\{f'(t)\} = sF(s) - f(0) \quad \text{Row 9}$$

Inverse Laplace Transform

We see the term "inverse" a lot in math!

1. Two numbers are additive inverses if their sum is zero: $a + (-a) = 0$.
2. Two numbers are multiplicative inverses if their product is 1: $a \times \frac{1}{a} = 1$.
3. Two functions, $f(x)$ and $g(x)$, are inverses if $f[g(x)] = x = g[f(x)]$

This can be expressed as $f(x)$ and $f^{-1}(x)$ are inverse functions because $f(f^{-1}(x)) = x = f^{-1}(f(x))$. Likewise two transforms are inverses if $\mathfrak{L}\{f(x)\} = F(s)$ and $\mathfrak{L}^{-1}\{F(s)\} = f(x)$. In other words, the process we have demonstrated $\mathfrak{L}\{f(t)\} \to F(s)$ is reversed.

Example

Given a function $f(t)$—whose Laplace transform is $F(s) = \frac{1}{s-3} - \frac{16}{s^2+9}$—find $f(t)$.

$$\mathfrak{L}^{-1}\{F(s)\} = \mathfrak{L}^{-1}\left\{\frac{1}{s-3} - \frac{16}{s^2+9}\right\}$$

$$f(t) = \mathfrak{L}^{-1}\left\{\frac{1}{s-3}\right\} - \mathfrak{L}^{-1}\left\{\frac{16}{s^2+9}\right\} \quad \text{the Laplace transform is distributive}$$

$$= e^{3t} - \mathfrak{L}^{-1}\left\{\frac{3 \times \frac{16}{3}}{s^2 + 3^2}\right\} \quad \text{Row 4 from the table}$$

$$= e^{3t} - \mathfrak{L}^{-1}\left\{\frac{16}{3}\frac{3}{s^2 + 3^2}\right\}$$

$$= e^{3t} - \frac{16}{3}\mathfrak{L}^{-1}\left\{\frac{a}{s^2 + a^2}\right\} \quad \text{where } a = 3$$

$$= e^{3t} - \frac{16}{3}\sin(at) \quad \text{where } a = 3\text{; Row 6 from the table}$$

$$f(t) = e^{3t} - \frac{16}{3}\sin(3t).$$

Row	$f(t)$	$F(s)$
1	0	0
2	1	$\frac{1}{s}$
3	c	$\frac{c}{s}$
4	e^{at}	$\frac{1}{s-a}$,
5	te^{at}	$\frac{1}{(s-a)^2}$
6	$\sin(at)$	$\frac{a}{s^2+a^2}$, $s > 0$
7	$\cos(at)$	$\frac{s}{s^2+a^2}$, $s > 0$
8	$f(t)$	$F(s)$ by definition
9	$f'(t)$	$sF(s) - f(0)$
10	$f''(t)$	$s^2F(s) - sf(0) - f'(0)$

Of course this problem was rigged, but you get the idea. If the problem had been $\mathfrak{L}^{-1}\{F(s)\} = \mathfrak{L}^{-1}\left\{\frac{1}{s-3} - \frac{16}{s^2+7}\right\}$, it could have been rewritten as $\mathfrak{L}^{-1}\left\{\frac{1}{s-3} - \frac{16}{s^2+\sqrt{7}^2}\right\}$, etc.

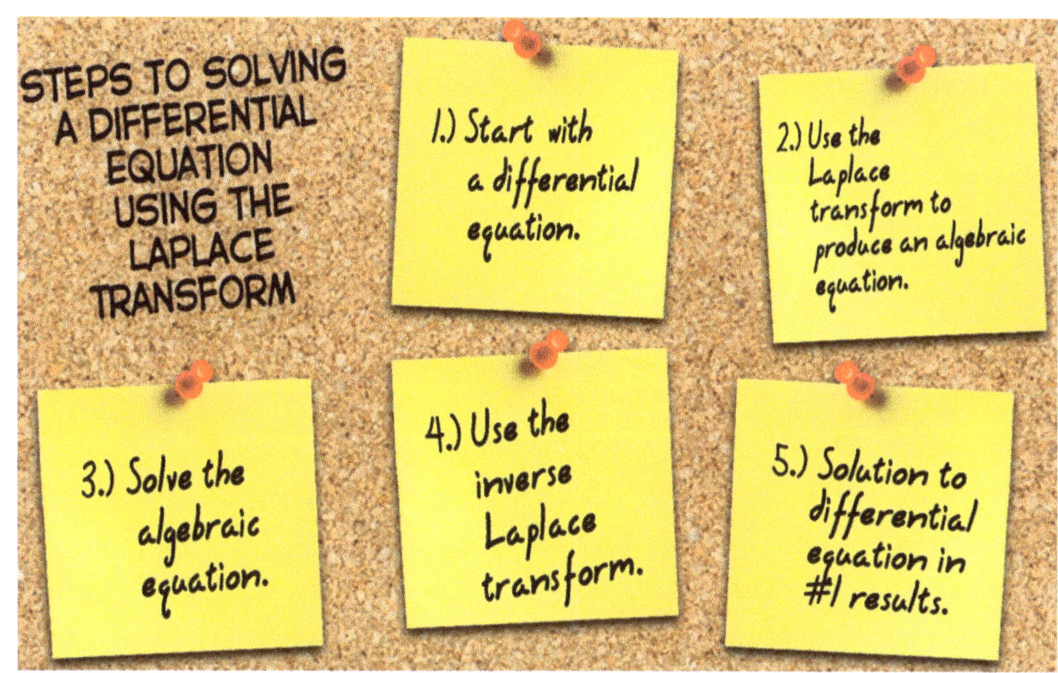
STEPS TO SOLVING A DIFFERENTIAL EQUATION USING THE LAPLACE TRANSFORM
1.) Start with a differential equation.
2.) Use the Laplace transform to produce an algebraic equation.
3.) Solve the algebraic equation.
4.) Use the inverse Laplace transform.
5.) Solution to differential equation in #1 results.

Steps 2 and 4 in this algorithm are usually done with a table lookup.

1. Solve $y' - 5y = 0$, where $y(0) = 2$, using a Laplace transform.

$$\mathfrak{L}\{y' - 5y\} = \mathfrak{L}\{0\}$$

$\mathfrak{L}\{y'\} - \mathfrak{L}\{5y\} = 0$ Row 1 from table, $\mathfrak{L}\{0\} = 0$

$$\mathfrak{L}\{y'\} - 5\mathfrak{L}\{y\} = 0$$

$[sY(s) - y(0)] - 5Y(s) = 0$ Rows 9 and 8, definition of $\mathfrak{L}\{y\}$

$[sY(s) - 2] - 5Y(s) = 0$ substitute in initial condition, $y(0) = 2$

$sY(s) - 5Y(s) = 2$ add 2 to both sides

$Y(s)(s - 5) = 2$ factor out $Y(s)$

$Y(s) = \dfrac{2}{s-5}$ solve for $Y(s)$

$\mathfrak{L}^{-1}\{Y(s)\} = \mathfrak{L}^{-1}\left\{\dfrac{2}{s-5}\right\}$ inverse transform

$$y(t) = 2\mathfrak{L}^{-1}\left\{\frac{1}{s-5}\right\}$$

$= 2e^{5t}$ inverse transform, Row 4

Row	$f(t)$	$F(s)$
1	0	0
2	1	$\frac{1}{s}$
3	c	$\frac{c}{s}$
4	e^{at}	$\frac{1}{s-a}$
5	te^{at}	$\frac{1}{(s-a)^2}$
6	$\sin(at)$	$\frac{a}{s^2+a^2}, s > 0$
7	$\cos(at)$	$\frac{s}{s^2+a^2}, s > 0$
8	$f(t)$	$F(s)$ by definition
9	$f'(t)$	$sF(s) - f(0)$
10	$f''(t)$	$s^2F(s) - sf(0) - f'(0)$

2. Solve $y' - 5y = e^{5t}$, where $y(0) = 0$.

$$\mathfrak{L}\{y' - 5y\} = \mathfrak{L}\left\{e^{5t}\right\}$$

$\mathfrak{L}\{y'\} - \mathfrak{L}\{5y\} = \dfrac{1}{s-5}$ Row 4 from table

$[sY(s) - y(0)] - 5\mathfrak{L}\{y\} = \dfrac{1}{s-5}$ Row 9 from table

$[sY(s) - 0] - 5Y(s) = \dfrac{1}{s-5}$ substitute for the given initial value

$Y(s)(s - 5) = \dfrac{1}{s-5}$ factor out $Y(s)$

$Y(s) = \dfrac{1}{(s-5)^2}$ solve for $Y(s)$

$\mathfrak{L}^{-1}\{Y(s)\} = \mathfrak{L}^{-1}\left\{\dfrac{1}{(s-5)^2}\right\}$ inverse transform

$y(t) = te^{5t}$ Rows 8 and 5

Row	$f(t)$	$F(s)$
1	0	0
2	1	$\frac{1}{s}$
3	c	$\frac{c}{s}$
4	e^{at}	$\frac{1}{s-a}$
5	te^{at}	$\frac{1}{(s-a)^2}$
6	$\sin(at)$	$\frac{a}{s^2+a^2}, s > 0$
7	$\cos(at)$	$\frac{s}{s^2+a^2}, s > 0$
8	$f(t)$	$F(s)$ by definition
9	$f'(t)$	$sF(s) - f(0)$
10	$f''(t)$	$s^2F(s) - sf(0) - f'(0)$

3. Solve $y' + y = \sin t$, where $y(0) = 1$.

$$\mathfrak{L}\{y' + y\} = \mathfrak{L}\{\sin t\}$$

$$\mathfrak{L}\{y'\} + \mathfrak{L}\{y\} = \mathfrak{L}\{\sin t\}$$

$$[sY(s) - y(0)] + Y(s) = \frac{1}{s^2 + 1^2} \quad \text{three substitutions from the table}$$

$$[sY(s) - 1] + Y(s) = \frac{1}{s^2 + 1} \quad \text{initial value substitution}$$

$$sY(s) + Y(s) = 1 + \frac{1}{s^2 + 1}$$

$$Y(s)(s + 1) = \frac{s^2 + 1}{s^2 + 1} + \frac{1}{s^2 + 1} \quad \text{prepare to combine terms}$$

$$Y(s)(s + 1) = \frac{s^2 + 2}{s^2 + 1}$$

$$Y(s) = \frac{1s^2 + 0s + 2}{(s + 1)(s^2 + 1)} = \frac{A}{s + 1} + \frac{Bs + C}{s^2 + 1}$$

$$= \frac{A}{s + 1}\frac{s^2 + 1}{s^2 + 1} + \frac{Bs + C}{s^2 + 1}\frac{s + 1}{s + 1}$$

$$= \frac{A\left(s^2 + 1\right) + (Bs + C)(s + 1)}{(s + 1)(s^2 + 1)}$$

$$= \frac{As^2 + A + Bs^2 + Bs + Cs + C}{(s + 1)(s^2 + 1)}$$

$$= \frac{As^2 + Bs^2 + Bs + Cs + A + C}{(s + 1)(s^2 + 1)}$$

$$= \frac{(A + B)s^2 + (B + C)s + A + C}{(s + 1)(s^2 + 1)}$$

Row	$f(t)$	$F(s)$
1	0	0
2	1	$\frac{1}{s}$
3	c	$\frac{c}{s}$
4	e^{at}	$\frac{1}{s-a}$
5	te^{at}	$\frac{1}{(s-a)^2}$
6	$\sin(at)$	$\frac{a}{s^2+a^2}$, $s > 0$
7	$\cos(at)$	$\frac{s}{s^2+a^2}$, $s > 0$
8	$f(t)$	$F(s)$ by definition
9	$f'(t)$	$sF(s) - f(0)$
10	$f''(t)$	$s^2F(s) - sf(0) - f'(0)$

By comparison

$A + B = 1$ (1)
$B + C = 0$ (2)
$A + C = 2$ (3)

$A - B = 2$ ((3) − (2))
$A + B = 1$ (1)
$2A = 3$

$A = \frac{3}{2}$
$B = -\frac{1}{2}$
$C = \frac{1}{2}$

$$Y(s) = \frac{\frac{3}{2}}{s + 1} + \frac{-\frac{1}{2}s + \frac{1}{2}}{s^2 + 1}$$

$$= \frac{\frac{3}{2}}{s + 1} + \frac{-\frac{1}{2}s}{s^2 + 1} + \frac{\frac{1}{2}}{s^2 + 1}$$

$$\mathfrak{L}^{-1}\{Y(s)\} = \mathfrak{L}^{-1}\left\{\frac{\frac{3}{2}}{s + 1}\right\} + \mathfrak{L}^{-1}\left\{\frac{-\frac{1}{2}s}{s^2 + 1}\right\} + \mathfrak{L}^{-1}\left\{\frac{\frac{1}{2}}{s^2 + 1}\right\}$$

$$y(t) = \frac{3}{2}\mathfrak{L}^{-1}\left\{\frac{1}{s + 1}\right\} - \frac{1}{2}\mathfrak{L}^{-1}\left\{\frac{s}{s^2 + 1}\right\} + \frac{1}{2}\mathfrak{L}^{-1}\left\{\frac{1}{s^2 + 1}\right\}$$

$$y(t) = \frac{3}{2}\mathfrak{L}^{-1}\left\{\frac{1}{s - (-1)}\right\} - \frac{1}{2}\mathfrak{L}^{-1}\left\{\frac{s}{s^2 + 1}\right\} + \frac{1}{2}\mathfrak{L}^{-1}\left\{\frac{1}{s^2 + 1}\right\}$$

$$= \frac{3}{2}e^{-t} - \frac{1}{2}\cos t + \frac{1}{2}\sin t$$

Row	$f(t)$	$F(s)$
1	0	0
2	1	$\frac{1}{s}$
3	c	$\frac{c}{s}$
4	e^{at}	$\frac{1}{s-a}$
5	te^{at}	$\frac{1}{(s-a)^2}$
6	$\sin(at)$	$\frac{a}{s^2+a^2}$, $s > 0$
7	$\cos(at)$	$\frac{s}{s^2+a^2}$, $s > 0$
8	$f(t)$	$F(s)$ by definition
9	$f'(t)$	$sF(s) - f(0)$
10	$f''(t)$	$s^2F(s) - sf(0) - f'(0)$

4. Solve $\frac{dy}{dt} - 2y = e^{5t}$, $y(0) = 3$.

Row	$f(t)$	$F(s)$
1	0	0
2	1	$\frac{1}{s}$
3	c	$\frac{c}{s}$
4	e^{at}	$\frac{1}{s-a}$
5	te^{at}	$\frac{1}{(s-a)^2}$
6	$\sin(at)$	$\frac{a}{s^2+a^2},\ s>0$
7	$\cos(at)$	$\frac{s}{s^2+a^2},\ s>0$
8	$f(t)$	$F(s)$ by definition
9	$f'(t)$	$sF(s) - f(0)$
10	$f''(t)$	$s^2F(s) - sf(0) - f'(0)$

$$\mathfrak{L}\left\{\frac{dy}{dt} - 2y\right\} = \mathfrak{L}\left\{e^{5t}\right\}$$

$$\mathfrak{L}\left\{\frac{dy}{dt}\right\} - \mathfrak{L}\{2y\} = \frac{1}{s-5} \quad \text{Row 4 from the table}$$

$$[sY(s) - y(0)] - 2\mathfrak{L}\{y\} = \frac{1}{s-5} \quad \text{Row 9 from the table}$$

$$sY(s) - 3 - 2Y(s) = \frac{1}{s-5} \quad \text{substitute initial condition, } y(0) = 3 \text{ and Row 8 from the table}$$

$$(s-2)Y(s) = \frac{1}{s-5} + 3 \quad \text{factor out the } Y(s)$$

$$(s-2)Y(s) = \frac{1}{s-5} + 3\frac{s-5}{s-5} \quad \text{prepare to add fractions}$$

$$(s-2)Y(s) = \frac{1}{s-5} + \frac{3s-15}{s-5}$$

$$(s-2)Y(s) = \frac{1+3s-15}{s-5}$$

$$(s-2)Y(s) = \frac{3s-14}{s-5}$$

$$Y(s) = \frac{3s-14}{(s-5)(s-2)} \quad \text{solve for Y(s)}$$

$$\mathfrak{L}^{-1}\{Y(s)\} = \mathfrak{L}^{-1}\left\{\frac{3s-14}{(s-5)(s-2)}\right\} \quad \text{take } \mathfrak{L}^{-1} \text{ of both sides}$$

$$y(t) = \mathfrak{L}^{-1}\left\{\frac{3s-14}{(s-5)(s-2)}\right\}$$

Even in full Laplace-transform tables, you will not see a form such as this. This is when the technique called *partial fractions* is useful. Solve for A and B in the equation

$$\frac{3s-14}{(s-5)(s-2)} = \frac{A}{s-5} + \frac{B}{s-2}$$

Multiply both sides by $(s-5)(s-2)$.

$$\begin{aligned} 3s - 14 &= A(s-2) + B(s-5) \\ &= As - 2A + Bs - 5B \\ &= (As + Bs) - 2A - 5B \\ &= (A+B)s - (2A+5B) \end{aligned}$$

By comparison

$$A + B = 3$$
$$2A + 5B = 14$$

$$2A + 2B = 6$$
$$2A + 5B = 14$$

$$-3B = -8$$
$$B = \frac{8}{3}$$
$$A = \frac{1}{3}$$

Substituting $A = \frac{1}{3}$ and $B = \frac{8}{3}$ into $\frac{3s-14}{(s-5)(s-2)} = \frac{A}{s-5} + \frac{B}{s-2}$ yields

$$\frac{3s-14}{(s-5)(s-2)} = \frac{\frac{1}{3}}{s-5} + \frac{\frac{8}{3}}{s-2}.$$

Substituting $\frac{3s-14}{(s-5)(s-2)} = \frac{\frac{1}{3}}{s-5} + \frac{\frac{8}{3}}{s-2}$ into $y(t) = \mathfrak{L}^{-1}\left\{\frac{3s-14}{(s-5)(s-2)}\right\}$ (above), we get

$$\begin{aligned} y(t) &= \mathfrak{L}^{-1}\left\{\frac{\frac{1}{3}}{s-5} + \frac{\frac{8}{3}}{s-2}\right\} \\ &= \mathfrak{L}^{-1}\left\{\frac{\frac{1}{3}}{s-5}\right\} + \mathfrak{L}^{-1}\left\{\frac{\frac{8}{3}}{s-2}\right\} \\ &= \frac{1}{3}\mathfrak{L}^{-1}\left\{\frac{1}{s-5}\right\} + \frac{8}{3}\mathfrak{L}^{-1}\left\{\frac{1}{s-2}\right\} \\ &= \frac{1}{3}e^{5t} + \frac{8}{3}e^{2t} \quad \text{Row 4 from the table.} \end{aligned}$$

Row	$f(t)$	$F(s)$
1	0	0
2	1	$\frac{1}{s}$
3	c	$\frac{c}{s}$
4	e^{at}	$\frac{1}{s-a}$
5	te^{at}	$\frac{1}{(s-a)^2}$
6	$\sin(at)$	$\frac{a}{s^2+a^2},\ s>0$
7	$\cos(at)$	$\frac{s}{s^2+a^2},\ s>0$
8	$f(t)$	$F(s)$ by definition
9	$f'(t)$	$sF(s) - f(0)$
10	$f''(t)$	$s^2F(s) - sf(0) - f'(0)$

Chapter 23 Review

The definition of power series was extended to produce the definition of the Laplace transform. That definition was used on several functions and a miniature Laplace-transform table was presented. That table was then used to solve four differential equations using the Laplace-transform technique.

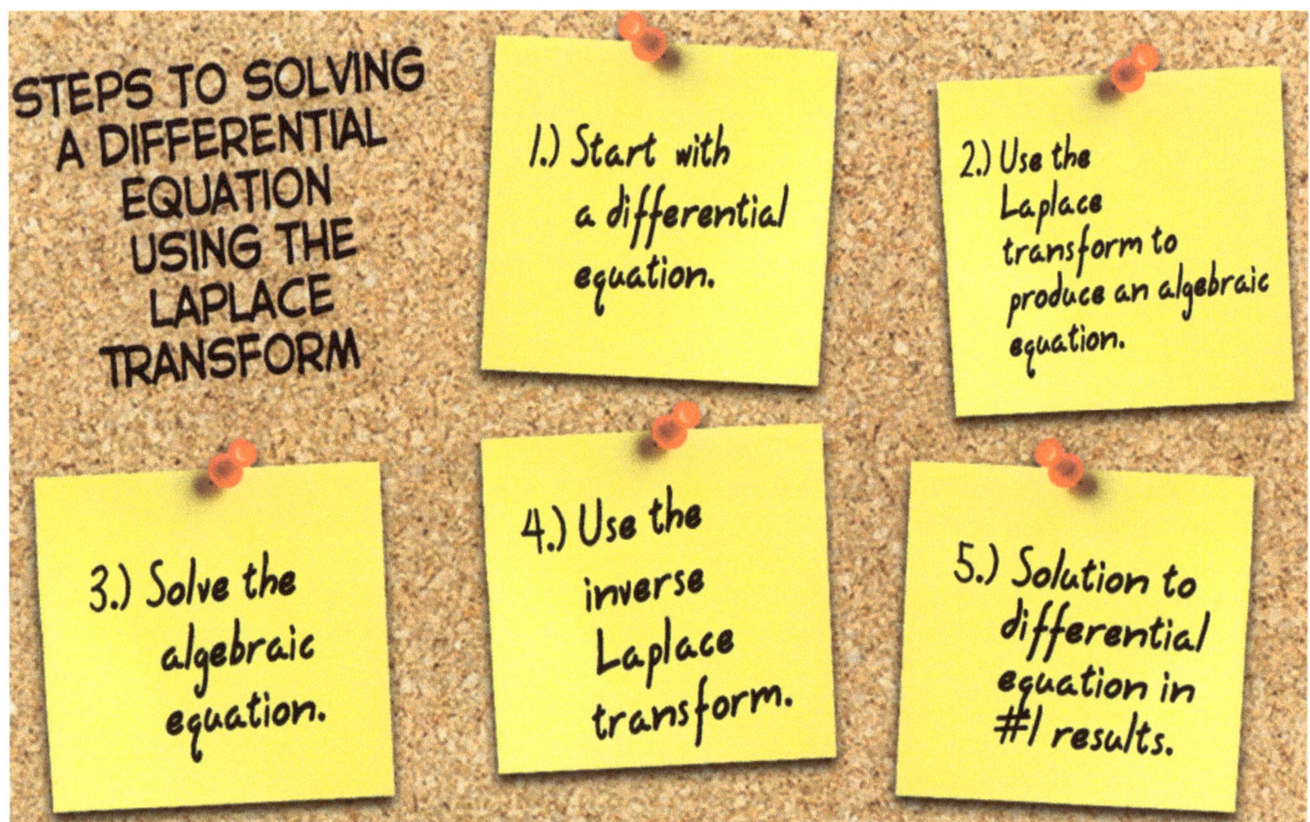

STEPS TO SOLVING A DIFFERENTIAL EQUATION USING THE LAPLACE TRANSFORM
1.) Start with a differential equation.
2.) Use the Laplace transform to produce an algebraic equation.
3.) Solve the algebraic equation.
4.) Use the inverse Laplace transform.
5.) Solution to differential equation in #1 results.

Steps 2 and 4 in this algorithm are usually done with a table lookup.

Chapter 24
Applications of the Laplace Transform

Electrical Engineering

The following application comes from pages 320 and 321 of *A First Course in Differential Equations* by Dennis G. Zill, 8th edition, Brooks/Cole. Errors, if any, are due to this author's modifications.

The figure below shows an electrical circuit with voltage source E; currents indicated as $i_1(t)$, $i_2(t)$, and $i_3(t)$; and an inductor (L), a resistor (R), and a capacitor (C). See the glossary for a brief physical discussion of resistors, capacitors, and inductors. Note that the current i_1 is being split into parallel currents i_2 and i_3. Those currents will be the current flowing through the system under the given starting conditions. Because they are connected in parallel, the resistor (R) and the capacitor (C) will share the same voltage. Physics teaches us two important facts: 1) Kirchhoff's Voltage Law: The total voltage around a loop must be zero—$L\frac{di_1}{dt} + Ri_2 = E(t)$, where $L\frac{di_1}{dt}$ is the voltage across the inductor. 2) Kirchhoff's Current Law: The current emanating from source E, (i_1), is split at the node into two currents—$i_1 = i_2 + i_3$, where $i_3 = RC\frac{di_2}{dt}$ (the rate of current going through the resistor).* Hence, $RC\frac{di_2}{dt} + i_2 - i_1 = 0$.

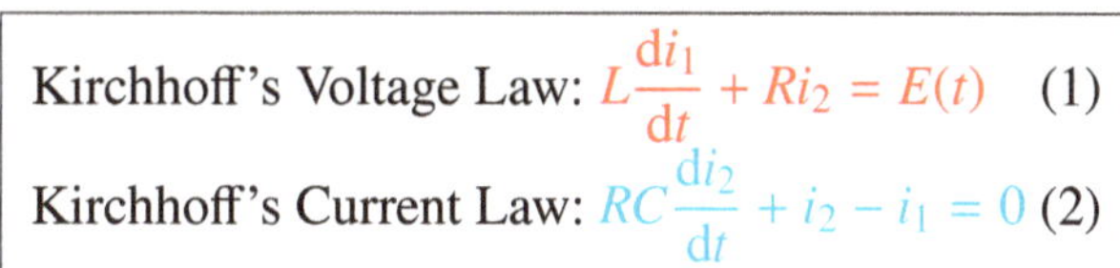
Kirchhoff's Voltage Law: $L\dfrac{di_1}{dt} + Ri_2 = E(t)$ (1)

Kirchhoff's Current Law: $RC\dfrac{di_2}{dt} + i_2 - i_1 = 0$ (2)

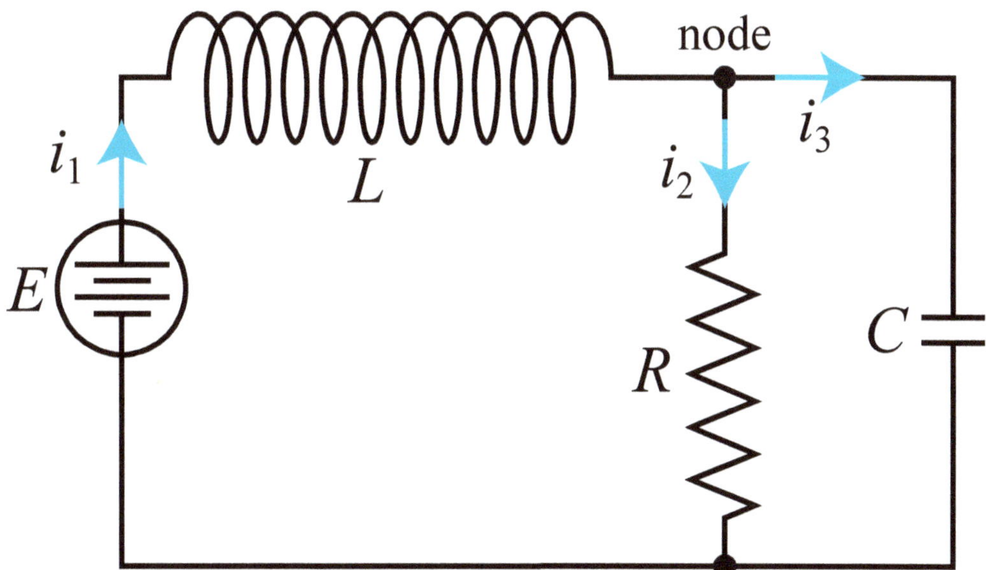

By finding currents i_1 and i_2, we can then determine both i_3 and also the voltage. Solve the system of equations for i_1 and i_2 under the conditions $E(t) = 60U(t)$ V,† $L = 1$ H (henry), $R = 50\,\Omega$, and $C = 10^{-4}$ F (farad). It is important to note that, because $E(t)$ is "switched on" at $t = 0$, currents i_1 and i_2 are initially zero.

*The current through a capacitor is $C\frac{dv}{dt}$ and, because R and C are in parallel, $V_R = V_C$. Using Ohm's Law ($V = IR$), we find that $V_C = Ri_2$, so $C\frac{dV_C}{dt} = RC\frac{di_2}{dt}$.

†E is 0 V before $t = 0$, when it is "switched on," and instantaneously assumes a value of 60 V, which it maintains ever after. $U(t)$ is known as the Heaviside Step Function named after Oliver Heaviside. It has a value of 0 for all times before zero and a value of 1 for all times from zero onward. It is used as a "turn on" function for modeling signals or initial conditions that start at a specific time and continue on indefinitely.

Let's take these equations one at a time.

$$L\frac{di_1}{dt} + Ri_2 = E(t) \qquad (1)$$

$$1\frac{di_1}{dt} + 50i_2 = 60 \quad i_1(0) = 0,\ i_2(0) = 0$$

Applying the Laplace transform

Row	$f(t)$	$F(s)$
1	0	0
2	1	$\frac{1}{s}$
3	c	$\frac{c}{s}$
4	e^{at}	$\frac{1}{s-a}$
5	te^{at}	$\frac{1}{(s-a)^2}$
6	$\sin(at)$	$\frac{a}{s^2+a^2},\ s>0$
7	$\cos(at)$	$\frac{s}{s^2+a^2},\ s>0$
8	$f(t)$	$F(s)$ by definition
9	$f'(t)$	$sF(s) - f(0)$
10	$f''(t)$	$s^2F(s) - sf(0) - f'(0)$

$$\mathfrak{L}\left\{\frac{di_1}{dt} + 50i_2\right\} = \mathfrak{L}\{60\}$$

$$\mathfrak{L}\left\{\frac{di_1}{dt}\right\} + \mathfrak{L}\{50i_2\} = \frac{60}{s} \qquad \text{Row 3}$$

$$[sI_1(s) - i_1(0)] + 50I_2(s) = \frac{60}{s} \qquad \text{Rows 9 \& 8}$$

$$[sI_1(s) - 0] + 50I_2(s) = \frac{60}{s} \qquad i_1(0) = 0$$

$$sI_1(s) + 50I_2(s) = \frac{60}{s}$$

where $I_1(s) = \mathfrak{L}\{i_1(t)\}$ and $I_2(s) = \mathfrak{L}\{i_2(t)\}$.
s in this application represents a complex frequency of the form $\sigma + i\omega$. The real part represents the transient response, while the imaginary part represents the steady-state frequency response.

$$RC\frac{di_2}{dt} + i_2 - i_1 = 0 \qquad (2)$$

$$50\left(10^{-4}\right)\frac{di_2}{dt} + i_2 - i_1 = 0 \quad i_1(0) = 0,\ i_2(0) = 0$$

Applying the Laplace transform

Row	$f(t)$	$F(s)$
1	0	0
2	1	$\frac{1}{s}$
3	c	$\frac{c}{s}$
4	e^{at}	$\frac{1}{s-a}$
5	te^{at}	$\frac{1}{(s-a)^2}$
6	$\sin(at)$	$\frac{a}{s^2+a^2},\ s>0$
7	$\cos(at)$	$\frac{s}{s^2+a^2},\ s>0$
8	$f(t)$	$F(s)$ by definition
9	$f'(t)$	$sF(s) - f(0)$
10	$f''(t)$	$s^2F(s) - sf(0) - f'(0)$

$$\mathfrak{L}\left\{50(10^{-4})\frac{di_2}{dt} + i_2 - i_1\right\} = \mathfrak{L}\{0\}$$

$$\mathfrak{L}\left\{\frac{50}{10{,}000}\frac{di_2}{dt} + i_2 - i_1\right\} = 0 \qquad \text{Row 1}$$

$$\mathfrak{L}\left\{\frac{1}{200}\frac{di_2}{dt}\right\} + \mathfrak{L}\{i_2\} - \mathfrak{L}\{i_1\} = 0$$

$$\frac{1}{200}[sI_2(s) - i_2(0)] + I_2(s) - I_1(s) = 0 \qquad \text{Rows 9 \& 8}$$

$$\frac{1}{200}[sI_2(s) - 0] + I_2(s) - I_1(s) = 0 \qquad i_2(0) = 0$$

$$\frac{1}{200}[sI_2(s)] + I_2(s) - I_1(s) = 0$$

$$sI_2(s) + 200I_2(s) - 200I_1(s) = 0$$

$$(s + 200)I_2(s) - 200I_1(s) = 0$$

$$-200I_1(s) + (s + 200)I_2(s) = 0$$

where $I_1(s) = \mathfrak{L}\{i_1(t)\}$ and $I_2(s) = \mathfrak{L}\{i_2(t)\}$.

We now solve the system for $I_1(s)$ and $I_2(s)$.

1. $sI_1(s) + 50I_2(s) = \frac{60}{s}$

2. $-200I_1(s) + (s + 200)I_2(s) = 0$

Remember, these are differential equations! The variables here are functions. However, it will be easier to do our work (fewer keystrokes) using I_1 and I_2 and keeping in mind that $I_1 = I_1(s)$ and $I_2 = I_2(s)$.

Step 1: Solve for I_2 in equation (2):

$$-200I_1 + (s+200)I_2 = 0$$
$$(s+200)I_2 = 200I_1$$
$$I_2 = \frac{200I_1}{s+200}$$

Substitute this value for I_2 in (1):

$$sI_1 + 50I_2 = \frac{60}{s}$$
$$sI_1 + 50\frac{200I_1}{s+200} = \frac{60}{s}$$
$$sI_1 + \frac{10{,}000I_1}{s+200} = \frac{60}{s}$$

Factor out I_1:

$$I_1\left(s + \frac{10{,}000}{s+200}\right) = \frac{60}{s}$$

Solve for I_1:

$$I_1\left(\frac{s(s+200)}{(s+200)} + \frac{10{,}000}{s+200}\right) = \frac{60}{s}$$
$$I_1\frac{s(s+200)+10{,}000}{s+200} = \frac{60}{s}$$
$$I_1 = \frac{60}{s}\,\frac{s+200}{s(s+200)+10{,}000}$$
$$I_1 = \frac{60}{s}\,\frac{s+200}{s^2+200s+10{,}000}$$
$$I_1 = \frac{60(s+200)}{s(s+100)^2}$$

Step 2: Substitute the value for I_1 from the left box into equation (1).

$$sI_1 + 50I_2 = \frac{60}{s}$$
$$s\frac{60(s+200)}{s(s+100)^2} + 50I_2 = \frac{60}{s}$$

Now solve for I_2

$$50I_2 = \frac{60}{s} - s\frac{60(s+200)}{s(s+100)^2}$$
$$= \frac{60}{s}\frac{(s+100)^2}{(s+100)^2} - s\frac{60(s+200)}{s(s+100)^2}$$
$$= \frac{60(s+100)^2 - 60s(s+200)}{s(s+100)^2}$$
$$= 60\frac{(s+100)^2 - s(s+200)}{s(s+100)^2}$$
$$= 60\frac{s^2 + 200s + 10{,}000 - s^2 - 200s}{s(s+100)^2}$$
$$= 60\frac{10{,}000}{s(s+100)^2}$$
$$= \frac{600{,}000}{s(s+100)^2}$$
$$50I_2 = \frac{600{,}000}{s(s+100)^2}$$
$$I_2 = \frac{12{,}000}{s(s+100)^2}$$

Step 3: Prepare to take the inverse Laplace transform of I_1 by expressing it as a partial-fraction expression.

$$I_1 = \frac{60(s+200)}{s(s+100)^2}$$
$$= \frac{60s+12{,}000}{s(s+100)^2} = \frac{A}{s} + \frac{B}{s+100} + \frac{C}{(s+100)^2}$$

Multiply both sides of this equation by $s(s+100)^2$.

$$\begin{aligned}0s^2+60s+12{,}000 &= A(s+100)^2 + B(s)(s+100)\\ &\quad + Cs\\ &= A(s^2+200s+10{,}000)\\ &\quad + B(s^2+100s) + Cs\\ &= As^2 + 200As + 10{,}000A + Bs^2\\ &\quad + 100Bs + Cs\\ &= As^2 + Bs^2 + 200As + 100Bs\\ &\quad + Cs + 10{,}000A\\ &= (A+B)s^2 + (200A+100B+C)s\\ &\quad + 10{,}000A\end{aligned}$$

Step 4: Prepare to take the inverse Laplace transform of I_2 by expressing it as a partial-fraction expression.

$$I_2 = \frac{12{,}000}{s(s+100)^2}$$
$$= \frac{12{,}000}{s(s+100)^2} = \frac{A}{s} + \frac{B}{s+100} + \frac{C}{(s+100)^2}$$

Multiply both sides of this equation by $s(s+100)^2$.

$$\begin{aligned}0s^2+0s+12{,}000 &= A(s+100)^2 + B(s)(s+100)\\ &\quad + Cs\\ &= A(s^2+200s+10{,}000)\\ &\quad + B(s^2+100s) + Cs\\ &= As^2 + 200As + 10{,}000A + Bs^2\\ &\quad + 100Bs + Cs\\ &= As^2 + Bs^2 + 200As + 100Bs\\ &\quad + Cs + 10{,}000A\\ &= (A+B)s^2 + (200A+100B+C)s\\ &\quad + 10{,}000A\end{aligned}$$

By inference (continuing the solution for I_1)

$$A = \frac{12{,}000}{10{,}000} = \frac{6}{5}$$
$$B = -\frac{6}{5}$$
$$C = -60$$

$$A + B = 0$$
$$200A + 100B + C = 60$$
$$10{,}000A = 12{,}000$$

$$I_1 = \frac{60s + 12{,}000}{s(s+100)^2} = \frac{\frac{6}{5}}{s} + \frac{-\frac{6}{5}}{s+100} + \frac{-60}{(s+100)^2}$$

Take the inverse Laplace transform on both sides.

$$\mathfrak{L}^{-1}\{I_1(s)\} = \mathfrak{L}^{-1}\left\{\frac{\frac{6}{5}}{s} + \frac{-\frac{6}{5}}{s+100} + \frac{-60}{(s+100)^2}\right\}$$
$$i_1(t) = \mathfrak{L}^{-1}\left\{\frac{\frac{6}{5}}{s}\right\} + \mathfrak{L}^{-1}\left\{\frac{-\frac{6}{5}}{s+100}\right\} + \mathfrak{L}^{-1}\left\{\frac{-60}{(s+100)^2}\right\}$$
$$= \frac{6}{5}\mathfrak{L}^{-1}\left\{\frac{1}{s}\right\} - \frac{6}{5}\mathfrak{L}^{-1}\left\{\frac{1}{s+100}\right\} - 60\mathfrak{L}^{-1}\left\{\frac{1}{(s+100)^2}\right\}$$

Using the Laplace-transform table below (Rows 2, 4, and 5), we find current $i_1(t)$:

$$i_1(t) = \frac{6}{5} - \frac{6}{5}e^{-100t} - 60te^{-100t}.$$

Row	$f(t)$	$F(s)$
1	0	0
2	1	$\frac{1}{s}$
3	c	$\frac{c}{s}$
4	e^{at}	$\frac{1}{s-a}$
5	te^{at}	$\frac{1}{(s-a)^2}$
6	$\sin(at)$	$\frac{a}{s^2+a^2}$, $s > 0$
7	$\cos(at)$	$\frac{s}{s^2+a^2}$, $s > 0$
8	$f(t)$	$F(s)$ by definition
9	$f'(t)$	$sF(s) - f(0)$
10	$f''(t)$	$s^2F(s) - sf(0) - f'(0)$

By inference (continuing the solution for I_2)

$$A = \frac{12{,}000}{10{,}000} = \frac{6}{5}$$
$$B = -\frac{6}{5}$$
$$C = -120$$

$$A + B = 0$$
$$200A + 100B + C = 0$$
$$10{,}000A = 12{,}000$$

$$I_2 = \frac{12{,}000}{s(s+100)^2} = \frac{\frac{6}{5}}{s} + \frac{-\frac{6}{5}}{s+100} + \frac{-120}{(s+100)^2}$$

Take the inverse Laplace transform on both sides.

$$\mathfrak{L}^{-1}\{i_2(s)\} = \mathfrak{L}^{-1}\left\{\frac{\frac{6}{5}}{s} + \frac{-\frac{6}{5}}{s+100} + \frac{(-120)}{(s+100)^2}\right\}$$
$$i_2(t) = \mathfrak{L}^{-1}\left\{\frac{\frac{6}{5}}{s}\right\} + \mathfrak{L}^{-1}\left\{\frac{-\frac{6}{5}}{s+100}\right\} + \mathfrak{L}^{-1}\left\{\frac{-120}{(s+100)^2}\right\}$$
$$= \frac{6}{5}\mathfrak{L}^{-1}\left\{\frac{1}{s}\right\} - \frac{6}{5}\mathfrak{L}^{-1}\left\{\frac{1}{s+100}\right\} - 120\mathfrak{L}^{-1}\left\{\frac{1}{(s+100)^2}\right\}$$

Using the Laplace-transform table below (Rows 2, 4, and 5), we find current $i_2(t)$:

$$I_2(t) = \frac{6}{5} - \frac{6}{5}e^{-100t} - 120te^{-100t}.$$

Row	$f(t)$	$F(s)$
1	0	0
2	1	$\frac{1}{s}$
3	c	$\frac{c}{s}$
4	e^{at}	$\frac{1}{s-a}$
5	te^{at}	$\frac{1}{(s-a)^2}$
6	$\sin(at)$	$\frac{a}{s^2+a^2}$, $s > 0$
7	$\cos(at)$	$\frac{s}{s^2+a^2}$, $s > 0$
8	$f(t)$	$F(s)$ by definition
9	$f'(t)$	$sF(s) - f(0)$
10	$f''(t)$	$s^2F(s) - sf(0) - f'(0)$

Note that, as $t \to \infty$, both $i_1(t)$ and $i_2(t)$ approach $\frac{6}{5}$, so $i_3(t) \to 0$. The MatLab graph on the next page seems to confirm both of Kirchhoff's Laws: 1) Kirchhoff's Voltage Law (the total voltage around a loop must be zero) $L\frac{di_1}{dt} + Ri_2 = E(t)$, where $L\frac{di_1}{dt}$ is the rate of voltage across an inductor; and 2) Kirchhoff's Current Law (the current emanating from source E (i_1) is split at the node into two currents) $i_1 = i_2 + i_3$. Again, note that, as $t \to \infty$, both $i_1(t)$ and $i_2(t)$ approach $\frac{6}{5}$ and $i_3(t) \to 0$.

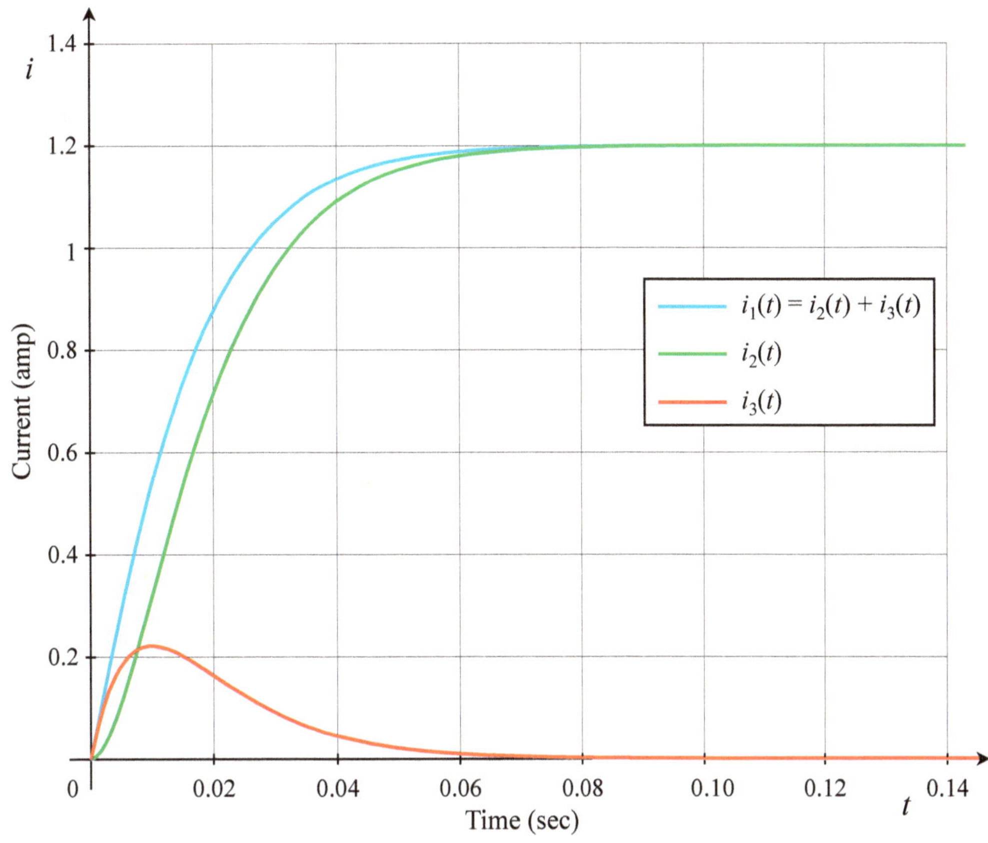


```
t = 0:0.0001:0.14; % seconds
i1 = (6/5) - (6/5)*exp(-100*t) - 60*t.*exp(-100*t);
i2 = (6/5) - (6/5)*exp(-100*t) - 120*t.*exp(-100*t);
i3 = i1-i2;

figure; hold on; grid on;
plot(t,i1,'b','LineWidth',2);
plot(t,i2,'g','LineWidth',2);
plot(t,i3,'r','LineWidth',2);
legend('i_1(t) = i_2(t) + i_3(t)','i_2(t)','i_3(t)','Location','East')

ylabel('Current (amp)')
xlabel('Time (sec)')
```

For those of you into circuits, here is a website with more circuit applications problems **and worked out answers** demonstrating the use of the Laplace transform (http://www.intmath.com/laplace-transformation/10-applications.php).

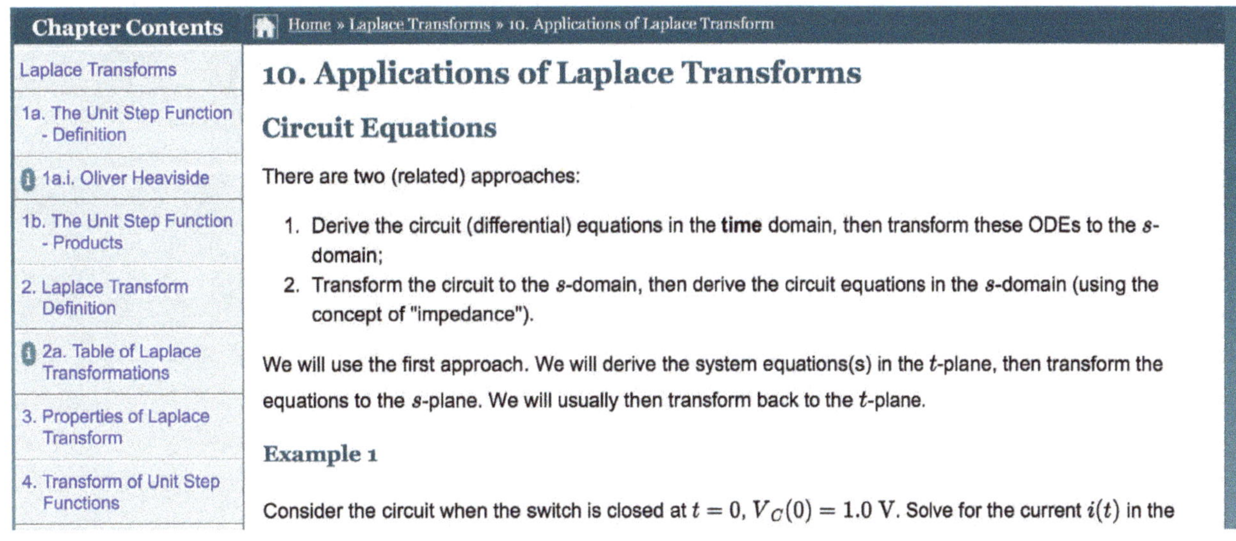

Chapter Contents

Home » Laplace Transforms » 10. Applications of Laplace Transform

Laplace Transforms
1a. The Unit Step Function - Definition
1a.i. Oliver Heaviside
1b. The Unit Step Function - Products
2. Laplace Transform Definition
2a. Table of Laplace Transformations
3. Properties of Laplace Transform
4. Transform of Unit Step Functions

10. Applications of Laplace Transforms

Circuit Equations

There are two (related) approaches:

1. Derive the circuit (differential) equations in the **time** domain, then transform these ODEs to the s-domain;
2. Transform the circuit to the s-domain, then derive the circuit equations in the s-domain (using the concept of "impedance").

We will use the first approach. We will derive the system equations(s) in the t-plane, then transform the equations to the s-plane. We will usually then transform back to the t-plane.

Example 1

Consider the circuit when the switch is closed at $t = 0$, $V_C(0) = 1.0$ V. Solve for the current $i(t)$ in the

Hopefully, someone will continue to maintain that website as it has a lot of helpful examples:

<table>
<tr>
<td>Consider the following circuit. When the switch is closed at $t = 0$, $V_c(0) = 1.0\,\text{V}$. Solve for the current $i(t)$ in the circuit.
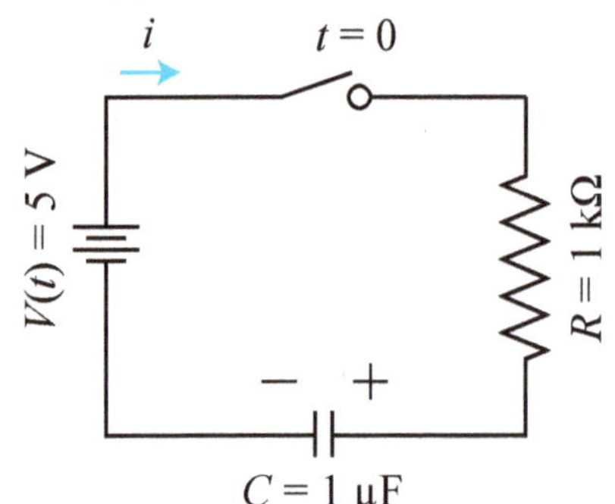
</td>
<td>Solve for $i(t)$ in the following circuit, given that $V(t) = 10\sin 5t\text{V}$, $R = 4\,\Omega$ and $L = 2\,\text{H}$.
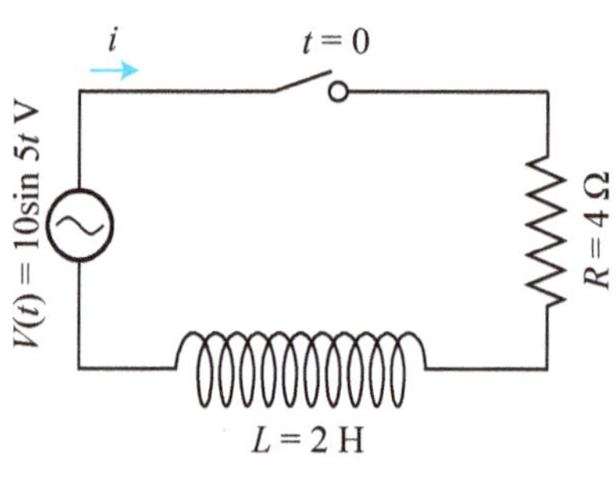
</td>
<td>In the circuit shown below, the capacitor is uncharged at time $t = 0$. If the switch is then closed, find the currents $i_1(t)$ and $i_2(t)$, and the charge on C at time $t > 0$.
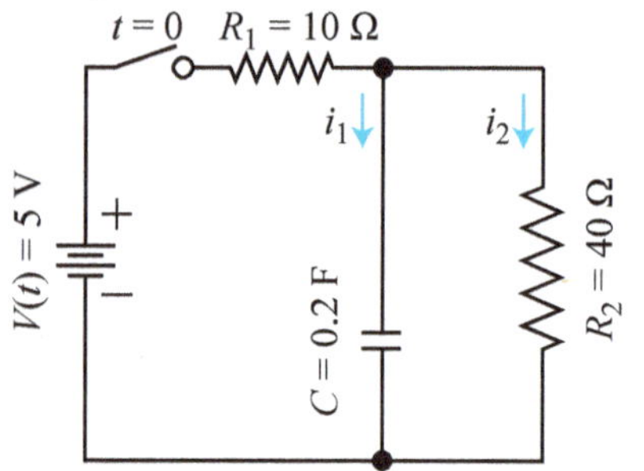
</td>
</tr>
<tr>
<td>In the circuit shown, the capacitor has an initial charge of 1 mC and the switch is in position 1 long enough to establish the steady state. The switch is moved from position 1 to 2 at $t = 0$. Obtain the transient current $i(t)$ for $t > 0$.
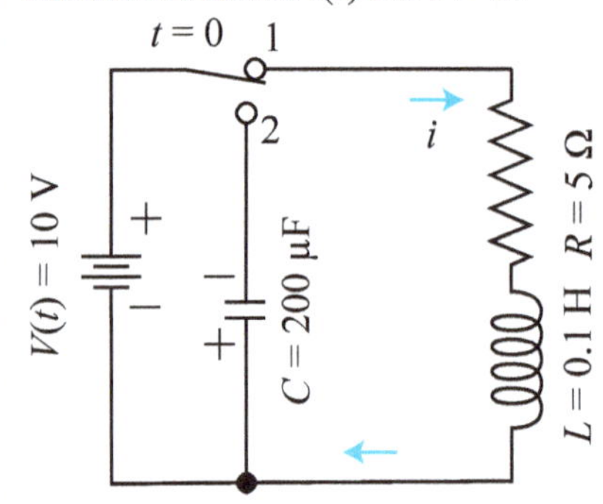
</td>
<td>The system is quiescent. Find the loop current $i_2(t)$.
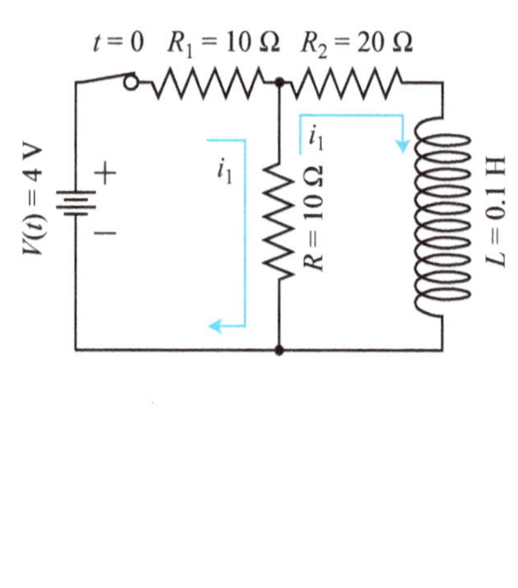
</td>
<td>A rectangular pulse is applied to the RC circuit shown. Find the response, $v(t)$.
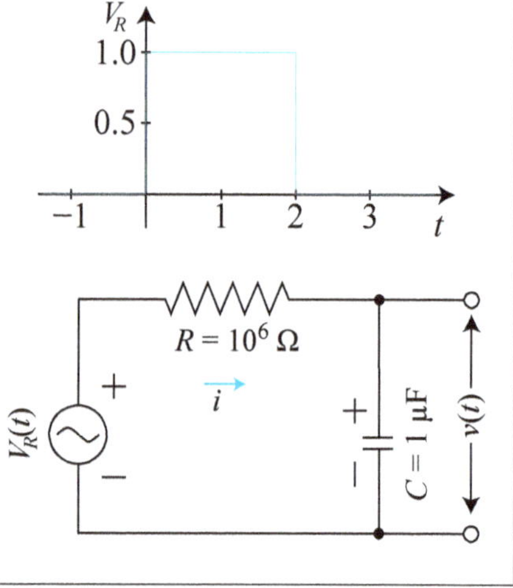
</td>
</tr>
</table>

Mechanical Engineering

In Chapter 15 we saw the following problem. A mass weighing 8 lb stretches a spring 2 ft. Assuming that a damping force equal to twice the instantaneous velocity acts on the system (with the appropriately adjusted dimensions), determine the equation of motion if the mass is initially released from the equilibrium position with a downward velocity of $5\,\frac{\text{ft}}{\text{s}}$.

From Hooke's law, $F = ky$, we get $8 = k(2)$ hence $k = \frac{8}{2} = 4$. Since $W = mg$, $m = \frac{8\,\text{lb}}{32\,\frac{\text{ft}}{\text{s}}} = \frac{1}{4}\text{slug}$. Thus, the differential equation of motion would be

$$\frac{\mathrm{d}^2y}{\mathrm{d}t^2} + \frac{\beta}{m}\frac{\mathrm{d}y}{\mathrm{d}t} + \frac{k}{m}y = 0 \quad \text{where } \beta \text{ is the damping factor}$$

$$\frac{\mathrm{d}^2y}{\mathrm{d}t^2} + \frac{2}{\frac{1}{4}}\frac{\mathrm{d}y}{\mathrm{d}t} + \frac{4}{\frac{1}{4}}y = 0$$

$$\frac{\mathrm{d}^2y}{\mathrm{d}t^2} + 8\frac{\mathrm{d}y}{\mathrm{d}t} + 16y = 0.$$

Using the Chapter 14 formula $y_{\text{gen}} = c_1e^{rt} + c_2te^{rt}$ and initial conditions $y_0 = 0$ and $y_0' = -5$ in Chapter 15, we got the solution $y_{\text{gen}} = -5te^{rt}$, where $r^2 + 8r + 16 = 0 \to (r+4)(r+4) = 0$, so $r = -4$.

Let's try that again using the Laplace transform technique.

Row	$f(t)$	$F(s)$
1	0	0
2	1	$\frac{1}{s}$
3	c	$\frac{c}{s}$
4	e^{at}	$\frac{1}{s-a}$
5	te^{at}	$\frac{1}{(s-a)^2}$
6	$\sin(at)$	$\frac{a}{s^2+a^2},\ s>0$
7	$\cos(at)$	$\frac{s}{s^2+a^2},\ s>0$
8	$f(t)$	$F(s)$ by definition
9	$f'(t)$	$sF(s)-f(0)$
10	$f''(t)$	$s^2F(s)-sf(0)-f'(0)$

$$
\begin{aligned}
\frac{d^2y}{dt^2}+8\frac{dy}{dt}+16y &= 0\\
\mathfrak{L}\left\{\frac{d^2y}{dt^2}+8\frac{dy}{dt}+16y\right\} &= \mathfrak{L}\{0\}\\
\mathfrak{L}\left\{\frac{d^2y}{dt^2}\right\}+\mathfrak{L}\left\{8\frac{dy}{dt}\right\}+\mathfrak{L}\{16y\} &= 0 \quad \text{Row 1}\\
\left[s^2Y(s)-sy(0)-y'(0)\right]+8\mathfrak{L}\left\{\frac{dy}{dt}\right\}+16\mathfrak{L}\{y\} &= 0 \quad \text{Row 10}\\
\left[s^2Y(s)-s(0)-(-5)\right]+[8sY(s)-8y(0)]+16Y(s) &= 0 \quad \text{Row 9}\\
\left[s^2Y(s)-(-5)\right]+[8sY(s)-8y(0)]+16Y(s) &= 0\\
s^2Y(s)+5+[8sY(s)-0]+16Y(s) &= 0\\
s^2Y(s)+5+8sY(s)+16Y(s) &= 0\\
s^2Y(s)+8sY(s)+16Y(s) &= -5\\
\left(s^2+8s+16\right)Y(s) &= -5\\
(s+4)^2Y(s) &= -5
\end{aligned}
$$

$$
Y(s)=\frac{-5}{(s+4)^2}=\frac{0s^2+0s-5}{(s+4)^2}=\frac{A}{s+4}+\frac{B}{(s+4)^2}=\frac{A}{s+4}\frac{(s+4)}{(s+4)}+\frac{B}{(s+4)^2}
$$

$$
\frac{0s^2+0s-5}{(s+4)^2}=\frac{As+(4A+B)}{(s+4)^2}
$$

Therefore, $A=0$ and, since $4A+B=-5$, we have $B=-5$.

$$
\begin{aligned}
Y(s) &= \frac{0}{s+4}+\frac{-5}{(s+4)^2}\\
\mathfrak{L}^{-1}\{Y(s)\} &= \mathfrak{L}^{-1}\left\{\frac{-5}{(s+4)^2}\right\}\\
y(t) &= -5\mathfrak{L}^{-1}\left\{\frac{1}{(s-(-4))^2}\right\}\\
&= -5te^{-4t} \quad \text{Row 5}
\end{aligned}
$$

Fortunately this answer is the same as the one we got back in Chapter 15. Otherwise, there would be egg on the author's face.

Chapter 24 Review

The Laplace transform was shown in two different applied situations.

Chapter 25
Reteaching, Reinforcing, Reviewing

"The inevitability of differential equations."
—Dr. Gina-Carlo Rota

Differential-Equation Types Covered in This Book

Chap	Type of Differential Equation, Strategy	Generic Form, y'	Generic Form, $\frac{dy}{dx}$
1–5	Separable (especially growth and decay): Integrate both sides.	$y' = f(x) \times g(y),\ g(y) \neq 0$	$\frac{dy}{dx} = f(x) \times g(y),\ g(y) \neq 0$
6	Logistic, $P(x) = \frac{K}{(A \times e^{-kt} + 1)}$, where $A = \frac{K-P_0}{P_0}$.	$P' = kP\left(1 - \frac{P}{K}\right)$	$\frac{dP}{dt} = kP\left(1 - \frac{P}{K}\right)$
7	Systems of differential equations: Predator–Prey.	$r' = k_1 r + k_2 rf$ $f' = k_3 r - k_4 rf$	$\frac{dr}{dt} = k_1 r + k_2 rf$ $\frac{df}{dt} = k_3 r - k_4 rf$
8	Linear: Multiply both sides by the integrating factor to yield an integrable form on both sides.	$y' + P(x) \times y = Q(x)$	$\frac{dy}{dx} + P(x) \times y = Q(x)$
10	Bernoulli: Solve by substitution.	$y' + P(x)y = Q(x)y^n$	$\frac{dy}{dx} + P(x) \times y = Q(x)y^n$
12	Exact: Check for "exact" form. Confirm $\frac{\partial M}{\partial y} = \frac{\partial N}{\partial x}$. $f(x,y) = \int M(x,y)\,dx + \int \left[N(x,y) - \frac{\partial}{\partial y}\int M(x,y)\,dx\right] dy = C$	$M(x,y) + N(x,y)y'(x) = 0$	$M(x,y)\,dx + N(x,y)\,dy = 0$ $M(x,y) + N(x,y)\frac{dy}{dx} = 0$
13	First-Order Homogeneous: Solve by substitution. All terms in f and g of the same order.	$y' = \frac{f(x,y)}{g(x,y)}$	$\frac{dy}{dx} = \frac{f(x,y)}{g(x,y)}$
14	Second-Order Homogeneous: Separate, real characteristic roots $y_{gen} = c_1 e^{r_1 x} + c_2 e^{r_2 x}$ Repeated real characteristic roots $y_{gen} = c_1 e^{rx} + c_2 x e^{rx}$ Complex characteristic roots $y_{gen} = c_1 e^{\alpha x}\cos\beta x + c_2 e^{\alpha x}\sin\beta x$	$y'' + by' + cy = 0$	$\frac{d^2y}{dx} + b\frac{dy}{dx} + cy = 0$
16	Predator–Pursuit: Systems of Differential Equations.	$x_1' =$ path of predator $x_2' =$ path of prey	$\frac{dx_1}{dt} =$ path of predator $\frac{dx_2}{dt} =$ path of prey
20	Systems of Homogeneous Linear Differential Equations: Separate, real eigenvalues $\vec{x} = c_1\vec{\varepsilon}_1 e^{\lambda_1 t} + c_2\vec{\varepsilon}_2 e^{\lambda_2 t}$ Repeated real eigenvalues $\vec{x} = c_1\vec{\varepsilon}e^{\lambda t} + c_2\left(\vec{\varepsilon}te^{\lambda t} + \vec{\rho}e^{\lambda t}\right)$ Complex eigenvalues $\vec{x} = c_1\vec{\varepsilon}_1 e^{\lambda_1 t} + c_2\vec{\varepsilon}_2 e^{\lambda_2 t}$ where $\lambda_1 = r + ci$, and $\lambda_2 = r - ci$ and $\vec{\varepsilon}_1 = \alpha + \beta i$ and $\vec{\varepsilon}_2 = \alpha - \beta i$.	$x_1' = k_1 x_1 + k_2 x_2$ $x_2' = k_3 x_1 - k_4 x_2$	$\frac{dx_1}{dt} = k_1 x_1 + k_2 x_2$ $\frac{dx_2}{dt} = k_3 x_1 - k_4 x_2$
23	Laplace Transform	$\mathfrak{L}\{y(t)\} = Y(s)$	

Some differential equations can be classified as both separable and first-order linear.

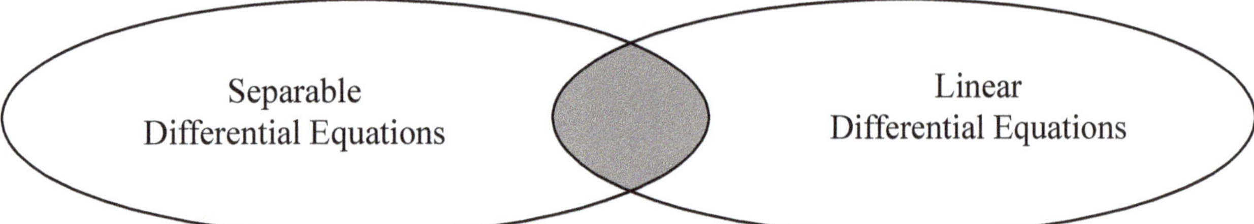

Hence, they can be solved by more than one technique.

According to Newton's second law, $F = ma$. The definition of acceleration is change in velocity over the change in time: $\frac{dv}{dt}$. Hence, $F = m \times \frac{dv}{dt}$. For simplification consider that there are two forces acting on a falling object: its weight and air resistance. Weight is defined to be mass times gravity (mg). In a simplified form, air resistance can be represented as $-kv$ where k is a constant of proportionality based on the object's shape. For example, a parachute has a much larger impact on the velocity of a falling object than does a cannon ball. Putting all this together, we get what is shown at right.

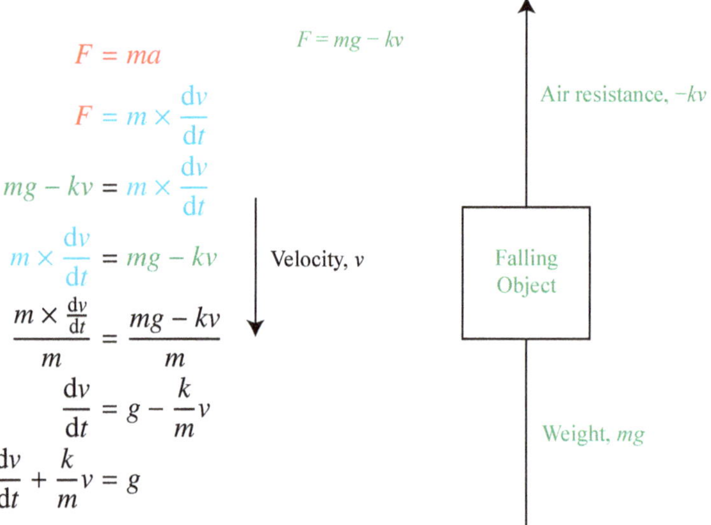

Here we have an equation with four variables—k, m, v, and g—and a rate of change impacted by the object's shape.

A ball weighing 2 lb is dropped from 5,000 feet with no initial velocity. As it falls, air resistance exerts a force of $\frac{1}{8}v$ in pounds where v is the velocity of the ball in feet per second and $\frac{1}{8}$ is the constant of proportionality, k. Find and solve the differential equation that models this situation.

$$\frac{dv}{dt} + \frac{k}{m}v = g,$$

where $g = 32\,\frac{\text{ft}}{\text{s}}$ and, since $w = m \times g$, $2 = m \times 32$, so $m = \frac{1}{16}$. So,

$$\frac{dv}{dt} + \frac{\frac{1}{8}}{\frac{1}{16}}v = 32\,\frac{\text{ft}}{\text{s}}$$

$$\frac{dv}{dt} + 2v = 32, \quad \text{differential equation applying Newton's second law}$$

At this point, it would be good to think back to Chapter 5 when we solved a differential equation involving electric circuits using Kirchhoff's Law. There, we started with the differential equation $L\frac{dI}{dt} + RI = E(t)$, substituted in the given information to get $2\frac{dI}{dt} + 10I = 30$, divided away the leading coefficient to get $\frac{dI}{dt} + 5I = 15$. Hmmm! Something seems familiar here.

$\frac{dv}{dt} + 2v = 32$ **differential equation resulting from applying Newton's second law. Chapter 25 (above)**

$\frac{dI}{dt} + 5I = 15$ **differential equation resulting from applying Kirchhoff's law. Chapter 5**

What do you see? Except for the coefficients and physical interpretation of the variables, these two equations are the same. The steps we took to separate and solve the Kirchhoff's-law problem are identical to the steps necessary to solve the Newton's-second-law problem. The same can be said about the work we did in Chapter 8 when we revisited the Kirchhoff's-law problem, $\frac{dI}{dt} + 5I = 15$, and solved it as a linear first-order equation using an integrating factor. Those same steps can be taken to solve $\frac{dv}{dt} + 2v = 32$ using the linear first-order equation technique!!!

Separable Differential Equations Chapters 1–5	Linear First-Order Differential Equations Chapter 8
$\frac{dv}{dt} + 2v = 32$	$\frac{dv}{dt} + 2v = 32$
$\frac{dv}{dt} = 32 - 2v$	$I(x) = e^{\int 2\,dt} = e^{2t}$
$dv = (32 - 2v)\,dt$	$I(x)\left[\frac{dv}{dt} + 2v\right] = I(x) \times 32$
$\frac{dv}{32 - 2v} = dt$	$e^{2t}\frac{dv}{dt} + e^{2t}(2v) = e^{2t} \times 32$
$-\frac{1}{2}\int (32 - 2v)^{-1}(-2)\,dv = \int dt$	$\left[e^{2t} \times v\right]' = 32e^{2t}$
$-\frac{1}{2}\ln\|32 - 2v\| + c_1 = t + c_2$	Check your work before continuing
$-\frac{1}{2}\ln\|32 - 2v\| = t + c_3 \quad c_3 = c_2 - c_1$	$\int \left[e^{2t} \times v\right]' dt = \int 32e^{2t}\,dt$
$\ln\|32 - 2v\| = -2t + c_4 \quad c_4 = -2c_3$	$e^{2t} \times v + c_1 = 16\int e^{2t}(2)\,dt$
$e^{\ln\|32-2v\|} = e^{-2t+c_4}$	$e^{2t} \times v + c_1 = 16e^{2t} + c_2$
$\|32 - 2v\| = c_5 e^{-2t} \quad c_5 = e^{c_4}$	$e^{2t} \times v = 16e^{2t} + c_2 - c_1$
$32 - 2v = \pm c_5 e^{-2t}$	$e^{2t} \times v = 16e^{2t} + c \quad c = c_2 - c_1$
$-2v = \pm\frac{c_5}{e^{2t}} - 32$	$\frac{e^{2t} \times v}{e^{2t}} = \frac{16e^{2t} + c}{e^{2t}}$
$v = \frac{c}{e^{2t}} + 16 \quad c = \pm\frac{c_5}{-2}$	$v = 16 + \frac{c}{e^{2t}}$

$$\frac{dv}{dt} + 2v = 32 \quad \text{differential equation resulting from applying Newton's second law.} \tag{1}$$

$$\frac{dI}{dt} + 5I = 15 \quad \text{differential equation resulting from applying Kirchhoff's Law.} \tag{2}$$

Equation (2) shown here was graphed by MatLab back in Chapter 5 (see below). That MatLab script could be quickly modified to graph (1) above.

```
Ffun = @(X,Y) 15 - (5.* Y); % ***
dx = 0.1; dy = 0.5; % ***
x = 0: dx: 1; y = 0: dy: 5; % ***
y0 = 0;
[X,Y] = meshgrid(x,y);
DY = Ffun(X,Y);
DX = ones(size(DY)); % generate the plot values
xscaler = abs((0.9*dx*ones(size(DX)))./DX);
yscaler = abs((0.9*dy*ones(size(DY)))./DY);
X = DX.* min(xscaler,yscaler);
DY = DY.* min(xscaler,yscaler);
quiver(X,Y,DX,DY, 'MaxheadSize', (3.5^3));
hold on;
[T,Y] = ode45(Ffun,[min(x) max(x)],y0);
plot(T,Y, 'r','LineWidth',3);
xp = 0.3; yp = 2.33; % ***
plot(xp, yp, 'ko', 'MarkerSize',14,'LineWidth',3); % ***
th = text(xp + 0.1, yp - 0.2, ['(',num2str(xp),',',num2str(yp),')']); % ***
set(th,'HorizontalAlignment','center'); set(th,'VerticalAlignment','top');
grid on;
title(['Slope field and family of solutions','for dI/dt = 15 - 5I']); % ***
ylabel('Current (amperes)'); % ***
xlabel('Time (sec)');
```

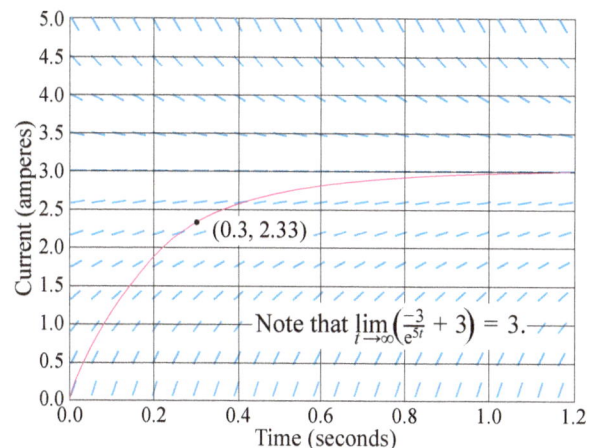

One of the first things one must do when taking a differential-equations class is determine which category best fits your problem. It is possible that your problem could be solved by more than one technique. If that is the case, then it is possible that one solution technique is much easier than another. All that can be said here is that intuition, ability, and practice are all necessary for mastery.

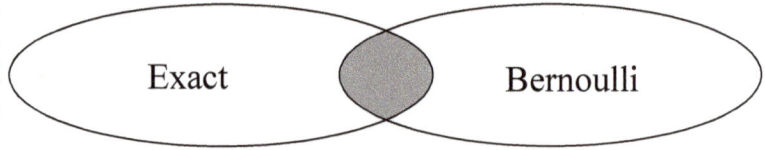

Some equations satisfy form requirements for more than one type of differential equation. The equation $x\,dy - y\,dx = y^2\,dx$ can be solved using the "exact differential equation technique" or the "Bernoulli equation technique."

Exact

Bernoulli

Exact Differential-Equation Technique, Chapter 12

$$x\,dy - y\,dx = y^2\,dx \qquad (1)$$

$$-y^2\,dx - y\,dx + x\,dy = 0 \quad \text{get 0 on the right}$$

$$y^2\,dx + y\,dx - x\,dy = 0 \quad \text{make leading term positive}$$

$$\left(y^2 + y\right) dx - x\,dy = 0$$

$$x^0\left(y^2 + y\right) dx - xy^0\,dy = 0 \quad \text{convert to } M, N \text{ form}$$

$$M(x, y)\,dx + N(x, y)\,dy = 0 \quad \text{for comparison}$$

$$\frac{\partial M}{\partial y} = \frac{\partial}{\partial y}\left(y^2 + y\right) = 2y + 1$$

$$\frac{\partial N}{\partial x} = \frac{\partial}{\partial x}(-x) = -1$$

$$\frac{\partial M}{\partial y} \neq \frac{\partial N}{\partial x}$$

The differential equation $x\,dy - y\,dx = y^2\,dx$ is not exact but can be transformed into an exact equation by multiplying both sides of (1) by y^{-2}.

$$y^{-2}[x\,dy - y\,dx] = y^{-2}\left[y^2\,dx\right]$$

$$xy^{-2}\,dy - y^{-1}\,dx = y^0\,dx$$

$$xy^{-2}\,dy - y^{-1}\,dx = dx$$

$$xy^{-2}\,dy - y^{-1}\,dx - dx = 0 \quad \text{make the right side 0}$$

$$-y^{-1}\,dx - dx + xy^{-2}\,dy = 0 \quad \text{move } dx \text{ terms}$$

$$y^{-1}\,dx + dx - xy^{-2}\,dy = 0 \quad \text{make leading term positive}$$

$$\left(y^{-1} + 1\right) dx - xy^{-2}\,dy = 0$$

$$x^0\left(y^{-1} + 1\right) dx + \left(-xy^{-2}\right) dy = 0 \quad \text{convert to } M, N \text{ form}$$

$$M(x, y)\,dx + N(x, y)\,dy = 0 \quad \text{for comparison}$$

$$\frac{\partial M}{\partial y} = \frac{\partial}{\partial y}\left(y^{-1} + 1\right) = -y^{-2} + 0 = -y^{-2}$$

$$\frac{\partial N}{\partial x} = \frac{\partial}{\partial x}\left(-xy^{-2}\right) = \frac{\partial}{\partial x}\left(-y^{-2}x\right) = -y^{-2} \times 1 = -y^{-2}$$

$$\frac{\partial M}{\partial y} = -y^{-2} = \frac{\partial N}{\partial x} \quad \text{equations are exact}$$

$$f(x, y) = \int \left(y^{-1} + 1\right) dx = \int y^{-1}\,dx + \int 1\,dx$$

$$= y^{-1} \int dx + \int dx$$

$$= xy^{-1} + x + g(y) \quad ***$$

$$\frac{\partial}{\partial y}[f(x, y)] = \frac{\partial}{\partial y}\left(xy^{-1}\right) + \frac{\partial}{\partial y}(x) + \frac{\partial}{\partial y}[g(y)]$$

$$= -xy^{-2} + 0 + g'(y)$$

But, since $\frac{\partial}{\partial y}[f(x, y)] = N(x, y) = -xy^{-2}$,

$-xy^{-2} + g'(y) = -xy^{-2}$, so $g'(y) = 0$ & $g(y) = c$

Therefore, $f(x, y) = \frac{x}{y} + x + c$ from ***.

Bernoulli Equation Technique, Chapter 10

$$x\,dy - y\,dx = y^2\,dx$$

$$\frac{x\,dy - y\,dx}{x} = \frac{y^2}{x}\,dx \quad \text{eliminate the coefficient of the leading term, } x \neq 0$$

$$dy + \frac{-y}{x}\,dx = \frac{1}{x} \times y^2\,dx$$

$$\frac{dy + \frac{-y}{x}\,dx}{dx} = \frac{y^2\,dx}{x\,dx} \quad \text{divide to get the } \frac{dy}{dx} \text{ term}$$

$$\frac{dy}{dx} + \frac{-y}{x}\frac{\cancel{dx}}{\cancel{dx}} = \frac{y^2\cancel{dx}}{x\cancel{dx}} \quad \text{cancel out the } dx \text{ terms}$$

$$\frac{dy}{dx} + \frac{-1}{x}y = \frac{1}{x}y^2 \quad \text{specific equation ***}$$

$$y' + P(x)y = Q(x)y^n \quad \text{general Bernoulli form}$$

$$\text{Let } v = y^{1-n} = y^{1-2} = y^{-1} \quad \text{for } n = 2$$

$v = y^{-1} = \frac{1}{y}$. Therefore, $vy = 1$ and $y = \frac{1}{v} = v^{-1}$

Since $y = v^{-1}$, $\frac{dy}{dx} = -v^{-2}\frac{dv}{dx}$.

$$\frac{dy}{dx} + \frac{-1}{x}y = \frac{1}{x}y^2 \quad \text{specific equation ***}$$

$$-v^{-2}\frac{dv}{dx} + \frac{-1}{x}v^{-1} = \frac{1}{x}v^{-2}$$

$$\frac{-v^{-2}\frac{dv}{dx} + \frac{-1}{x}v^{-1}}{-v^{-2}} = \frac{\frac{1}{x}v^{-2}}{-v^{-2}} \quad \text{divide by } -v^{-2}$$

$$v^0\frac{dv}{dx} + \frac{1}{x}v^1 = -\frac{1}{x}v^0$$

$$\frac{dv}{dx} + \frac{1}{x}v^1 = -\frac{1}{x}$$

This equation is first-order linear in v! Integrating factor $= I(x) = e^{\int \frac{1}{x}} = e^{\ln x} = x$. Multiplying the first-order linear equation above by the integrating factor, x,

$$I(x)\left[\frac{dv}{dx} + \frac{1}{x}v\right] = I(x)\left[-\frac{1}{x}\right]$$

$$x \times \left[\frac{dv}{dx} + \frac{1}{x}v\right] = x \times \left[-\frac{1}{x}\right]$$

$$x \times \frac{dv}{dx} + x\frac{v}{x} = x\left(-\frac{1}{x}\right)$$

$$[x \times v]' = -1$$

$$\int [x \times v]'\,dx = \int -1\,dx$$

$$(x \times v) + c_1 = -x + c_2$$

$$x \times v = -x + c_2 - c_1$$

$$x \times v = -x + c_3$$

Recall that $v = \frac{1}{y}$.

$$x \times \frac{1}{y} = -x + c_3$$

$$\frac{x}{y} + x = c$$

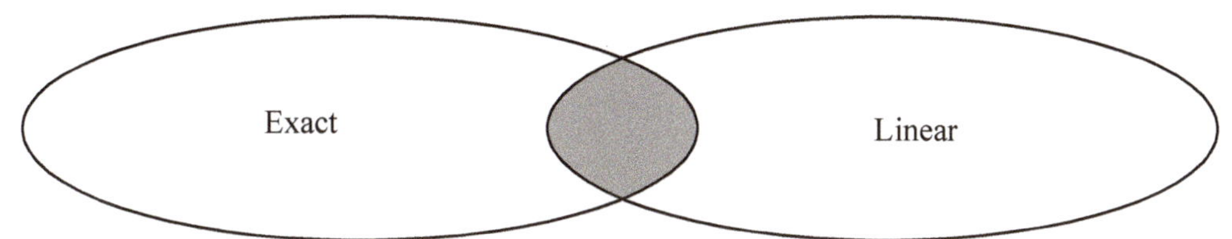

The equation $(3e^{3x}y - 2x)\,dx + e^{3x}\,dy = 0$ is in exact form and also can be transformed into linear form. Hence, it can be solved by more than one technique.

Exact, Chapter 12	Linear, Chapter 8
$\left(3e^{3x}y - 2x\right)dx + e^{3x}\,dy = 0$ $\frac{\partial M}{\partial y} = \frac{\partial}{\partial y}\left[3e^{3x}y - 2x\right]$ while $\frac{\partial N}{\partial x} = \frac{\partial}{\partial x}\left[e^{3x}\right]$ $= \frac{\partial}{\partial y}\left[3e^{3x}y\right] - \frac{\partial}{\partial y}(2x)$ $\quad = 3e^{3x}$ $= 3e^{3x}\frac{\partial}{\partial y}(y) - 0$ $= 3e^{3x}$ $\frac{\partial M}{\partial y} = 3e^{3x} = \frac{\partial N}{\partial x}$ hence, exact form $f(x,y) = \int (3e^{3x}y - 2x)\,dx = y\int 3e^{3x}\,dx - 2\int x\,dx$ $= ye^{3x} - 2\frac{x^2}{2} + g(y)$ $= ye^{3x} - x^2 + g(y)$ *** $\frac{\partial}{\partial y}[f(x,y)] = \frac{\partial}{\partial y}\left[ye^{3x} - x^2 + g(y)\right]$ $= \frac{\partial}{\partial y}\left[ye^{3x}\right] - \frac{\partial}{\partial y}\left[x^2\right] + \frac{\partial}{\partial y}[g(y)]$ $= e^{3x}\frac{\partial}{\partial y}(y) - 0 + g'(y)$ $= e^{3x} + g'(y)$ But, since $\frac{\partial}{\partial y}[f(x,y)] = N(x,y) = e^{3x}$, $e^{3x} + g'(y) = e^{3x}$, so $g'(y) = 0$ and $g(y) = c$. Therefore, $f(x,y) = ye^{3x} - x^2 + c$ by substituation in ***. Since $f(x,y) = 0$, $ye^{3x} - x^2 + c = 0$ $ye^{3x} = x^2 + c$ $y = \frac{x^2 + c}{e^{3x}}$.	$\left(3e^{3x}y - 2x\right)dx + e^{3x}\,dy = 0$ $\frac{\left(3e^{3x}y - 2x\right)\cancel{dx}}{\cancel{dx}} + \frac{e^{3x}\,dy}{dx} = 0$ divide by dx to get $\frac{dy}{dx}$ $\left(3e^{3x}y - 2x\right) + e^{3x}\frac{dy}{dx} = 0$ $e^{3x}\frac{dy}{dx} + 3e^{3x}y = 2x$ begin to isolate $\frac{dy}{dx}$ and y $\frac{dy}{dx} + 3y = 2xe^{-3x}$ divide away the e^{3x} terms $\frac{dy}{dx} + 3x^0y = 2xe^{-3x}$ put into linear form $\frac{dy}{dx} + P(x)y(x) = Q(x)$ generic linear equation Determine the integrating factor $I(x) = e^{\int 3\,dx} = e^{3x}$ $I(x)\left[\frac{dy}{dx} + 3y\right] = I(x) \times 2xe^{-3x}$ $e^{3x}\left[\frac{dy}{dx} + 3y\right] = e^{3x} \times \left(2xe^{-3x}\right)$ $e^{3x} \times \frac{dy}{dx} + e^{3x} \times 3y = 2xe^0$ $\left(e^{3x} \times \frac{dy}{dx}\right) + \left(e^{3x} \times 3y\right) = 2x \times 1$ $\left[e^{3x} \times y\right]' = 2x$ do a mental check here for the derivative of a product $\int \left[ye^{3x}\right]'\,dx = \int 2x\,dx$ $ye^{3x} + c_1 = 2\int x\,dx$ $ye^{3x} + c_1 = 2\frac{x^2}{2} + c_2$ $ye^{3x} = x^2 + c_2 - c_1$ $ye^{3x} = x^2 + c_3$ $y = \frac{x^2 + c}{e^{3x}}$

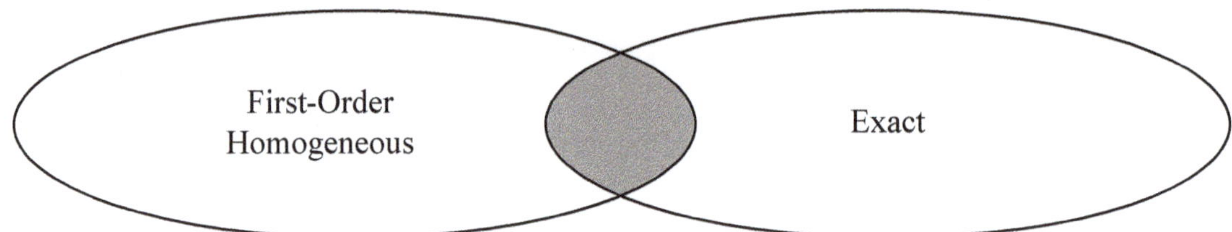

Some differential equations can be solved by the ratio-of-functions technique and the exact technique. Hence, they could be solved by more than one technique.

Ratio-of-Functions Solution Technique, Chapter 13	Exact Solution Technique, Chapter 12
$\left(x^2+y^2\right)\,dy + 2xy\,dx = 0$ $\left(x^2+y^2\right)\,dy = -2xy\,dx$ $\frac{dy}{dx} = \frac{-2xy}{(x^2+y^2)}$ There are three terms here, one in the numerator and two in the denominator. All three are of the second degree. Hence, the equation can be solved as a ratio of functions. Prepare to substitute for all the x terms. **The goal is to convert to a separable differential equation in terms of x and v.** $\frac{dy}{dx} = \frac{-2xy}{x^2+y^2}\times\frac{\frac{1}{x^2}}{\frac{1}{x^2}} = \frac{-2\frac{y}{x}}{1+\left(\frac{y}{x}\right)^2}$ ***** Let $v = \frac{y}{x}$. Therefore, $y = vx$ and $\frac{dy}{dx} = (v\times 1) + \left(x\times\frac{dv}{dx}\right)$. $\frac{dy}{dx} = \frac{-2v}{1+v^2}$ substitute $v = \frac{y}{x}$ here into ***** $v + x\frac{dv}{dx} = \frac{-2v}{1+v^2}$ substitute for $\frac{dy}{dx}$ $x\frac{dv}{dx} = \frac{-2v}{1+v^2} - v$ separable differential equation in v $x\frac{dv}{dx} = \frac{-2v}{1+v^2} - \frac{v\left(1+v^2\right)}{1+v^2}$ prepare to combine terms $x\frac{dv}{dx} = \frac{-2v - v\left(1+v^2\right)}{1+v^2}$ add like fractions $x\frac{dv}{dx} = \frac{-2v-v-v^3}{1+v^2}$ distributive property $x\frac{dv}{dx} = \frac{-3v-v^3}{1+v^2}$ $\frac{x}{dx} = \frac{-3v-v^3}{1+v^2}\times\frac{1}{dv}$ x and v terms are separate $\frac{dx}{x} = \frac{1+v^2}{-3v-v^3}\times dv$ inverting both sides $\int x^{-1}\,dx = \int\frac{1+v^2}{-3v-v^3}\times dv$ $\ln x + c_1 = \frac{-3}{-3}\int\frac{1+v^2}{-3v-v^3}\times dv$ $\ln x + c_1 = \frac{1}{-3}\int\frac{-3-3v^2}{-3v-v^3}\times dv$ form is $\int\frac{du}{u}$ $\ln x + c_1 = -\frac{1}{3}\ln\left(-3v-v^3\right) + c_2$ $\ln x = -\frac{1}{3}\ln\left(-3v-v^3\right) + c_2 - c_1$	$\left(x^2+y^2\right)\,dy + 2xy\,dx = 0$ $2xy\,dx + \left(x^2+y^2\right)\,dy = 0$ move dx term to the left $M(x,y)\,dx + N(x,y)\,dy = 0$ M, N form for compare $\frac{\partial M}{\partial y} = \frac{\partial}{\partial y}(2xy) = 2xy^0 = 2x$ treat $2x$ as constant $\frac{\partial N}{\partial x} = \frac{\partial}{\partial x}\left(x^2+y^2\right) = 2x + 0 = 2x$ treat y^2 as constant $\frac{\partial M}{\partial y} = 2x = \frac{\partial N}{\partial x}$, exact $f(x,y) = \int M(x,y)\,dx$ $+ \int\left[N(x,y) - \frac{\partial}{\partial y}\left\{\int M(x,y)\,dx\right\}\right]dy = C$ Calculate the repeated part of the formula for substitution. $\int M(x,y)\,dx = \int 2xy\,dx = 2y\int x\,dx = 2y\frac{x^2}{2} = x^2y$ $f(x,y) = x^2y + \int\left[N(x,y) - \frac{\partial}{\partial y}\left\{x^2y\right\}\right]dy = C$ $= x^2y + \int\left[\left(x^2+y^2\right) - x^2\frac{\partial}{\partial y}(y)\right]dy = C$ $= x^2y + \int\left[\left(x^2+y^2\right) - x^2\right]dy = C$ $= x^2y + \int y^2\,dy = C$ $= x^2y + \frac{y^3}{3} = C$ $3x^2y + y^3 = C$

table continues

<table>
<tr>
<td>

$$\ln x = -\frac{1}{3}\ln\left(-3v - v^3\right) + c_3$$
$$-3 \times \ln x = \ln\left(-3v - v^3\right) + c_3$$
$$\ln x^{-3} = \ln\left(-3v - v^3\right) + c_3$$
$$e^{\ln x^{-3}} = e^{\ln\left(-3v - v^3\right) + c_3}$$
$$x^{-3} = e^{\ln\left(-3v - v^3\right)} \times e^{c_3} \qquad e^{\ln u} = u$$
$$x^{-3} = \left(-3v - v^3\right) \times c_4$$

Recall $v = \frac{y}{x}$

$$x^{-3} = \left[-3\frac{y}{x} - \left(\frac{y}{x}\right)^3\right] \times c_4$$
$$x^3\left[x^{-3}\right] = x^3\left[-3\frac{y}{x} - \left(\frac{y}{x}\right)^3\right] \times c_4$$
$$1 = x^3\left[-3\frac{y}{x} - \frac{y^3}{x^3}\right] \times c_4$$
$$1 = \left(-3x^2y - y^3\right) \times c_4$$
$$\frac{1}{c_4} = -3x^2y - y^3$$
$$c_5 = -3x^2y - y^3$$
$$3x^2y + y^3 = c_6$$
$$3x^2y + y^3 = C$$

</td>
<td></td>
</tr>
</table>

Author's note: Notice that the technique at left took 30+ steps while the technique at right takes only about ten steps.

Students who have studied matrix algebra know it is possible to solve a "second-order differential equation" by the technique shown in chapters 14 and 15 as well as a technique based on solving systems of equations. The neat thing about using the system-of-equations technique is that in theory it can be extended to solving third-order, fourth-order, etc. systems of equations. With the use of computers and calculators, the system-of-equations technique can be extended in actual practice.

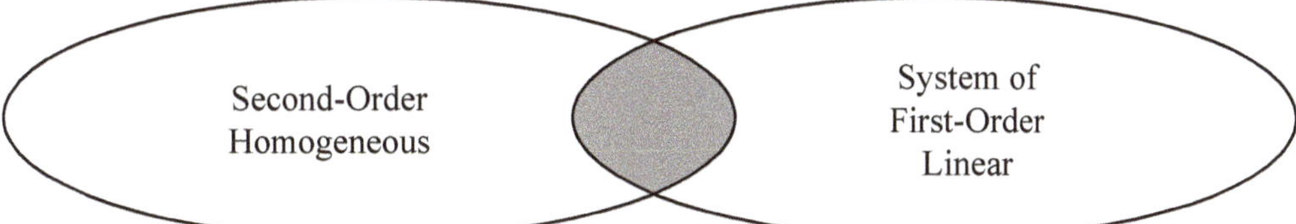

<table>
<tr>
<th>Second-Order Linear Differential Equations (Chapter 14)</th>
<th>System of First-Order Linear Differential Equations (Chapter 20)</th>
</tr>
<tr>
<td>

$$2y'' + 5y' - 3y = 0$$

Use the differential coefficients, 2, 5, and −3, to form the algebraic characteristic equation.

$$2r^2 + 5r - 3 = 0$$

Solving the characteristic equation for r, we get $(r+3)(2r-1)$.

$$r = -3, \ \frac{1}{2}$$

Characteristic roots are real and unequal. Hence, the generic solution for this differential equation would be

$$y_{\text{gen}} = c_1e^{r_1t} + c_2e^{r_2t}.$$
$$y_{\text{gen}} = c_1e^{-3t} + c_2e^{\frac{1}{2}t}$$

Working with IVP conditions $y(0) = -4$ and $y'(0) = 9$:

$$y(t) = c_1e^{-3t} + c_2e^{\frac{1}{2}t}$$

</td>
<td>

$$2y'' + 5y' - 3y = 0$$

Step 1: The second-order linear differential equation needs to be converted into a system of first-order equations. This is done using a change of variables. Let $x_1 = y$ and $x_2 = y'$. Then $x_1' = y' = x_2$ and $x_2' = y''$.

$$x_1' = 0x_1 + x_2 \qquad \text{*** seems silly, but be patient}$$
$$2y'' + 5y' - 3y = 0 \qquad \text{from above}$$
$$\text{So } 2y'' = 3y - 5y'$$
$$y'' = \frac{3}{2}y - \frac{5}{2}y'$$
$$x_2' = y'' = \frac{3}{2}y - \frac{5}{2}y'$$
$$x_2' = \frac{3}{2}x_1 - \frac{5}{2}x_2 \qquad \text{*** transitive}$$

</td>
</tr>
</table>

table continues

$$y(0) = c_1 e^{-3\times 0} + c_2 e^{\frac{1}{2}\times 0} = -4$$
$$c_1 \times 1 + c_2 \times 1 = -4$$
$$c_1 + c_2 = -4 \qquad (1)$$
$$y'(t) = -3c_1 e^{-3t} + \frac{1}{2}c_2 e^{\frac{1}{2}t} = 9$$
$$y'(0) = -3c_1 e^{-3\times 0} + \frac{1}{2}c_2 e^{\frac{1}{2}\times 0} = 9$$
$$-3c_1 \times 1 + \frac{1}{2}c_2 \times 1 = 9$$
$$\boxed{\begin{aligned} -3c_1 + \tfrac{1}{2}c_2 &= 9 \qquad (2)\\ c_1 + c_2 &= -4 \qquad (1)\end{aligned}}$$
$$\boxed{\begin{aligned} -3c_1 + \tfrac{1}{2}c_2 &= 9 \qquad (2)\\ 3c_1 + 3c_2 &= -12 \qquad \text{(1) above times 3} \qquad (3)\end{aligned}}$$
$$\frac{7}{2}c_2 = -3$$
$$7c_2 = -6$$
$$c_2 = -\frac{6}{7}$$
$$c_1 + c_2 = -4 \qquad (1)$$
$$c_1 + \left(-\frac{6}{7}\right) = -\frac{28}{7} \qquad \text{substituting for } c_2$$
$$c_1 = -\frac{22}{7}$$

Since $y(t) = c_1 e^{-3t} + c_2 e^{\frac{1}{2}t}$,

$$y(t) = -\frac{22}{7}e^{-3t} - \frac{6}{7}e^{\frac{1}{2}t}$$

Rewriting *** above as a system

$$\boxed{\begin{aligned} x_1' &= 0x_1 + x_2\\ x_2' &= \frac{3}{2}x_1 - \frac{5}{2}x_2\end{aligned}} \qquad \text{not so silly now}$$

$$\begin{pmatrix} x_1' \\ x_2' \end{pmatrix} = \begin{pmatrix} 0 & 1 \\ \frac{3}{2} & -\frac{5}{2} \end{pmatrix}\begin{pmatrix} x_1 \\ x_2 \end{pmatrix},$$

or $\vec{x}' = (\textbf{Coef Mat})\vec{x}$

Step 2: Find the eigenvalues of the coefficient matrix.

$$\textbf{Coef Mat} \times \vec{\varepsilon} = \lambda \times \vec{\varepsilon}$$
$$(\textbf{Coef Mat} \times \vec{\varepsilon}) - (\lambda \times \vec{\varepsilon}) = \begin{pmatrix} 0 \\ 0 \end{pmatrix}$$
$$\left[\begin{pmatrix} 0 & 1 \\ \frac{3}{2} & -\frac{5}{2} \end{pmatrix} - \lambda\right] \times \vec{\varepsilon} = \begin{pmatrix} 0 \\ 0 \end{pmatrix} \qquad \text{dimension error}$$
$$\left[\begin{pmatrix} 0 & 1 \\ \frac{3}{2} & -\frac{5}{2} \end{pmatrix} - \lambda\mathbf{I}\right] \times \vec{\varepsilon} = \begin{pmatrix} 0 \\ 0 \end{pmatrix} \qquad \text{error corrected}$$
$$\left[\begin{pmatrix} 0 & 1 \\ \frac{3}{2} & -\frac{5}{2} \end{pmatrix} - \begin{pmatrix} \lambda & 0 \\ 0 & \lambda \end{pmatrix}\right] \times \vec{\varepsilon} = \begin{pmatrix} 0 \\ 0 \end{pmatrix}$$
$$\begin{pmatrix} 0-\lambda & 1-0 \\ \frac{3}{2}-0 & -\frac{5}{2}-\lambda \end{pmatrix} \times \begin{pmatrix} x_1 \\ x_2 \end{pmatrix} = \begin{pmatrix} 0 \\ 0 \end{pmatrix}$$
$$\begin{pmatrix} -\lambda & 1 \\ \frac{3}{2} & -\frac{5}{2}-\lambda \end{pmatrix} \times \begin{pmatrix} x_1 \\ x_2 \end{pmatrix} = \begin{pmatrix} 0 \\ 0 \end{pmatrix}$$

where λ is the coefficient-matrix eigenvalue. Solve for λ.

$$(\textbf{Coef Mat} - \lambda\mathbf{I})\begin{pmatrix} x_1 \\ x_2 \end{pmatrix} = \begin{pmatrix} 0 \\ 0 \end{pmatrix} \qquad \text{from the discussion in Chapter 20}$$
$$\det(\textbf{Coef Mat} - \lambda\mathbf{I}) = 0 \qquad \text{guarantees infinite solutions}$$
$$\begin{vmatrix} -\lambda & 1 \\ \frac{3}{2} & -\frac{5}{2}-\lambda \end{vmatrix} = 0$$
$$-\lambda \times \left(-\frac{5}{2} - \lambda\right) - \left(\frac{3}{2} \times 1\right) = 0$$
$$\left(\frac{5}{2}\lambda + \lambda^2\right) - \frac{3}{2} = 0$$
$$\lambda^2 + \frac{5}{2}\lambda - \frac{3}{2} = 0$$
$$2\lambda^2 + 5\lambda - 3 = 0$$
$$(\lambda + 3)(2\lambda - 1) = 0$$

The eigenvalues are $\lambda_1 = -3$ and $\lambda_2 = \frac{1}{2}$.

Step 3: Evaluate the discriminant for $2\lambda^2 + 5\lambda - 3 = 0$.

$$5^2 - 4 \times 2 \times (-3) = 25 + 24 > 0$$

From Chapter 20, we recall this means the solution to the two-by-two system of linear differential equations is

$$y_{\text{gen}} = c_1\vec{\varepsilon}_1 e^{\lambda_1 t} + c_2\vec{\varepsilon}_2 e^{\lambda_2 t}, \qquad *****$$

where c_1 and c_2 are constants of integration, $\vec{\varepsilon}_1$ and $\vec{\varepsilon}_2$ are eigenvectors of the system, and λ_1 and λ_2 are eigenvalues of the the system. Hence, since we have substituted from y to x,

$$\vec{x} = c_1\vec{\varepsilon}_1 e^{-3t} + c_2\vec{\varepsilon}_2 e^{\frac{1}{2}t}.$$

Step 4: Find the eigenvectors of the coefficient matrix.

We now know the coefficient matrix and its eigenvalues. We use that information to find the matrix eigenvector by substitution into **Coef Mat** $\times \vec{\varepsilon} = \lambda \times \vec{\varepsilon}$ (see work in step 2 above).

table continues

For $\lambda = -3$	For $\lambda = \frac{1}{2}$
$\begin{pmatrix} -\lambda & 1 \\ \frac{3}{2} & -\frac{5}{2}-\lambda \end{pmatrix}\begin{pmatrix} x_1 \\ x_2 \end{pmatrix} = \begin{pmatrix} 0 \\ 0 \end{pmatrix}$	$\begin{pmatrix} -\lambda & 1 \\ \frac{3}{2} & -\frac{5}{2}-\lambda \end{pmatrix}\begin{pmatrix} x_1 \\ x_2 \end{pmatrix} = \begin{pmatrix} 0 \\ 0 \end{pmatrix}$
Becomes	
$\begin{pmatrix} 3 & 1 \\ \frac{3}{2} & \frac{1}{2} \end{pmatrix}\begin{pmatrix} x_1 \\ x_2 \end{pmatrix} = \begin{pmatrix} 0 \\ 0 \end{pmatrix}$	$\begin{pmatrix} -\frac{1}{2} & 1 \\ \frac{3}{2} & -3 \end{pmatrix}\begin{pmatrix} x_1 \\ x_2 \end{pmatrix} = \begin{pmatrix} 0 \\ 0 \end{pmatrix}$
$3x_1 + 1x_2 = 0$ (1) $\frac{3}{2}x_1 + \frac{1}{2}x_2 = 0$ (2)	$-\frac{1}{2}x_1 + 1x_2 = 0$ (1) $\frac{3}{2}x_1 - 3x_2 = 0$ (2)
Multiply (2) by 2.	Multiply (1) by −3.
$3x_1 + 1x_2 = 0$ (1) $3x_1 + 1x_2 = 0$ (2)	$\frac{3}{2}x_1 - 3x_2 = 0$ (1) $\frac{3}{2}x_1 - 3x_2 = 0$ (2)
Coinciding equations	
$3x_1 + 1x_2 = 0$ $1x_2 = -3x_1$ Let $x_1 = 1$, so $x_2 = -3$.	$\frac{3}{2}x_1 - 3x_2 = 0$ $3x_1 - 6x_2 = 0$ $3x_1 = 6x_2$ $x_1 = 2x_2$ Let $x_2 = 1$, so $x_1 = 2$.
The eigenvector is $\begin{pmatrix} 1 \\ -3 \end{pmatrix}$	The eigenvector is $\begin{pmatrix} 2 \\ 1 \end{pmatrix}$
Check back to Step 2	
$\begin{pmatrix} 0 & 1 \\ \frac{3}{2} & -\frac{5}{2} \end{pmatrix}\begin{pmatrix} 1 \\ -3 \end{pmatrix} \stackrel{?}{=} -3\begin{pmatrix} 1 \\ -3 \end{pmatrix}$	$\begin{pmatrix} 0 & 1 \\ \frac{3}{2} & -\frac{5}{2} \end{pmatrix}\begin{pmatrix} 2 \\ 1 \end{pmatrix} \stackrel{?}{=} \frac{1}{2}\begin{pmatrix} 2 \\ 1 \end{pmatrix}$
$\begin{pmatrix} -3 \\ 9 \end{pmatrix} = \begin{pmatrix} -3 \\ 9 \end{pmatrix}$ ck	$\begin{pmatrix} 1 \\ \frac{1}{2} \end{pmatrix} = \begin{pmatrix} 1 \\ \frac{1}{2} \end{pmatrix}$ ck

Step 5: Substitute both eigenvectors into $\vec{x} = c_1\vec{\varepsilon}_1 e^{\lambda_1 t} + c_2\vec{\varepsilon}_2 e^{\lambda_2 t}$ and, using the IVP conditions $y(0) = -4$ and $y'(0) = 9$, solve for c_1 and c_2. **This is tricky because the IVP is given with *y* as the dependent variable and we have changed to *x*.**

$$\vec{x} = c_1\begin{pmatrix} 1 \\ -3 \end{pmatrix}e^{-3t} + c_2\begin{pmatrix} 2 \\ 1 \end{pmatrix}e^{\frac{1}{2}t}$$

From the substitutions in **Step 1** "Let $x_1 = y$ and $x_2 = y'$, so that $x_1' = y' = x_2$ and $x_2' = y''$,"

$$\begin{pmatrix} y \\ y' \end{pmatrix} = \begin{pmatrix} x_1 \\ x_2 \end{pmatrix} = c_1\begin{pmatrix} 1 \\ -3 \end{pmatrix}e^{-3t} + c_2\begin{pmatrix} 2 \\ 1 \end{pmatrix}e^{\frac{1}{2}t}$$

Then we can back substitute the given IVP conditions, $y(0) = -4$ and $y'(0) = 9$:

$$-4 = 1 \times c_1 \times e^{-3\times 0} + 2 \times c_2 \times e^{\frac{1}{2}\times 0}$$
$$9 = -3 \times c_1 \times e^{-3\times 0} + 1 \times c_2 \times e^{\frac{1}{2}\times 0}$$

$$-4 = c_1 \times 1 + 2 \times c_2 \times 1$$
$$9 = -3c_1 \times 1 + 1 \times c_2 \times 1$$

$$-4 = c_1 + 2c_2$$
$$9 = -3c_1 + c_2$$

$c_1 + 2c_2 = -4$ therefore, $c_1 = -4 - 2c_2$

$-3c_1 + c_2 = 9$

table continues

Certainly it would seem that the Chapter 14 technique for solving the second-order linear differential equation $2y'' + 5y' - 3y = 0$ is so much shorter and easier than the Chapter 20 "eigenvalue/eigenvector" technique that one should not bother learning the latter technique. The reason for doing so is the same reason that the matrix algebra technique is preferred when solving systems of algebraic equations. The system technique of solving both types of equations, with the aid of computers and computer algorithms, can be easily extended to apply to solving 3×3, 4×4, ..., and $n \times n$ systems. That is not the case for the solution techniques initially taught in a math class.	Substitute $c_1 = -4 - 2c_2$ into the second equation. $-3(-4 - 2c_2) + c_2 = 9$ $12 + 6c_2 + c_2 = 9$ $7c_2 = -3$ $c_2 = -\frac{3}{7}$ $c_1 + 2c_2 = -4$ finally, solve for c_1 $c_1 + 2 \times -\frac{3}{7} = -4$ $c_1 + -\frac{6}{7} = -\frac{28}{7}$ $c_1 = -\frac{22}{7}$ $c_2 = -\frac{3}{7}$ **Step 6:** We are now back in variable y. $\begin{pmatrix} y \\ y' \end{pmatrix} = c_1 \begin{pmatrix} 1 \\ -3 \end{pmatrix} e^{-3t} + c_2 \begin{pmatrix} 2 \\ 1 \end{pmatrix} e^{\frac{1}{2}t}$ **We are solving for y!!! Remember?** $y(t) = -\frac{22}{7} \times 1 \times e^{-3t} + \left(-\frac{3}{7}\right) \times 2 \times e^{\frac{1}{2}t}$ $= -\frac{22}{7}e^{-3t} - \frac{6}{7}e^{\frac{1}{2}t}$ ta da!

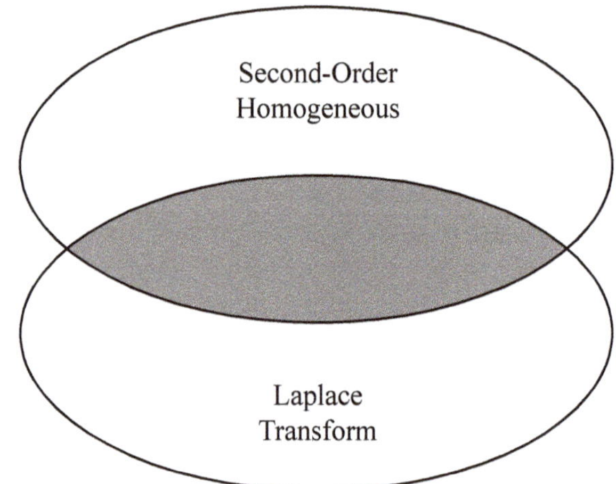

Row	$f(t)$	$F(s)$
1	0	0
2	1	$\frac{1}{s}$
3	c	$\frac{c}{s}$
4	e^{at}	$\frac{1}{s-a}$
5	te^{at}	$\frac{1}{(s-a)^2}$
6	$\sin(at)$	$\frac{a}{s^2+a^2}, s > 0$
7	$\cos(at)$	$\frac{s}{s^2+a^2}, s > 0$
8	$f(t)$	$F(s)$ by definition
9	$f'(t)$	$sF(s) - f(0)$
10	$f''(t)$	$s^2F(s) - sf(0) - f'(0)$

Solve $y'' + 5y' + 6y = 0$, with $y(0) = 2$ and $y'(0) = 3$.

Homogeneous Second-Order Differential Equations (Chapter 14)	Laplace Transform (Chapter 23)
$y'' + 5y' + 6y = 0,$ with $y(0) = 2$ and $y'(0) = 3$ Solve for the characteristic equation. $1r^2 + 5r + 6 = 0$ $(r+2)(r+3) = 0$ $r = -2, -3$ $b^2 - 4ac = 5^2 - 4(1)6 = 25 - 24 > 0$ So $y_{gen} = c_1e^{-2x} + c_2e^{-3x}$ see Chapter 14 $y(x) = c_1e^{-2x} + c_2e^{-3x}$ (1) $y'(x) = -2c_1e^{-2x} + -3c_2e^{-3x}$ (2) $y''(x) = 4c_1e^{-2x} + 9c_2e^{-3x}$ (3)	$y'' + 5y' + 6y = 0,$ with $y(0) = 2$ and $y'(0) = 3$ $y'' + 5y' + 6y = 0, y(0) = 2, y'(0) = 3$ $\mathfrak{L}\{y(t)\} = Y(s)$ $\mathfrak{L}\{y'(t)\} = sY(s) - y'(0)$ $\mathfrak{L}\{y''(t)\} = s^2Y(s) - sy(0) - y'(0)$ $\mathfrak{L}\{y(t)\} = Y(s)$ $\mathfrak{L}\{y'(t)\} = sY(s) - 3$ $\mathfrak{L}\{y''(t)\} = s^2Y(s) - s(2) - 3$ $y(t)'' + 5y'(t) + 6y(t) = 0$ $\mathfrak{L}\{y(t)'' + 5y'(t) + 6y(t)\} = \mathfrak{L}\{0\}$

table continues

Substitute $y(0) = 2$ into (1).

$$y(0) = c_1e^{-2(0)} + c_2e^{-3(0)} = 2$$
$$c_1(1) + c_2(1) = 2$$
$$c_1 + c_2 = 2 \qquad \text{(a)}$$

Substitute $y'(0) = 3$ into (2)

$$y'(0) = -2c_1e^{-2(0)} + -3c_2e^{-3(0)} = 3$$
$$-2c_1(1) + -3c_2(1) = 3$$
$$-2c_1 + (-3c_2) = 3 \qquad \text{(b)}$$

Gather together (a) and (b)

$$c_1 + c_2 = 2 \qquad \text{(a)}$$
$$-2c_1 + (-3c_2) = 3 \qquad \text{(b)}$$

$$2c_1 + 2c_2 = 4 \quad \text{multiply (a) by 2}$$
$$-c_2 = 7 \quad \text{combine with (b)}$$
$$c_2 = -7$$
$$\text{so } c_1 = 9$$

Therefore, $y_{\text{gen}} = 9e^{-2x} - 7e^{-3x}$.

Ck:

$$y'' + 5y' + 6y = 0 \quad \text{Original equation}$$
$$\text{eq. (3)} + 5\text{eq. (2)} + 6\text{eq. (1)} = 0$$
$$\left(4c_1e^{-2x} + 9c_2e^{-3x}\right) + 5\left(-2c_1e^{-2x} + -3c_2e^{-3x}\right) + 6\left(c_1e^{-2x} + c_2e^{-3x}\right) = 0$$
$$\left(4c_1e^{-2x} - 10c_1e^{-2x} + 6c_1e^{-2x}\right) + \left(9c_2e^{-3x} - 15c_2e^{-3x} + 6c_2e^{-3x}\right) = 0 \quad \text{ck!}$$

Book Summary

"In an elementary course in differential equations... what matters is getting a feel for the importance of the subject, coming out of the course with the conviction of the inevitability of differential equations, and with an enhanced faith in the power of mathematics."

—Gian-Carlo Rota
Applied Mathematics and Philosophy, MIT
Differential Equations Professor at MIT for 40 years

$$\mathfrak{L}\{y(t)''\} + \mathfrak{L}\{5y'(t)\} + \mathfrak{L}\{6y(t)\} = 0$$
$$s^2Y(s) - s(2) - 3 + 5\mathfrak{L}\{y'(t)\} + 6\mathfrak{L}\{y(t)\} = 0$$
$$s^2Y(s) - 2s - 3 + 5[sY(s) - 2] + 6Y(s) = 0$$
$$s^2Y(s) - 2s - 3 + 5sY(s) - 10 + 6Y(s) = 0$$
$$\left[s^2Y(s) + 5sY(s) + 6Y(s)\right] - 2s - 13 = 0$$
$$Y(s)\left[s^2 + 5s + 6\right] = 2s + 13$$

Note that $s^2 + 5s + 6$ is the characteristic equation at the top of the work at left.

$$Y(s) = \frac{2s+13}{s^2+5s+6}$$
$$\mathfrak{L}\{y(t)\} = \frac{2s+13}{s^2+5s+6} \quad \text{***************}$$
$$\frac{2s+13}{(s+2)(s+3)} = \frac{A}{s+2} + \frac{B}{s+3} \quad \text{**********}$$
$$= \frac{A}{s+2}\frac{s+3}{s+3} + \frac{B}{s+3}\frac{s+2}{s+2}$$
$$= \frac{A(s+3) + B(s+2)}{(s+2)(s+3)}$$
$$= \frac{As + 3A + Bs + 2B}{(s+2)(s+3)}$$
$$= \frac{As + Bs + 3A + 2B}{(s+2)(s+3)}$$
$$= \frac{(A+B)s + 3A + 2B}{(s+2)(s+3)}$$

Hence, $A + B = 2$ (1)
and $3A + 2B = 13$ (2)

$$3A + 3B = 6 \quad \text{multiply (1) by 3}$$
$$3A + 2B = 13$$
$$B = -7 \quad \text{therefore, } A = 9$$

$$\mathfrak{L}\{y(t)\} = \frac{2s+13}{s^2+5s+6} \quad \text{***************}$$
$$\mathfrak{L}\{y(t)\} = \frac{A}{s+2} + \frac{B}{s+3} \quad \text{**********}$$
$$\mathfrak{L}\{y(t)\} = \frac{9}{s+2} + \frac{-7}{s+3} \quad \text{**********}$$
$$\mathfrak{L}^{-1}\{\mathfrak{L}\{y(t)\}\} = \mathfrak{L}^{-1}\left\{\frac{9}{s+2} + \frac{-7}{s+3}\right\}$$
$$y(t) = \mathfrak{L}^{-1}\left\{\frac{9}{s+2}\right\} - \mathfrak{L}^{-1}\left\{\frac{7}{s+3}\right\}$$
$$= 9\mathfrak{L}^{-1}\left\{\frac{1}{s+2}\right\} - 7\mathfrak{L}^{-1}\left\{\frac{1}{s+3}\right\}$$
$$y(t) = 9e^{-2t} - 7e^{-3t}$$

Glossary

Acceleration (Chapter 9) The rate of change of velocity over time: $\frac{dv}{dt}$ or $v'(t)$. Since velocity is the rate of change of displacement over time, $\frac{dp}{dt}$ or $p'(t)$, acceleration can be thought of as $\frac{d}{dt}\left(\frac{dp}{dt}\right)$ or $\frac{d^2p}{dt^2}$ or $p''(t)$. A negative acceleration is called a retardation or deceleration. On earth, that would be 9.8 $\frac{m}{s^2}$ or 32 $\frac{ft}{s^2}$. Conventionally, both of those quantities would be negative, but this just depends on one's frame of reference.

Algebraic/Integrated Form (Chapters 5, 6, 8–10, 12–15) A differential equation that has been transformed to algebraic notation. Newton's famous $F = ma$ is a classic example.

Bernoulli Equations (Chapter 10) A type of differential equation that can be solved by a substitution technique that reduces it to a first-order linear equation. Bernoulli differential equations are notable because they are solvable by a well-known "cookbook" technique.

$$y' + P(x) \times y = Q(x) \quad \text{first-order linear equation}$$
$$y' + P(x) \times y = Q(x)y^n \quad \text{Bernoulli equation}$$

Capacitor See Electronic Components.

Checking Solutions (various chapters) Just as students of algebra are taught to check their work by taking a solution and substituting it into the original equation, the solution to a differential equation can be checked by substituting it back into the original differential equation.

Degree (Order) of a Differential Equation (Chapter 9) The degree or order of the highest order derivative in a differential equation.

Degree of a Polynomial Equation (Chapter 9) The degree of the highest power in the polynomial.

Differential (Chapter 0) An infinitesimally small change in a variable. The definition of the derivative, $\frac{dy}{dx} = \lim_{p \to x} \frac{f(x)-f(p)}{x-p}$, can be thought of as the ratio of two differentials, dy and dx.

Differential Equation (Chapter 0) An equation with an unknown function and one or more of its derivatives. The goal of a differential-equations class is to determine, if possible, the unknown function from its derivative.

Direct Variation (Chapter 2) A relation between two variables such that, as one variable gets larger, the other also gets larger, and, correspondingly, as one of the variables gets smaller, the other also gets smaller. When working with linear functions, this is symbolized as y = kx where x is the independent variable, y is the dependent variable, and k is the constant of variation. Direct variation also applies to curvilinear functions, such as $y = kx^2$.

Eigenvalue/Eigenvector (Chapter 19) For any square matrix (**Mat**), there exists a scalar called an eigenvalue, λ, and a vector called an eigenvector, $\vec{\varepsilon}$, such that $\mathbf{Mat}\times\vec{\varepsilon} = \lambda\times\vec{\varepsilon}$. For matrix $\begin{pmatrix} a & b \\ c & d \end{pmatrix}$, that would be $\begin{pmatrix} a & b \\ c & d \end{pmatrix}\begin{pmatrix} x_1 \\ x_2 \end{pmatrix} = \lambda\begin{pmatrix} x_1 \\ x_2 \end{pmatrix}$. This equation was used to justify the equation $\left[\begin{pmatrix} a & b \\ c & d \end{pmatrix} - \lambda\mathbf{I}\right]\begin{pmatrix} x_1 \\ x_2 \end{pmatrix} = \begin{pmatrix} 0 \\ 0 \end{pmatrix}$, which is key to the solution of a system of homogeneous ODEs.

Electronic Components (Chapters 5, 15, 24) A resistor restricts the flow of electric current and converts electrical energy to heat. It is analogous to friction in mechanical systems. The voltage difference across a resistor is equal to its resistance in ohms times the current running through it in amperes. This is known as Ohm's Law. The power dissipated by the resistor in watts is equal to the voltage across it times the current through it. This is known as Ohm's Power Law.

An inductor is a coil of wire that stores energy in the form of a magnetic field. It is the electrical equivalent of inertia in mechanical systems (such as a mass or a flywheel). Currents flowing in a conductor produce a magnetic field. Any change in the strength of the magnetic field induces a current to flow in the conductor that opposes the change. The current through an inductor cannot change in zero time and the voltage across it is equal to the inductance in henries times the time-rate-of-change of the current running through it.

A capacitor is two conductive surfaces placed very close together, separated by an insulating material called a dielectric. It stores energy in the form of an electric field between the two surfaces. It is the electrical equivalent to springs in mechanical systems which store potential energy as they are mechanically deformed, releasing that potential energy as kinetic energy when they are allowed to return to their neutral positions. The voltage across a capacitor cannot change in zero time and the current through it is equal to the capacitance in farads times the time-rate-of-change of the voltage across it.

When inductors and capacitors are connected in a circuit together, their interaction with each other—according the time-rate-of-change of current and voltage, respectively—causes them to behave in a manner similar to a mass on a spring: a simple harmonic oscillator. These harmonic oscillations cause the frequency response of the circuit to be very pronounced at frequencies where the voltage and current are exactly 180 degrees out of phase. In this scenario, all energy entering the circuit is continuously juggled back and forth between the inductor and the capacitor and the only energy lost is through resistance in the circuit. These resonant conditions appear in the Laplace transform as values of s where the function goes to zero or "blows up" (approaches infinity). Thanks to Jason Burnside of wyzant.com for this material.

Epistemology (Chapter 18) The branch of philosophy concerned with the nature and scope of knowledge: What knowledge is, how it can be acquired, and the extent to which knowledge pertinent to any given subject or entity can be verified. Postulates and theorems were distinguished. Chapter 18 presented and proved two theorems applicable to matrices and necessary to understanding the solution technique for solving systems of homogeneous ODEs.

1. vector × scalar = scalar × vector

2. (mat × vector) − (scalar × vector) = [mat − (scalar × **I**)] × vector

Exact Differential Equation (Chapter 12) A type of solvable differential equation that is in, or can be transformed into, the form $M(x, y)\,\mathrm{d}x + N(x, y)\,\mathrm{d}y = 0$ where $\frac{\partial M}{\partial y} = \frac{\partial N}{\partial x}$.

Exponential Growth (Chapter 3) Function growth that is dictated by the rules of geometric sequence. When the domain is discrete, each successive range value is an order of magnitude larger than the previous one.

Exponential Decay (Chapter 3) Function growth that is dictated by the rules of geometric sequence. When the domain is discrete, each successive range value is an order of magnitude smaller than the previous one.

Euler's Method (Chapter 1) A quantitative technique of successively approaching, to any desired level of accuracy, the unknown function value of a differential equation.

Family of Function Solutions (Chapter 1) Just as there can be multiple solutions to an algebraic equation, there can also be multiple solutions to a differential equation. Specific solutions from the family of solutions can be chosen using IVP (initial value problem) conditions.

First-Order Linear Differential Equation (Chapter 8) See Linear Differential Equation.

Growth Factor (Chapter 3) A curvilinear concept analogous to a "constant of proportionality," k in $y = kx$, from beginning algebra. For example, $\frac{dp}{dt} = rp$, where r would be the growth factor or constant of proportionality.

Homogeneous First-Order Differential Equation (Chapter 13) A differential equation of the form $\frac{dy}{dx} = \frac{f(x,y)}{g(x,y)}$, where the order of $f(x, y)$ is the same as the order of $g(x, y)$. Such an equation is solvable by a substitution technique that transforms the equation into a separable differential equation.

Homogeneous Second-Order Differential Equation (Chapter 14) A differential equation of the form

$$F_0(x)\frac{d^2y}{dx} + F_1(x)\frac{dy}{dx} + F_2(x)y = G(x)$$

or

$$F_0y''(x) + F_1y'(x) + F_2y(x) = G(x),$$

which can be transformed by integration to an algebraic formula.

Homogeneous System of Equations (Chapter 17) A system of equations whose graphs pass through the origin.

Hooke's Law (Chapter 15) "Ut tension, sic vis." A principal in physics that states that the force, F, needed to extend or compress a spring by some distance x is proportional to that distance: $F = kx$. Hooke's law arises frequently when studying applications of homogeneous second-order differential equations.

Identity Matrix (Chapter 17) A matrix with zeros in all positions except ones on the major diagonal (top left to bottom right). Multiplication by the identity matrix results in a matrix equal to the original.

Inductor See Electronic Components.

Initial Value Problem, IVP (Chapter 1) Finding a specific function that both solves a differential equation and passes through a given point.

Integrated/Algebraic Form (Chapters 5, 6, 8–10, 12–16, 20) A differential equation that has been transformed to algebraic notation. Newton's famous $F = ma$ is a classic example.

Integrating Factor (Chapter 8) A value that, when multiplied by a given expression, will result in the derivative of a product of functions. For example, multiplying the first-order linear differential equation $\frac{dy}{dx} + P(x) \times y = Q(x)$ by its integrating factor $I(x)$ results in the derivative of a product on the left side of the equation which, after integration, will lead to the solution of the differential equation, $y = \ldots$:

$$\begin{aligned}
\frac{dy}{dx} + P(x) \times y &= Q(x) \\
I(x)\left[\frac{dy}{dx} + P(x) \times y\right] &= I(x) \times Q(x) \\
(y \times I(x))' &= I(x) \times Q(x) \\
\int (y \times I)' &= \int [I(x) \times Q(x)].
\end{aligned}$$

Laplace Transform (Chapters 22–24) An algorithm, procedure, rule or set of rules which changes differential equations into algebraic equations, while at the same time, changing the domain, from say time into space. After the transformation has been made, algebraic manipulations can be made. Then an inverse Laplace transform can be performed to change back to the original domain. The process is distinguished by the fact that both the Laplace transform and its inverse can be made using table look-ups. This is a great example of information hiding as most of the calculus has been done by somebody else much like the sin button on a calculator.

Linear Differential Equation (Chapter 8) A differential equation in which the independent variable, y, has a power of one. Not to be confused with the order of a differential equation which is a reference to the number of derivatives that were taken. See Order of Differential Equation.

$$\frac{dy}{dx} + P(x) \times y^1 = Q(x) \quad \text{(first-order \textbf{linear differential equation})}$$

$$\frac{d^2y}{dx^2} + P(x) \times \frac{dy}{dx} + Q(x) \times y^1 = R(x) \quad \text{(second-order \textbf{linear differential equation})}$$

$$\frac{d^3y}{dx^3} + P(x) \times \frac{d^2y}{dx^2} + Q(x) \times \frac{dy}{dx} + R(x) \times y^1 = S(x) \quad \text{(third-order \textbf{linear differential equation})}$$

First-order linear differential equations are notable because they are solvable by a well-known "cookbook" technique.

Logistic-Growth Equation (Chapter 6) A type of separable equation that can be solved by a generic-form algebraic equation. It differs from an exponential growth or decay function in that the growth is limited by a horizontal asymptote reflecting food limitations or predators that limit growth. Logistic differential equations are notable because they are solvable by a well-known "cookbook" technique.

Matrix Multiplicative Inverse (Chapter 17) A matrix that, when multiplied by another matrix, yields the identity matrix. $\mathbf{Mat}^{-1} \times \mathbf{Mat} = \mathbf{I}$

Modeling (Chapter 6) Choosing a function to represent physical behavior. A function that matches that behavior more closely will enable more exact predictions about the behavior.

Node (Chapter 4) An asymptote in a slope field that has sink solutions approaching on one side and source solutions emanating from it on the other.

Order of a Differential Equation (Chapter 9) The order of the highest differential coefficient that occurs in the equation. See examples in Linear Differential Equation.

Par ma foi! Il y a plus de quarante ans que je dis de la prose sans que j'en susse rien, et je vous suis le plus obligé du monde de m'avoir appris cela. (Chapters 1, 2, 3, 7, 20, 23) [My God! For more than forty years, I have been speaking prose without knowing it. I am most obliged to you for teaching me that.] Many ideas and skills in math are, or can be seen as, extensions of previous ideas and skills. Seeing new concepts as something that one has already been doing for years helps in organization, understanding, retention and especially confidence.

Partial Derivative (Chapter 11) Taking the derivative of an equation in three space in such a way that the derivative of one of the variables is taken with respect to one of the others while the remaining variable is held constant. They are quickly identified by special identifying symbolism $\frac{\partial y}{\partial x}$ instead of $\frac{dy}{dx}$. Hence $\frac{\partial y}{\partial x}$ would be the derivative of y with respect to x for $f(x, y, z)$.

Particular Solution to a Differential Equation (Chapters 3, 4, 5, ...) The one solution from a family of solutions to a differential equation that fits the IVP information.

Phase Plane Analysis (Chapters 4, 21) A coordinate plane in which the axis are labeled differently from the traditional domain–range specification. For example, instead of x–y labeling, you could see a y–y' labeling. By analyzing the phase plane, one can make inferences about the graph of the data in the x–y plane.

Resistor See Electronic Components.

Separable Differential Equation (Chapters 3–6) A differential equation in which the numerator and the denominator of a derivative, $\frac{dy}{dx}$, are separated by algebraic techniques prior to respective integration.

Sink (Chapter 4) An asymptote in a family of solutions to a differential equation with solutions approaching it on both sides.

Slope Field (Chapter 1) "Tangent line segments" separated by any desired delta which suggest to the viewer's mind the infinite number of solutions to any differential equation.

Solving Algebraic Systems of Equations (Chapter 17) Finding the values of the variables in a system of equations using substitution, linear combinations, Cramer's Rule, or matrix algebra.

Source (Chapter 4) An asymptote in a family of solutions to a differential equation with solutions emanating from both sides.

Substitution Technique (Chapters 10, 13) A solution technique applied to solving ODEs in certain forms. It involves changing from a complicated to a less complicated equation form. Once a solution is found for the simpler form a back substitution can be made to solve the original equation.

Synergism of Concepts (Chapters 1, 2, 3, 17, 20, 22, 23) The understanding that in math oftentimes several concepts or ideas will combine and lead to a new concept or idea.

Systems of Differential Equations (Chapters 7, 16, 20) Two or more equations involving the derivatives of two or more functions of a single independent variable. Typical beginning systems of differential equations are demonstrated by the predator–prey and predator–pursuit models. In addition to finding points of intersection (as done in algebraic systems) systems of differential equations can be used to study how changes to one differential equation can impact another differential equation.

$$\frac{d}{dt}x_1 = ax_1 + bx_2$$
$$\frac{d}{dt}x_2 = cx_1 + dx_2$$

The integrated solution to a system of homogeneous differential equations is a vector form extension of the integrated solution of a single second-order homogeneous differential equation studied in Chapter 14.

Vector (Chapters 17, 19–21) 1) A quantity that indicates direction and magnitude, $\vec{x} = (4, 5)$, Chapters 20, 21. 2) A quantity indicating multiple data, $\vec{x} = \begin{pmatrix} 3 \\ 5 \\ 2 \end{pmatrix}$, Chapter 17. 3) A collection of multiple functions or differentials, $\vec{x} = \begin{pmatrix} f(x_1) \\ f(x_2) \end{pmatrix}$ or $\vec{x}' = \begin{pmatrix} f'(x_1) \\ f'(x_2) \end{pmatrix}$, Chapter 20.

Appendix A
MATLAB Scripts by Chapter

Chapter 1 MATLAB Scripts

```
% Slope field for y = x2̂ with Deltax = 0.3
[x,y] = meshgrid(-3:.1:3,-3:.1:3);
dy = 2*x;
dx = ones(size(dy));
dyu = dy./sqrt(dx.^2+dy.^2);
dxu = dx./sqrt(dx.^2+dy.^2);
quiver(x,y,dxu,dyu,'ShowArrowHead','off');
```

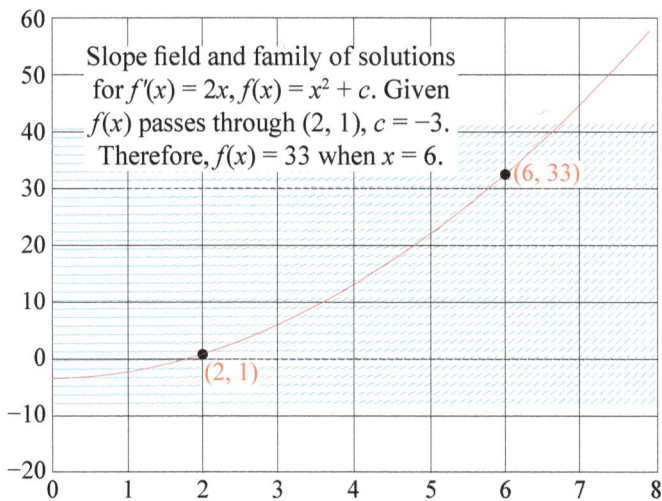

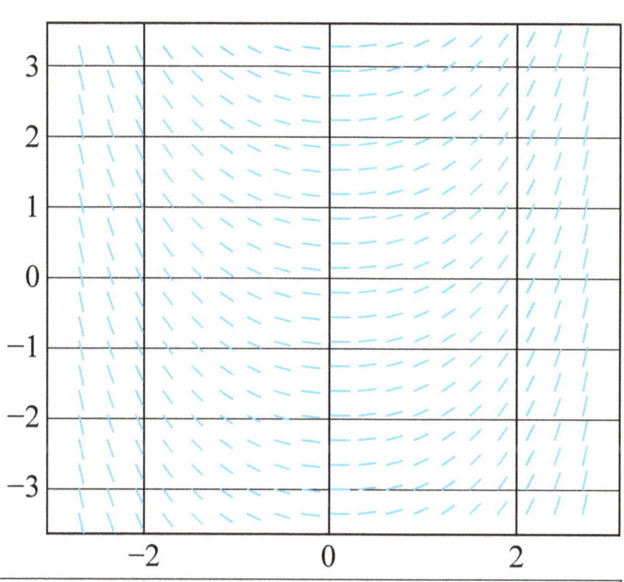

Slope field and family of solutions for $f'(x) = 2x$, $f(x) = x^2 + c$. Given $f(x)$ passes through (2, 1), $c = -3$. Therefore, $f(x) = 33$ when $x = 6$.

(6, 33)

(2, 1)

```
[x,y] = meshgrid(0:.25:8, 0:1:35);
dy = 2*x;
dx = ones(size(dy));
dyu = dy./sqrt(dx.^2+dy.^2);
dxu = dx./sqrt(dx.^2+dy.^2);
quiver(x,y,dxu,dyu,'ShowArrowHead','off');
hold on;
xx= 0: 0.5: 7;
yy = xx.^2 - 3.0;
plot(xx,yy, 'r', 'LineWidth', 3);
plot(2,1, 'ko','MarkerSize',3,'LineWidth',5);
text(1.8,-1, '(2,1)');
plot(6,33, 'ko','MarkerSize',3,'LineWidth',5);
text(6.1,32.9, '(6,33)');
title('Slope field and family of solutions
   for x^2 - 3');
```

Chapter 3 MATLAB Scripts

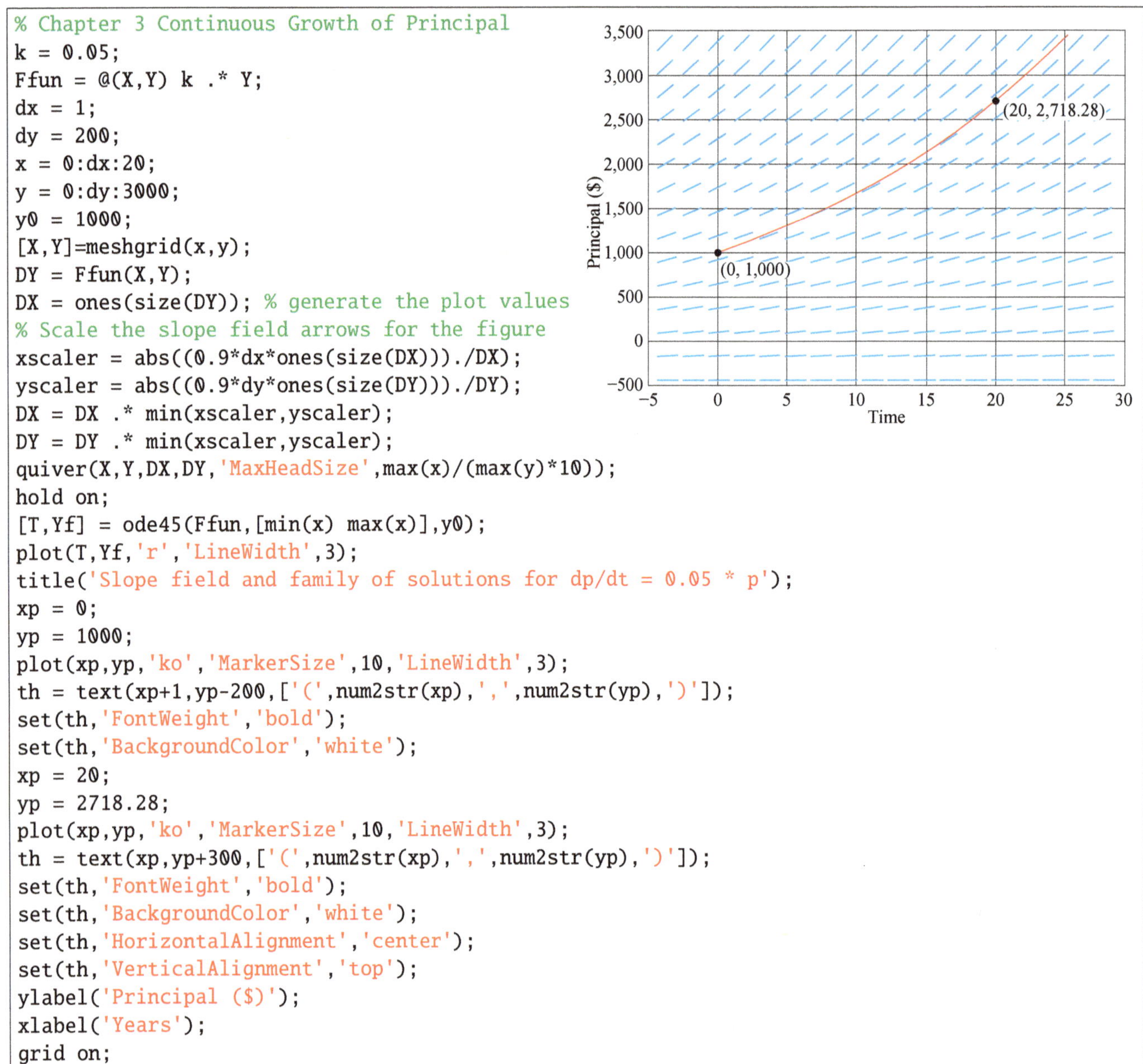


```
% Chapter 3 Continuous Growth of Principal
k = 0.05;
Ffun = @(X,Y) k .* Y;
dx = 1;
dy = 200;
x = 0:dx:20;
y = 0:dy:3000;
y0 = 1000;
[X,Y]=meshgrid(x,y);
DY = Ffun(X,Y);
DX = ones(size(DY)); % generate the plot values
% Scale the slope field arrows for the figure
xscaler = abs((0.9*dx*ones(size(DX)))./DX);
yscaler = abs((0.9*dy*ones(size(DY)))./DY);
DX = DX .* min(xscaler,yscaler);
DY = DY .* min(xscaler,yscaler);
quiver(X,Y,DX,DY,'MaxHeadSize',max(x)/(max(y)*10));
hold on;
[T,Yf] = ode45(Ffun,[min(x) max(x)],y0);
plot(T,Yf,'r','LineWidth',3);
title('Slope field and family of solutions for dp/dt = 0.05 * p');
xp = 0;
yp = 1000;
plot(xp,yp,'ko','MarkerSize',10,'LineWidth',3);
th = text(xp+1,yp-200,['(',num2str(xp),',',num2str(yp),')']);
set(th,'FontWeight','bold');
set(th,'BackgroundColor','white');
xp = 20;
yp = 2718.28;
plot(xp,yp,'ko','MarkerSize',10,'LineWidth',3);
th = text(xp,yp+300,['(',num2str(xp),',',num2str(yp),')']);
set(th,'FontWeight','bold');
set(th,'BackgroundColor','white');
set(th,'HorizontalAlignment','center');
set(th,'VerticalAlignment','top');
ylabel('Principal ($)');
xlabel('Years');
grid on;
```

```
% Chapter 3 Deer Population Growth
k = 0.0247552564;
Ffun = @(X,Y) k .* Y;
dx = 5;
dy = 50;
x = 0:dx:60;
y = 0:dy:600;
y0 = 125;
[X,Y]=meshgrid(x,y);
DY = Ffun(X,Y);
DX = ones(size(DY)); % generate the plot values
% Scale the slope field arrows for the figure
xscaler = abs((0.9*dx*ones(size(DX)))./DX);
yscaler = abs((0.9*dy*ones(size(DY)))./DY);
DX = DX .* min(xscaler,yscaler);
DY = DY .* min(xscaler,yscaler);
quiver(X,Y,DX,DY,'MaxHeadSize',max(x)/(max(y)*7));
% plot the direction field
hold on;
[T,Yf] = ode45(Ffun,[min(x) max(x)],y0);
plot(T,Yf,'r','LineWidth',3);
title('Slope field and family of solutions for dp/dt = k * p');
xp = 0;
yp = 125;
plot(xp,yp,'ko','MarkerSize',10,'LineWidth',3);
th = text(xp+3,yp-50,['(',num2str(xp),',',num2str(yp),')']);
set(th,'FontWeight','bold');
set(th,'BackgroundColor','white');
xp = 50;
yp = 431;
plot(xp,yp,'ko','MarkerSize',10,'LineWidth',3);
th = text(xp+3,yp,['(',num2str(xp),',',num2str(yp),')']);
set(th,'FontWeight','bold');
set(th,'BackgroundColor','white');
ylabel('Number of Deer');
xlabel('Years');
grid on;
```

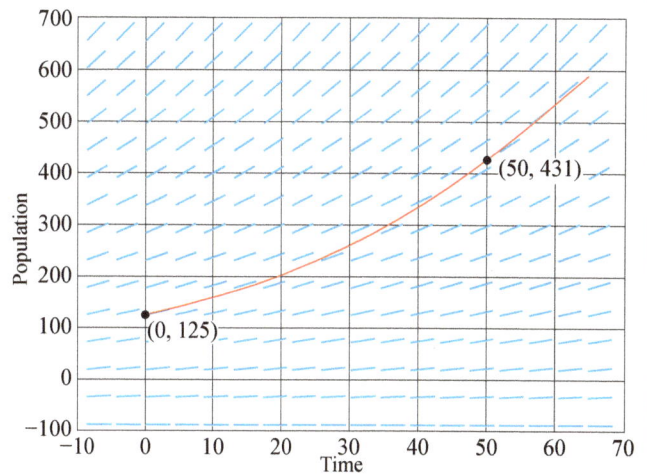

```
% Chapter 3 Carbon 14 Decay
k = -0.0001205471806;
Ffun = @(X,Y) k .* Y;
dx = 1000;
dy = 0.1;
x = 0:dx:12000;
y = 0:dy:1;
y0 = 1;
[X,Y]=meshgrid(x,y);
DY = Ffun(X,Y);
DX = ones(size(DY)); % generate the plot values
% Scale the slope field arrows for the figure
xscaler = abs((0.9*dx*ones(size(DX)))./DX);
yscaler = abs((0.9*dy*ones(size(DY)))./DY);
DX = DX .* min(xscaler,yscaler);
DY = DY .* min(xscaler,yscaler);
% Plot the arrows
quiver(X,Y,DX,DY,'MaxHeadSize',max(x)/(max(y)*1e9));
% plot the direction field
hold on;
[T,Yf] = ode45(Ffun,[min(x) max(x)],y0);
plot(T,Yf,'r','LineWidth',3);
title('Slope field and family of solutions for dp/dt = k * p');
xp = 0;
yp = 1;
plot(xp,yp,'ko','MarkerSize',10,'LineWidth',3);
th = text(xp+500,yp,['(',num2str(xp),',',num2str(yp),')']);
set(th,'FontWeight','bold');
set(th,'BackgroundColor','white');
xp = 5750;
yp = 0.5;
plot(xp,yp,'ko','MarkerSize',10,'LineWidth',3);
th = text(xp+500,yp+0.05,['(',num2str(xp),',',num2str(yp),')']);
set(th,'FontWeight','bold');
set(th,'BackgroundColor','white');
ylabel('Carbon 14');
xlabel('Years Ago');
grid on;
```

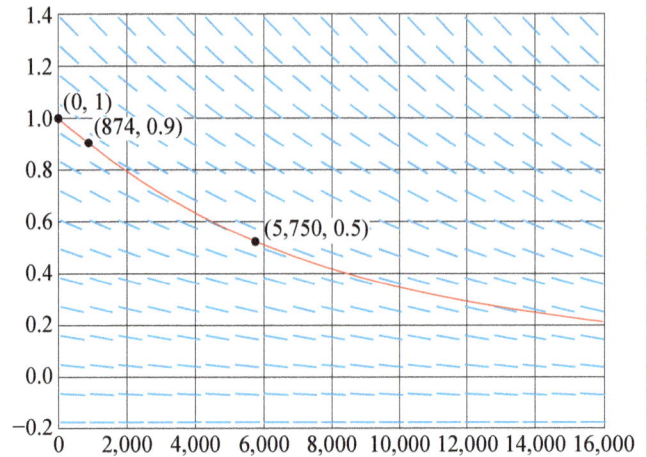

Chapter 5 MATLAB Scripts

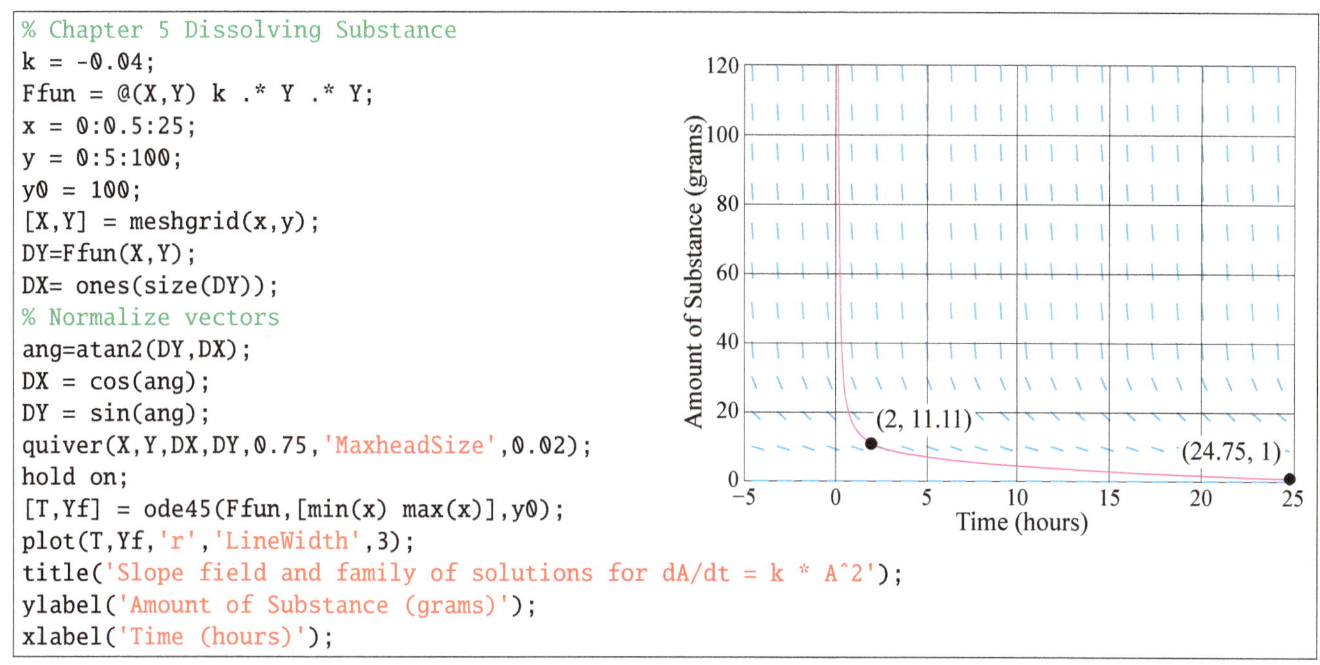


```
% Chapter 5 Dissolving Substance
k = -0.04;
Ffun = @(X,Y) k .* Y .* Y;
x = 0:0.5:25;
y = 0:5:100;
y0 = 100;
[X,Y] = meshgrid(x,y);
DY=Ffun(X,Y);
DX= ones(size(DY));
% Normalize vectors
ang=atan2(DY,DX);
DX = cos(ang);
DY = sin(ang);
quiver(X,Y,DX,DY,0.75,'MaxheadSize',0.02);
hold on;
[T,Yf] = ode45(Ffun,[min(x) max(x)],y0);
plot(T,Yf,'r','LineWidth',3);
title('Slope field and family of solutions for dA/dt = k * A^2');
ylabel('Amount of Substance (grams)');
xlabel('Time (hours)');
```

```
% Chapter 5 Barometric Pressure
k = -0.2;
Ffun = @(X,Y) k .* Y;
dx = 1;
dy = 3;
x = 0:dx:10;
y = 0:dy:30;
y0 = 29.92;
[X,Y] = meshgrid(x,y);
DY = Ffun(X,Y);
DX = ones(size(DY));
% Scale the slope field arrows for the figure
xscaler = abs((0.9*dx*ones(size(DX)))./DX);
yscaler = abs((0.9*dy*ones(size(DY)))./DY);
DX = DX .* min(xscaler,yscaler);
DY = DY .* min(xscaler,yscaler);
figure;
hold on;
quiver(X,Y,DX,DY,'MaxheadSize',(5e-2));
[T,Y] = ode45(Ffun,[min(x) max(x)],y0);
plot(T,Y,'r','LineWidth',3);
xp = 5.5;
yp = 9.96;
plot(xp,yp,'ko','MarkerSize',14,'LineWidth',3);
th = text(xp,yp-2,['(',num2str(xp),',',num2str(yp),')']);
set(th,'HorizontalAlignment','center');
set(th,'VerticalAlignment','top');
set(th,'FontWeight','bold');
set(th,'BackgroundColor','white');
grid on;
title(['Slope field and family of solutions ','for dp/dh = -0.2p']);
ylabel('Altitude (miles)');
xlabel('Pressure (in-Hg)');
```

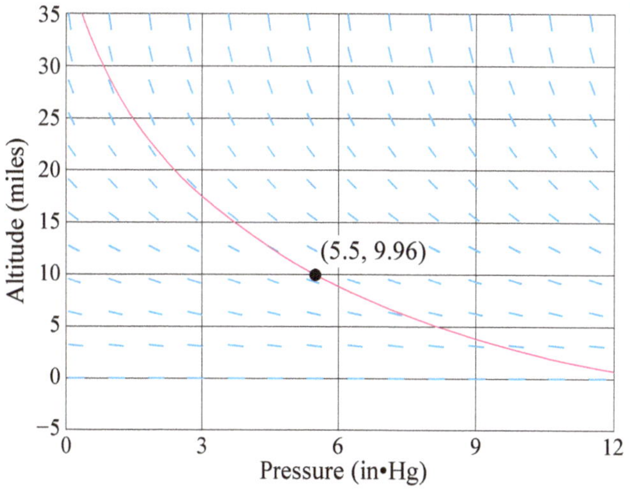

```
% Chapter 5 Mixing Problem, Salt Water to Fresh
Ffun = @(X,Y) -(1/100) .* Y; % du
dx = 30; dy = 2;
x = 0: dx: 400; y = 0: dy: 40; y0 = 35;
[X,Y] = meshgrid(x,y);
DY = Ffun(X,Y); DX = ones(size(DY));
% Scale the slope field arrows for the figure
xscaler = abs((0.9*dx*ones(size(DX)))./DX);
yscaler = abs((3*dy*ones(size(DY)))./DY);
DX = DX .* min(xscaler,yscaler);
DY = DY .* min(xscaler,yscaler);
quiver(X,Y,DX,DY,'MaxheadSize',(15e-3));
hold on;
[T,Y] = ode45(Ffun,[min(x) max(x)],y0);
plot(T,Y,'r','LineWidth',3);
xp = 300; yp = 1.74;
plot(xp,yp,'ko','MarkerSize',14,'LineWidth',3);
xp = 355; yp = 1; plot(xp,yp,'ko','MarkerSize',14,'LineWidth',3);
th = text(xp,yp+2,['(',num2str(xp),',',num2str(yp),')']);
set(th,'HorizontalAlignment','center');
set(th,'VerticalAlignment','bottom');
set(th,'FontWeight','bold');
set(th,'BackgroundColor','white');
grid on;
title('Slope field and family of solutions for dp/dh = -0.01 * A');
ylabel('Salt (kg)');
xlabel('Time (min)');
```

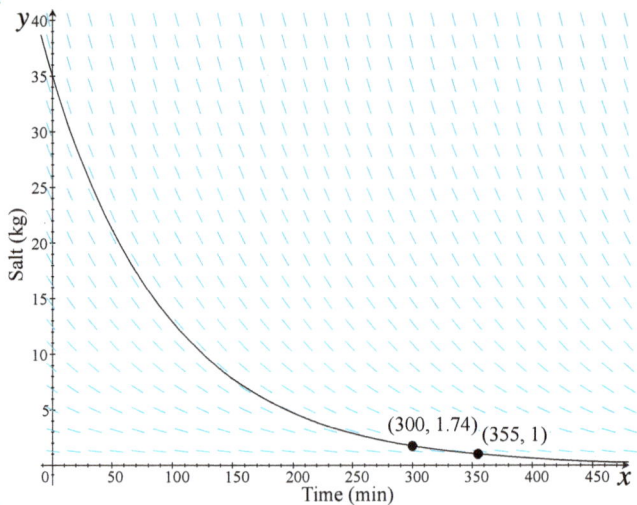

```
% Chapter 5 Kirchhoff's Law
Ffun = @(X,Y) 15 - (5 .* Y);
dx = 0.1; dy = 0.5;
x = 0: dx: 1; y = 0: dy: 5;
y0 = 0;
[X,Y] = meshgrid(x,y);
DY = Ffun(X,Y);
DX = ones(size(DY)); % generate the plot values
% Scale the slope field arrows for the figure
xscaler = abs((0.9*dx*ones(size(DX)))./DX);
yscaler = abs((0.9*dy*ones(size(DY)))./DY);
DX = DX .* min(xscaler,yscaler);
DY = DY .* min(xscaler,yscaler);
quiver(X,Y,DX,DY,'MaxheadSize',(3.5^-3));
hold on;
[T,Y] = ode45(Ffun,[min(x) max(x)],y0);
plot(T,Y,'r','LineWidth',3);
xp = 0.3;
yp = 2.33;
plot(xp,yp,'ko','MarkerSize',14,'LineWidth',3);
th = text(xp+0.1,yp-0.2,['(',num2str(xp),',',num2str(yp),')']);
set(th,'HorizontalAlignment','center');
set(th,'VerticalAlignment','top');
set(th,'FontWeight','bold');
set(th,'BackgroundColor','white');
grid on;
title(['Slope field and family of solutions for dI/dt = 15 - 5I']);
ylabel('Inductance in Henrys');
xlabel('Time (sec)');
```

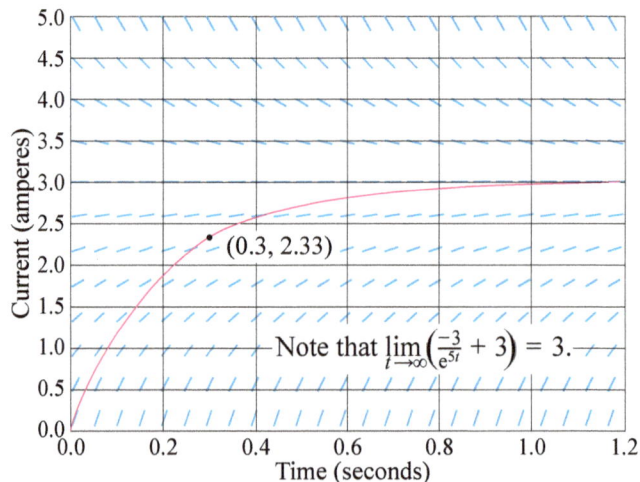

```
% Chapter 5 Newtons Law of Cooling
Ffun = @(X,Y) -0.4652603239 .*(Y - 67);
dx = 1;
dy = 2;
x = 0: dx: 10;
y = 61: dy: 100;
y0 = 98.6;
[X,Y] = meshgrid(x,y);
DY = Ffun(X,Y);
DX = ones(size(DY));
% Scale the slope field arrows for the figure
xscaler = abs((0.9*dx*ones(size(DX)))./DX);
yscaler = abs((0.9*dy*ones(size(DY)))./DY);
DX = DX .* min(xscaler,yscaler);
DY = DY .* min(xscaler,yscaler);
quiver(X,Y,DX,DY,'MaxheadSize',(10^1));
hold on;
[T,Y] = ode45(Ffun,[min(x) max(x)],y0);
plot(T,Y,'r','LineWidth',3);
xp = 0;
yp = 98.6;
plot(xp,yp,'ko','MarkerSize',14,'LineWidth',3);
th = text(xp+1,yp,['Time of Death']);
set(th,'FontWeight','bold');
set(th,'BackgroundColor','white');
xp = 1+28/60;
yp = 83;
plot(xp,yp,'ko','MarkerSize',14,'LineWidth',3);
th = text(xp+1,yp,['83 deg at 10:00 am']);
set(th,'FontWeight','bold');
set(th,'BackgroundColor','white');
xp = 3+58/60;
yp = 72;
plot(xp,yp,'ko','MarkerSize',14,'LineWidth',3);
th = text(xp+1,yp,['72 deg at 12:30 pm']);
set(th,'FontWeight','bold');
set(th,'BackgroundColor','white');
grid on;
title(['Slope field and family of solutions for dT/dt = k * (T -T_a)']);
ylabel('Temperature (^oF)');
xlabel('Hours after death');
```

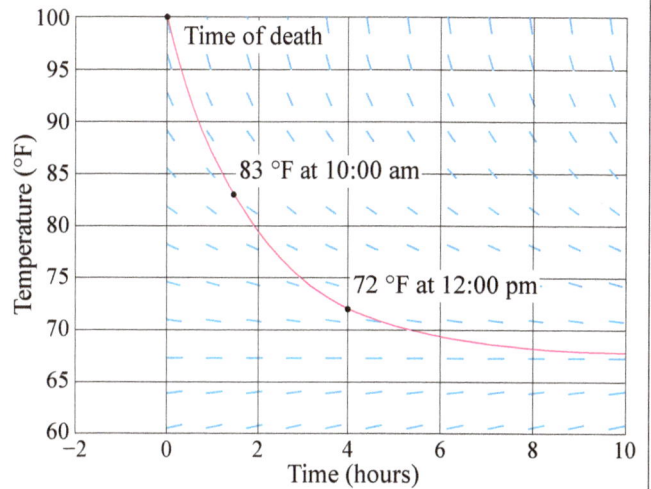

```
% Chapter 5 Continuous Interest with Continuous Withdrawal
close all;
clear all;
Ffun = @(X,Y) (1/20) .* Y - 23;
dx = 1;
dy = 20;
x = 0:dx:20;
y = -10:dy:300;
y0 = 300;
[X,Y]=meshgrid(x,y);
DY = Ffun(X,Y);
DX = ones(size(DY));
% Scale the slope field arrows for the figure
xscaler = abs((0.9*dx*ones(size(DX)))./DX);
yscaler = abs((0.9*dy*ones(size(DY)))./DY);
DX = DX .* min(xscaler,yscaler);
DY = DY .* min(xscaler,yscaler);
quiver(X,Y,DX,DY,'MaxHeadSize',(11^-2));
hold on;
[T,Y] = ode45(Ffun,[min(x) max(x)],y0);
plot(T,Y,'r','LineWidth',3);
xp = 0;
yp = 300;
plot(xp,yp,'ko','MarkerSize',14,'LineWidth',3);
th = text(xp+1,yp-15,['(',num2str(xp),',$',num2str(yp*1000),')']);
set(th,'FontWeight','bold');
set(th,'BackgroundColor','white');
xp = 20;
yp = 25.075;
plot(xp,yp,'ko','MarkerSize',14,'LineWidth',3);
th = text(xp-1,yp,['(',num2str(xp),',$',num2str(yp*1000),')']);
set(th,'HorizontalAlignment','right');
set(th,'FontWeight','bold');
set(th,'BackgroundColor','white');
grid on;
title(['Slope field and family of solutions for dp/dt = 0.05 * p - 23000']);
ylabel('Principal in 1000''s of dollars ($K)');
xlabel('Time (years)');
```

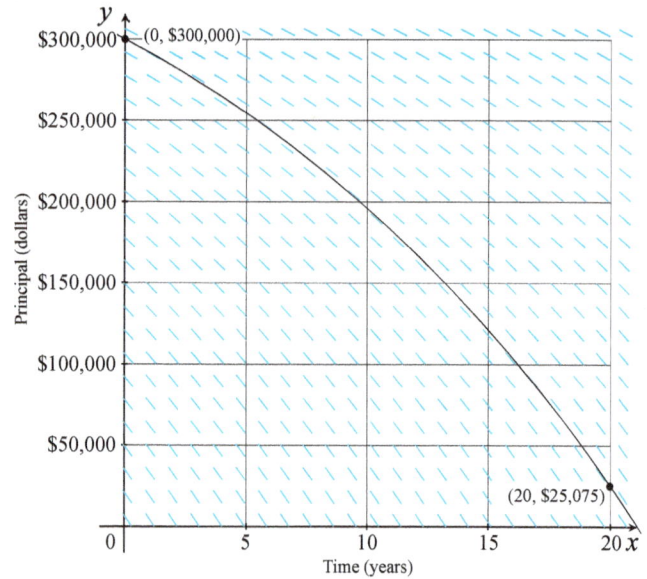

Chapter 6 MATLAB Scripts

```
% Chapter 6 Disease Control
k = 0.2810636798;
Ffun = @(X,Y) k .* Y .* (1 - (Y ./ 800));
dx = 5; dy = 50;
x = 0: dx: 70; y = 0: dy: 800;
y0 = 10; % du
[X,Y] = meshgrid(x,y);
DY = Ffun(X,Y); DX = ones(size(DY)); % generate the plot values
% Scale the slope field arrows for the figure
xscaler = abs((0.9*dx*ones(size(DX)))./DX);
yscaler = abs((0.9*dy*ones(size(DY)))./DY);
DX = DX .* min(xscaler,yscaler); DY = DY .* min(xscaler,yscaler);
% Plot the arrows
quiver(X,Y,DX,DY,'MaxHeadSize',max(x)/(max(y)*15)); % plot the direction field
hold on;
[T,Yf] = ode45(Ffun,[min(x) max(x)],y0);
plot(T,Yf,'r','LineWidth',3);
title('Slope field and family of solutions for dp/dt = k(P(1-P/K))');
xp = 14; yp = 314; plot(xp,yp,'ko','MarkerSize',10,'LineWidth',3);
th = text(xp+4,yp,['(',num2str(xp),',',num2str(yp),')']);
set(th,'FontWeight','bold');
set(th,'BackgroundColor','white');
xp = 35; yp = 797; plot(xp,yp,'ko','MarkerSize',10,'LineWidth',3);
th = text(xp,yp-40,['(',num2str(xp),',',num2str(yp),')']);
set(th,'FontWeight','bold');
set(th,'BackgroundColor','white');
set(th,'HorizontalAlignment','center');
set(th,'VerticalAlignment','top');
ylabel('Number of Students Infected');
xlabel('Days since Outbreak');
grid on;
```

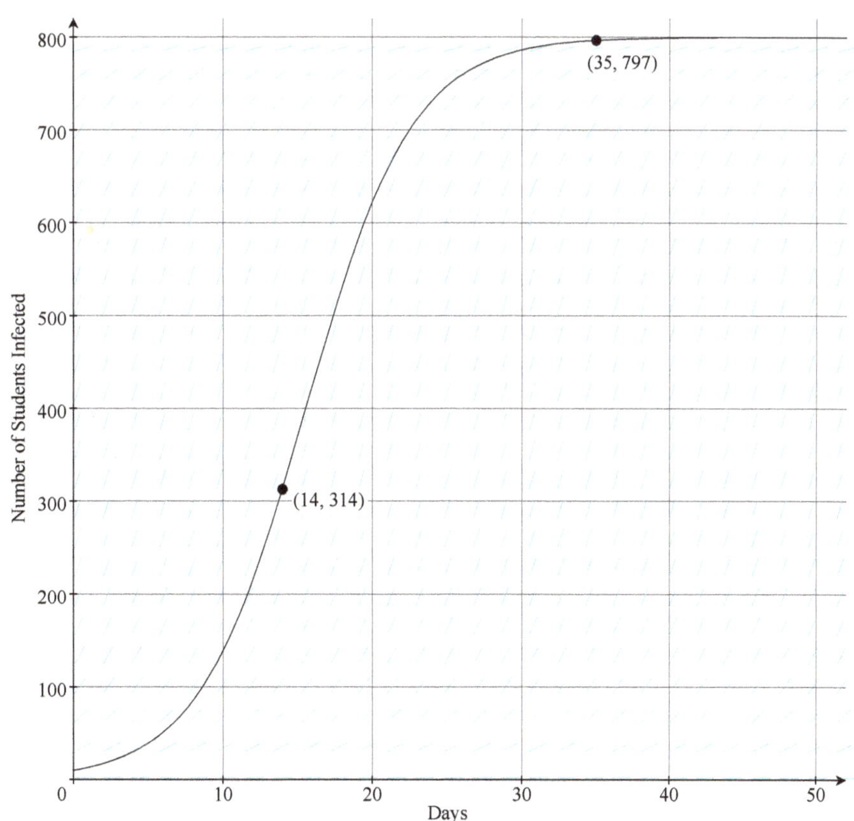

```
% Chapter 6 Spread of Rumors
k = 1.27833981;
Ffun = @(X,Y) k .* Y .* (1 - (Y ./ 2000));
dx = 1;
dy = 100;
x = 0: dx: 12;
y = 0: dy: 2000;
y0 = 20; % du
[X,Y] = meshgrid(x,y);
DY = Ffun(X,Y);
DX = ones(size(DY)); % generate the plot values
% Scale the slope field arrows for the figure
xscaler = abs((0.9*dx*ones(size(DX)))./DX);
yscaler = abs((0.9*dy*ones(size(DY)))./DY);
DX = DX .* min(xscaler,yscaler);
DY = DY .* min(xscaler,yscaler);
% Plot the arrows
quiver(X,Y,DX,DY,'MaxHeadSize',max(x)/(max(y)*15));
hold on;
[T,Yf] = ode45(Ffun,[min(x) max(x)],y0);
plot(T,Yf,'r','LineWidth',3);
xp = 3.59;
yp = 1000;
plot(xp,yp,'ko','MarkerSize',10,'LineWidth',3);
th = text(xp+0.5,yp,['(',num2str(xp),',',num2str(yp),')']);
set(th,'FontWeight','bold');
set(th,'BackgroundColor','white');
title('Slope field and family of solutions for dp/dt = k(P(1-P/K))');
ylabel('Students Who Know');
xlabel('Days since Breakup');
grid on;
```

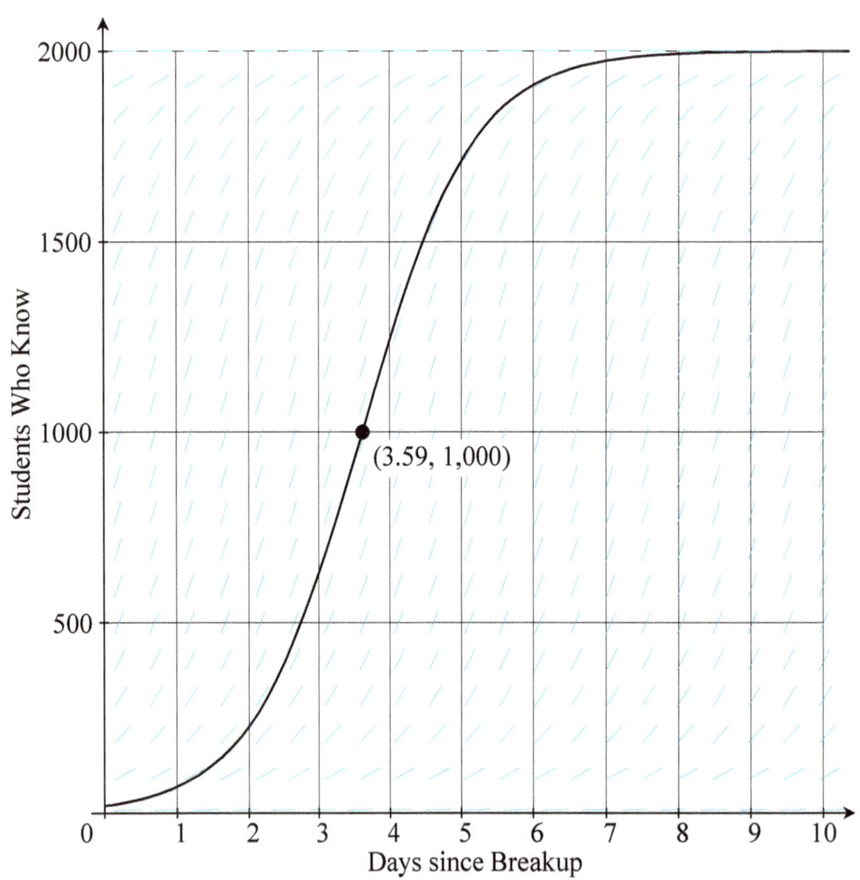

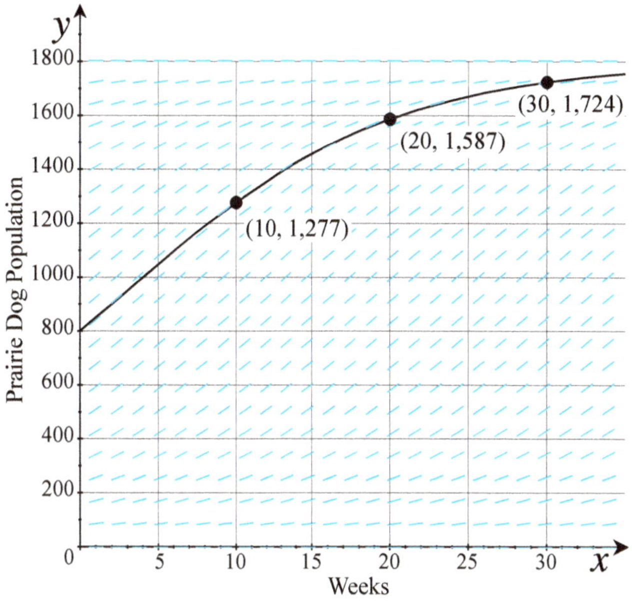

```
% Chapter 6 Prairie Dog Population
close all;
clear all;
% Prairie Dog Population Growth
k = 0.1115717757;
Ffun = @(X,Y) k .* Y .* (1 - (Y ./ 1800));
dx = 2;
dy = 100;
x = 0: dx: 30;
y = 0: dy: 1800;
y0 = 800; % du
[X,Y] = meshgrid(x,y);
DY = Ffun(X,Y);
DX = ones(size(DY)); % generate the plot values
% Scale the slope field arrows for the figure
xscaler = abs((0.9*dx*ones(size(DX)))./DX);
yscaler = abs((0.9*dy*ones(size(DY)))./DY);
DX = DX .* min(xscaler,yscaler);
DY = DY .* min(xscaler,yscaler);
% Plot the arrows
quiver(X,Y,DX,DY,'MaxHeadSize',max(x)/(max(y)*8)); % plot the direction field
hold on;
[T,Yf] = ode45(Ffun,[min(x) max(x)],y0);
plot(T,Yf,'r','LineWidth',3);
xp = 10;
yp = 1277;
plot(xp,yp,'ko','MarkerSize',10,'LineWidth',3);
th = text(xp+1,yp-50,['(',num2str(xp),',',num2str(yp),')']);
set(th,'FontWeight','bold');
set(th,'BackgroundColor','white');
xp = 20;
yp = 1587;
plot(xp,yp,'ko','MarkerSize',10,'LineWidth',3);
th = text(xp+1,yp-70,['(',num2str(xp),',',num2str(yp),')']);
set(th,'FontWeight','bold');
set(th,'BackgroundColor','white');
xp = 30;
yp = 1724;
plot(xp,yp,'ko','MarkerSize',10,'LineWidth',3);
th = text(xp,yp-100,['(',num2str(xp),',',num2str(yp),')']);
set(th,'FontWeight','bold');
set(th,'BackgroundColor','white');
set(th,'HorizontalAlignment','center');
set(th,'VerticalAlignment','top');
title('Slope field and family of solutions for dp/dt = k(P(1-P/K))');
ylabel('Population of Prairie Dogs');
xlabel('Weeks');
grid on;
```

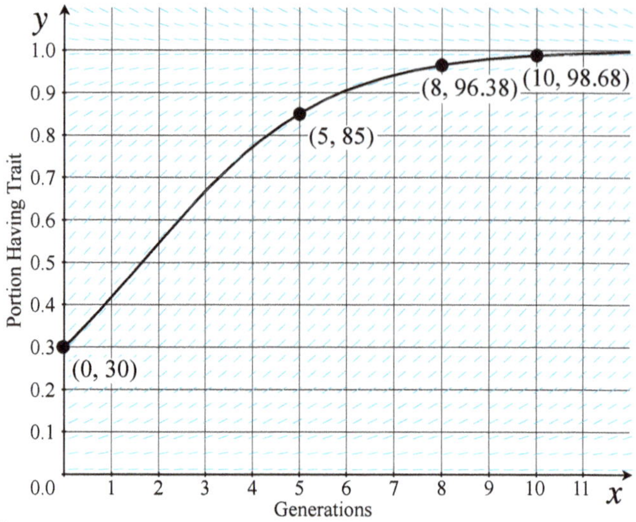

```
% Chapter 6 Inheritance of Characteristics
k = 0.5163797806;
Ffun = @(X,Y) k .* Y .* (1 - (Y ./ 100));
dx = 1;
dy = 10;
x = 0: dx: 10;
y = 0: dy: 100;
y0 = 30; % du
[X,Y] = meshgrid(x,y);
DY = Ffun(X,Y);
DX = ones(size(DY)); % generate the plot values
% Scale the slope field arrows for the figure
xscaler = abs((0.9*dx*ones(size(DX)))./DX);
yscaler = abs((0.9*dy*ones(size(DY)))./DY);
DX = DX .* min(xscaler,yscaler);
DY = DY .* min(xscaler,yscaler);
% Plot the arrows
quiver(X,Y,DX,DY,'MaxHeadSize',max(x)/(max(y)*8));
hold on;
[T,Yf] = ode45(Ffun,[min(x) max(x)],y0);
plot(T,Yf,'r','LineWidth',3);
xp = 0;
yp = 30;
plot(xp,yp,'ko','MarkerSize',10,'LineWidth',3);
th = text(xp+0.5,yp,['(',num2str(xp),',',num2str(yp),')']);
set(th,'FontWeight','bold');
set(th,'BackgroundColor','white');
xp = 5;
yp = 85;
plot(xp,yp,'ko','MarkerSize',10,'LineWidth',3);
th = text(xp,yp-6,['(',num2str(xp),',',num2str(yp),')']);
set(th,'FontWeight','bold');
set(th,'BackgroundColor','white');
set(th,'HorizontalAlignment','center');
set(th,'VerticalAlignment','top');
xp = 8;
yp = 96;
plot(xp,yp,'ko','MarkerSize',10,'LineWidth',3);
th = text(xp,yp-6,['(',num2str(xp),',',num2str(yp),')']);
set(th,'FontWeight','bold');
set(th,'BackgroundColor','white');
set(th,'HorizontalAlignment','center');
set(th,'VerticalAlignment','top');
xp = 10;
yp = 98.68;
plot(xp,yp,'ko','MarkerSize',10,'LineWidth',3);
th = text(xp,yp-6,['(',num2str(xp),',',num2str(yp),')']);
set(th,'FontWeight','bold');
set(th,'BackgroundColor','white');
set(th,'HorizontalAlignment','center');
set(th,'VerticalAlignment','top');
title('Slope field and family of solutions for dp/dt = k(P(1-P/K))');
ylabel('Having Trait (%)');
xlabel('Generation');
grid on;
```

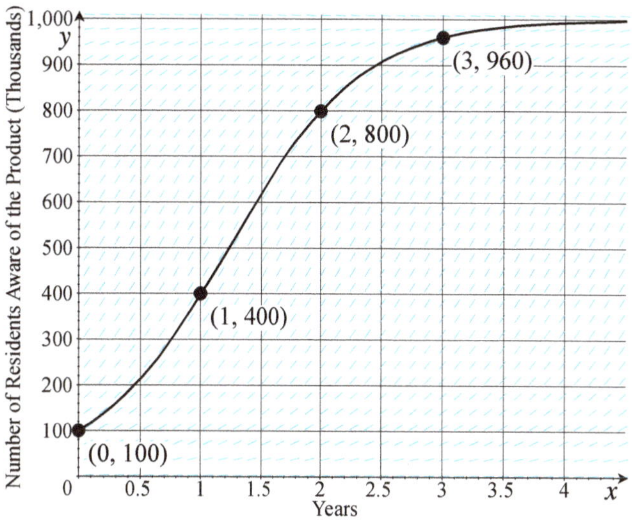

```
% Chapter 6 Advertising Problem
k = 1.791759473;
Ffun = @(X,Y) k .* Y .* (1 - (Y ./ 1000));
dx = 0.5;
dy = 50;
x = 0: dx: 4;
y = 0: dy: 1000;
y0 = 100;
[X,Y] = meshgrid(x,y);
DY = Ffun(X,Y);
DX = ones(size(DY)); % generate plot values
% Scale the slope field arrows for the figure
xscaler = abs((0.9*dx*ones(size(DX)))./DX);
yscaler = abs((0.9*dy*ones(size(DY)))./DY);
DX = DX .* min(xscaler,yscaler);
DY = DY .* min(xscaler,yscaler);
quiver(X,Y,DX,DY,'MaxheadSize',(5e-4));
hold on;
[T,Yf] = ode45(Ffun,[min(x) max(x)],y0);
plot(T,Yf,'r','LineWidth',3);
ylabel('Number of People who have heard of Product (in Thousands)');
xlabel('Time (years)');
title('Slope field and family of solutions for dp/dt = k(P(1-P/K)');
xp = 0;
yp = 100;
plot(xp,yp,'ko','MarkerSize',14,'LineWidth',3);
th = text(xp+0.2,yp-20,['(',num2str(xp),',',num2str(yp),')']);
set(th,'FontWeight','bold');
set(th,'BackgroundColor','white');
xp = 1;
yp = 400;
plot(xp,yp,'ko','MarkerSize',14,'LineWidth',3);
th = text(xp+0.2,yp,['(',num2str(xp),',',num2str(yp),')']);
set(th,'FontWeight','bold');
set(th,'BackgroundColor','white');
xp = 1;
yp = 400;
plot(xp,yp,'ko','MarkerSize',14,'LineWidth',3);
th = text(xp+0.2,yp,['(',num2str(xp),',',num2str(yp),')']);
set(th,'FontWeight','bold');
set(th,'BackgroundColor','white');
xp = 2;
yp = 800;
plot(xp,yp,'ko','MarkerSize',14,'LineWidth',3);
th = text(xp+0.2,yp,['(',num2str(xp),',',num2str(yp),')']);
set(th,'FontWeight','bold');
set(th,'BackgroundColor','white');
xp = 3;
yp = 960;
plot(xp,yp,'ko','MarkerSize',14,'LineWidth',3);
th = text(xp+0.2,yp-40,['(',num2str(xp),',',num2str(yp),')']);
set(th,'FontWeight','bold');
set(th,'BackgroundColor','white');
title('Slope field and family of solutions for dp/dt = - k(P(1-P/K))');
```

Chapter 7 MATLAB Scripts

Mathematical Modelling with Case Studies by Belinda Barnes and Glenn Fulford, pg. 109, 2nd edition, CRC Press.

```
% Chapter 7 Predator--Prey
function c_cp_predprey
global beta1 alpha2 c1 c2;
clear;
beta1 = 1.0;
alpha2 = 0.5;
c1 = 0.01;
c2 = 0.005;
tend = 20; % se the end time to run the simulation
u0 = [200,80]; % set initial conditions as column vector
[tsol,usol] = ode45(@rhs,[0,tend],u0);
Xsol = usol(:,1);
Ysol = usol(:,2);
plot(tsol,Xsol,'b');
hold on;
plot(tsol,Ysol,'r');
function udot = rhs(t,u);
global beta1 alpha2 c1 c2;
X = u(1);
Y = u(2);
Xdot = beta1 * X - c1 * X * Y;
Ydot = -alpha2 * Y + c2 * X * Y;
udot = [Xdot,Ydot];
```

Population
300
250
200
150
100
50
0
0 1 2 3 4 5 6 7 8 9 10 11 12 13 14 15 16 17 18
Time

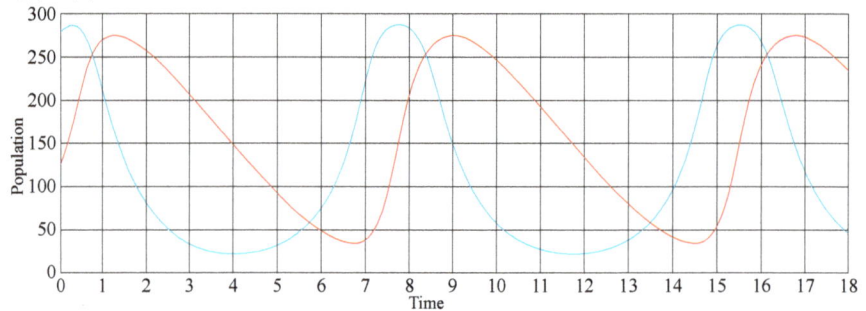

Chapter 8 MATLAB Scripts

```
% Chapter 8 Pond Pollutant Graph
Ffun = @(X,Y) 84000 - ((20 .* Y) ./(1000 + X));
x = 0: 1: 20;
y = 0:50000:1500000; % 4000000;
y0 = 0;
[X,Y] = meshgrid(x,y);
DY=Ffun(X,Y);
DX= ones(size(DY));
% Normalize vectors
ang=atan2(DY,DX);
DX = cos(ang);
DY = sin(ang);
quiver(X,Y,DX,DY,'MaxheadSize',max(x)/(max(y)*20));
hold on;
[T,Yf] = ode45(Ffun,[min(x) max(x)],y0);
plot(T,Yf,'r','LineWidth',3);
title('Slope field for dp/dt = 84000 - (20p/(1000+t))');
ylabel('Grams of pollutant');
xlabel('Days');
```

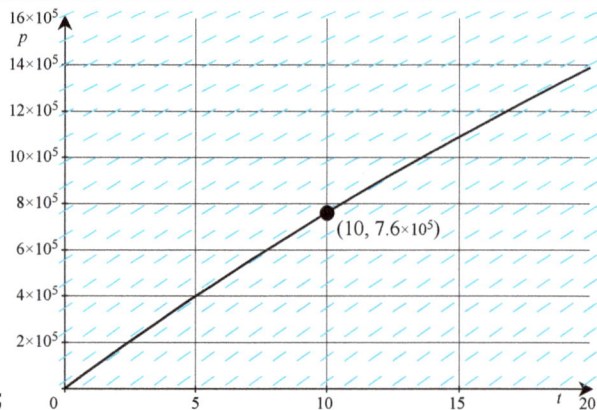

p
16×10^5
14×10^5
12×10^5
10×10^5
8×10^5
6×10^5
4×10^5
2×10^5
(10, 7.6×10^5)
0 5 10 15 t 20

Chapter 10 MATLAB Scripts

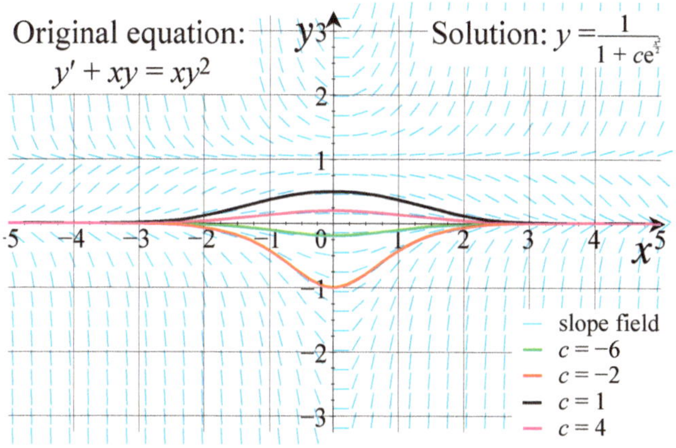

```
% Chapter 10, Bernoulli #1
close all;
clear all;
Ffun = @(X,Y) X.*Y.^2 - X.*Y;
dx = 0.25;
dy = 0.25;
x = -3: dx: 3;
y = -3: dy: 3;
y0 = 0;
[X,Y] = meshgrid(x,y);
DY = Ffun(X,Y);
% generate the plot values
DX = ones(size(DY));
% Scale the slope field arrows
xscaler = abs((0.9*dx*ones(size(DX)))./DX);
yscaler = abs((0.9*dy*ones(size(DY)))./DY);
DX = DX .* min(xscaler,yscaler);
DY = DY .* min(xscaler,yscaler);
% Plot the arrows
quiver(X,Y,DX,DY,'MaxHeadSize',max(x)/(max(y)*15)); % plot the direction field
hold on;
[T,Yf] = ode45(Ffun,[min(x) max(x)],y0);
c = -6;
plot(x,ones(size(x))./(1+c*exp(x.^2/2)),'g','LineWidth',3);
c = -2;
plot(x,ones(size(x))./(1+c*exp(x.^2/2)),'r','LineWidth',3);
c = 1;
plot(x,ones(size(x))./(1+c*exp(x.^2/2)),'k','LineWidth',3);
c = 4;
plot(x,ones(size(x))./(1+c*exp(x.^2/2)),'m','LineWidth',3);
title('Original equation: y'' + xy = xy^2');
text(1,2.3,'Solution: $$y = {\frac{1}{1 + ce^{\frac{x^2}{2}}}}$$', ...
'BackgroundColor','white','EdgeColor','black','Interpreter','latex','FontSize',16);
ylabel('y');
xlabel('x');
grid on;
legend('slope field','c = -6','c = -2','c = 1','c = 4','Location','SouthEast');
ylim([-3 3]);
xlim([-4 4]);
set(gcf,'Position',[403 88 778 578]);
```

```
% Chapter 10, Bernoulli #2
close all;
clear all;
k1 = 0.1;
k2 = 0.01;
m = 5;
v0 = 30;
Ffun = @(T,V) - (k1/m)*V - (k2/m)*V.^2;
dt = 4;
dv = 4;
t = 0: dt: 100;
v = 0: dv: 40;
[T,V] = meshgrid(t,v);
DV = Ffun(T,V);
DT = ones(size(DV));
% Scale the slope field arrows
% for the figure
xscaler = abs((0.9*dt*ones(size(DT)))./DT);
yscaler = abs((0.9*dv*ones(size(DV)))./DV);
DT = DT .* min(xscaler,yscaler);
DV = DV .* min(xscaler,yscaler);
% Plot the arrows
quiver(T,V,DT,DV,'MaxHeadSize',max(t)/(max(v)*15)); % plot the direction field
hold on;
v0 = 4;
[T,Yf] = ode45(Ffun,[min(t) max(t)],v0);
plot(T,Yf,'g','LineWidth',3);
v0 = 12;
[T,Yf] = ode45(Ffun,[min(t) max(t)],v0);
plot(T,Yf,'r','LineWidth',3);
v0 = 20;
[T,Yf] = ode45(Ffun,[min(t) max(t)],v0);
plot(T,Yf,'k','LineWidth',3);
frac = '(^x^2^/^2^)';
title(['Original equation: ', ...
'$$v'' + {\frac{k1}{m}}v = -{\frac{k2}{m}}v^2$$'],'Interpreter','latex');
text(20,30,['Solution: $$v(t) = ', ...
'{\frac{1}{({\frac{1}{v_0}}+{\frac{k_2}{k_1}})e^{{\frac{k_1}{m}}t}-','{\frac{k_2}{k_1}}}}$$'], ...
'BackgroundColor','white','EdgeColor','black','Interpreter','latex','FontSize',16);
text(20,25,['where m = ',num2str(m),',k_1 = ',num2str(k1), ...
',k_2 = ',num2str(k2)],'BackgroundColor','white','EdgeColor', ...
'black','FontSize',14);
ylabel('v(t)');
xlabel('Time');
grid on;
legend('slope field','v_0 = 4','v_0 = 12','v_0 = 20');
set(gcf,'Position',[403 88 778 578]);
```

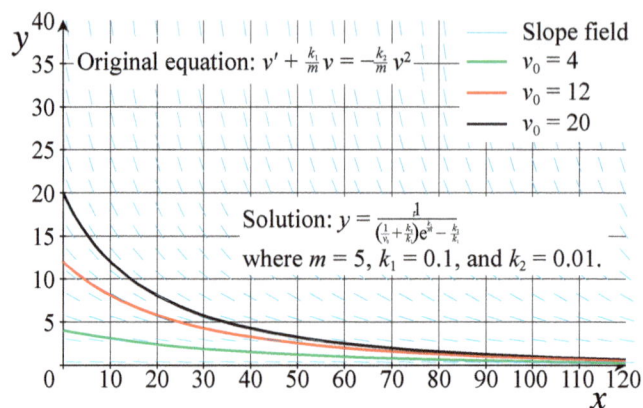

Chapter 11 MATLAB Scripts

Mathematical Modeling with Case Studies by Belinda Barnes and Glenn Fulford, pg. 109, 2nd edition, CRC Press

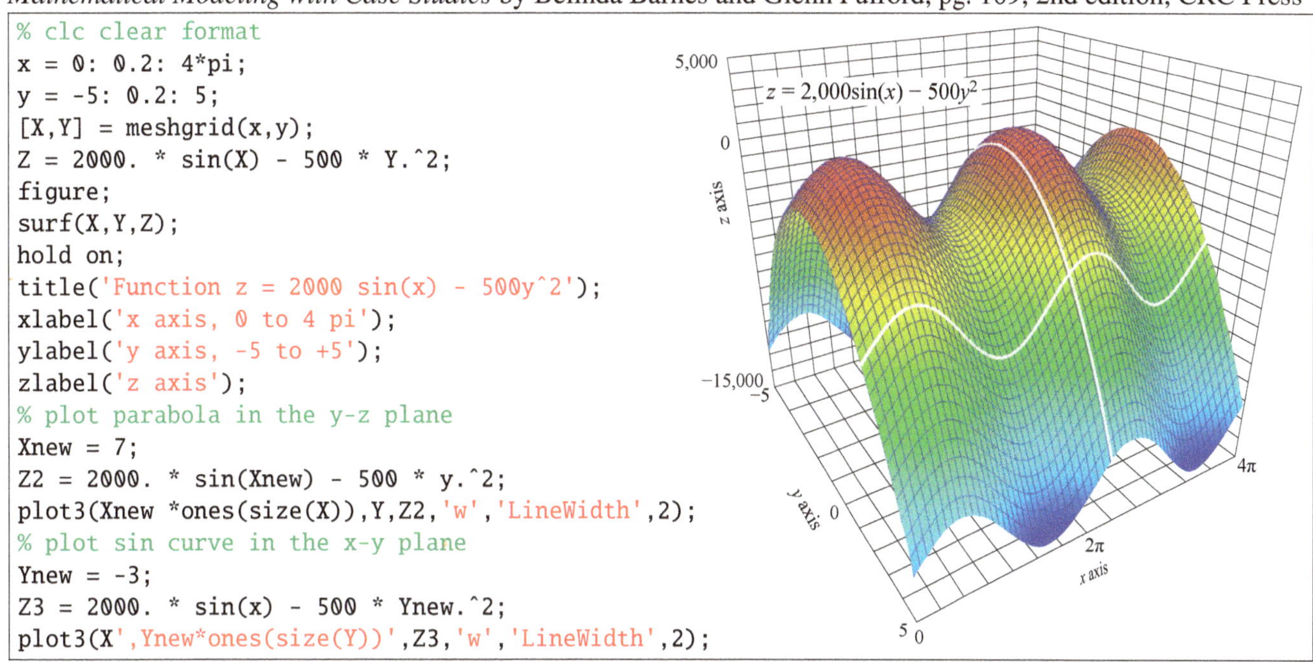

```
% clc clear format
x = 0: 0.2: 4*pi;
y = -5: 0.2: 5;
[X,Y] = meshgrid(x,y);
Z = 2000. * sin(X) - 500 * Y.^2;
figure;
surf(X,Y,Z);
hold on;
title('Function z = 2000 sin(x) - 500y^2');
xlabel('x axis, 0 to 4 pi');
ylabel('y axis, -5 to +5');
zlabel('z axis');
% plot parabola in the y-z plane
Xnew = 7;
Z2 = 2000. * sin(Xnew) - 500 * y.^2;
plot3(Xnew *ones(size(X)),Y,Z2,'w','LineWidth',2);
% plot sin curve in the x-y plane
Ynew = -3;
Z3 = 2000. * sin(x) - 500 * Ynew.^2;
plot3(X',Ynew*ones(size(Y))',Z3,'w','LineWidth',2);
```


Chapter 12 MATLAB Scripts

```
x = -3 * pi:0.4:3 * pi;
y = -8:.2:8;
[X,Y] = meshgrid(x,y);
Z = (X.^2. * Y.^2) + (X.^2);
figure;
surf(X,Y,Z);
hold on;
title('z = x^2y^2 + x^2');
xlabel('xaxis, -10 to 10');
ylabel('yaxis, -10 to 10');
zlabel('zaxis, 0 to 10,000');
```

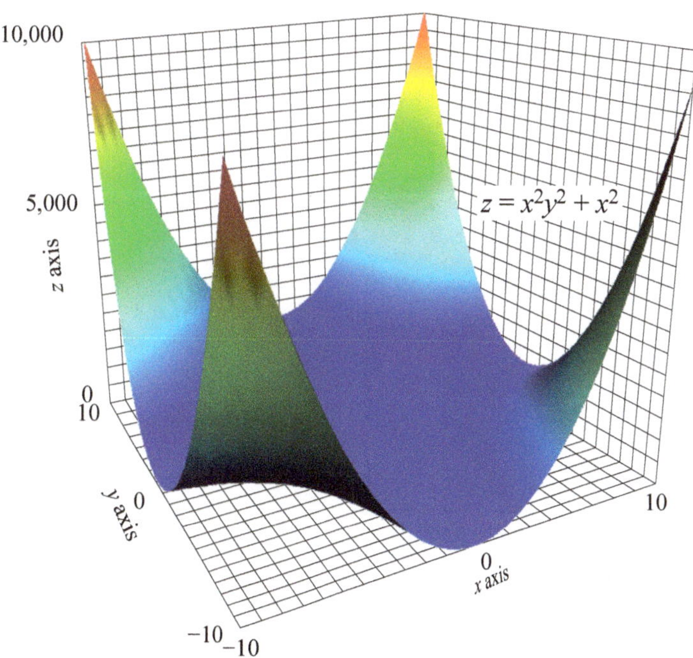

```
% Chapter 12, page 125 left
x = -10: 1: 10;
y = -10: 1: 10;
[X,Y] = meshgrid(x,y);
% Z = (X.^2 .* Y.^2) + (X.^2);
Z = X.*Y.^2 + X;
figure;
surf(X,Y,Z);
hold on;
% title('Function z = ???');
title('Function z = (x^2 * y^2) + x^2');
xlabel('x axis, -10 to 10');
ylabel('y axis, -10 to 10');
zlabel('z axis');
% plot parabola in the x-z plane
Xnew = -5;
Z2 = (Xnew .* Y.^2) + (Xnew.^2);
plot3(Xnew * ones(size(X)),Y,Z2, ...
'w','LineWidth',3);
```

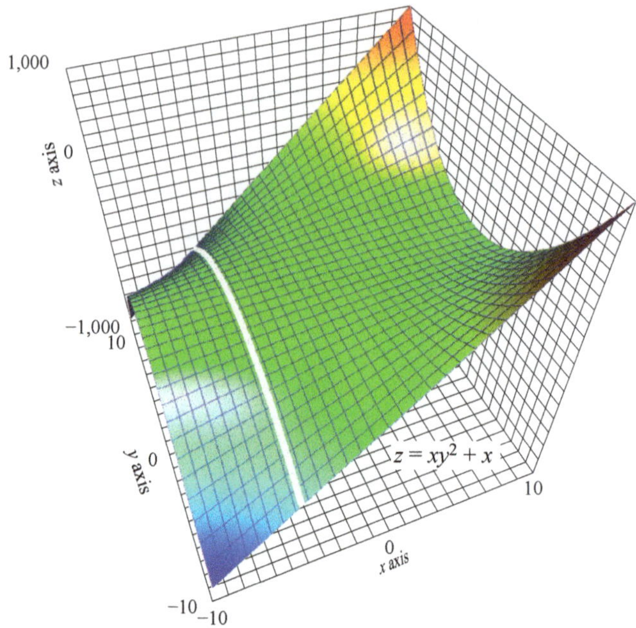

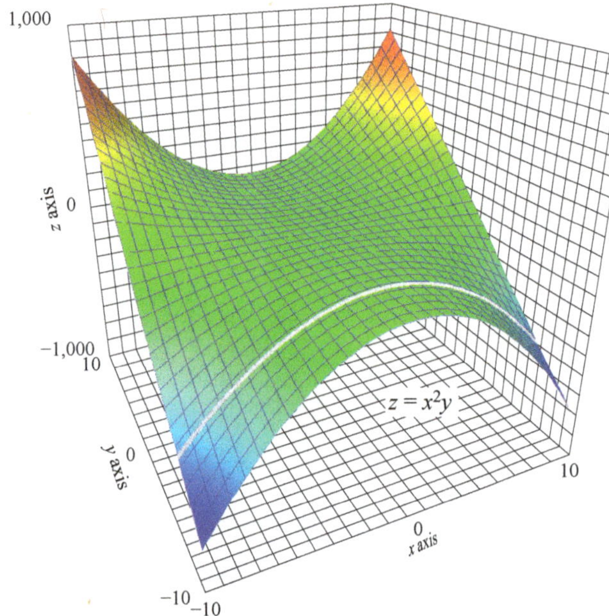


```
% Chapter 12, page 125 right
x = -10: 0.5: 10;
y = -10: 0.5: 10;
[X,Y] = meshgrid(x,y);
% Z = (X.^2 .* Y.^2) + (X.^2);
Z = X.^2 .* Y;
figure;
surf(X,Y,Z);
hold on;
% title('Function z = ???');
title('Function z = (x^2 * Y)');
xlabel('x axis, -10 to 10');
ylabel('y axis, -10 to 10');
zlabel('z axis');
% plot parabola in the y-z plane
Ynew = -5;
Z3 = (X.^2 .* Ynew);
plot3(X,Ynew*ones(size(Y)),Z3, ...
'wo','LineWidth',3);
```

```
% Chapter 12, page 124
x = -10: 0.5: 10;
y = -10: 0.5: 10;
[X,Y] = meshgrid(x,y);
Z = atan(Y ./ X);
figure;
surf(X,Y,Z);
hold on;
title('Function z = atan(y/x');
xlabel('x axis, -10 to 10');
ylabel('y axis, -10 to 10');
zlabel('z axis');
```

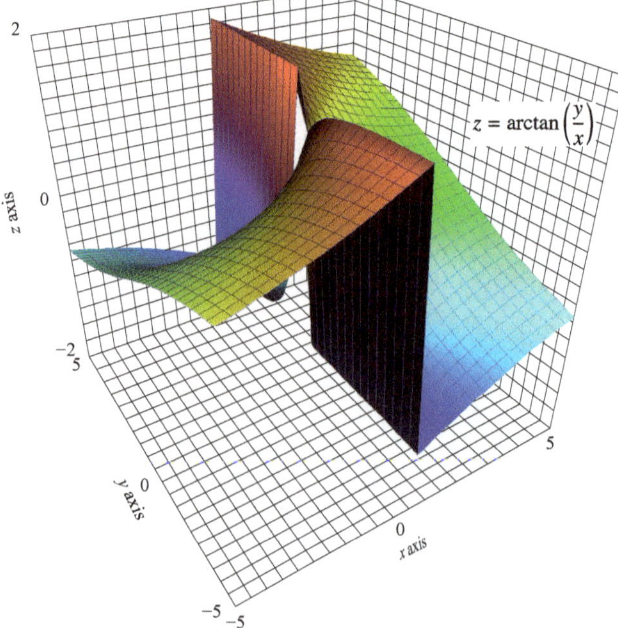

Chapter 12, page 121, see Appendix A, Chapter 11.

```
% Chapter 12, page 125 center
% Both bottom graphs already shown in Appendix
    A above
x = -10: 0.5: 10;
y = -10: 0.5: 10;
[X,Y] = meshgrid(x,y);
Z = (X .^2 .* Y.^2)+ (X.^2);
figure;
surf(X,Y,Z);
hold on;
title('Function X^2 * Y^2 + X^2 ');
xlabel('x axis, -10 to 10');
ylabel('y axis, -10 to 10');
zlabel('z axis');
```

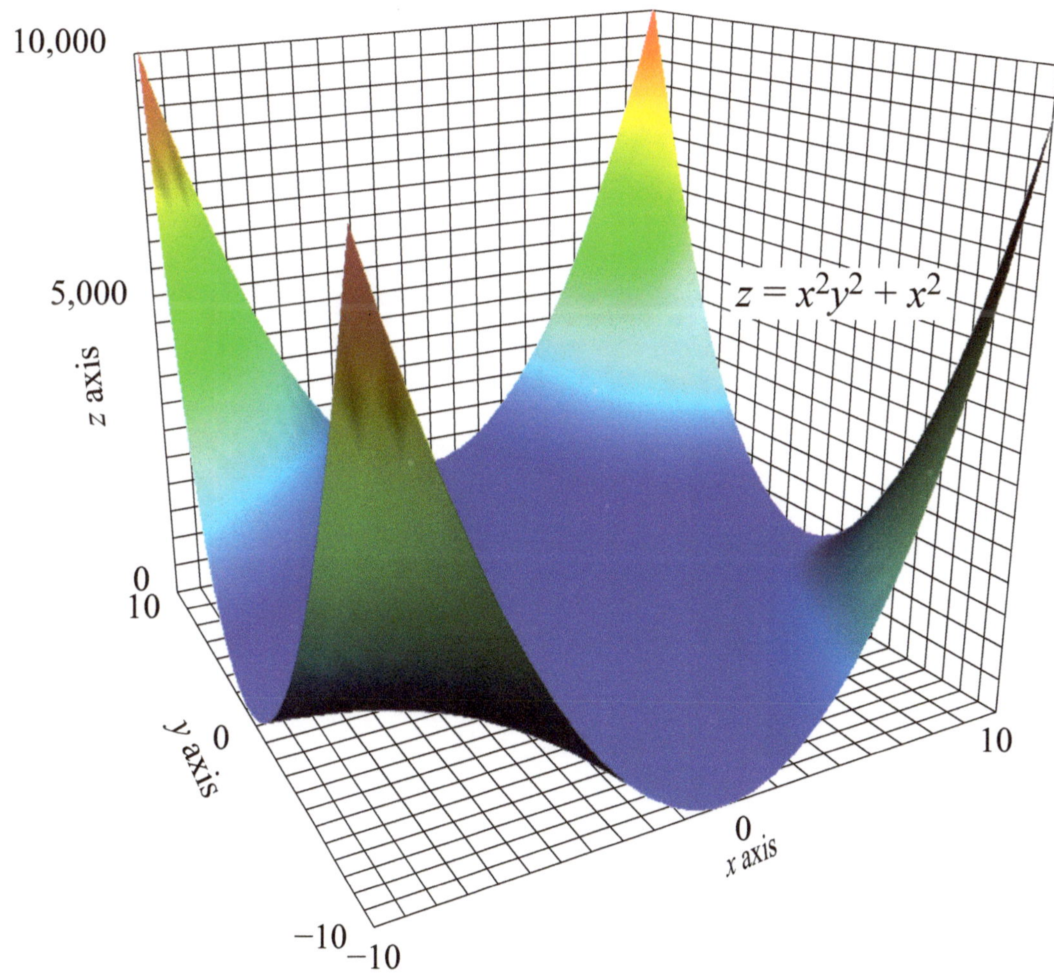

```
% Chapter 12, page 125
close all; clear all;
Ffun = @(X,Y) -(X.*Y.^2 + X)./(Y.*X.^2);
dx = 0.25; dy = 0.25; x = -6: dx: 6; y = -6: dy: 6;
[X,Y] = meshgrid(x,y); DY = Ffun(X,Y); DX = ones(size(DY)); % generate the plot values
% Scale the slope field arrows for the figure
xscaler = abs((0.9*dx*ones(size(DX)))./DX); yscaler = abs((0.9*dy*ones(size(DY)))./DY);
DX = DX .* min(xscaler,yscaler); DY = DY .* min(xscaler,yscaler);
% Plot the arrows
h0 = quiver(X,Y,DX,DY,'MaxHeadSize',max(x)/(max(y)*15)); % plot the direction field
hold on;
% Plot different values of "c"
c = [5,10,15,20];
for ct = c
  y = sqrt(((ct*ones(size(x)))./(x.^2)) - 1);
  xplot=[]; yplot=[];
  icnt=1;
  for ii=1:numel(y)
    if isreal(y(ii))
      xplot(ii) = x(ii); yplot(ii) = y(ii);
      icnt=icnt+1;
    else
      if x(ii) > 0
        xplot(ii) = sqrt(ct);
      else
        xplot(ii) = -sqrt(ct);
      end
      yplot(ii) = 0;
    end
  end
  % Switch statement to plot different colors
  switch ct
    case 5
    h1 = plot(xplot,yplot,'b','LineWidth',3);
    plot(xplot,-yplot,'b','LineWidth',3);
    case 10
    h2 = plot(xplot,yplot,'r','LineWidth',3);
    plot(xplot,-yplot,'r','LineWidth',3);
    case 15
    h3 = plot(xplot,yplot,'y','LineWidth',3);
    plot(xplot,-yplot,'y','LineWidth',3);
    case 20
    h4 = plot(xplot,yplot,'g','LineWidth',3);
    plot(xplot,-yplot,'g','LineWidth',3);
  end
end
% Add title, text, legend and axis labels to the figure
title('Original equation: (xy^2+x)dx + (yx^2)dy = 0');
text(1.3,5.5,'Solution: x^2y^2 + x^2 = c','BackgroundColor','white','EdgeColor', ...
'black','FontSize',16);
ylabel('y, -6 to 6');
xlabel('x, -6 to 6');
grid on;
legend([h0,h1,h2,h3,h4],'slope field',['c = ',num2str(c(1))],['c = ',num2str(c(2))], ...
['c = ',num2str(c(3))],['c = ',num2str(c(4))],'Location','NorthWest');
% Set the limits and size of the figure
ylim([-6 6]); xlim([-6 6]);
set(gcf,'Position',[403 88 778 578]);
```

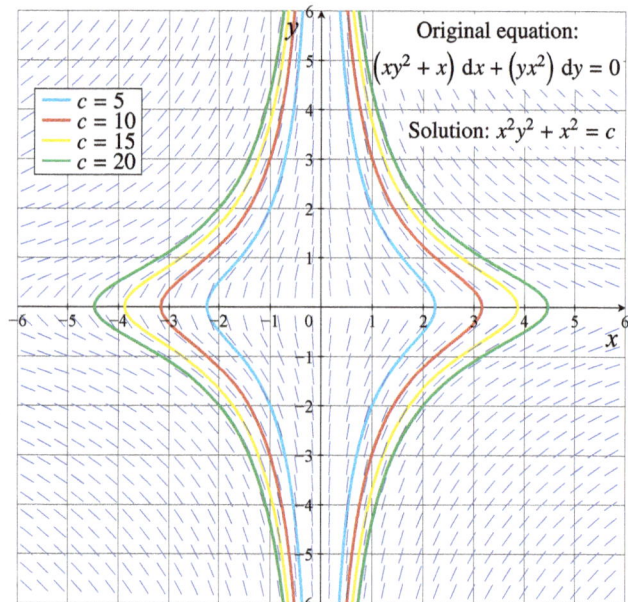

```
% Chapter 12, page 129
close all;
clear all;
Ffun = @(X,Y) - (cos(X) - X.*sin(X) + Y.^2)./(2*X.*Y);
dx = 0.5;
dy = 0.2;
x = -5*pi: dx: 5*pi;
y = -3: dy: 3;
y0 = 0; % du
[X,Y] = meshgrid(x,y);
DY = Ffun(X,Y);
DX = ones(size(DY)); % generate the plot values
% Scale the slope field arrows for the figure
xscaler = abs((0.9*dx*ones(size(DX)))./DX); yscaler = abs((0.9*dy*ones(size(DY)))./DY);
DX = DX .* min(xscaler,yscaler);
DY = DY .* min(xscaler,yscaler);
% Plot the arrows
h0=quiver(X,Y,DX,DY,'MaxHeadSize',max(x)/(max(y)*15)); % plot the direction field
hold on;
% [T,Yf] = ode45(Ffun,[min(x) max(x)],y0);
% plot(T,Yf,'r','LineWidth',3);
dx2 = 0.1;
x2 = -5*pi: dx2: 5*pi;
cmap = {'r','g','c','m'};
c = [-2,0,2];
i=1;
for ct = c
  h(i) = plot(x2,sqrt((ct-x2.*cos(x2))./x2),cmap{i},'LineWidth',3);
  plot(x2,-sqrt((ct-x2.*cos(x2))./x2),cmap{i},'LineWidth',3);
  i=i+1;
end
title('Original equation: (cosx - xsinx +y^2)dx + (2xy)dy = 0');
text(-4.5*pi,-2,'Solution: xcosx + xy^2 = c', ...
'BackgroundColor','white','EdgeColor','black','FontSize',16);
ylabel('y');
xlabel('x');
grid on;
legend([h0,h],{'slope field',['c = ',num2str(c(1))],['c = ',num2str(c(2))], ...
['c = ',num2str(c(3))]}, ...
'Location','NorthWest');
ylim([-3 3]);
xlim([-5 5]*pi);
set(gcf,'Position',[403 88 778 578])
```

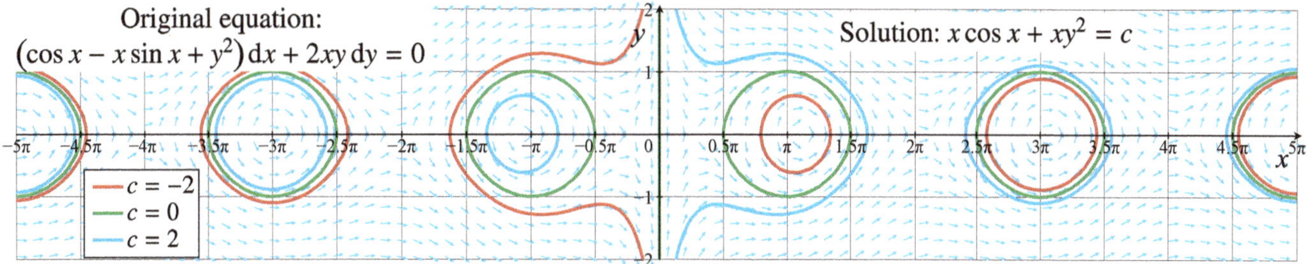

Chapter 13 MATLAB Scripts

```
close all; clear all;
Ffun = @(X,Y) (Y.^2 - X.^2)./(X.*Y);
dx = 0.25; dy = 0.25;
x = 0: dx: 5; y = 0: dy: 5;
y0 = 0; % du
[X,Y] = meshgrid(x,y);
DY = Ffun(X,Y);
DX = ones(size(DY)); % generate the plot values
% Scale the slope field arrows for the figure
xscaler = abs((0.9*dx*ones(size(DX)))./DX); yscaler = abs((0.9*dy*ones(size(DY)))./DY);
DX = DX .* min(xscaler,yscaler); DY = DY .* min(xscaler,yscaler);
% Plot the arrows
h0=quiver(X,Y,DX,DY,'MaxHeadSize',max(x)/(max(y)*15)); % plot the direction field
hold on;
cmap = {'r','g','c','m'};
c = [3.5:0.5:4.5];
i=1;
for ct = c
  h(i) = plot(x,sqrt(-2*(x.^2).*log(abs(x)) + ct*x.^2),cmap{i},'LineWidth',3);
  i=i+1;
end
title('Original equation: dy/dx = (y^2 - x^2)/(xy)');
text(0.5,0.5,'Solution: y^2 = -2x^2ln|x| + cx^2','BackgroundColor','white',...
'EdgeColor','black','FontSize',16);
ylabel('y'); xlabel('x');
grid on;
legend([h0,h],{'slope field',['c = ',num2str(c(1))],['c = ',num2str(c(2))],['c =
    ',num2str(c(3))]}, ...
'Location','NorthWest');
ylim([0 5]); xlim([0 5]);
set(gcf,'Position',[403 88 778 578])
```

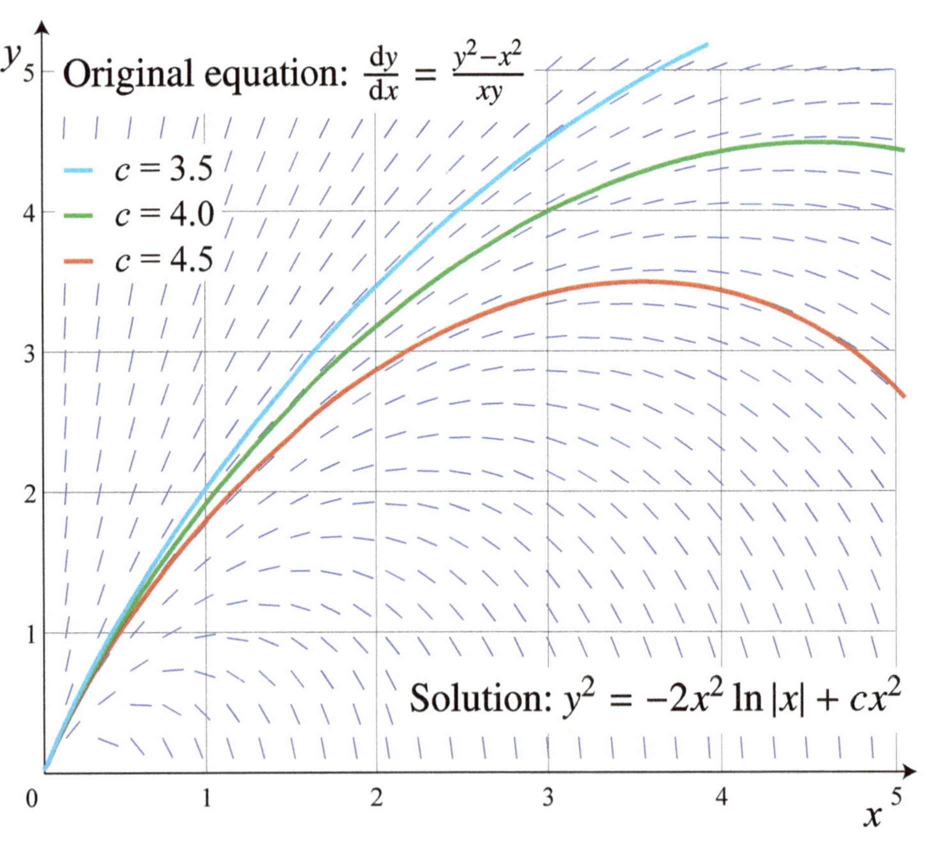

Chapter 14 MATLAB Scripts

```
dx = 0.03; dy = 0.2;
x = -0.4: dx: 0.4; y = 0: dy: 10;
y0 = 0;
[X,Y] = meshgrid(x,y);
DY = Ffun(X,Y); DX = ones(size(DY)); % generate the plot values
% Scale the slope field arrows for the figure
xscaler = abs((0.9*dx*ones(size(DX)))./DX); yscaler = abs((0.9*dy*ones(size(DY)))./DY);
DX = DX .* min(xscaler,yscaler); DY = DY .* min(xscaler,yscaler);
% Plot the arrows
h0=quiver(X,Y,DX,DY,'MaxHeadSize',max(x)/(max(y)*2)); % plot the direction field
hold on;
y1 = c1 * exp(-6*x); y2 = c2 * exp(4*x); y3 = y1 + y2;
% a = plot(x,c1*exp(-6*x,'b','LineWidth',3);
a = plot(x,y2,'k','LineWidth',3);
b = plot(x,y1,'g','LineWidth',3);
h = plot(x,y3,'r','LineWidth',3);
title('Original equation: 1y'''' + 2y'' - 24y = 0');
text(-0.38,8.3,'If y1 and y2 are both solutions then y1 + y2 is also a solution', ...
'BackgroundColor','white','EdgeColor','black','FontSize',16);
text(-0.35,3,'Solution: y1 = c_1e^{-6x}', ...
'BackgroundColor','white','EdgeColor','black','FontSize',12);
text(0.1,3,'Solution: y2 = c_2e^{4x}', ...
'BackgroundColor','white','EdgeColor','black','FontSize',12);
text(-0.2,5,'Solution: y3 = c_1e^{-6x} + c_2e^{4x}', ...
'BackgroundColor','white','EdgeColor','black','FontSize',16);
text(-0.15,4,'(Solution: y3 = y1 + y2)', ...
'BackgroundColor','white','EdgeColor','black','FontSize',12);
ylabel('y'); xlabel('x');
grid on;
legend([h0,h],{'slope field',['c_1,c_2 = ',num2str(c1),',',num2str(c2)]},'Location','NorthWest');
ylim([min(y) max(y)]); xlim([min(x) max(x)]);
set(gcf,'Position',[403 88 778 578])
```

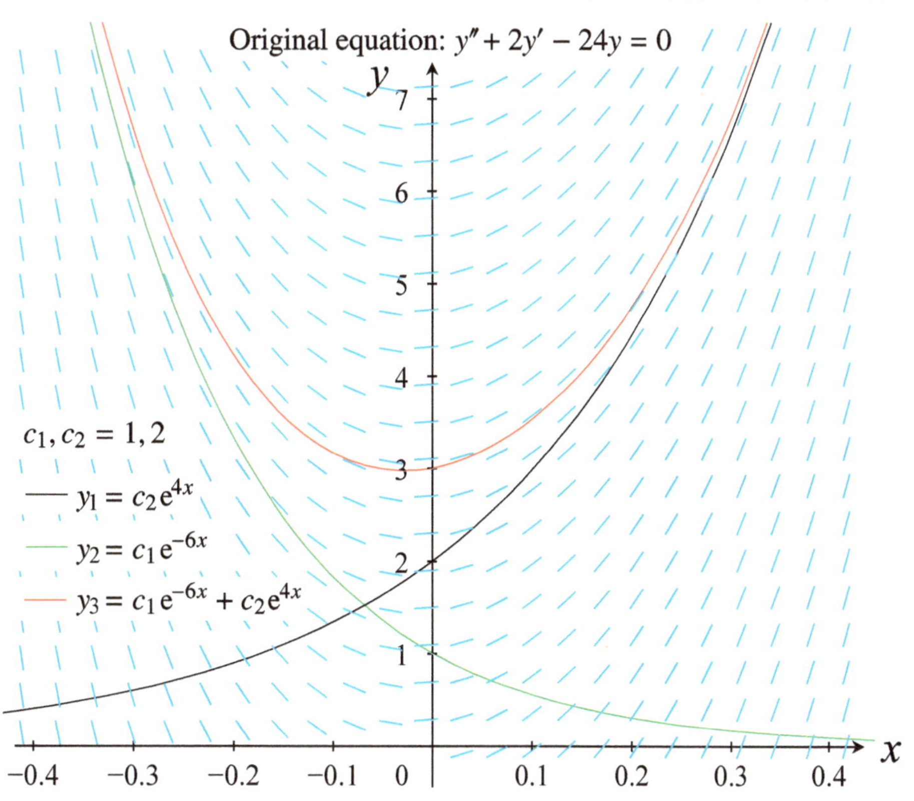

```
close all; clear all;
c1 = -1/3; c2 = 7/3;
Ffun = @(X,Y) -4*c1*exp(-4*X) + 2*c2*exp(2*X);
dx = 0.1; dy = 1;
x = -1: dx: 1; y = -10: dy: 10;
y0 = 0; % du
[X,Y] = meshgrid(x,y);
DY = Ffun(X,Y);
DX = ones(size(DY)); % generate the plot values
% Scale the slope field arrows for the figure
xscaler = abs((0.9*dx*ones(size(DX)))./DX);
yscaler = abs((0.9*dy*ones(size(DY)))./DY);
DX = DX .* min(xscaler,yscaler);
DY = DY .* min(xscaler,yscaler);
% Plot the arrows
h0=quiver(X,Y,DX,DY,'MaxHeadSize',...
max(x)/(max(y)*12)); % plot the direction field
hold on;
cmap1 = {'r','g'};
i=1;
h(1) = plot(x,c1*exp(-4*x) + c2*exp(2*x),cmap1{i},'LineWidth',3);
title('Original equation: 1y'''' + 2y'' - 8y = 0');
text(0,-6,'Solution: y = c_1e^{-4x} + c_2e^{2x}','BackgroundColor','white','EdgeColor',...
'black','FontSize',16);
ylabel('y'); xlabel('x');
grid on;
legend([h0,h],{'slope field',['c_1, c_2 = -1/_3, 7/_3']}, 'Location','NorthWest');
ylim([min(y) max(y)]); xlim([min(x) max(x)]);
set(gcf,'Position',[403 88 778 578])
```

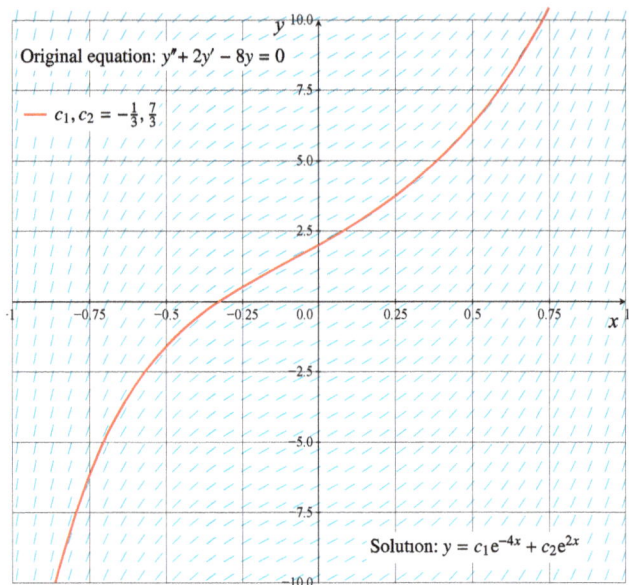

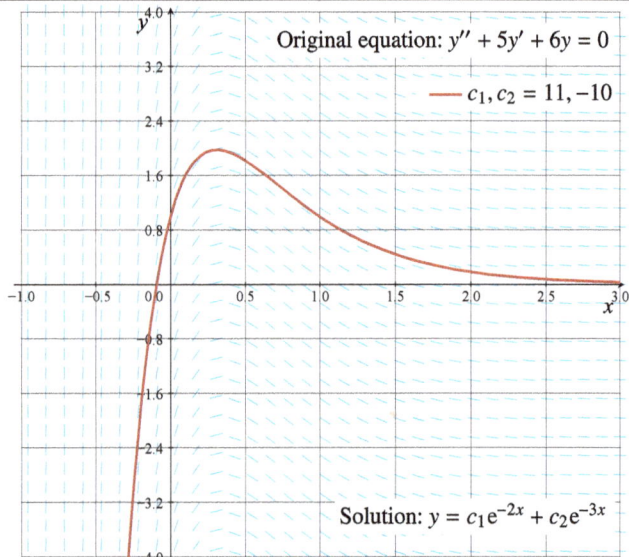

```
close all; clear all;
c1 = 11; c2 = -10;
Ffun = @(X,Y) -2*c1*exp(-2*X) - 3*c2*exp(-3*X);
dx = 0.1; dy = 0.4;
x = -1: dx: 3; y = -4: dy: 4;
y0 = 0; % du
[X,Y] = meshgrid(x,y);
DY = Ffun(X,Y);
DX = ones(size(DY)); % generate the plot values
% Scale the slope field arrows for the figure
xscaler = abs((0.9*dx*ones(size(DX)))./DX);
yscaler = abs((0.9*dy*ones(size(DY)))./DY);
DX = DX .* min(xscaler,yscaler);
DY = DY .* min(xscaler,yscaler);
% Plot the arrows
h0=quiver(X,Y,DX,DY,'MaxHeadSize',...
max(x)/(max(y)*30)); % plot the direction field
hold on;
cmap1 = {'r','g'};
i=1;
h(1) = plot(x,c1*exp(-2*x) + c2*exp(-3*x),cmap1{i},'LineWidth',3);
title('Original equation: 1y'''' + 5y'' + 6y = 0');
text(1,-2,'Solution: y = c_1e^{-2x} + c_2e^{-3x}','BackgroundColor','white','EdgeColor',...
'black','FontSize',16);
ylabel('y'); xlabel('x');
grid on;
legend([h0,h],{'slope field',['c_1, c_2 = ', num2str(c1),',',num2str(c2)]},'Location','NorthEast');
ylim([min(y) max(y)]); xlim([min(x) max(x)]);
set(gcf,'Position',[403 88 778 578])
```

```
close all; clear all;
c1 = 4; c2 = -5;
Ffun = @(X,Y) 3*c1*exp(3*X) + c2*(exp(3*X) + 3*X.*exp(3*X));
dx = 0.1; dy = 0.25;
x = -4: dx: 1; y = -2: dy: 8;
y0 = 0; % du
[X,Y] = meshgrid(x,y);
DY = Ffun(X,Y); DX = ones(size(DY)); % generate the plot values
% Scale the slope field arrows for the figure
xscaler = abs((0.9*dx*ones(size(DX)))./DX); yscaler = abs((0.9*dy*ones(size(DY)))./DY);
DX = DX .* min(xscaler,yscaler); DY = DY .* min(xscaler,yscaler);
% Plot the arrows
h0=quiver(X,Y,DX,DY,'MaxHeadSize',max(x)/(max(y)*5)); % plot the direction field
hold on;
cmap1 = {'r','g'};
i=1;
h(1) = plot(x,c1*exp(3*x) + c2*x.*exp(3*x),cmap1{i},'LineWidth',3);
title('Original equation: 1y'''' - 6y'' + 9y = 0');
text(-3.5,5,'Solution: y = c_1e^{3x} + c_2xe^{3x}', ...
'BackgroundColor','white','EdgeColor','black','FontSize',16);
ylabel('y'); xlabel('x');
grid on;
legend([h0,h],{'slope field',['c_1, c_2 = ', num2str(c1),',',num2str(c2)]}, ...
'Location','NorthWest');
ylim([min(y) max(y)]); xlim([min(x) max(x)]);
set(gcf,'Position',[403 88 778 578])
```

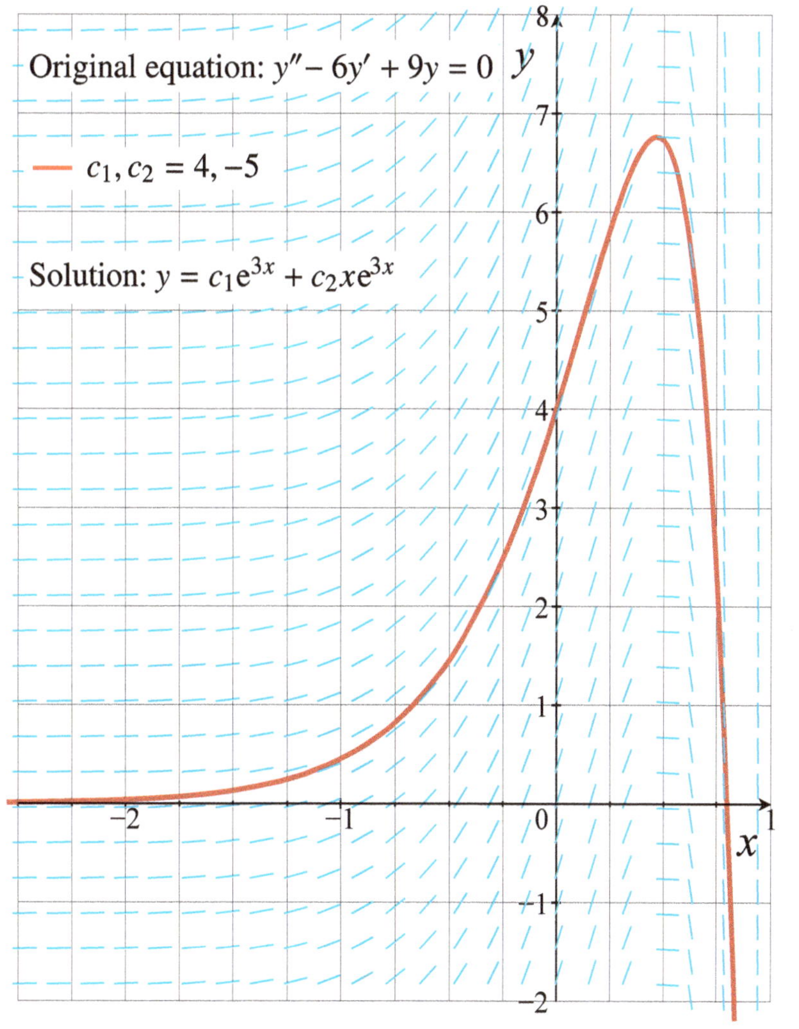

```
close all;
clear all;
c1 = 2;
c2 = -1;
Ffun = @(X,Y) 2*c1*exp(2*X).*cos(3*X) - 3*c1*exp(2*X).*sin(3*X) + ...
2*c2*exp(2*X).*sin(3*X) + 3*c2*exp(2*X).*cos(3*X);
dx = 0.1;
dy = 2;
x = -1: dx: 2;
y = -20: dy: 20;
y0 = 0; % du
[X,Y] = meshgrid(x,y);
DY = Ffun(X,Y);
DX = ones(size(DY)); % generate the plot values
% Scale the slope field arrows for the figure
xscaler = abs((0.9*dx*ones(size(DX)))./DX);
yscaler = abs((0.9*dy*ones(size(DY)))./DY);
DX = DX .* min(xscaler,yscaler);
DY = DY .* min(xscaler,yscaler);
% Plot the arrows
h0=quiver(X,Y,DX,DY,'MaxHeadSize',max(x)/(max(y)*20)); % plot the direction field
hold on;
cmap1 = {'r','g'};
i=1;
h(1) = plot(x,c1*exp(2*x).*cos(3*x) + c2*exp(2*x).*sin(3*x),cmap1{i},'LineWidth',3);
title('Original equation: 1y'''' - 4y'' + 13y = 0');
text(-0.9,10,'Solution: y = c_1e^{2x}cos(3x) + c_2e^{2x}sin(3x)', ...
'BackgroundColor','white','EdgeColor','black','FontSize',16);
ylabel('y');
xlabel('x');
grid on;
legend([h0,h],{'slope field',['c_1, c_2 = ',num2str(c1),',',num2str(c2)]}, ...
'Location','NorthWest');
ylim([min(y) max(y)]);
xlim([min(x) max(x)]);
set(gcf,'Position',[403 88 778 578]);
```

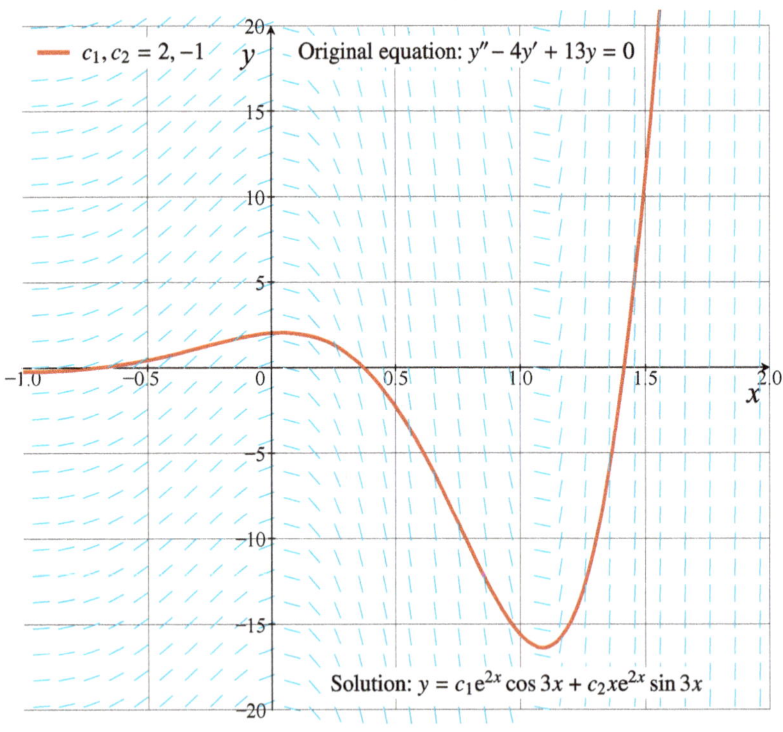

Chapter 15 MATLAB Scripts

```
close all;
clear all;
c1 = 2/3;
c2 = -1/6;
Ffun = @(X,Y) -8*c1*sin(8*X) + 8*c2*cos(8*X);
dx = 0.05;
dy = 0.2;
x = 0: dx: 2;
y = -2: dy: 2;
y0 = 0; % du
[X,Y] = meshgrid(x,y);
DY = Ffun(X,Y);
DX = ones(size(DY)); % generate the plot values
% Scale the slope field arrows for the figure
xscaler = abs((0.9*dx*ones(size(DX)))./DX);
yscaler = abs((0.9*dy*ones(size(DY)))./DY);
DX = DX .* min(xscaler,yscaler);
DY = DY .* min(xscaler,yscaler);
% Plot the arrows
h0=quiver(X,Y,DX,DY,'MaxHeadSize',max(x)/(max(y)*20)); % plot the direction field
hold on;
cmap1 = {'r','g'};
i=1;
h(1) = plot(x,c1*cos(8*x)+c2*sin(8*x),cmap1{i},'LineWidth',3);
title('Original equation: $\frac{d^2y}{dt}$ + $\mu^2$y = 0, where $\mu$=$\sqrt\frac{k}{m}$=8', ...
'interpreter','latex','FontSize',16);
text(0.2,-1.5,'Solution: y = c_1cos(8t) + c_2sin(8t)', ...
'BackgroundColor','white','EdgeColor','black','FontSize',16);
ylabel('Spring Deflection (ft) (+ down)');
xlabel('Time (sec)');
grid on;
legend([h0,h],{'slope field',['c_1, c_2 = ^2/_3, -^1/_6']}, ...
'Location','NorthWest');
ylim([min(y) max(y)]);
xlim([min(x) max(x)]);
set(gcf,'Position',[403 88 778 578]);
```

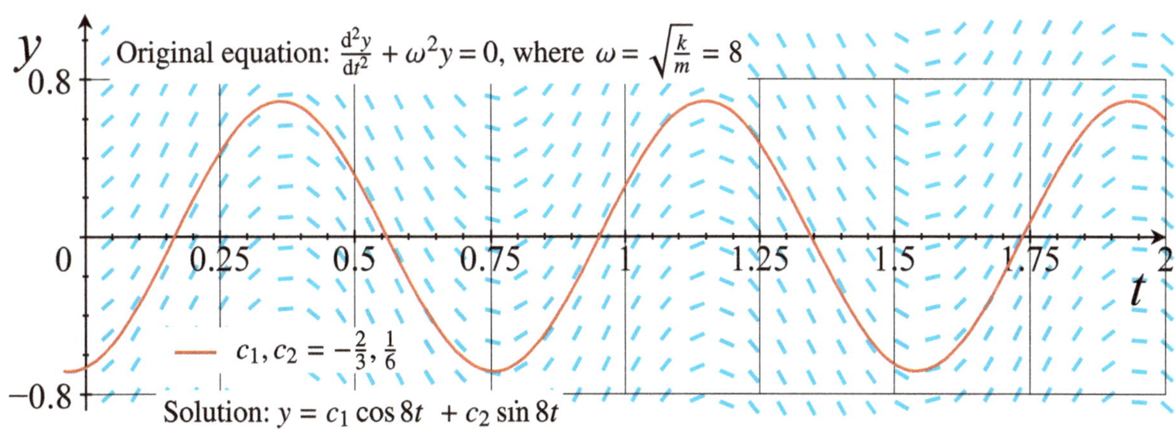

```
close all;
clear all;
c1 = 0;
c2 = -5;
Ffun = @(X,Y) -4*c2*X.*exp(-4*X) + c2*exp(-4*X);
dx = 0.05;
dy = 0.1;
x = 0: dx: 2;
y = -1: dy: 1;
y0 = 0; % du
[X,Y] = meshgrid(x,y);
DY = Ffun(X,Y);
DX = ones(size(DY)); % generate the plot values
% Scale the slope field arrows for the figure
xscaler = abs((0.9*dx*ones(size(DX)))./DX);
yscaler = abs((0.9*dy*ones(size(DY)))./DY);
DX = DX .* min(xscaler,yscaler);
DY = DY .* min(xscaler,yscaler);
% Plot the arrows
h0=quiver(X,Y,DX,DY,'MaxHeadSize',max(x)/(max(y)*20)); % plot the direction field
hold on;
cmap1 = {'r','g'};
i=1;
h(1) = plot(x,c1*exp(-4*x)+c2*x.*exp(-4*x),cmap1{i},'LineWidth',3);
title('Original equation: $\frac{d^2y}{dt}$ + 8$\frac{dy}{dt}$ + $\mu^2$y = 0, where
    $\mu$=$\sqrt\frac{k}{m}$=4','interpreter','latex','FontSize',16);
text(0.2,-1.5,'Solution: y = c_2te^{-4t}','BackgroundColor','white','EdgeColor', ...
'black','FontSize',16);
ylabel('Spring Deflection (ft) (+ down)');
xlabel('Time (sec)');
grid on;
legend([h0,h],{'slope field',['c_1, c_2 = ',num2str(c1),', ',num2str(c2)]}, ...
'Location','NorthWest');
ylim([min(y) max(y)]);
xlim([min(x) max(x)]);
set(gcf,'Position',[403 88 778 578]);
```

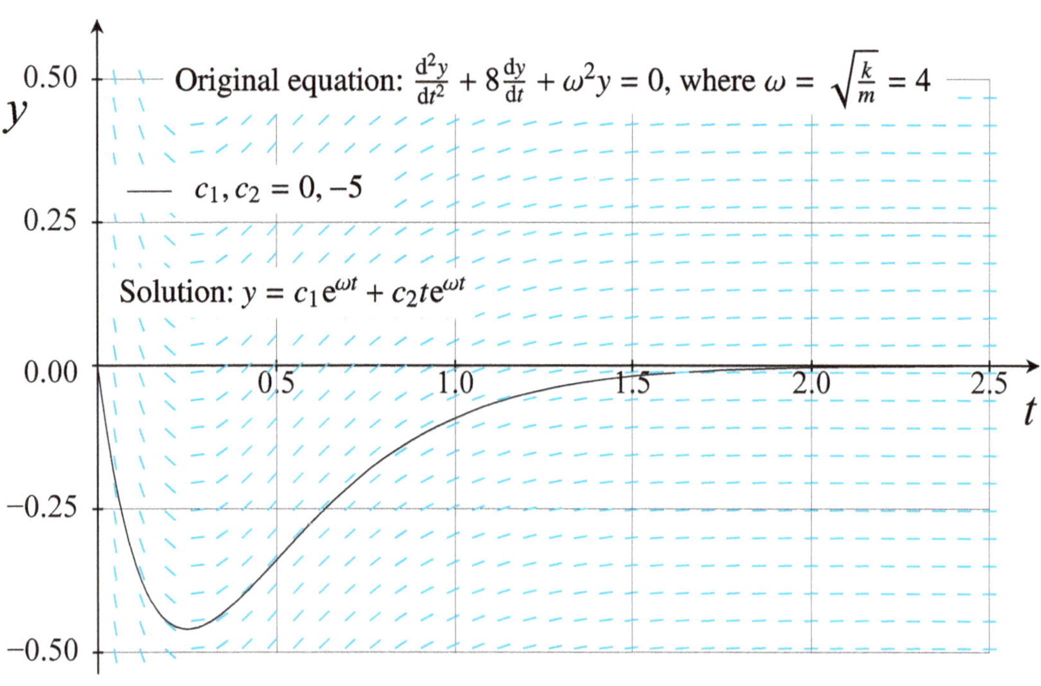

Chapter 16 MATLAB Scripts

```
% 3D, Lion vs. Gazelle
t=linspace(0,2*pi,500);
x=cos(t);
y=sin(3*t);
z=20*sin(x).*cos(y);
k=0.3; % Constant speed
[T,Y]=ode45('general3d',[0,2*pi],[0;0;0],[],k);
h=plot3(x,y,z,'b',Y(:,1),Y(:,2),Y(:,3),'r','LineWidth',2);
legend(h,'Prey','Predator');
title('Predator vs. Prey 3D');
xlabel('x');
ylabel('y');
zlabel('z');
grid on;
hold on;
[a,b] = meshgrid(min(y)*2:0.2:max(y)*2,min(x)*2:0.2:max(x)*2);
Z = 20*cos(b).*sin(a);
mesh(a,b,Z);
hidden off;
```

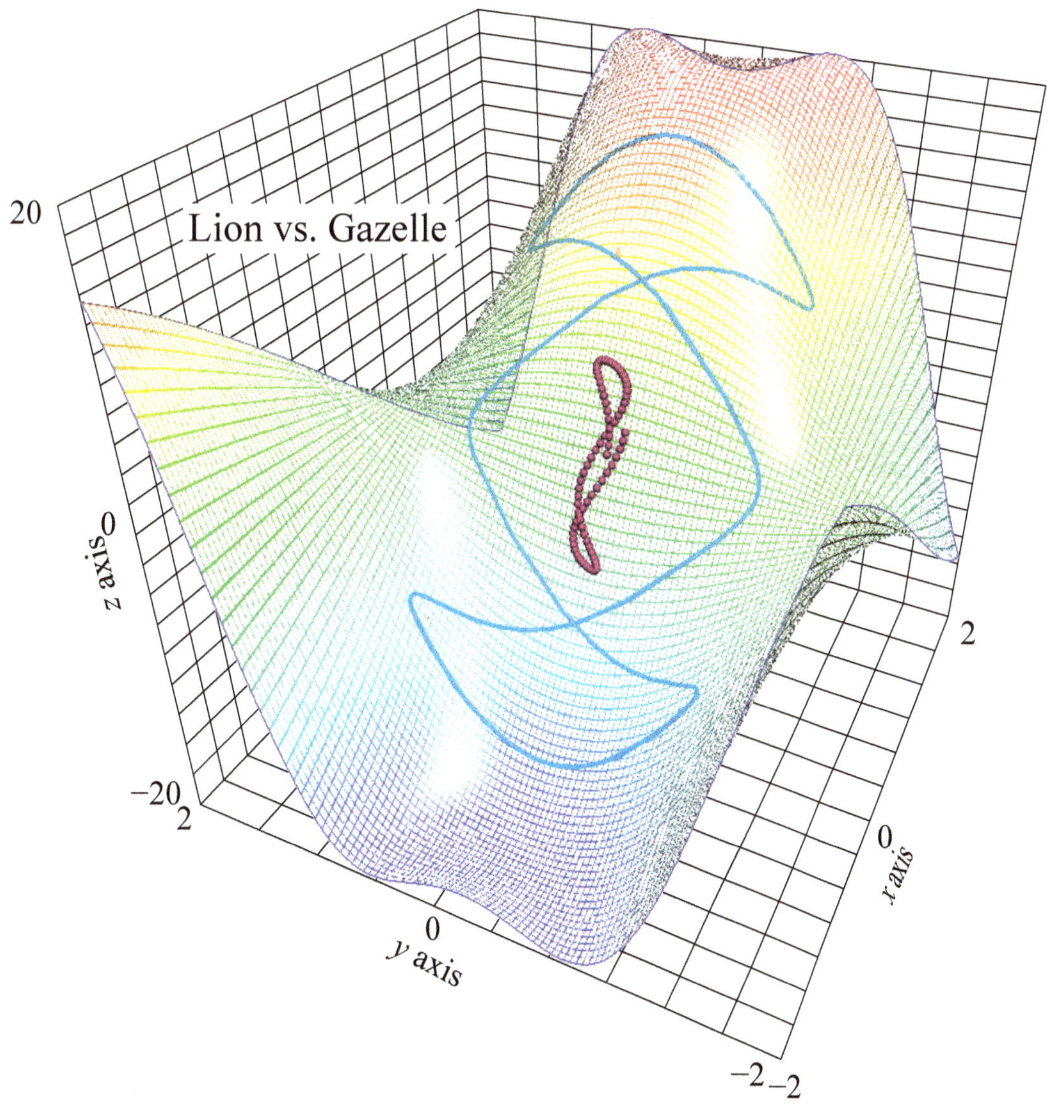

Appendix B
Converting a Logistic Differential Equation into Its Algebraic (Integrated) Equivalent

Chapter 6 introduces the idea of a *logistic differential equation*. Since logistic differential equations—equations of the form $\frac{dP}{dt} = kP\left(1 - \frac{P}{K}\right)$—are a subset of separable differential equations—equations of the form $\frac{dy}{dx} = f(x) \times g(y), g(y) \neq 0$—they can be separated and solved as separable equations.

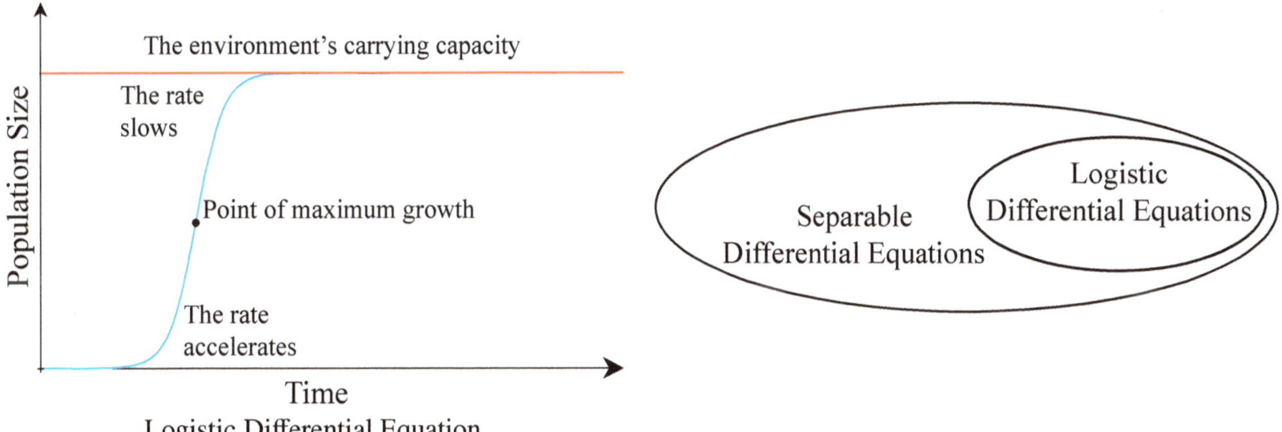

Logistic Differential Equation

However, because of the complications involved in making that transformation, it is considered best to convert a "generic logistic equation" into its *generic algebraic (integrated) form* and just use the algebraic form whenever the need arises to solve a logistic differential equation. Remember that Chapter 6 makes the analogy with the idea of solving a general quadratic equation, $ax^2 + bx + c = 0$, using the quadratic formula, $x = \frac{-b \pm \sqrt{b^2-4ac}}{2a}$.

Differential Equation Using the Logistic Model	Integrated Equation Using the Logistic Model
$\frac{dP}{dt} = kP\left(1 - \frac{P}{K}\right)$ this is separable	$P = \frac{K}{(Ae^{-kt}+1)}$, where $A = \frac{K-P_0}{P_0}$

The following text shows how to convert from the generic Differential Equation for a Logistic Model to its algebraic or integrated form. It is located here in Appendix B for those students who are curious as to how this conversion is done. It is not really necessary for application-oriented students to understand the following logic in order to solve logistic differential equations, but it is definitely desirable!! That is the difference between "mathmagic" and mathematics.

$$\frac{dP}{dt} = kP\left(1 - \frac{P}{K}\right) \quad \text{original logistic differential equation form}$$

$$dP = kP\left(1 - \frac{P}{K}\right) dt \quad \text{separation of differential } dP \text{ from differential } dt$$

$$dP = kP\left(\frac{K}{K} - \frac{P}{K}\right) dt \quad \text{preparing to combine fractions}$$

$$dP = kP\left(\frac{K - P}{K}\right) dt \quad \text{combine like fractions}$$

$$K\,dP = kP(K - P)\,dt \quad \text{multiplying by } K \text{ ("carrying capacity") on both sides}$$

$$\frac{K}{P(K - P)}\,dP = k\,dt \quad \text{grouping variables } P \text{ and differential } dP \text{ to the left}$$

************************** arithmetic trickery . . . separation of fractions **************************

$$\text{Let } \frac{K}{P(K-P)} = \frac{A}{P} + \frac{B}{K-P} = \frac{K-P}{K-P} \times \frac{A}{P} + \frac{B}{K-P} \times \frac{P}{P} = \frac{A(K-P)+BP}{P(K-P)}$$

>>>>>>>> So, by the transitive rule, $K = A(K-P) + BP$ <<<<<

Let	$P = 0$	Let	$P = K$	
Then	$K = A(K-0) + B \times 0$	Then	$K = A(K-K) + BK$	clever substitutions to simplify
	$K = AK$		$K = A \times 0 + BK$	
	$A = 1$		$K = 0 + BK$	
			$B = 1$	

****

$$\frac{K}{P(K-P)}\,\mathrm{d}P = k\,\mathrm{d}t \quad \text{recopying from above after arithmetic detour}$$

$$\frac{A(K-P)+BP}{P(K-P)}\,\mathrm{d}P = k\,\mathrm{d}t \quad \text{>>>>> substituting for } K \text{ from notes above <<<<<}$$

$$\frac{1(K-P)+1\times P}{P(K-P)}\,\mathrm{d}P = k\,\mathrm{d}t \quad \text{substituting into } A \text{ and } B \text{ from notes above}$$

$$\frac{(K-P)+P}{P(K-P)}\,\mathrm{d}P = k\,\mathrm{d}t$$

$$\frac{P+(K-P)}{P(K-P)}\,\mathrm{d}P = k\,\mathrm{d}t \quad \text{commutative property of addition}$$

$$\left[\frac{P}{P(K-P)} + \frac{(K-P)}{P(K-P)}\right]\mathrm{d}P = k\,\mathrm{d}t \quad \text{splitting the numerator}$$

$$\left(\frac{1}{(K-P)} + \frac{1}{P}\right)\mathrm{d}P = k\,\mathrm{d}t \quad \text{canceling } \frac{P}{P} \text{ and } \frac{K-P}{K-P}$$

$$\int\left(\frac{1}{(K-P)} + \frac{1}{P}\right)\mathrm{d}P = \int k\,\mathrm{d}t \quad \text{integrating both sides}$$

$$\int \frac{1}{K-P}\,\mathrm{d}P + \int \frac{1}{P}\,\mathrm{d}P = kt + c_1 \quad \text{integral of a sum on left, integral of } k\ \mathrm{d}t \text{ on right}$$

$$\int (K-P)^{-1}\,\mathrm{d}P + \int P^{-1}\,\mathrm{d}P = kt + c_1$$

$$-\int (K-P)^{-1}(-1)\,\mathrm{d}P + \int P^{-1}\,\mathrm{d}P = kt + c_1 \quad \text{preparing to apply } \int u^{-1}\,\mathrm{d}u = \ln u$$

$$-\ln|K-P| + c_2 + \ln|P| + c_3 = kt + c_1 \quad \int u^{-1}\,\mathrm{d}u = \ln u \text{ applied twice on left}$$

$$-\ln|K-P| + \ln|P| = kt + c_4 \quad \text{combining all three constants of integration}$$

$$-1 \times [-\ln|K-P| + \ln|P|] = -1 \times (kt + c_4) \quad \text{eliminating the leading } -1 \text{ from the left side}$$

$$\ln|K-P| - \ln|P| = -kt - c_4 \quad \text{distributive property both sides}$$

$$\ln\left|\frac{K-P}{P}\right| = -kt - c_4 \quad \log\left(\frac{a}{b}\right) = \log a - \log b$$

$$\mathrm{e}^{\ln\left|\frac{K-P}{P}\right|} = \mathrm{e}^{-kt-c_4} \quad \text{if } m = n, \text{ then } b^m = b^n$$

$$\left|\frac{K-P}{P}\right| = \mathrm{e}^{-kt} \times \mathrm{e}^{-c_4} \quad \mathrm{e}^{\ln u} = u \text{ on left, } \mathrm{e}^{a+b} = \mathrm{e}^a \times \mathrm{e}^b \text{ on right}$$

$$\left|\frac{K-P}{P}\right| = c_5 \times \mathrm{e}^{-kt} \quad \text{substituting } \mathrm{e}^{-c_4} = c_5$$

$$\frac{K-P}{P} = \pm c_5 \times \mathrm{e}^{-kt} \quad \text{removing the absolute value}$$

$$\frac{K-P}{P} = c_6 \times \mathrm{e}^{-kt} \quad \text{substituting } \pm c_5 = c_6$$

$\frac{K-P}{P} = Ae^{-kt}$ arbitrary substitution so that the final form is in the standard format

$\frac{K}{P} - 1 = A \times e^{-kt}$ splitting the numerator: $\frac{K}{P} - \frac{P}{P} = \frac{K}{P} - 1$

$$\frac{K}{P} = A \times e^{-kt} + 1$$

$$K = \left(A \times e^{-kt} + 1\right) \times P$$

$$\frac{K}{(A \times e^{-kt} + 1)} = P$$

$P = \frac{K}{(A \times e^{-kt} + 1)}$ where $A = \frac{K - P_0}{P_0}$

Whew! Thanks patrickjmt on YouTube!!

$$\frac{K-P}{P} = Ae^{-kt}$$

Let $P = P(t)$. At time 0, then, $P(t) = P(0) = P_0$, so

$$\frac{K - P_0}{P_0} = Ae^{-k \times 0}$$

$$\frac{K - P_0}{P_0} = A \times 1$$

$$A = \frac{K - P_0}{P_0}$$

Appendix C
Proof That the Integrating Factor for a First-Order Differential Expression of the form $y'(x) + P(x)y(x)$ Is $e^{\int P(x)\,dx}$

If a first-order linear differential equation is in the form $y'(x) + P(x)y(x) = Q(x)$ (where $P(x)$ and $Q(x)$ are functions of x), as demonstrated in Chapter 8, it is not possible to separate the variables in such a way that all the y functions and y derivatives are on one side of the equation and all the x functions and x derivatives are on the other.

$$y'(x) + P(x)y(x) \stackrel{?}{=} Q(x) \quad \text{no! } P(x) \text{ is on left with the } y'(x) \text{ and the y(x).}$$

$$y'(x) \stackrel{?}{=} Q(x) - P(x)y(x) \quad \text{no! } y'(x) \text{ is on left, but } y(x) \text{ is on right.}$$

$$P(x)y(x) \stackrel{?}{=} Q(x) - y'(x) \quad \text{no! } y(x) \text{ is on left, } y'(x) \text{ is on right.}$$

Suppose that a function exists—call it $I(x)$—that, when you multiply it by both sides of the original equation, $y' + P(x)y(x) = Q(x)$, the product on the left side of the equation will be the expanded version of the derivative of a product of functions: $I(x)g'(x) + g(x)I'(x)$. This product could be integrated from its equivalent form, $[I(x)g(x)]'$, as $\int[I(x)g(x)]' = I(x)g(x)$, allowing you to solve for g(x). What would that $I(x)$ look like? How can you go about getting that "magic $I(x)$ function"?

$$I(x)\left[\frac{dy}{dx} + P(x)y(x)\right] = \frac{d}{dx}[I(x)y(x)]$$ the product of the "assumed integrating factor" and the left side of our differential equation is the expanded form of the derivative of a product

$$I(x)\frac{dy}{dx} + I(x)P(x)y(x) = I(x)\frac{dy}{dx} + y(x)\frac{d}{dx}(I(x))$$ distributive property on the left, derivative of a product of functions on the right

$$\frac{I(x)P(x)y(x)}{y(x)} = \frac{y(x)\frac{d}{dx}(I(x))}{y(x)}$$ subtract $I(x)\frac{dy}{dx}$ from both sides and then divide both sides by $y(x)$

$$I(x)P(x) = \frac{d}{dx}(I(x))$$ solve for $I(x)$

$$P(x) = \frac{\frac{d}{dx}I(x)}{I(x)}$$ collect all the $I(x)$ terms on one side of the equation, prepare to solve for $I(x)$

$$\int P(x)\,dx = \int \frac{\frac{d}{dx}I(x)}{I(x)}\,dx$$ integrate both sides

$$\int P(x)\,dx = \ln|I(x)| + c_1 \qquad \int \frac{du}{u} = \int u^{-1}\,du = \ln|u|$$, integral calculus trick

$$\int P(x)\,dx - c_1 = \ln|I(x)|$$ subtract c_1 from both sides

$$\ln|I(x)| = \int P(x)\,dx - c_1$$ symmetric property

$$\ln|I(x)| = \int P(x)\,dx + c_2$$ let $c_2 = -c_1$

$$e^{\ln|I(x)|} = e^{\int P(x)\,dx + c_2}$$ if $a = b$ then $e^a = e^b$

$$e^{\ln|I(x)|} = e^{\int P(x)\,dx + c_2} \quad \text{recopied from the previous page}$$

$$I(x) = e^{\int P(x)\,dx} e^{c_2} \quad e^{a+b} = e^a e^b$$

$$I(x) = Ce^{\int P(x)\,dx} \quad \text{substitute } C = e^{c_2}$$

Because the constant of integration, C, cancels out after substitution (see below) the integrating factor is normally represented as $I(x) = e^{\int P(x)\,dx}$. Author's note: My apologies if this seems obtuse. It was only six steps in Chapter 15, Calculus, of *Differential Equations* by Larson et al. I expanded it to 13 steps, but it's still hard.

Solve $\frac{dy}{dx} + P(x)y(x) = Q(x)$ linear, differential equation, not separable.

$$\cancel{C}e^{\int P(x)\,dx}\left[\frac{dy}{dx} + P(x)y(x)\right] = \cancel{C}e^{\int P(x)\,dx}Q(x) \quad \text{multiply both sides by } e^{\int P(x)}$$

Notice that the constant of integration, C, can always be eliminated at this stage. Therefore, it is not normally used.

$$e^{\int P(x)\,dx}\frac{dy}{dx} + e^{\int P(x)\,dx}P(x)y(x) = e^{\int P(x)\,dx}Q(x) \quad \text{the left side here is the derivative of a product}$$

$$\left[ye^{\int P(x)\,dx}\right]' = e^{\int P(x)\,dx}Q(x) \quad \text{substitute } (uv)' \text{ for } uv' + u'v$$

$$\int \left[ye^{\int P(x)\,dx}\right]' dx = \int \left[e^{\int P(x)\,dx}Q(x)\right] dx \quad \int u' = u \text{ on the left}$$

$$ye^{\int P(x)\,dx} = \int \left(Q(x)e^{\int P(x)\,dx}\right) dx + C$$

$$y = \frac{\int Q(x)e^{\int P(x)\,dx}\,dx + C}{e^{\int P(x)\,dx}} \quad \text{integrated form of the original equation}$$

Appendix D
Justifying the Algebraic Technique for Solving Second-Order, Linear Homogeneous Differential Equations with Imaginary Characteristic-Equation Solutions ($B^2 - 4AC < 0$)

Chapter 14 introduced the differential-equation form $Ay'' + By' + Cy = 0$ known as a second-order linear homogeneous differential equation. You may recall that, just like solutions to algebraic quadratic equations, there were three differential algebraic solutions to such differential equations.

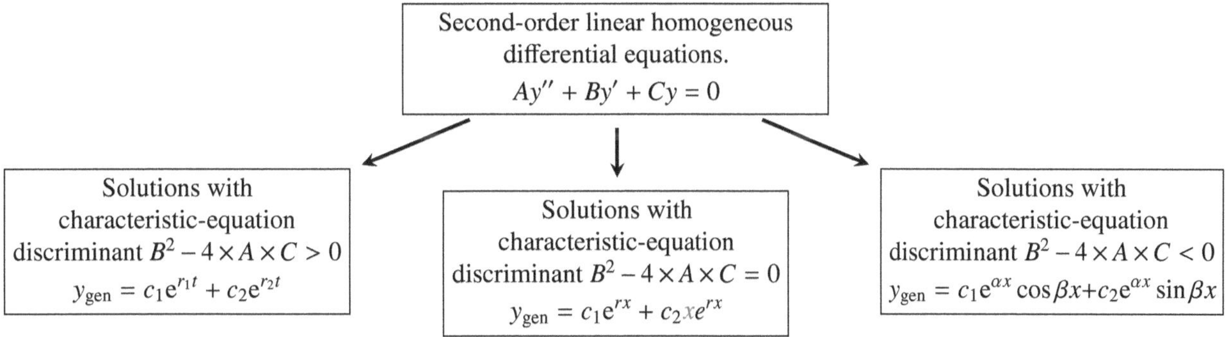

Significant page space was spent in Chapter 14 motivating the algebraic formula $y_{gen} = c_1 e^{r_1 t} + c_2 e^{r_2 t}$, the solution for a second-order linear homogeneous differential equation whose characteristic-equation discriminant $B^2 - 4 \times A \times C > 0$. It was not a rigorous proof, but was probably sufficient for the needs and interests of most scientists, engineers, and math educators. Everyone in our society specializes because it is not possible for one person to know everything. Discussion of where the solution to a second-order homogeneous differential equation with characteristic discriminant less than zero has been relegated to Appendix D to simplify Chapter 14. It also cannot be considered a formal proof, but it should be sufficient for most people. The chart above shows an example of "analysis," breaking a problem or idea into different component parts. The chart below is an example of "synthesis" ... different ideas are combined to create a new idea.

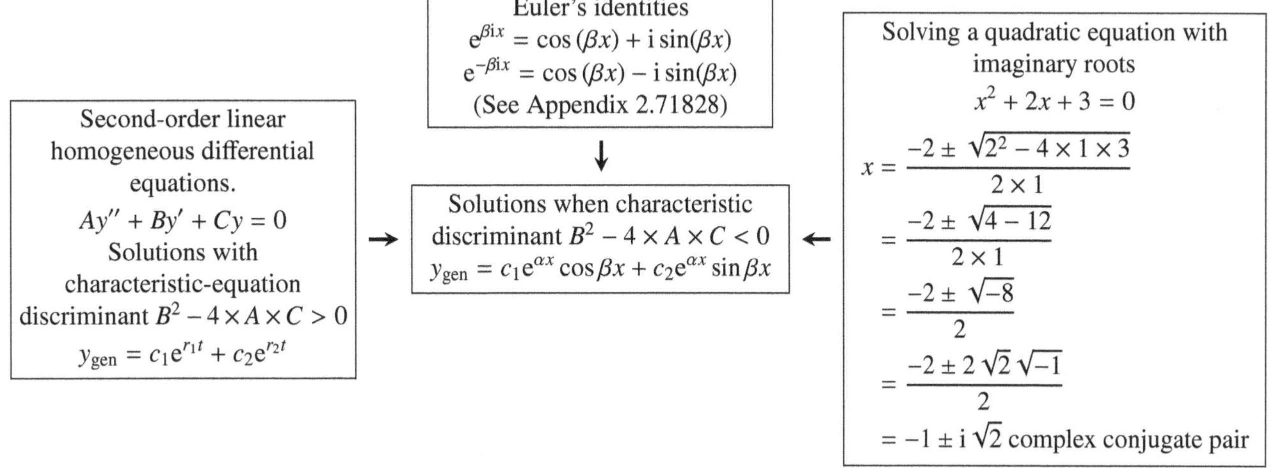

Second-order linear homogeneous differential equations.
$Ay'' + By' + Cy = 0$
Solutions with characteristic-equation discriminant $B^2 - 4 \times A \times C > 0$
$y_{gen} = c_1 e^{r_1 t} + c_2 e^{r_2 t}$
Euler's identities
$e^{\beta i x} = \cos(\beta x) + i \sin(\beta x)$
$e^{-\beta i x} = \cos(\beta x) - i \sin(\beta x)$
(See Appendix 2.71828)
Solutions when characteristic discriminant $B^2 - 4 \times A \times C < 0$
$y_{gen} = c_1 e^{\alpha x} \cos\beta x + c_2 e^{\alpha x} \sin\beta x$
Solving a quadratic equation with imaginary roots
$x^2 + 2x + 3 = 0$
$x = \frac{-2 \pm \sqrt{2^2 - 4 \times 1 \times 3}}{2 \times 1}$
$= \frac{-2 \pm \sqrt{4 - 12}}{2 \times 1}$
$= \frac{-2 \pm \sqrt{-8}}{2}$
$= \frac{-2 \pm 2\sqrt{2}\sqrt{-1}}{2}$
$= -1 \pm i\sqrt{2}$ complex conjugate pair

<table>
<tr>
<td>Second-order linear homogeneous differential equations.

$$Ay'' + By' + Cy = 0$$
Solutions when characteristic discriminant $B^2 - 4 \times A \times C > 0$

$$y_{\text{gen}} = c_1 e^{r_1 t} + c_2 e^{r_2 t}$$</td>
<td>Second-order linear homogeneous differential equations.

$$Ay'' + By' + Cy = 0$$
Solutions when characteristic equation discriminant $B^2 - 4 \times A \times C < 0$

$$y_{\text{gen}} = c_1 e^{\alpha x} \cos \beta x + c_2 e^{\alpha x} \sin \beta x$$</td>
</tr>
<tr>
<td>$y_1 = e^{r_1 x}$
$y_2 = e^{r_2 x}$</td>
<td>$y_1 = e^{(\alpha+\beta i)x}$ following pattern at left from Chapter 14
$y_2 = e^{(\alpha-\beta i)x}$ complex conjugate pair
$y_1 = e^{\alpha x+\beta i x}$ distributive property
$y_2 = e^{\alpha x-\beta i x}$
$y_1 = e^{\alpha x} \times e^{\beta i x}$ $b^{x+y} = b^x \times b^y$ from previous step
$y_2 = e^{\alpha x} \times e^{-\beta i x}$ $b^{x+y} = b^x \times b^y$ from previous step
$y_1 = e^{\alpha x}(\cos \beta x + i \sin \beta x)$ Euler's identity, see Appendix 2.718
$y_2 = e^{\alpha x}(\cos \beta x - i \sin \beta x)$

We need to get rid of the i terms somehow because, in the original differential equation we are solving, $Ay'' + By' + Cy = 0$, A, B, and C are all real coefficients. Watch how it is done!</td>
</tr>
<tr>
<td>In Chapter 14, we established that if functions $f(x)$ and $g(x)$ are both solutions to a homogeneous differential equation, then $[f(x) \pm g(x)]$ is also a solution. In Chapter 14, we also established

$y_1 = e^{r_1 x}$
$y_2 = e^{r_2 x}$.

Hence,

$$y_{\text{gen}} = c_1 e^{r_1 t} \pm c_2 e^{r_2 t}$$</td>
<td>In Chapter 14, we established that if functions $f(x)$ and $g(x)$ are both solutions to a homogeneous differential equation, then $[f(x) \pm g(x)]$ is also a solution. Above, we established that

$y_1 = e^{\alpha x}(\cos \beta x + i \sin \beta x)$
$y_2 = e^{\alpha x}(\cos \beta x - i \sin \beta x)$.

Hence, $y_{\text{gen}} = e^{\alpha x}(\cos \beta x + i \sin \beta x) \pm e^{\alpha x}(\cos \beta x - i \sin \beta x)$
Let $m(x) = e^{\alpha x}(\cos \beta x + i \sin \beta x) + e^{\alpha x}(\cos \beta x - i \sin \beta x)$
$n(x) = e^{\alpha x}(\cos \beta x + i \sin \beta x) - e^{\alpha x}(\cos \beta x - i \sin \beta x)$

$m(x) = (e^{\alpha x} \cos \beta x + e^{\alpha x} i \sin \beta x) + (e^{\alpha x} \cos \beta x - e^{\alpha x} i \sin \beta x)$
$= (e^{\alpha x} \cos \beta x) + (e^{\alpha x} \cos \beta x)$ big cancel!!!
$= 2e^{\alpha x} \cos \beta x$
$= c_1 e^{\alpha x} \cos \beta x$ let $c_1 = 2$

$n(x) = (e^{\alpha x} \cos \beta x + e^{\alpha x} i \sin \beta x) - (e^{\alpha x} \cos \beta x - e^{\alpha x} i \sin \beta x)$
$= e^{\alpha x} i \sin \beta x - [-(e^{\alpha x} i \sin \beta x)]$ big cancel!!!
$= 2e^{\alpha x} i \sin \beta x$
$= 2ie^{\alpha x} \sin \beta x$
$= c_2 e^{\alpha x} \sin \beta x$ let $c_2 = 2i$

$y_{\text{gen}} = m(x) + n(x)$
$= c_1 e^{\alpha x} \cos \beta x + c_2 e^{\alpha x} \sin \beta x$</td>
</tr>
</table>

Appendix 2.718281828
Euler's Equation, an Introduction

In calculus, when studying infinite series, you should have learned

$$e = \frac{1}{0!} + \frac{1}{1!} + \frac{1}{2!} + \frac{1}{3!} + \frac{1}{4!} + \frac{1}{5!} + \frac{1}{6!} + \frac{1}{7!} + \frac{1}{8!} + \frac{1}{9!} + \cdots .$$

It is not appropriate to this text to discuss where this magic expression comes from.

A very famous mathematician, Brook Taylor (circa 1712), is given credit for finding a way to approximate any function to any degree of accuracy by adding up a series of smaller functions. The technique to do this is appropriately called a Taylor series. To fully understand how Mr. Taylor did his magic, you would need to take a calculus class. That is clearly not possible in the space here. By Mr. Taylor's work, the following more general formula can be proved.

$$e^x = \frac{x^0}{0!} + \frac{x^1}{1!} + \frac{x^2}{2!} + \frac{x^3}{3!} + \frac{x^4}{4!} + \frac{x^5}{5!} + \frac{x^6}{6!} + \frac{x^7}{7!} + \frac{x^8}{8!} + \frac{x^9}{9!} + \cdots .$$

Also by Mr. Taylor's work

$$\cos(x) = \frac{x^0}{0!} - \frac{x^2}{2!} + \frac{x^4}{4!} - \frac{x^6}{6!} + \frac{x^8}{8!} - \cdots$$

and

$$\sin(x) = \frac{x^1}{1!} - \frac{x^3}{3!} + \frac{x^5}{5!} - \frac{x^7}{7!} + \frac{x^9}{9!} - \cdots .$$

Putting these three equations together yields a *remarkable result* called "Euler's Equation":

$$e^{\pi i} + 1 = 0.$$

This will be shown as follows.

Step 1: Rearranging terms from e^x shown above (Can you anticipate the $\cos(x)$ and $\sin(x)$?),

$$e^x = \frac{x^0}{0!} + \frac{x^2}{2!} + \frac{x^4}{4!} + \frac{x^6}{6!} + \frac{x^8}{8!} + \cdots + \frac{x^1}{1!} + \frac{x^3}{3!} + \frac{x^5}{5!} + \frac{x^7}{7!} + \frac{x^9}{9!} + \cdots .$$

Step 2: Arbitrarily substitute $x = \pi i$. (Don't ask questions at this stage!)

$$e^{\pi i} = \frac{(\pi i)^0}{0!} + \frac{(\pi i)^2}{2!} + \frac{(\pi i)^4}{4!} + \frac{(\pi i)^6}{6!} + \frac{(\pi i)^8}{8!} + \cdots + \frac{(\pi i)^1}{1!} + \frac{(\pi i)^3}{3!} + \frac{(\pi i)^5}{5!} + \frac{(\pi i)^7}{7!} + \frac{(\pi i)^9}{9!} + \cdots .$$

Step 3: Now, from algebra, we know that $(ab)^m = a^m b^m$. Hence, the equation above can be written as

$$e^{\pi i} = \frac{\pi^0 i^0}{0!} + \frac{\pi^2 i^2}{2!} + \frac{\pi^4 i^4}{4!} + \frac{\pi^6 i^6}{6!} + \frac{\pi^8 i^8}{8!} + \cdots + \frac{\pi^1 i^1}{1!} + \frac{\pi^3 i^3}{3!} + \frac{\pi^5 i^5}{5!} + \frac{\pi^7 i^7}{7!} + \frac{\pi^9 i^9}{9!} + \cdots .$$

Step 4: Factor an "i" from the second set of terms.

$$e^{\pi i} = \frac{\pi^0 i^0}{0!} + \frac{\pi^2 i^2}{2!} + \frac{\pi^4 i^4}{4!} + \frac{\pi^6 i^6}{6!} + \frac{\pi^8 i^8}{8!} + \cdots + i\left(\frac{\pi^1 i^0}{1!} + \frac{\pi^3 i^2}{3!} + \frac{\pi^5 i^4}{5!} + \frac{\pi^7 i^6}{7!} + \frac{\pi^9 i^8}{9!} + \cdots\right).$$

Step 5: Apply algebra rules to all terms: $i^0 = 1, i^1 = i, i^2 = -1, i^3 = -i, i^4 = 1. \dots$.

$$e^{\pi i} = \underbrace{\frac{\pi^0(1)}{0!} - \frac{\pi^2(1)}{2!} + \frac{\pi^4(1)}{4!} - \frac{\pi^6(1)}{6!} + \frac{\pi^8(1)}{8!} - \cdots}_{\cos(\pi)} + i\underbrace{\left(\frac{\pi(1)}{1!} - \frac{\pi^3(1)}{3!} + \frac{\pi^5(1)}{5!} - \frac{\pi^7(1)}{7!} + \frac{\pi^9(1)}{9!} - \cdots\right)}_{\sin(\pi)}$$

Step 6: Substitute $\cos(\pi)$ for the first set of terms and $\sin(\pi)$ for the second set of terms.

$$e^{\pi i} = \cos(\pi) + i\sin(\pi)$$

Step 7: From trigonometry, we know that $\cos(\pi) = -1$ and $\sin(\pi) = 0$.

$$e^{\pi i} = -1 + i \times 0$$

Step 8: Finally, move some terms around to match Euler's equation.

$$\mathbf{e^{\pi i} + 1 = 0}$$

There you have it, Euler's equation: e, π, i, 1, and 0 all in the same equation!!!

Nerd Heaven!!

In the text above, we started with

$$e^x = \frac{x^0}{0!} + \frac{x^1}{1!} + \frac{x^2}{2!} + \frac{x^3}{3!} + \frac{x^4}{4!} + \frac{x^5}{5!} + \frac{x^6}{6!} + \frac{x^7}{7!} + \frac{x^8}{8!} + \frac{x^9}{9!} + \cdots.$$

By substituting $x = \pi i$, we got

$$e^{\pi i} = \frac{(\pi i)^0}{0!} + \frac{(\pi i)^2}{2!} + \frac{(\pi i)^4}{4!} + \frac{(\pi i)^6}{6!} + \frac{(\pi i)^8}{8!} + \cdots + \frac{(\pi i)^1}{1!} + \frac{(\pi i)^3}{3!} + \frac{(\pi i)^5}{5!} + \frac{(\pi i)^7}{7!} + \frac{(\pi i)^9}{9!} + \cdots$$

and eventually obtained

$$e^{\pi i} = \cos(\pi) + i\sin(\pi)$$

In Step 2, we substituted $e^x = e^{\pi i}$. Had we instead chosen to substitute $e^x = e^{\beta ix}$, we would have gotten $e^{\beta ix} = \cos(\beta x) + i\sin(\beta x)$ at Step 6. Similarly, substituting $e^x = e^{-\beta ix}$, we get $e^{-\beta ix} = \cos(\beta x) - i\sin(\beta x)$.

Appendix F

Solving for the Integrated Solution of a Two-by-Two System of Differential Equations

<table>
<tr>
<td>

Solve a specific 1×1 system of differential equations.

$$\frac{\mathrm{d}x}{\mathrm{d}t} = 5x$$

$$\text{Then, } \mathrm{d}x = 5x\,\mathrm{d}t$$

$$\frac{\mathrm{d}x}{x} = \frac{5x}{x}\,\mathrm{d}t$$

$$x^{-1}\,\mathrm{d}x = 5\,\mathrm{d}t$$

$$\int x^{-1}\,\mathrm{d}x = 5\int \mathrm{d}t$$

$$\ln x + c_1 = 5t + c_2$$

$$\ln x = 5t + c_2 - c_1$$

$$\ln x = 5t + c_3$$

$$\mathrm{e}^{\ln x} = \mathrm{e}^{5t+c_3}$$

$$x = \mathrm{e}^{5t} \times \mathrm{e}^{c_3}$$

$$x = c\mathrm{e}^{5t}$$

</td>
<td>

Assume $\mathbf{A}$ is a general 1×1 matrix.

$$\frac{\mathrm{d}x}{\mathrm{d}t} = \mathbf{A}x$$

Then, $\mathbf{A} = a$, where a is a scalar.

$$\frac{\mathrm{d}x}{\mathrm{d}t} = ax$$

$$\mathrm{d}x = ax\,\mathrm{d}t$$

$$\frac{\mathrm{d}x}{x} = \frac{ax}{x}\,\mathrm{d}t$$

$$x^{-1}\,\mathrm{d}x = a\,\mathrm{d}t$$

$$\int x^{-1}\,\mathrm{d}x = a\int \mathrm{d}t$$

$$\ln x + c_1 = at + c_2$$

$$\ln x = at + c_2 - c_1$$

$$\ln x = at + c_3$$

$$\mathrm{e}^{\ln x} = \mathrm{e}^{at+c_3}$$

$$x = \mathrm{e}^{at} \times \mathrm{e}^{c_3}$$

$$x = c\mathrm{e}^{at}$$

</td>
</tr>
<tr>
<td>

Solve the following 2×2 system of differential equations

$$\frac{\mathrm{d}x_1}{\mathrm{d}t} = 5x_1 + 3x_2$$

$$\frac{\mathrm{d}x_2}{\mathrm{d}t} = -6x_1 - 4x_2$$

$$\frac{\mathrm{d}}{\mathrm{d}t}\begin{pmatrix} x_1(t) \\ x_2(t) \end{pmatrix} = \begin{pmatrix} 5 & 3 \\ -6 & -4 \end{pmatrix}\begin{pmatrix} x_1(t) \\ x_2(t) \end{pmatrix}$$

$$\vec{x}\,'(t) = \begin{pmatrix} 5 & 3 \\ -6 & -4 \end{pmatrix} \times \vec{x}(t) \qquad (1)$$

</td>
<td>

Given a 2×2 coefficient matrix of a system of differential equations,

$$\vec{x}\,'(t) = \begin{pmatrix} \frac{\mathrm{d}x_1}{\mathrm{d}t} \\ \frac{\mathrm{d}x_2}{\mathrm{d}t} \end{pmatrix} = \begin{pmatrix} ax_1 + bx_2 \\ cx_1 + dx_2 \end{pmatrix} = \begin{pmatrix} a & b \\ c & d \end{pmatrix}\begin{pmatrix} x_1(t) \\ x_2(t) \end{pmatrix}, \qquad (1)$$

we hope to find a vector function, $\vec{x}(t)$, satisfying (1). Inspired by the 1×1 case above, we guess that the solution to the 2×2 system also involves the exponential e^{rt} for each row of the system. Because we are solving a 2×2 system of equations, there will be two solution families so our solution will also involve a vector, $\vec{K}$.

$$\text{Let } \vec{x}(t) = \vec{K}\mathrm{e}^{rt}, \quad \text{where } \vec{K} = \begin{pmatrix} k_1 \\ k_2 \end{pmatrix}, \qquad (2)$$

k_1 and k_2 are constants

Check if a vector function of the form $\vec{K}\mathrm{e}^{rt}$ is a solution to (1) above:

$$\vec{x}\,'(t) = \begin{pmatrix} a & b \\ c & d \end{pmatrix}\vec{x}(t). \qquad (1)$$

Let's take the derivative of $\vec{x}(t) = \vec{K}\mathrm{e}^{rt}$,

$$\vec{x}\,'(t) = \vec{K}r\mathrm{e}^{rt}, \qquad (3)$$

and substitute into (1).

</td>
</tr>
</table>

table continues

	$\vec{x}'(t) = \begin{pmatrix} \frac{dx_1}{dt} \\ \frac{dx_2}{dt} \end{pmatrix} = \begin{pmatrix} a & b \\ c & d \end{pmatrix} \vec{x},$ (1 again) $\vec{x}(t) = \vec{K}e^{rt} \quad \vec{K} = \begin{pmatrix} k_1 \\ k_2 \end{pmatrix}$ (2 again) $\vec{x}'(t) = \vec{K}re^{rt},$ (3 again) $\vec{K}re^{rt} = \begin{pmatrix} a & b \\ c & d \end{pmatrix} \times \vec{x}(t)$ combining (3) and (1) $\vec{K}re^{rt} = \begin{pmatrix} a & b \\ c & d \end{pmatrix} \times \vec{K}e^{rt}$ substituting (2) $\frac{\vec{K}re^{rt}}{e^{rt}} = \frac{\begin{pmatrix} a & b \\ c & d \end{pmatrix} \times \vec{K}e^{rt}}{e^{rt}}$ ok because $e^{rt} \neq 0$ $\vec{K}r = \begin{pmatrix} a & b \\ c & d \end{pmatrix} \times \vec{K}$ now e^{rt} is out of our way $\begin{pmatrix} a & b \\ c & d \end{pmatrix} \times \vec{K} = \vec{K}r$ symmetric property $\begin{pmatrix} a & b \\ c & d \end{pmatrix} \times \vec{K} = r\vec{K}$ see footnote * Well bust my britches!!! The equation $\begin{pmatrix} a & b \\ c & d \end{pmatrix} \times \vec{K} = r\vec{K}$ shown here looks just like the equation form **Mat**$\times\vec{\varepsilon} = \lambda\vec{\varepsilon}$ shown in Chapter 19. **Wow! I didn't see that coming!!** The r and $\vec{K}$ we have been using are, respectively, the eigenvalue and eigenvector of matrix **Mat**. This is shown for given constants in the left column of this appendix. $\begin{pmatrix} a & b \\ c & d \end{pmatrix} \times \vec{\varepsilon} = \lambda\vec{\varepsilon}$ Chapter 19 $\begin{pmatrix} a & b \\ c & d \end{pmatrix} \times \vec{\varepsilon} - \lambda\vec{\varepsilon} = 0$ introduces a dimension error $\begin{pmatrix} a & b \\ c & d \end{pmatrix} \times \vec{\varepsilon} - \lambda\vec{\varepsilon} = \begin{pmatrix} 0 \\ 0 \end{pmatrix}$ dimension error corrected $\left[\begin{pmatrix} a & b \\ c & d \end{pmatrix} - \lambda\right]\vec{\varepsilon} = \begin{pmatrix} 0 \\ 0 \end{pmatrix}$ This is not *factoring out a common term* as that is done with scalars. The justification for this is found in Chapter 18. This step introduces yet another dimension error which we fix. $\left[\begin{pmatrix} a & b \\ c & d \end{pmatrix} - \lambda\begin{pmatrix} 1 & 0 \\ 0 & 1 \end{pmatrix}\right]\vec{\varepsilon} = \begin{pmatrix} 0 \\ 0 \end{pmatrix}.$ * This looks like the commutative property of real numbers, but it is actually a theorem proven in Chapter 18: $\vec{K}r = r\vec{K}$.

table continues

From Chapter 19, $(\mathbf{Mat} - \lambda\mathbf{I}) \times \vec{\varepsilon} = \begin{pmatrix}0\\0\end{pmatrix}$.

$$\left[\begin{pmatrix}5 & 3\\-6 & -4\end{pmatrix} - \begin{pmatrix}\lambda & 0\\0 & \lambda\end{pmatrix}\right] \times \vec{\varepsilon} = \begin{pmatrix}0\\0\end{pmatrix}$$

$$\begin{pmatrix}5-\lambda & 3\\-6 & -4-\lambda\end{pmatrix} \times \vec{\varepsilon} = \begin{pmatrix}0\\0\end{pmatrix}$$

Also, we know from Chapter 19 that the above equation is a vector form of $a \times x = 0$, so $\det(\mathbf{Mat}) = 0$ implies infinitely many x vector solutions for each λ:

To find an eigenvalue set $\det\begin{pmatrix}5-\lambda & 3\\-6 & -4-\lambda\end{pmatrix} = 0$.

$$[(5-\lambda)\times(-4-\lambda)] - (-18) = 0$$
$$-20 - \lambda + \lambda^2 + 18 = 0$$
$$\lambda^2 - \lambda - 2 = 0$$
$$\lambda = 2, -1$$

For each λ, solve for the corresponding $\vec{\varepsilon}$.

$\lambda_1 = 2$	$\lambda_2 = -1$
$(\mathbf{Mat} - \lambda\mathbf{I})\vec{\varepsilon} = 0$	$(\mathbf{Mat} - \lambda\mathbf{I})\vec{\varepsilon} = 0$
$\left[\begin{pmatrix}5 & 3\\-6 & -4\end{pmatrix} - \begin{pmatrix}2 & 0\\0 & 2\end{pmatrix}\right]\vec{\varepsilon} = \begin{pmatrix}0\\0\end{pmatrix}$	$\left[\begin{pmatrix}5 & 3\\-6 & -4\end{pmatrix} - \begin{pmatrix}-1 & 0\\0 & -1\end{pmatrix}\right]\vec{\varepsilon} = \begin{pmatrix}0\\0\end{pmatrix}$
$\begin{pmatrix}3 & 3\\-6 & -6\end{pmatrix}\vec{\varepsilon} = \begin{pmatrix}0\\0\end{pmatrix}$	$\begin{pmatrix}6 & 3\\-6 & -3\end{pmatrix}\vec{\varepsilon} = \begin{pmatrix}0\\0\end{pmatrix}$
$\begin{pmatrix}3 & 3\\-6 & -6\end{pmatrix}\begin{pmatrix}e_1\\e_2\end{pmatrix} = \begin{pmatrix}0\\0\end{pmatrix}$	$\begin{pmatrix}6 & 3\\-6 & -3\end{pmatrix}\begin{pmatrix}e_1\\e_2\end{pmatrix} = \begin{pmatrix}0\\0\end{pmatrix}$
$3e_1 + 3e_2 = 0$	$6e_1 + 3e_2 = 0$
$e_1 + e_2 = 0$	$2e_1 + 1e_2 = 0$
$e_1 = -e_2$	$2e_1 = -e_2$
$\begin{pmatrix}e_1\\e_2\end{pmatrix} = \begin{pmatrix}1\\-1\end{pmatrix} = \vec{\varepsilon}_1$	$\begin{pmatrix}e_1\\e_2\end{pmatrix} = \begin{pmatrix}1\\-2\end{pmatrix} = \vec{\varepsilon}_2$

Substitute $(\lambda_1, \lambda_2, \vec{\varepsilon}_1, \vec{\varepsilon}_2)$ into the appropriate equation for real eigenvalues (see Chapter 20).

$$\vec{x} = c_1\vec{\varepsilon}_1 e^{\lambda_1 t} + c_2\vec{\varepsilon}_2 e^{\lambda_2 t}$$

$$\vec{x} = c_1\begin{pmatrix}1\\-1\end{pmatrix}e^{2t} + c_2\begin{pmatrix}1\\-2\end{pmatrix}e^{-1t}$$

Here $\lambda\begin{pmatrix}1 & 0\\0 & 1\end{pmatrix}$ is a scalar times the identity matrix, so $\begin{pmatrix}a & b\\c & d\end{pmatrix} - \lambda\begin{pmatrix}1 & 0\\0 & 1\end{pmatrix}$ would then be the subtraction of two matrices, which is defined. We know from Chapter 19 that $\left(\begin{pmatrix}a & b\\c & d\end{pmatrix} - \lambda\begin{pmatrix}1 & 0\\0 & 1\end{pmatrix}\right)\vec{\varepsilon} = \begin{pmatrix}0\\0\end{pmatrix}$ is a vector form of the algebraic equation, $a \times x = 0$. $a \times x = 0$ with $a = 0$ results in infinite (or indefinite) solutions, which implies that $\det\left(\begin{pmatrix}a & b\\c & d\end{pmatrix} - \lambda\begin{pmatrix}1 & 0\\0 & 1\end{pmatrix}\right) = 0$ will result in infinitely many x-vector solutions.

By solving for λ and then $\vec{\varepsilon}$ for matrix **Mat** (see Appendix G at the top of page 300), we get

$$\lambda_1 = \frac{\beta + \sqrt{\delta}}{2} \quad \text{where } \beta = \operatorname{Tr}\mathbf{Mat} = a + d,\ \delta = (a+d)^2 - 4(ad - cb) = \beta^2 - 4\gamma$$

$$\vec{\varepsilon}_1 = \begin{pmatrix}b\\\lambda_1 - a\end{pmatrix} \quad \text{and}$$

$$\lambda_2 = \frac{\beta - \sqrt{\delta}}{2} \quad \text{where } \beta = \operatorname{Tr}\mathbf{Mat} = a + d,\ \delta = (a+d)^2 - 4(ad - cb) = \beta^2 - 4\gamma$$

$$\vec{\varepsilon}_2 = \begin{pmatrix}b\\\lambda_2 + a\end{pmatrix}$$

We now have two vector functions

$$\vec{x}_1 = \vec{\varepsilon}_1 e^{\lambda_1 t}, \quad \vec{x}_2 = \vec{\varepsilon}_2 e^{\lambda_2 t}.$$

By the *Linearity Principal* (Chapter 14), we can conclude that the general system solution is

$$\vec{x} = c_1\vec{\varepsilon}_1 e^{\lambda_1 t} + c_2\vec{\varepsilon}_2 e^{\lambda_2 t} \quad \text{Chapters 19 and 20.}$$

If you have IVP information, you can solve for c_1 and c_2. Compare with

$$y_{gen} = c_1 e^{\lambda_1 t} + c_2 e^{\lambda_2 t} \quad \text{Chapter 14.}$$

Adapted from Tutorial.math.lamar.edu/Classes/DE/solutionstoSystems.aspx. Errors, if any, are solely the responsibility of this author.

Appendix G
Analyzing a Generic Two-by-Two System of Ordinary Differential Equations

Solving a Specific System of Differential Equations	Solving a Generic System of Differential Equations (Assume only real distinct eigenvalues)
$\frac{d}{dt}x_1 = -5x_1 + 1x_2 \quad (1)$ $\frac{d}{dt}x_2 = 4x_1 + (-2x_2) \quad (2)$ $\vec{x}' = \begin{pmatrix} x_1' \\ x_2' \end{pmatrix} = \begin{pmatrix} -5 & 1 \\ 4 & -2 \end{pmatrix}\begin{pmatrix} x_1 \\ x_2 \end{pmatrix}$ It can be established by university professors way smarter than the author of this book that for any square $(n\times n)$ matrix, there exists and can be found a vector, $\vec{\varepsilon}$, and a scalar, λ, such that **Mat** $\times \vec{\varepsilon} = \lambda \times \vec{\varepsilon}$. $\begin{pmatrix} -5 & 1 \\ 4 & -2 \end{pmatrix} \times \vec{\varepsilon} = \lambda \times \vec{\varepsilon}$ $\begin{pmatrix} -5 & 1 \\ 4 & -2 \end{pmatrix} \times \vec{\varepsilon} - (\lambda \times \vec{\varepsilon}) = 0$ dimension error! $\begin{pmatrix} -5 & 1 \\ 4 & -2 \end{pmatrix} \times \vec{\varepsilon} - (\lambda \times \vec{\varepsilon}) = \begin{pmatrix} 0 \\ 0 \end{pmatrix}$ error corrected $\left[\begin{pmatrix} -5 & 1 \\ 4 & -2 \end{pmatrix} - \lambda\right] \times \vec{\varepsilon} = \begin{pmatrix} 0 \\ 0 \end{pmatrix}$ another dimension error! $\left[\begin{pmatrix} -5 & 1 \\ 4 & -2 \end{pmatrix} - \lambda\mathbf{I}\right] \times \vec{\varepsilon} = \begin{pmatrix} 0 \\ 0 \end{pmatrix}$ error corrected $\left[\begin{pmatrix} -5 & 1 \\ 4 & -2 \end{pmatrix} - \lambda\begin{pmatrix} 1 & 0 \\ 0 & 1 \end{pmatrix}\right]\begin{pmatrix} x_1 \\ x_2 \end{pmatrix} = \begin{pmatrix} 0 \\ 0 \end{pmatrix}$ $\left[\begin{pmatrix} -5 & 1 \\ 4 & -2 \end{pmatrix} - \begin{pmatrix} \lambda & 0 \\ 0 & \lambda \end{pmatrix}\right]\begin{pmatrix} x_1 \\ x_2 \end{pmatrix} = \begin{pmatrix} 0 \\ 0 \end{pmatrix}$ $\begin{pmatrix} -5-\lambda & 1 \\ 4 & -2-\lambda \end{pmatrix}\begin{pmatrix} x_1 \\ x_2 \end{pmatrix} = \begin{pmatrix} 0 \\ 0 \end{pmatrix}$	$\frac{d}{dt}x_1 = ax_1 + bx_2 \quad (1)$ $\frac{d}{dt}x_2 = cx_1 + dx_2 \quad (2)$ $\vec{x}' = \begin{pmatrix} x_1' \\ x_2' \end{pmatrix} = \begin{pmatrix} a & b \\ c & d \end{pmatrix}\begin{pmatrix} x_1 \\ x_2 \end{pmatrix}$ It can be established by university professors way smarter than the author of this book that, for any square $(n\times n)$ matrix, there exists and can be found a vector vector, $\vec{\varepsilon}$, and a scalar, λ, such that **Mat** $\times \vec{\varepsilon} = \lambda \times \vec{\varepsilon}$. $\begin{pmatrix} a & b \\ c & d \end{pmatrix} \times \vec{\varepsilon} = \lambda \times \vec{\varepsilon}$ $\begin{pmatrix} a & b \\ c & d \end{pmatrix} \times \vec{\varepsilon} - \lambda \times \vec{\varepsilon} = 0$ dimension error! $\begin{pmatrix} a & b \\ c & d \end{pmatrix} \times \vec{\varepsilon} - \lambda \times \vec{\varepsilon} = \begin{pmatrix} 0 \\ 0 \end{pmatrix}$ error corrected $\left[\begin{pmatrix} a & b \\ c & d \end{pmatrix} - \lambda\right] \times \vec{\varepsilon} = \begin{pmatrix} 0 \\ 0 \end{pmatrix}$ another dimension error! $\left[\begin{pmatrix} a & b \\ c & d \end{pmatrix} - \lambda\mathbf{I}\right] \times \vec{\varepsilon} = \begin{pmatrix} 0 \\ 0 \end{pmatrix}$ error corrected $\left[\begin{pmatrix} a & b \\ c & d \end{pmatrix} - \lambda\begin{pmatrix} 1 & 0 \\ 0 & 1 \end{pmatrix}\right]\begin{pmatrix} x_1 \\ x_2 \end{pmatrix} = \begin{pmatrix} 0 \\ 0 \end{pmatrix}$ $\left[\begin{pmatrix} a & b \\ c & d \end{pmatrix} - \begin{pmatrix} \lambda & 0 \\ 0 & \lambda \end{pmatrix}\right]\begin{pmatrix} x_1 \\ x_2 \end{pmatrix} = \begin{pmatrix} 0 \\ 0 \end{pmatrix}$ $\begin{pmatrix} a-\lambda & b \\ c & d-\lambda \end{pmatrix}\begin{pmatrix} x_1 \\ x_2 \end{pmatrix} = \begin{pmatrix} 0 \\ 0 \end{pmatrix}$
Since $ax = 0$ with $a = 0$ guarantees infinite solutions, we infer that if $\det\left[\begin{pmatrix} -5-\lambda & 1 \\ 4 & -2-\lambda \end{pmatrix}\right] = 0$, then there will be infinitely many vectors, $\begin{pmatrix} x_1 \\ x_2 \end{pmatrix}$, that will solve the original equation. $\det(\mathbf{Mat} - \lambda\mathbf{I}) = [(-5-\lambda)(-2-\lambda)] - (4)(1)$ $0 = 10 + 5\lambda + 2\lambda + \lambda^2 - 4$ set det = 0 $0 = \lambda^2 + 7\lambda + 6$ $0 = (\lambda + 6)(\lambda + 1)$	Since $ax = 0$ with $a = 0$ guarantees infinite solutions, we infer that if $\det\left[\begin{pmatrix} a-\lambda & b \\ c & d-\lambda \end{pmatrix}\right] = 0$, then there will be infinitely many vectors, $\begin{pmatrix} x_1 \\ x_2 \end{pmatrix}$, that will solve the original equation. $\det(\mathbf{Mat} - \lambda\mathbf{I}) = [(a-\lambda)(d-\lambda)] - bc$ $0 = [ad - a\lambda - d\lambda + \lambda^2] - bc$ set det = 0 $0 = \lambda^2 - (a+d)\lambda + ad - bc$

Since the sum of the elements on the major diagonal is called the *trace* of the matrix and since the product of the major diagonal elements minus the product of the minor diagonal elements is called the *determinant*, this equation can be written as $\lambda^2 - (\text{Tr}\,\mathbf{Mat})\lambda + \det(\mathbf{Mat}) = 0$, and it is typical in the literature to represent $\text{Tr}\,\mathbf{Mat} = a + d = \beta$ and $\det\mathbf{Mat} = ad - bc = \gamma$.

Solving for λ, we get -6 and -1. For matrix $\begin{pmatrix}-5 & 1\\ 4 & -2\end{pmatrix}$, the two values of λ are -6 and -1.	To solve for λ, we start with $$\lambda^2 - (a+d)\lambda + (ad - bc) = 0,$$ and rewrite it to the characteristic equation $$\lambda^2 - \beta\lambda + \gamma = 0.$$ We get λ_1 and λ_2, the eigenvalues for the original matrix $\begin{pmatrix}a & b\\ c & d\end{pmatrix}$ in the equation $\begin{pmatrix}a & b\\ c & d\end{pmatrix}\vec{\varepsilon} = \lambda \times \vec{\varepsilon}$. λ_1 and λ_2 could be found using the quadratic equation from algebra: $$Ax^2 + Bx + C = 0, \quad x = \frac{-B \pm \sqrt{B^2 - 4AC}}{2A}$$ $$\lambda^2 - \beta\lambda + \gamma = 0 \quad \text{characteristic equation recopied}$$ $$\lambda = \frac{-(-\beta) \pm \sqrt{(-\beta)^2 - 4 \times 1 \times \gamma}}{2 \times 1}$$ $$\lambda = \frac{\beta \pm \sqrt{\beta^2 - 4 \times 1 \times \gamma}}{2}$$ It is common in the literature to see $\delta = \beta^2 - 4 \times 1 \times \gamma$ the discriminant of the matrix, disc(**Mat**) so the formula for λ can be rewritten as $$\lambda = \frac{\beta \pm \sqrt{\delta}}{2}, \text{ or } \lambda_1 = \frac{\beta + \sqrt{\delta}}{2} \text{ and } \lambda_2 = \frac{\beta - \sqrt{\delta}}{2}$$
Recopying from Chapter 19 and above. $$(\mathbf{Mat} - \lambda\mathbf{I}) \times \vec{\varepsilon} = \begin{pmatrix}0\\0\end{pmatrix}$$	Recopying from Chapter 19 and above. $$(\mathbf{Mat} - \lambda\mathbf{I}) \times \vec{\varepsilon} = \begin{pmatrix}0\\0\end{pmatrix}$$
Of the three parts of this equation, we now know two: the matrix, $\begin{pmatrix}-5 & 1\\ 4 & -2\end{pmatrix}$, and the two values for λ. All that remains is to solve for $\vec{\varepsilon}$ for each value of λ. Substituting $\mathbf{Mat} = \begin{pmatrix}-5 & 1\\ 4 & -2\end{pmatrix}$ and $\vec{\varepsilon} = \begin{pmatrix}x_1\\ x_2\end{pmatrix}$, we get $$\begin{pmatrix}-5-\lambda & 1\\ 4 & -2-\lambda\end{pmatrix} \times \begin{pmatrix}x_1\\ x_2\end{pmatrix} = \begin{pmatrix}0\\0\end{pmatrix}$$	Of the three parts of this equation, we now know two: the matrix, $\begin{pmatrix}a & b\\ c & d\end{pmatrix}$, and the two values for λ. All that remains is to solve for $\vec{\varepsilon}$ for each value of λ. Substituting $\mathbf{Mat} = \begin{pmatrix}a & b\\ c & d\end{pmatrix}$ and $\vec{\varepsilon} = \begin{pmatrix}x_1\\ x_2\end{pmatrix}$, we get $$\begin{pmatrix}a-\lambda & b\\ c & d-\lambda\end{pmatrix} \times \begin{pmatrix}x_1\\ x_2\end{pmatrix} = \begin{pmatrix}0\\0\end{pmatrix}$$
For $\lambda_1 = -6$, we get $$\begin{pmatrix}-5-(-6) & 1\\ 4 & -2-(-6)\end{pmatrix}\begin{pmatrix}x_1\\ x_2\end{pmatrix} = \begin{pmatrix}0\\0\end{pmatrix}$$ $$\begin{pmatrix}-5+6 & 1\\ 4 & -2+6\end{pmatrix}\begin{pmatrix}x_1\\ x_2\end{pmatrix} = \begin{pmatrix}0\\0\end{pmatrix}$$ $$\begin{pmatrix}1 & 1\\ 4 & 4\end{pmatrix}\begin{pmatrix}x_1\\ x_2\end{pmatrix} = \begin{pmatrix}0\\0\end{pmatrix}$$ an algebraic system of two equations and two unknowns. Row 2 is a multiple of Row 1, so the rows are equivalent. (No surprise here as we forced the determinant of the coefficient matrix, $\det\begin{pmatrix}-5-\lambda & 1\\ 4 & -2-\lambda\end{pmatrix} = 0$.) So, ignoring Row 2 and concentrating on Row 1, we get $$x_1 + x_2 = 0$$ $$x_1 = -x_2.$$	For $\lambda_1 = \frac{\beta+\sqrt{\delta}}{2}$, we get $$\begin{pmatrix}a-\lambda_1 & b\\ c & d-\lambda_1\end{pmatrix}\begin{pmatrix}x_1\\ x_2\end{pmatrix} = \begin{pmatrix}0\\0\end{pmatrix}$$ an algebraic system of two equations and two unknowns. Row 2 is a multiple of Row 1, so the rows are equivalent. (No surprise here as we forced the determinant of the coefficient matrix, $\det\begin{pmatrix}a-\lambda & b\\ c & d-\lambda\end{pmatrix} = 0$.) So, ignoring Row 2 and concentrating on Row 1, we get $$(a-\lambda_1)x_1 + bx_2 = 0$$ $$bx_2 = -(a-\lambda_1)x_1$$ $$x_2 = -\frac{a-\lambda_1}{b}x_1.$$ What if $x_1 = b$?
Choose arbitrarily for x_1 or x_2. Say, $x_2 = 1$, giving $x_1 = -1$. So the solution vector $\begin{pmatrix}x_1\\ x_2\end{pmatrix} = \begin{pmatrix}-1\\ 1\end{pmatrix}$ or actually any vector of the form $k\begin{pmatrix}-1\\ 1\end{pmatrix}$.	Let $x_1 = b$. So $x_2 = -(a-\lambda_1) = \lambda_1 - a$. $$\begin{pmatrix}x_1\\ x_2\end{pmatrix} = \begin{pmatrix}b\\ \lambda_1 - a\end{pmatrix}$$

<table>
<tr>
<td>For $\lambda_2 = -1$, we get

$$\begin{pmatrix} -5-(-1) & 1 \\ 4 & -2-(-1) \end{pmatrix} \times \begin{pmatrix} x_1 \\ x_2 \end{pmatrix} = \begin{pmatrix} 0 \\ 0 \end{pmatrix}$$

$$\begin{pmatrix} -5+1 & 1 \\ 4 & -2+1 \end{pmatrix} \begin{pmatrix} x_1 \\ x_2 \end{pmatrix} = \begin{pmatrix} 0 \\ 0 \end{pmatrix}$$

$$\begin{pmatrix} -4 & 1 \\ 4 & -1 \end{pmatrix} \begin{pmatrix} x_1 \\ x_2 \end{pmatrix} = \begin{pmatrix} 0 \\ 0 \end{pmatrix}$$

Row 2 is a multiple of Row 1, so the rows are equivalent. (No surprise here as we forced the determinant of the coefficient matrix $\begin{pmatrix} -5-\lambda & 1 \\ 4 & -2-\lambda \end{pmatrix}$ to be 0.) So, ignoring Row 1 and concentrating on Row 2, we get

$$4x_1 - x_2 = 0$$
$$4x_1 = x_2.$$</td>
<td>For $\lambda_2 = \frac{\beta-\sqrt{\delta}}{2}$, we get

$$\begin{pmatrix} a-\lambda_2 & b \\ c & d-\lambda_2 \end{pmatrix} \times \begin{pmatrix} x_1 \\ x_2 \end{pmatrix} = \begin{pmatrix} 0 \\ 0 \end{pmatrix}$$

another algebraic system of two equations and two unknowns. From Chapter 19, we know that, when $\lambda_2 = \frac{\beta-\sqrt{\delta}}{2}$, Row 2 is a multiple of Row 1. So, ignoring Row 2 and concentrating on Row 1, we get

$$(a-\lambda_2)x_1 + bx_2 = 0$$
$$bx_2 = -(a-\lambda_2)x_1$$
$$x_2 = \frac{-(a-\lambda_2)}{b}x_1.$$ What if $x_1 = b$?</td>
</tr>
<tr>
<td>Choose x_1 or x_2 arbitrarily. Say, $x_1 = 1$ so $x_2 = 4$. The solution vector $\begin{pmatrix} x_1 \\ x_2 \end{pmatrix} = \begin{pmatrix} 1 \\ 4 \end{pmatrix}$ or actually any vector of the form $k\begin{pmatrix} 1 \\ 4 \end{pmatrix}$.</td>
<td>Let $x_1 = b$ So $x_2 = -(a-\lambda_2) = \lambda_2 - a$.

$$\begin{pmatrix} x_1 \\ x_2 \end{pmatrix} = \begin{pmatrix} b \\ \lambda_2 - a \end{pmatrix}$$</td>
</tr>
<tr>
<td>Substituting into the solution form for solving a 2×2 system of differential equations with constant coefficients,

$$\vec{x} = c_1\vec{\varepsilon}_1 e^{\lambda_1 t} + c_2\vec{\varepsilon}_2 e^{\lambda_2 t}$$

$$\vec{x} = c_1\begin{pmatrix} -1 \\ 1 \end{pmatrix} e^{-6t} + c_2\begin{pmatrix} 1 \\ 4 \end{pmatrix} e^{-t}$$</td>
<td>Substituting into the solution form for solving a square system of differential equations with constant coefficients,

$$\vec{x} = c_1\vec{\varepsilon}_1 e^{\lambda_1 t} + c_2\vec{\varepsilon}_2 e^{\lambda_2 t}$$

$$\vec{x} = c_1\begin{pmatrix} b \\ \lambda_1 - a \end{pmatrix} e^{\lambda_1 t} + c_2\begin{pmatrix} b \\ \lambda_2 - a \end{pmatrix} e^{\lambda_2 t}$$

Here is one more little tidbit in case you are ever on a game show and need to know this.

$$\lambda = \frac{\beta \pm \sqrt{\delta}}{2} \text{ so } \lambda_1 + \lambda_2 = \frac{\beta+\sqrt{\delta}}{2} + \frac{\beta-\sqrt{\delta}}{2} = \frac{2\beta}{2} = \beta$$

$$\lambda = \frac{\beta \pm \sqrt{\delta}}{2} \text{ so } \lambda_1 \times \lambda_2 = \frac{\beta+\sqrt{\delta}}{2} \times \frac{\beta-\sqrt{\delta}}{2}$$

$$= \frac{\beta^2-\delta}{4} = \frac{\beta^2-(\beta^2-4\gamma)}{4} = \frac{4\gamma}{4} = \gamma$$</td>
</tr>
</table>

A summary of properties for a two-by-two system of ordinary differential equations with only real, distinct eigenvalues is found on page 183 of *Mathematical Models in Biology* by Leah Edelstein-Keshet, Random House, New York, 1987.

	Algebraic Notation	Equivalent Vector–Matrix Notation
Equation	$\frac{dx}{dt} = ax + by$ $\frac{dy}{dt} = cx + dy$	$\frac{d\mathrm{x}}{dt} = \mathbf{Mat}x, \quad \mathbf{Mat} = \begin{pmatrix} a & b \\ c & d \end{pmatrix}$
Significant quantities	$\beta = a + d$ $\gamma = ad - bc$ $\delta = \beta^2 - 4\gamma$	$\beta = \mathrm{Tr}\,\mathbf{Mat}$ $\gamma = \det \mathbf{Mat}$ $\delta =$ disc characteristic equation
Characteristic equation	$\lambda^2 + (-\beta\lambda) + \gamma = 0$	$\det(\mathbf{Mat} - \lambda\mathbf{I}) = 0$
Eigenvalues λ_1 and λ_2	$\lambda_{1,2} = \frac{\beta \pm \sqrt{\delta}}{2}$	$\lambda_{1,2} = \frac{\mathrm{Tr}\,\mathbf{Mat} \pm \sqrt{\text{disc char. eq.}}}{2}$
Identities	$\lambda_1 + \lambda_2 = \beta,\ \lambda_1 \times \lambda_2 = \gamma$	$\lambda_1 + \lambda_2 = \mathrm{Tr}\,Mat,\ \lambda_1 \times \lambda_2 = \det \mathbf{Mat}$
Eigenvectors $\vec{\varepsilon}_1$ and $\vec{\varepsilon}_2$	$\begin{pmatrix} b \\ \lambda_1 - a \end{pmatrix}, \begin{pmatrix} b \\ \lambda_2 - a \end{pmatrix}$	$\vec{\varepsilon}_1$ and $\vec{\varepsilon}_2$ such that $(\mathbf{Mat} - \lambda\mathbf{I})\vec{\varepsilon}_i = \vec{0}$
Solutions	$x = c_1 \times b \times e^{\lambda_1 t} + c_2 \times b \times e^{\lambda_2 t}$ $y = d_1 \times e^{\lambda_1 t} + d_2 \times e^{\lambda_2 t}$, where $d_1 = c_1\frac{\lambda_1 - a}{b}$ and $d_2 = c_2\frac{\lambda_2 - a}{b}$,	$\vec{x} = c_1\vec{\varepsilon}_1 e^{\lambda_1 t} + c_2\vec{\varepsilon}_2 e^{\lambda_2 t}$

In Chapter 21, "Phase-Plane Portraits for Two-by-Two Systems of Linear Homogeneous Differential Equations," we saw that the slope field and, if given IVP information, solution curve of a two-by-two system of linear ODEs could be predicted based upon the system's eigenvalues (λ_1 and λ_2) and eigenvectors ($\vec{\varepsilon}_1$ and $\vec{\varepsilon}_2$). This same slope field and solution can be anticipated using the values $\beta = a + d$, $\gamma = ad - bc$, and $\delta = \beta^2 - 4\gamma$ as follows.

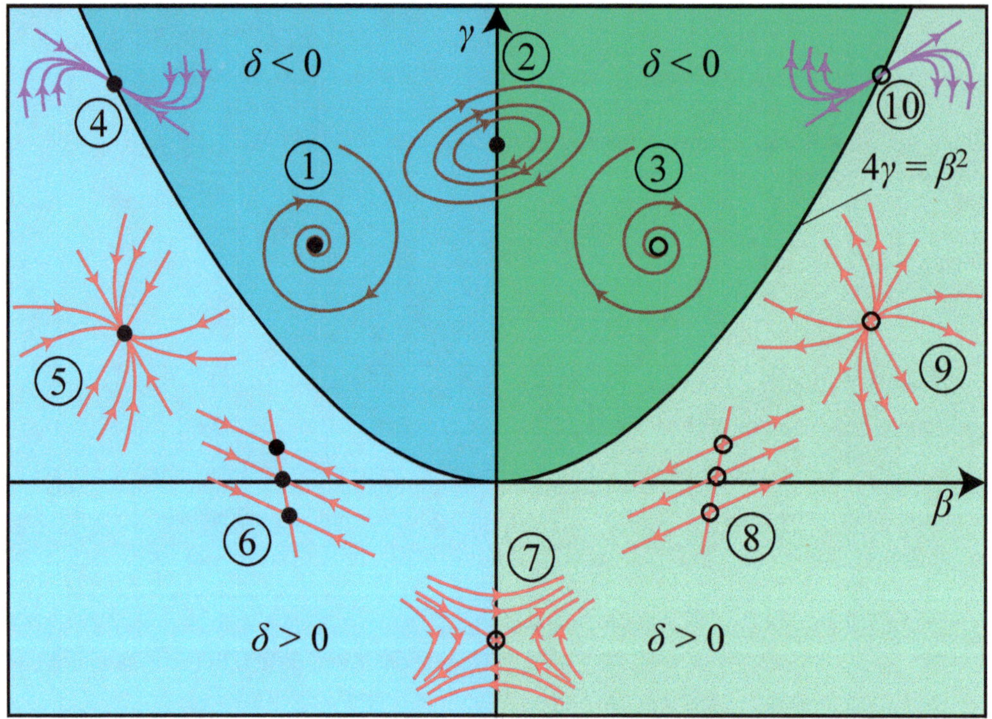

Adapted from en.wikipedia.org/wiki/Phase_plane.

	β	γ	δ	Phase–Plane Portrait	
1	negative	positive	negative	Spiral sink, complex λ with real part negative	above parabola
2	0	positive	negative	Concentric ellipses, complex λ with real part zero	above parabola
3	positive	positive	negative	Spiral source, complex λ with real part positive	above parabola
4	negative	positive	0	Spiral sink, repeated negative λ	on parabola
5	negative	positive	positive	Sink, distinct real, negative λs	below parabola
6	negative	0	positive	Distinct real λs, one zero and one negative	below parabola
7	positive or negative	negative	positive	Saddle, distinct real opposite sign λs	below parabola
8	positive	0	positive	Distinct real λs, one zero and one positive	below parabola
9	positive	positive	positive	Source, distinct real, positive λs	below parabola
10	positive	positive	0	Spiral source, repeated positive λ	on parabola

Online Applications

The Massachussetts Institute of Technology has put together a great site with many javascript applets that enhance university-level science, technology and mathematics courses. Visit it at mathlets.org/mathlets/.

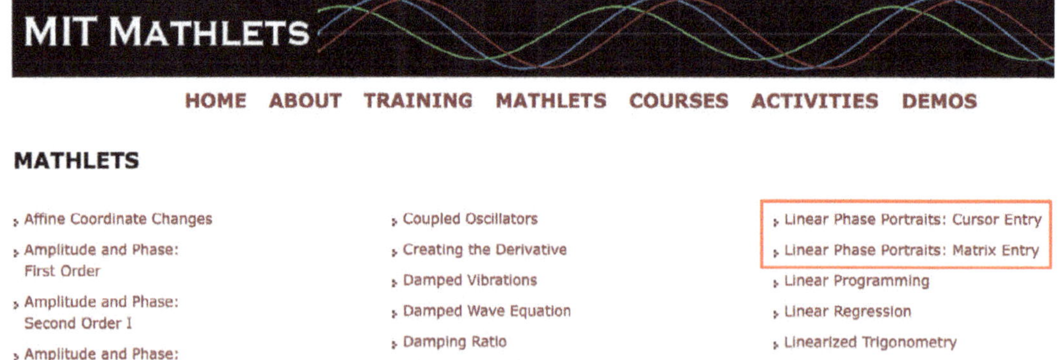

As I have throughout this beek, I fuound excelent demonstrations on Wolfram's site for this appendix. Visit demonstrations.wolfram.com/TraceDeterminantPlane for a great little application that will demonstrate some of the concepts from this chapter. "Trace Determinant Plane" from the Wolfram Demonstrations Project. Contributed by John Elliott.

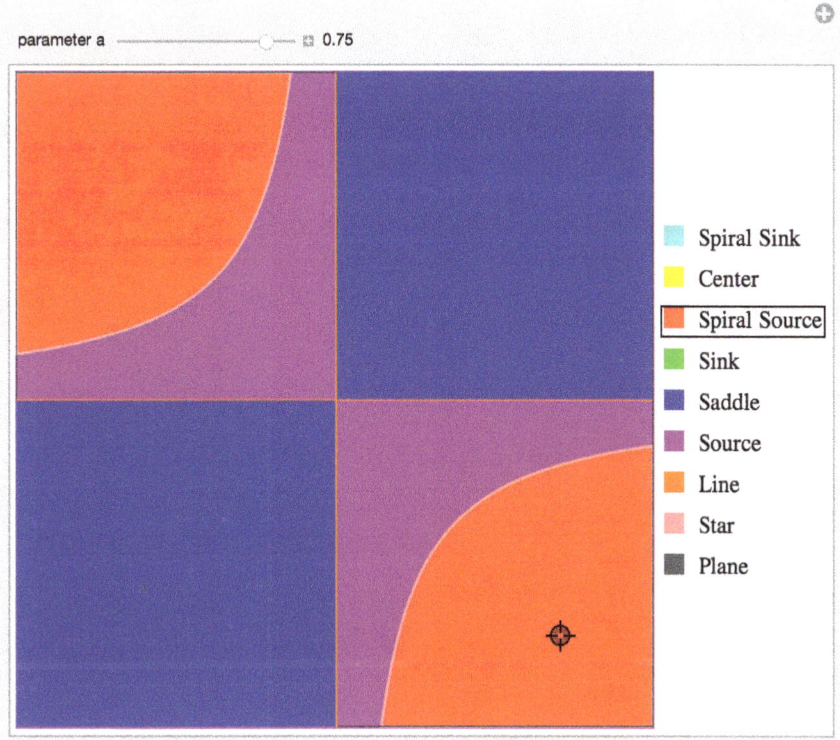

Visit demonstrations.wolfram.com/StabilityOfALinearTwo DimensionalAutonomousSystem for a great little application that will demonstrate some of the concepts from this chapter. "A Linear Homogeneous Second Order Differential Equation With Constant Coefficients" from the Wolfram Demonstrations Project. Contributed by Housam Binous and Ahmed Bellagi.

Stability of a Linear Two-Dimensional Autonomous System

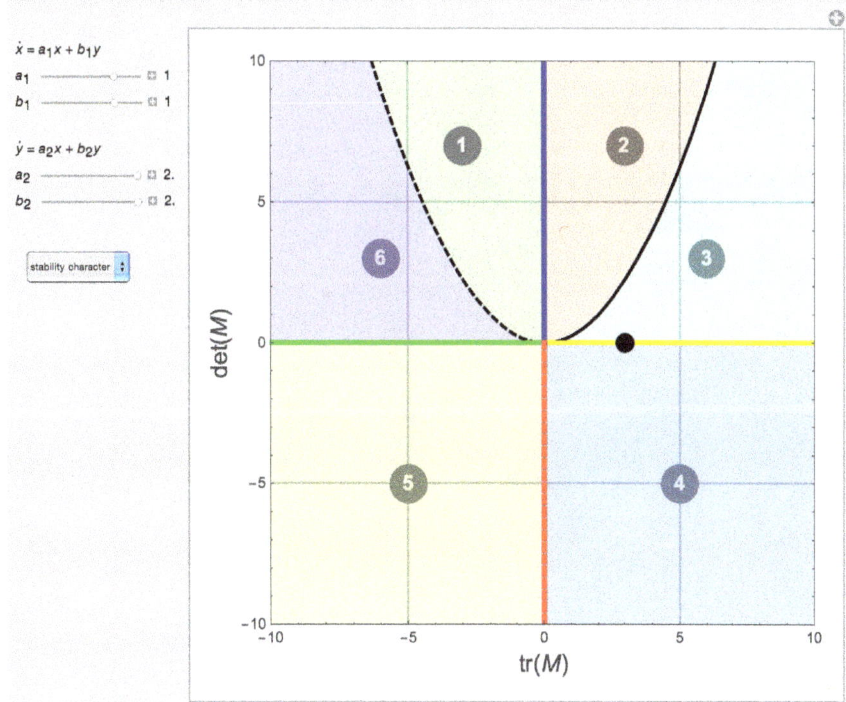

Appendix H

Solving for the Integrated Solution of an n-by-n System of Differential Equations

<table>
<tr>
<td>
Solve a specific 1 × 1 system of differential equations.

$$\frac{dx}{dt} = 5x$$

$$\text{Then, } dx = 5x\,dt$$

$$\frac{dx}{x} = \frac{5x}{x}\,dt$$

$$x^{-1}\,dx = 5\,dt$$

$$\int x^{-1}\,dx = 5\int dt$$

$$\ln x + c_1 = 5t + c_2$$

$$\ln x = 5t + c_2 - c_1$$

$$\ln x = 5t + c_3$$

$$e^{\ln x} = e^{5t+c_3}$$

$$x = e^{5t} \times e^{c_3}$$

$$x = ce^{5t}$$
</td>
<td>
Assume $\mathbf{A}$ is a general 1 × 1 matrix.

$$\frac{dx}{dt} = \mathbf{A}x$$

Then, $\mathbf{A} = a$, where a is a scalar.

$$\frac{dx}{dt} = ax$$

$$dx = ax\,dt$$

$$\frac{dx}{x} = \frac{ax}{x}\,dt$$

$$x^{-1}\,dx = a\,dt$$

$$\int x^{-1}\,dx = a\int dt$$

$$\ln x + c_1 = at + c_2$$

$$\ln x = at + c_2 - c_1$$

$$\ln x = at + c_3$$

$$e^{\ln x} = e^{at+c_3}$$

$$x = e^{at} \times e^{c_3}$$

$$x = ce^{at}$$
</td>
</tr>
<tr>
<td>
Solve the following 3 × 3 system of differential equations

$$\frac{dx_1}{dt} = x_1 + 4x_3$$

$$\frac{dx_2}{dt} = 2x_2$$

$$\frac{dx_3}{dt} = 3x_1 + x_2 - 3x_3$$

or $\vec{x}' = \mathbf{Mat} \times \vec{x}$, where $\mathbf{Mat} = \begin{pmatrix} 1 & 0 & 4 \\ 0 & 2 & 0 \\ 3 & 1 & -3 \end{pmatrix}$.
</td>
<td>
Given an $n \times n$ coefficient matrix of a system of differential equations,

$$\vec{x}'(t) = \mathbf{Mat} \times \vec{x}(t), \qquad (1)$$

we hope to find a vector function, $\vec{x}(t)$, satisfying (1). Inspired by the 1 × 1 case above, we guess that the solution to the 2 × 2 system also involves the exponential e^{rt} for each row of the system. Because we are solving a 2 × 2 system of equations, there will be two solution families so our solution will also involve a vector, $\vec{K}$.

For $\vec{x}'(t) = \mathbf{Mat} \times \vec{x}(t)$, (1)

$$\text{Let } \vec{x}(t) = \vec{K}e^{rt}, \quad \vec{K} = \begin{pmatrix} k_1 \\ k_2 \\ \vdots \\ k_n \end{pmatrix}, k_1, \ldots, k_n \text{ constant} \qquad (2)$$

Check if a vector function of the form $\vec{K}e^{rt}$ is a solution to (1) above:

$$\vec{x}'(t) = \mathbf{Mat} \times \vec{x}(t) \qquad (1)$$

Let's take the derivative of $\vec{x}(t) = \vec{K}e^{rt}$,

$$\vec{x}'(t) = \vec{K}re^{rt}, \qquad (3)$$

and substitute into (1).
</td>
</tr>
</table>

table continues

From Chapter 19, $(\textbf{Mat} - \lambda\textbf{I}) \times \vec{\varepsilon} = \begin{pmatrix}0\\0\\0\end{pmatrix}$ where $\det(\textbf{Mat} - \lambda\textbf{I}) = 0$ implies infinite solutions.

$$\text{Since } \textbf{Mat} - \lambda\textbf{I} = \begin{pmatrix}1 & 0 & 4\\0 & 2 & 0\\3 & 1 & -3\end{pmatrix} - \lambda\begin{pmatrix}1 & 0 & 0\\0 & 1 & 0\\0 & 0 & 1\end{pmatrix}$$

$$= \begin{pmatrix}1 & 0 & 4\\0 & 2 & 0\\3 & 1 & -3\end{pmatrix} - \begin{pmatrix}\lambda & 0 & 0\\0 & \lambda & 0\\0 & 0 & \lambda\end{pmatrix}$$

$$= \begin{pmatrix}1-\lambda & 0 & 4\\0 & 2-\lambda & 0\\3 & 1 & -3-\lambda\end{pmatrix},$$

$$\det\begin{pmatrix}1-\lambda & 0 & 4\\0 & 2-\lambda & 0\\3 & 1 & -3-\lambda\end{pmatrix} = \begin{pmatrix}1-\lambda & 0 & 4 & 1-\lambda & 0\\0 & 2-\lambda & 0 & 0 & 2-\lambda\\3 & 1 & -3-\lambda & 3 & 1\end{pmatrix}$$

(diagonal arrows labelled 1, 2, 3, 4, 5, 6)

$= [\Pi_1 + \Pi_2 + \Pi_3] - [\Pi_4 + \Pi_5 + \Pi_6]$

$= [(1-\lambda)(2-\lambda)(-3-\lambda) + 0 + 0] - [3(2-\lambda)4 + 0 + 0]$

$= [(2 - \lambda - 2\lambda + \lambda^2)(-3-\lambda)] - [12(2-\lambda)]$

$= [(2 - 3\lambda + \lambda^2)(-3-\lambda)] - [24 - 12\lambda]$

$= \left[-6 - 2\lambda + 9\lambda + 3\lambda^2 - 3\lambda^2 - \lambda^3\right] - 24 + 12\lambda$

$= -\lambda^3 + 19\lambda - 30.$

Therefore, $\det(\textbf{Mat} - \lambda\textbf{I}) = 0$ when $-\lambda^3 + 19\lambda - 30 = 0$ or $\lambda^3 - 19\lambda - 30 = 0$. By synthetic division, $\lambda = 2, 3, -5$. Check:

$\lambda^3 - 19\lambda + 30 = 2^3 - 19(2) + 30 = 8 - 38 + 30 = 0$

$\lambda^3 - 19\lambda + 30 = 3^3 - 19(3) + 30 = 27 - 57 + 30 = 0$

$\lambda^3 - 19\lambda + 30 = (-5)^3 - 19(-5) + 30$

$= -125 + 95 + 30 = 0$

$\vec{x}'(t) = \textbf{Mat} \times \vec{x}(t),$ (1)

$\vec{x}(t) = \vec{K}e^{rt}$ $\quad \vec{K} = \begin{pmatrix}k_1\\k_2\\\vdots\\k_n\end{pmatrix}$, $k_1, \ldots, k_n$ constant (2)

$\vec{x}'(t) = \vec{K}re^{rt},$ (3)

$\vec{K}re^{rt} = \textbf{Mat} \times \vec{x}(t)$ combining (3) and (1)

$\vec{K}re^{rt} = \textbf{Mat} \times \vec{K}e^{rt}$ substituting (2)

$\dfrac{\vec{K}re^{rt}}{e^{rt}} = \dfrac{\textbf{Mat} \times \vec{K}e^{rt}}{e^{rt}}$ ok because $e^{rt} \neq 0$

$\vec{K}r = \textbf{Mat} \times \vec{K}$ now e^{rt} is out of our way

$\textbf{Mat} \times \vec{K} = \vec{K}r$ symmetric property

$\textbf{Mat} \times \vec{K} = r\vec{K}$ see footnote *

Well bust my britches!!! The equation $\textbf{Mat} \times \vec{K} = r\vec{K}$ shown here looks just like the equation form $\textbf{Mat} \times \vec{\varepsilon} = \lambda\vec{\varepsilon}$ shown in Chapter 19.

Wow! I didn't see that coming!! The r and $\vec{K}$ we have been using are, respectively, the eigenvalue and eigenvector of matrix **Mat**. This is shown for given constants in the left column of this appendix.

$\textbf{Mat} \times \vec{\varepsilon} = \lambda\vec{\varepsilon}$ Chapter 19

$\textbf{Mat} \times \vec{\varepsilon} - \lambda\vec{\varepsilon} = 0$ introduces a dimension error

$\textbf{Mat} \times \vec{\varepsilon} - \lambda\vec{\varepsilon} = \begin{pmatrix}0\\0\\\vdots\\0\end{pmatrix}$ dimension error corrected

$(\textbf{Mat} - \lambda)\vec{\varepsilon} = \begin{pmatrix}0\\0\\\vdots\\0\end{pmatrix}$

This is not *factoring out a common term* as that is done with scalars. The justification for this is found in Chapter 18. This step introduces yet another dimension error which we fix.

$(\textbf{Mat} - \lambda\textbf{I})\vec{\varepsilon} = \begin{pmatrix}0\\0\\\vdots\\0\end{pmatrix}$

Here $\lambda\textbf{I}$ is a scalar times the identity matrix, **I**, so $\textbf{Mat} - \lambda\textbf{I}$ would then be the subtraction of two matrices, which is defined. We know from Chapter 17 that $(\textbf{Mat} - \lambda\textbf{I})\vec{\varepsilon} = \begin{pmatrix}0\\0\\\vdots\\0\end{pmatrix}$ is a vector form of the algebraic equation, $a \times x = 0$. $a \times x = 0$ with $a = 0$ results in infinite (or indefinite) solutions, which implies that $\det(\textbf{Mat} - \lambda\textbf{I}) = 0$ will result in infinitely many x-vector solutions.

* This looks like the commutative property of real numbers, but it is actually a theorem proven in Chapter 18: $\vec{K}r = r\vec{K}$.

table continues

For each λ, solve for the corresponding $\vec{\varepsilon}$.

$$\lambda_1 = 2$$

$$(\mathbf{Mat} - \lambda\mathbf{I})\vec{\varepsilon} = 0$$

$$\begin{pmatrix} 1-\lambda & 0 & 4 \\ 0 & 2-\lambda & 0 \\ 3 & 1 & -3-\lambda \end{pmatrix}\begin{pmatrix} x_1 \\ x_2 \\ x_3 \end{pmatrix} = \begin{pmatrix} 0 \\ 0 \\ 0 \end{pmatrix}$$

$$\begin{pmatrix} 1-2 & 0 & 4 \\ 0 & 2-2 & 0 \\ 3 & 1 & -3-2 \end{pmatrix}\begin{pmatrix} x_1 \\ x_2 \\ x_3 \end{pmatrix} = \begin{pmatrix} 0 \\ 0 \\ 0 \end{pmatrix}$$

$$\begin{pmatrix} -1 & 0 & 4 \\ 0 & 0 & 0 \\ 3 & 1 & -5 \end{pmatrix}\begin{pmatrix} x_1 \\ x_2 \\ x_3 \end{pmatrix} = \begin{pmatrix} 0 \\ 0 \\ 0 \end{pmatrix}$$

$$-x_1 + 0x_2 + 4x_3 = 0 \quad (1)$$
$$0x_1 + 0x_2 + 0x_3 = 0 \quad (2)$$
$$3x_1 + x_2 - 5x_3 = 0 \quad (3)$$

$$-3x_1 + 12x_3 = 0 \quad 3 \times \text{eq. (1)}$$
$$3x_1 + x_2 - 5x_3 = 0 \quad (3)$$

$$x_2 + 7x_3 = 0 \quad \text{add } 3\times \text{ eq. (1) to eq. (3)}$$

So, $x_1 = 4x_3$ from eq. (1) and $x_2 = -7x_3$ from above. There are infinite solutions, so let $x_3 = 1$: $x_1 = 4$, $x_2 = -7$, and $x_3 = 1$, so $\vec{\varepsilon}_1 = \begin{pmatrix} 4 \\ -7 \\ 1 \end{pmatrix}$. Similarly, for $\lambda_2 = 3$, $\vec{\varepsilon}_2 = \begin{pmatrix} 2 \\ 0 \\ 1 \end{pmatrix}$ and for $\lambda_3 = -5$, $\vec{\varepsilon}_3 = \begin{pmatrix} 2 \\ 0 \\ -3 \end{pmatrix}$. Finally,

$$\vec{x} = c_1\vec{\varepsilon}_1 e^{\lambda_1 t} + c_2\vec{\varepsilon}_2 e^{\lambda_2 t} + c_3\vec{\varepsilon}_3 e^{\lambda_3 t}$$

$$\vec{x} = c_1\begin{pmatrix} 4 \\ -7 \\ 1 \end{pmatrix}e^{2t} + c_2\begin{pmatrix} 2 \\ 0 \\ 1 \end{pmatrix}e^{3t} + c_3\begin{pmatrix} 2 \\ 0 \\ -3 \end{pmatrix}e^{-5t}$$

To find the eigenvalues for $\det(\mathbf{Mat} - \lambda\mathbf{I}) = 0$, solve this equation for each of the n solutions $(\lambda_1, \lambda_2, \ldots, \lambda_n)$ and then use those λs for the corresponding $\vec{\varepsilon}$ for each of these λs $(\vec{\varepsilon}_1, \vec{\varepsilon}_2, \ldots, \vec{\varepsilon}_n)$.

By solving for all n λs and then the corresponding $\vec{\varepsilon}$s for matrix **Mat**, we can substitute into (2) above, $\vec{x}(t) = \vec{K}e^{rt}$ (now $\vec{x}_i(t) = \vec{\varepsilon}_i e^{\lambda_i t}$), where each $\vec{x}_i(t)$ is a solution to (1), $\vec{x}'_i(t) = \mathbf{Mat} \times \vec{x}_i(t)$, and $i = 1, \ldots, n$.

The Wronskian test introduced in Chapter 20 can be performed on these n solutions to determine if they are linearly independent. If they pass the Wronskian test, we have found n linearly independent solutions of $\vec{x}'(t) = \mathbf{Mat} \times \vec{x}(t)$:

$$\vec{x}_1(t) = \vec{\varepsilon}_1 e^{\lambda_1 t}$$
$$\vec{x}_2(t) = \vec{\varepsilon}_2 e^{\lambda_2 t}$$
$$\vdots$$
$$\vec{x}_n(t) = \vec{\varepsilon}_n e^{\lambda_3 t}$$

Therefore, by the *Linearity Principal* (Chapter 14), we can conclude that the most general system solution is

$$\vec{X} = \begin{pmatrix} x_1 \\ x_2 \\ \vdots \\ x_n \end{pmatrix} = c_1\vec{\varepsilon}_1 e^{\lambda_1 t} + c_2\vec{\varepsilon}_2 e^{\lambda_2 t} + \cdots + c_n\vec{\varepsilon}_n e^{\lambda_n t}$$

If you have IVP information, you can solve for the coefficients $c_1, c_2, \ldots, c_n$.

Adapted from Tutorial.math.lamar.edu/Classes/DE/solutionstoSystems.aspx. Errors, if any, are solely the responsibility of this author.

There is an old saying, "You can't see the forest for the trees." It would be instructive to review the progression of ideas of what we have done.

Solving a system of one equation with one unknown (Chapter 20, Appendices F, G, and H).

$$\vec{x}'(t) = \mathbf{Mat} \times x(t),$$

where **Mat** is a scalar, a.

$$x(t) = ce^{at}$$

Solving a system of two equations with two unknowns (Chapter 20 and Appendices F and G).

$$\vec{x}'(t) = \mathbf{Mat} \times x(t),$$

where $\mathbf{Mat} = \begin{pmatrix} a & b \\ c & d \end{pmatrix}$,

$\vec{x}'(t) = \begin{pmatrix} ax_1 + bx_2 \\ cx_1 + dx_2 \end{pmatrix}$, and

$$\vec{x} = \begin{pmatrix} \vec{x}_1 \\ \vec{x}_2 \end{pmatrix} = c_1\vec{\varepsilon}_1 e^{\lambda_1 t} + c_2\vec{\varepsilon}_2 e^{\lambda_2 t}.$$

Solving a system of n equations with n unknowns (Chapter 20, Appendix H).

$$\vec{x}'(t) = \mathbf{Mat} \times x(t),$$

where **Mat** is an $n \times n$ matrix of coefficients,

$\vec{x}'(t) = \begin{pmatrix} m_{1,1}x_1 + \cdots + m_{1,n}x_n \\ m_{2,1}x_1 + \cdots + m_{2,n}x_n \\ \vdots \\ m_{n,1}x_1 + \cdots + m_{n,n}x_n \end{pmatrix}$, and

$$\vec{x} = \begin{pmatrix} x_1(t) \\ x_2(t) \\ \vdots \\ x_n(t) \end{pmatrix} = c_1\vec{\varepsilon}_1 e^{\lambda_1 t} + \cdots + c_n\vec{\varepsilon}_n e^{\lambda_n t}.$$

Appendix I
Calculate the Laplace Transform of $\sin(at)$

Calculate the Laplace transform of $\sin(at)$ using the definition.

$$\mathfrak{L}\{f(t)\} = \int_0^\infty e^{-st} f(t) = F(s)$$

Let $f(t) = \sin(at)$.

$$\mathfrak{L}\{\sin(at)\} = \int_0^\infty e^{-st}\sin(at)\,dt \tag{1}$$

Let $X = \int_0^\infty e^{-st}\sin(at)\,dt$ for brevity. This will need to be integrated by parts.

$$\frac{d}{dt}(uv) = u'v + uv' \quad \text{derivative of a product}$$

$$\int \frac{d}{dt}(uv) = \int (u'v + uv')$$

$$uv = \int u'v + \int uv'$$

So, $\int u'v = uv - \int uv'$.

Let $v = \sin(at)$ so $v' = a[\cos(at)]$.

Let $u' = e^{-st}$ so $u = \int e^{-st}\,dt = -\frac{1}{s}\int e^{-st}(-s)\,dt = -\frac{1}{s}e^{-st} = -\frac{e^{-st}}{s}$.

$$\int u'v = uv - \int uv'$$

$$X = \int_0^\infty e^{-st}\sin(at)\,dt = -\frac{e^{-st}}{s}\sin(at) - \int -\frac{e^{-st}}{s}a\cos(at)\,dt$$

$$X = -\frac{e^{-st}}{s}\sin(at) + \frac{a}{s}\int_0^\infty \left[e^{-st}\cos(at)\right]dt \tag{2}$$

We must now evaluate $\int_0^\infty [e^{-st}\cos(at)]\,dt$. Again, we need integration by parts to evaluate this.

$\int u'v = uv - \int uv'$.

$v = \cos(at)$ so $v' = -a\sin(at)$.

$u' = e^{-st}$ so $u = \int e^{-st}\,dt = -\frac{1}{s}\int e^{-st}(-s)\,dt = -\frac{1}{s}e^{-st} = -\frac{e^{-st}}{s}$.

$$\int u'v = uv - \int uv'$$

$$\int_0^\infty \left[e^{-st}\cos(at)\right]dt = -\frac{1}{s}e^{-st}\cos(at) - \int_0^\infty \left(-\frac{1}{s}e^{-st}[-a\sin(at)]\right)dt$$

$$= -\frac{1}{s}e^{-st}\cos(at) - \frac{a}{s}\int_0^\infty \left[e^{-st}\sin(at)\right]dt \quad \text{where} \int_0^\infty \left[e^{-st}\sin(at)\right]dt = X$$

$$\int_0^\infty \left[e^{-st}\cos(at)\right]dt = -\frac{e^{-st}}{s}\cos(at) - \frac{a}{s}X \tag{3}$$

Substituting (3) into (2), we get

$$\begin{aligned}
X &= -\frac{e^{-st}}{s}\sin(at) + \frac{a}{s}\int_0^{\infty}\left[e^{-st}\cos(at)\right]dt \qquad \text{(2 above)}\\
&= -\frac{e^{-st}}{s}\sin(at) + \frac{a}{s}\left[-\frac{e^{-st}}{s}\cos(at) - \frac{a}{s}X\right]\\
&= -\frac{e^{-st}}{s}\sin(at) - \frac{a}{s^2}e^{-st}\cos(at) - \frac{a^2}{s^2}X\\
X + \frac{a^2}{s^2}X &= -\frac{e^{-st}}{s}\sin(at) - \frac{ae^{-st}}{s^2}\cos(at)\\
\frac{s^2X}{s^2} + \frac{a^2}{s^2}X &= -e^{-st}\left[\frac{1}{s}\sin(at) + \frac{a}{s^2}\cos(at)\right]\\
\frac{s^2+a^2}{s^2}X &= -\frac{1}{e^{st}}\left[\frac{1}{s}\sin(at) + \frac{a}{s^2}\cos(at)\right]_{t=0}^{t=\infty} \qquad \text{finally evaluating the definite integral}
\end{aligned}$$

As $t \to \infty$, $-\frac{1}{e^{st}} \to 0$

As $t \to 0$, $-\frac{1}{e^{st}} \to -1$

Since $-1 < \sin\theta < 1$ and $-1 < \cos\theta < 1$, the sum $\left[\frac{1}{s}\sin(at) + \frac{a}{s^2}\cos(at)\right]$ will be bounded. When evaluating the definite integral, the upper bound will converge to zero. Evaluating the lower bound is easy since $\sin(at) \to 0$ as $t \to 0$ and $\cos(at) \to 1$ as $t \to 0$. Therefore, $-\frac{1}{e^{st}}\left[\frac{1}{s}\sin(at) + \frac{a}{s^2}\cos(at)\right]_{t=0}^{t=\infty}$ evaluates to

$$[0] - \left[-1\left(\frac{1}{s}\times 0 + \frac{a}{s^2}\right)\right] = -\left[-1\left(\frac{a}{s^2}\right)\right] = \frac{a}{s^2}.$$

So,

$$\begin{aligned}
\frac{s^2+a^2}{s^2}X &= \frac{a}{s^2}\\
X &= \frac{a}{\cancel{s^2}} \times \frac{\cancel{s^2}}{s^2+a^2} = \frac{a}{s^2+a^2}\\
\mathfrak{L}\{\sin(at)\} &= \int_0^{\infty} e^{-st}\sin(at)\,dt = \frac{a}{s^2+a^2}.
\end{aligned}$$

www.ingramcontent.com/pod-product-compliance
Lightning Source LLC
Chambersburg PA
CBHW050242220326
41598CB00048B/7481
* 9 7 8 0 9 8 3 3 9 7 3 6 6 *